Introduction to
CALCULUS AND ANALYSIS

Volume One

Richard Courant and Fritz John

Courant Institute of Mathematical Sciences
New York University

Interscience Publishers

A Division of John Wiley and Sons, Inc.
New York · London · Sydney

Preface

During the latter part of the seventeenth century the new mathematical analysis emerged as the dominating force in mathematics. It is characterized by the amazingly successful operation with infinite processes or limits. Two of these processes, differentiation and integration, became the core of the systematic Differential and Integral Calculus, often simply called "Calculus," basic for all of analysis.

The importance of the new discoveries and methods was immediately felt and caused profound intellectual excitement. Yet, to gain mastery of the powerful art appeared at first a formidable task, for the available publications were scanty, unsystematic, and often lacking in clarity. Thus, it was fortunate indeed for mathematics and science in general that leaders in the new movement soon recognized the vital need for writing textbooks aimed at making the subject accessible to a public much larger than the very small intellectual elite of the early days. One of the greatest mathematicians of modern times, Leonard Euler, established in introductory books a firm tradition and these books of the eighteenth century have remained sources of inspiration until today, even though much progress has been made in the clarification and simplification of the material.

After Euler, one author after the other adhered to the separation of differential calculus from integral calculus, thereby obscuring a key point, the reciprocity between differentiation and integration. Only in 1927 when the first edition of R. Courant's German *Vorlesungen über Differential und Integralrechnung*, appeared in the Springer-Verlag was this separation eliminated and the calculus presented as a unified subject.

From that German book and its subsequent editions the present work originated. With the cooperation of James and Virginia McShaue a greatly expanded and modified English edition of the "Calculus" was prepared and published by Blackie and Sons in Glasgow since 1934, and distributed in the United States in numerous reprintings by Inter-science-Wiley.

During the years it became apparent that the need of college and university instruction in the United States made a rewriting of this work desirable. Yet, it seemed unwise to tamper with the original versions which have remained and still are viable.

Instead of trying to remodel the existing work it seemed preferable to supplement it by an essentially new book in many ways related to the European originals but more specifically directed at the needs of the present and future students in the United States. Such a plan became feasible when Fritz John, who had already greatly helped in the preparation of the first English edition, agreed to write the new book together with R. Courant.

While it differs markedly in form and content from the original, it is animated by the same intention: To lead the student directly to the heart of the subject and to prepare him for active application of his knowledge. It avoids the dogmatic style which conceals the motivation and the roots of the calculus in intuitive reality. To exhibit the interaction between mathematical analysis and its various applications and to emphasize the role of intuition remains an important aim of this new book. Somewhat strengthened precision does not, as we hope, interfere with this aim.

Mathematics presented as a closed, linearly ordered, system of truths without reference to origin and purpose has its charm and satisfies a philosophical need. But the attitude of introverted science is unsuitable for students who seek intellectual independence rather than indoctrination; disregard for applications and intuition leads to isolation and atrophy of mathematics. It seems extremely important that students and instructors should be protected from smug purism.

The book is addressed to students on various levels, to mathematicians, scientists, engineers. It does not pretend to make the subject easy by glossing over difficulties, but rather tries to help the genuinely interested reader by throwing light on the interconnections and purposes of the whole.

Instead of obstructing the access to the wealth of facts by lengthy discussions of a fundamental nature we have sometimes postponed such discussions to appendices in the various chapters.

Numerous examples and problems are given at the end of various chapters. Some are challenging, some are even difficult; most of them supplement the material in the text. In an additional pamphlet more problems and exercises of a routine character will be collected, and moreover, answers or hints for the solutions will be given.

Many colleagues and friends have been helpful. Albert A. Blank not only greatly contributed incisive and constructive criticism, but he also played a major role in ordering, augmenting, and sifting of the problems and exercises, and moreover he assumed the main responsibility for the pamphlet. Alan Solomon helped most unselfishly and effectively in all phases of the preparation of the book. Thanks is also due to Charlotte John, Anneli Lax, R. Richtmyer, and other friends, including James and Virginia McShane.

The first volume is concerned primarily with functions of a single variable, whereas the second volume will discuss the more ramified theories of calculus for functions of several variables.

A final remark should be addressed to the student reader. It might prove frustrating to attempt mastery of the subject by studying such a book page by page following an even path. Only by selecting shortcuts first and returning time and again to the same questions and difficulties can one gradually attain a better understanding from a more elevated point.

An attempt was made to assist users of the book by marking with an asterisk some passages which might impede the reader at his first attempt. Also some of the more difficult problems are marked by an asterisk.

We hope that the work in the present new form will be useful to the young generation of scientists. We are aware of many imperfections and we sincerely invite critical comment which might be helpful for later improvements.

Richard Courant
Fritz John

June 1965

Contents

ix

Chapter *7* *Infinite Sums and Products* 510

1

Introduction

Since antiquity the intuitive notions of continuous change, growth, and motion, have challenged scientific minds. Yet, the way to the understanding of continuous variation was opened only in the seventeenth century when modern science emerged and rapidly developed in close conjunction with integral and differential calculus, briefly called calculus, and mathematical analysis.

The basic notions of Calculus are derivative and integral: the derivative is a measure for the rate of change, the integral a measure for the total effect of a process of continuous change. A precise understanding of these concepts and their overwhelming fruitfulness rests upon the concepts of limit and of function which in turn depend upon an understanding of the continuum of numbers. Only gradually, by penetrating more and more into the substance of Calculus, can one appreciate its power and beauty. In this introductory chapter we shall explain the basic concepts of number, function, and limit, at first simply and intuitively, and then with careful argument.

1.1 The Continuum of Numbers

The positive integers or *natural numbers* 1, 2, 3, ... are abstract symbols for indicating "how many" objects there are in a *collection* or *set* of discrete elements.

These symbols are stripped of all reference to the concrete qualities of the objects counted, whether they are persons, atoms, houses, or any objects whatever.

The natural numbers are the adequate instrument for counting elements of a collection or "set." However, they do not suffice for another equally important objective: to *measure* quantities such as the length of a curve and the volume or weight of a body. The question,

1

"how much?", cannot be answered immediately in terms of the natural numbers. The profound need for expressing measures of quantities in terms of what we would like to call numbers forces us to extend the number concept so that we may describe a continuous gradation of measures. This extension is called the *number continuum* or the system of "real numbers" (a nondescriptive but generally accepted name). The extension of the number concept to that of the continuum is so convincingly natural that it was used by all the great mathematicians and scientists of earlier times without probing questions. Not until the nineteenth century did mathematicians feel compelled to seek a firmer logical foundation for the real number system. The ensuing precise formulation of the concepts, in turn, led to further progress in mathematics. We shall begin with an unencumbered intuitive approach, and later on we shall give a deeper analysis of the system of real numbers.[1]

a. The System of Natural Numbers and Its Extension. Counting and Measuring

The Natural and the Rational Numbers.   The sequence of "natural" numbers 1, 2, 3, . . . is considered as given to us. We need not discuss how these abstract entities, the numbers, may be categorized from a philosophical point of view. For the mathematician, and for anybody working with numbers, it is important merely to know the rules or laws by which they may be combined to yield other natural numbers. These laws form the basis of the familiar rules for adding and multiplying numbers in the decimal system; they include the *commutative laws* $a + b = b + a$ and $ab = ba$, the *associative laws* $a + (b + c) = (a + b) + c$ and $a(bc) = (ab)c$, the *distributive law* $a(b + c) = ab + ac$, the cancellation law that $a + c = b + c$ implies $a = b$, etc.

The inverse operations, subtraction and division, are not always possible within the set of natural numbers; we cannot subtract 2 from 1 or divide 1 by 2 and stay within that set. To make these operations possible without restriction we are forced to extend the concept of number by inventing the number 0, the "negative" integers, and the fractions. The totality of all these numbers is called the class or set of *rational numbers*; they are all obtained from unity by using the "rational operations" of calculation, namely, addition, subtraction, multiplication, and division.[2]

A rational number can always be written in the form p/q, where p

---

[1] A more complete exposition is given in *What Is Mathematics?* by Courant and Robbins, Oxford University Press, 1962.

[2] The word "rational" here does not mean reasonable or logical but is derived from the word "ratio" meaning the relative proportion of two magnitudes.

and q are integers and $q \neq 0$. We can make this representation unique
by requiring that q is positive and that p and q have no common factor
larger than 1.

Within the domain of rational numbers all the *rational operations,*
addition, multiplication, subtraction, and division (except division by
zero), can be performed and produce again rational numbers. As we
know from elementary arithmetic, operations with rational numbers
obey the same laws as operations with natural numbers: thus the
rational numbers extend the system of positive integers in a completely straightforward way.

Graphical Representation of Rational Numbers. Rational numbers
are usually represented graphically by points on a straight line L,
the *number axis.* Taking an arbitrary point of L as the origin or point 0

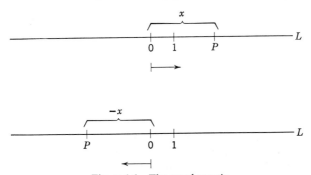

Figure 1.1   The number axis.

and another arbitrary point as the point 1, we use the distance between
these two points to serve as a scale or unit of measurement and define the
direction from 0 to 1 as "positive." The line with a direction thus
imposed is called a directed line. It is customary to depict L so that
the point 1 is to the right of the point 0 (Fig. 1.1). The location of any
point P on L is completely determined by two pieces of information:
the distance of P from the origin 0 and the direction from 0 to P (to the
right or left of 0). The point P on L representing a positive rational
number lies at distance x units to the right of 0. A negative rational
number x is represented by the point $-x$ units to the left of 0. In either
case the distance from 0 to the point which represents x is called the
absolute value of x, written $|x|$, and we have

$$|x| = \begin{cases} x, & \text{if } x \text{ is positive or zero,} \\ -x, & \text{if } x \text{ is negative.} \end{cases}$$

We note that $|x|$ is never negative and equals zero only when $x = 0$.

From elementary geometry we recall that with ruler and compass it is possible to construct a subdivision of the unit length into any number of equal parts. It follows that any rational length can be constructed and hence that the point representing a rational number x can be found by purely geometrical methods.

In this way we obtain a geometrical representation of rational numbers by points on L, the *rational points*. Consistent with our notation for the points 0 and 1, we take the liberty of denoting both the rational number and the corresponding point on L by the same symbol x.

The relation $x < y$ for two rational numbers means geometrically that the point x lies to the left of the point y. In that case the distance between the points is $y - x$ units. If $x > y$, the distance is $x - y$ units. In either case the distance between two rational points x, y of L is $|y - x|$ units and is again a rational number.

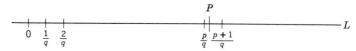

Figure 1.2

A segment on L with end points a, b where $a < b$ will be called an *interval*. The particular segment with end points 0, 1 is called the *unit interval*. If the end points are included in the interval, we say the interval is *closed*; if the end points are excluded, the interval is called *open*. The open interval, denoted by (a, b), consists of those points x for which $a < x < b$, that is, of those points that lie "between" a and b. The closed interval, denoted by $[a, b]$, consists of the points x for which $a \leq x \leq b$.[1] In either case the *length of the interval* is $b - a$.

The points corresponding to the integers 0, ± 1, ± 2, . . . subdivide the number axis into intervals of unit length. Every point on L is either an end point or interior point of one of the intervals of the subdivision. If we further subdivide every interval into q equal parts, we obtain a subdivision of L into intervals of length $1/q$ by rational points of the form p/q. Every point P of L is then either a rational point of the form p/q or lies between two successive rational points p/q and $(p + 1)/q$ (see Fig. 1.2). Since successive points of subdivision are $1/q$ units apart, it follows that we can find a rational point p/q whose distance from P does not exceed $1/q$ units. The number $1/q$ can be made as small as we please by choosing q as a sufficiently large positive integer. For example, choosing $q = 10^n$ (where n is any natural number) we can

---

[1] The relation $a \leq x$ (read "*a* less than or equal to *x*") is interpreted as "either $a < x$, or $a = x$." We interpret the double signs $\geq$ and $\pm$ in similar fashion.

find a "decimal fraction" $x = p/10^n$ whose distance from P is less than $1/10^n$. Although we do not assert that every point of L is a rational point we see at least that rational points can be found arbitrarily close to any point P of L.

Density

The arbitrary closeness of rational points to a given point P of L is expressed by saying: *The rational points are dense on the number axis.* It is clear that even smaller sets of rational numbers are dense, for example, the points $x = p/10^n$, for all natural numbers n and integers p.

Density implies that between any two distinct rational points a and b there are infinitely many other rational points. In particular, the point halfway between a and b, $c = \frac{1}{2}(a + b)$, corresponding to the arithmetic mean of the numbers a and b, is again rational. Taking the midpoints of a and c, of b and c, and continuing in this manner, we can obtain any number of rational points between a and b.

An arbitrary point P on L can be located to any degree of precision by using rational points. At first glance it might then seem that the task of locating P by a number has been achieved by introducing the rational numbers. After all, in physical reality quantities are never given or known with absolute precision but always only with a degree of uncertainty and therefore might just as well be considered as measured by rational numbers.

Incommensurable Quantities. Dense as the rational numbers are, they do not suffice as a theoretical basis of measurement by numbers. Two quantities whose ratio is a rational number are called *commensurable* because they can be expressed as integral multiples of a common unit. As early as in the fifth or sixth century B.C. Greek mathematicians and philosophers made the surprising and profoundly exciting discovery that there exist quantities which are not commensurable with a given unit. In particular, line segments exist which are not rational multiples of a given unit segment.

It is easy to give an example of a length incommensurable with the unit length: the diagonal l of a square with the sides of unit length. For, by the theorem of Pythagoras, the square of this length l must be equal to 2. Therefore, if l were a rational number and consequently equal to p/q, where p and q are positive integers, we should have $p^2 = 2q^2$. We can assume that p and q have no common factors, for such common factors could be canceled out to begin with. According to the above equation, p^2 is an even number; hence p itself must be even, say $p = 2p'$. Substituting $2p'$ for p gives us $4p'^2 = 2q^2$, or $q^2 = 2p'^2$; consequently, q^2

is even and so q is also even. This proves that p and q both have the factor 2. However, this contradicts our hypothesis that p and q have no common factor. Since the assumption that the diagonal can be represented by a fraction p/q leads to a contradiction, it is false.

This reasoning, a characteristic example of *indirect proof*, shows that the symbol $\sqrt{2}$ cannot correspond to any rational number. Another example is π, the ratio of the circumference of a circle to its diameter. The proof that π is not rational is much more complicated and was obtained only in modern times (Lambert, 1761). It is easy to find many incommensurable quantities (see Problem 1, p. 106); in fact, incommensurable quantities are in a sense far more common than the commensurable ones (see p. 99).

Irrational Numbers

Because the system of rational numbers is not sufficient for geometry, it is necessary to invent new numbers as measures of incommensurable quantities: these new numbers are called "irrational." The ancient Greeks did not emphasize the abstract number concept, but considered geometric entities, such as line segments, as the basic elements. In a purely geometrical way, they developed a logical system for dealing and operating with incommensurable quantities as well as commensurable (rational) ones. This important achievement, initiated by the Pythagoreans, was greatly advanced by Eudoxus and is expressed at length in Euclid's famous *Elements*. In modern times mathematics was recreated and vastly expanded on a foundation of number concepts rather than geometrical ones. With the introduction of analytic geometry a reversal of emphasis developed in the ancient relationship between numbers and geometrical quantities and the classical theory of incommensurables was all but forgotten or disregarded. It was assumed as a matter of course that to every point on the number axis there corresponds a rational or irrational number and that this totality of "real" numbers obeys the same arithmetical laws as the rational numbers do. Only later, in the nineteenth century, was the need for justifying such an assumption felt and was eventually completely satisfied in a remarkable booklet by Dedekind which makes fascinating reading even today.[1]

---

[1] R. Dedekind, "Nature and Meaning of Number" in *Essays on Number*, London and Chicago, 1901. (The first of these essays, "Continuity and Irrational Numbers," supplies a detailed account of the definition and laws of operation with real numbers.) Reprinted under title *Essays on the Theory of Numbers*, Dover, New York, 1964. The original of these translations appeared in 1887 under the title "Was sind und wass sollen die Zahlen?"

In effect, Dedekind showed that the "naive" approach practiced by all the great mathematicians from Fermat and Newton to Gauss and Riemann was on the right track: That the system of real numbers (as symbols for the lengths of segments, or otherwise defined) is a consistent and complete instrument for scientific measurement, and that in this system the rules of computation of the rational number system remain valid.

Without harm, one could leave it at that and turn directly to the substance of calculus. However, for a deeper understanding of the concept of real number, which is necessary for our later work, the following account as well as the Supplement to this chapter should be studied.

b. Real Numbers and Nested Intervals

For the moment let us think of the points on a line L as the basic elements of the continuum. We postulate that to each point on L there corresponds a "real number" x, its *coordinate*, and that for these numbers x, y the relationships just described for the rational numbers retain their meaning. In particular, the relationship $x < y$ indicates order on L and the expression $|y - x|$ means the distance between the point x and the point y. The basic problem is to relate these numbers (or measurements on the geometrically given continuum of points) to the rational numbers considered originally and hence ultimately to the integers. In addition, we have to explain how to operate with the elements of this "number-continuum" in the same way as with the rational numbers. Eventually, we shall formulate the concept of the continuum of numbers independently of the intuitive geometric concepts, but for the present we postpone some of the more abstract discussion to the Supplement.

How can we describe an irrational real number? For some numbers such as $\sqrt{2}$ or π, we can give a simple geometric characterization, but that is not always feasible. A method flexible enough to yield every real point consists in describing the value x by a sequence of rational approximations of greater and greater precision. Specifically, we shall approximate x simultaneously from the right and from the left with successively increasing accuracy and in such a way that the margin of error approaches zero. In other words, we use a "sequence" of rational intervals containing x, with each interval of the sequence containing the next one, such that the length of the interval, and with it the error of the approximation, can be made smaller than any specified positive number by taking intervals sufficiently far along in the sequence.

To begin, let x be confined to a closed interval $I_1 = [a_1, b_1]$, that is,

$$a_1 \leq x \leq b_1,$$

where a_1 and b_1 are rational (see Fig. 1.3). Within I_1 we consider a "subinterval" $I_2 = [a_2, b_2]$ containing x, that is,

$$a_1 \leq a_2 \leq x \leq b_2 \leq b_1,$$

where a_2 and b_2 are rational. For example, we may choose for I_2 one of the halves of I_1, for x must lie in one or both of the half-intervals. Within I_2 we consider a subinterval $I_3 = [a_3, b_3]$ which also contains x:

$$a_1 \leq a_2 \leq a_3 \leq x \leq b_3 \leq b_2 \leq b_1,$$

where a_3 and b_3 are rational, etc. We require that the length of the interval I_n *tends to zero* with increasing n; that is, that the length of I_n is less than any preassigned positive number for all sufficiently large n. A set of closed intervals $I_1, I_2, I_3, \ldots$ each containing the

Figure 1.3   A nested sequence of intervals.

next one and such that the lengths tend to zero will be called a "nested sequence of intervals." The point x is uniquely determined by the nested sequence; that is, no other point y can lie in all I_n, since the distance between x and y would exceed the length of I_n once n is sufficiently large. Since here we always choose rational points for the end points of the I_n and since every interval with rational end points is described by two rational numbers, we see that every point x of L, that is, every real number, can be precisely described with the help of infinitely many rational numbers. The converse statement is not so obvious; we shall accept it as a basic *axiom*.

POSTULATE OF NESTED INTERVALS. *If I_1, I_2, I_3, ... form a nested sequence of intervals with rational end points, there is a point x contained in all I_n.*[1]

As we shall see, this is an *axiom of continuity*: it guarantees that no gaps exist on the real axis. We shall use the axiom to characterize the real continuum and to justify all operations with limits which are

---

[1] It is important to emphasize for a nested sequence that the intervals I_n are *closed*. If, for example, I_n denotes the open interval $0 < x < 1/n$, then each I_n contains the following one and the lengths of the intervals tend to zero; but there is no x contained in all I_n.

basic for calculus and analysis. (There also are many other ways of formulating this axiom as we shall see later.)

c. Decimal Fractions. Bases Other than Ten

Infinite Decimal Fractions. One of the many ways of defining real numbers is the familiar description in terms of infinite decimals. It is entirely possible to take the infinite decimals as the basic objects rather than the points of the number axis, but we would rather proceed in a more suggestive geometrical way by defining the infinite decimal representation of real numbers in terms of nested sequences of intervals.

Let the number axis be subdivided into unit intervals by the points corresponding to integers. A point x either lies between two successive points of subdivision or is itself one of the dividing points. In either case there is at least one integer c_0 such that

$$c_0 \leq x \leq c_0 + 1,$$

so that x belongs to the closed interval $I_0 = [c_0, c_0 + 1]$. We divide I_0 into ten equal parts by points $c_0 + \frac{1}{10}, c_0 + \frac{2}{10}, \ldots, c_0 + \frac{9}{10}$. The point x must then belong to at least one of the closed subintervals of I_0 (possibly to two adjacent ones if x is one of the points of subdivision). In other words, there is a digit c_1 (that is, one of the integers 0, 1, 2, ..., 9) such that x belongs to the closed interval I_1 given by

$$c_0 + \tfrac{1}{10}c_1 \leq x \leq c_0 + \tfrac{1}{10}c_1 + \tfrac{1}{10}.$$

Dividing I_1 in turn into ten equal parts, we find a digit c_2 such that x lies in the interval I_2 given by

$$c_0 + \tfrac{1}{10}c_1 + \tfrac{1}{100}c_2 \leq x \leq c_0 + \tfrac{1}{10}c_1 + \tfrac{1}{100}c_2 + \tfrac{1}{100}.$$

We repeat this process. After n steps x is confined to an interval I_n given by

$$c_0 + \frac{1}{10}c_1 + \cdots + \frac{1}{10^n}c_n \leq x \leq c_0 + \frac{1}{10}c_1 + \cdots + \frac{1}{10^n}c_n + \frac{1}{10^n},$$

where $c_1, c_2, \ldots$ are all digits. The interval I_n has length $1/10^n$, which tends to zero for increasing n. It is clear that the I_n form a nested set of intervals, and hence that x is determined uniquely by the I_n. Since the I_n are known, once the numbers $c_0, c_1, c_2, \ldots$ are given we find that an arbitrary real number can be described completely by an infinite sequence of integers $c_0, c_1, c_2, \ldots$, where all except the first are digits, having values from zero to nine only. In ordinary decimal notation the connection between x and $c_0, c_1, c_2, \ldots$ is indicated by writing

$$x = c_0 + 0.c_1 c_2 c_3 \cdots.$$

(Usually, the integer c_0 itself is also written in decimal notation if c_0 is positive.) Conversely, by the axiom of continuity, every such expression denoting an infinite decimal fraction represents a real number.

It is possible that there are two different decimal representations of the same number; for example,

$$1 = 0.99999 \cdots = 1.00000 \cdots.$$

In our construction the integer c_0 is determined uniquely by x unless x itself is an integer. In that case we could choose either $c_0 = x$ or $c_0 = x - 1$. Once a choice has been made the digit c_1 is unique unless x is one of the new points subdividing I_0 into ten equal parts. Continuing we find that c_0 and all c_k are determined uniquely by x unless x occurs as a point of subdivision at some stage. If this should happen for the first time at the nth stage, then

$$x = c_0 + \frac{1}{10} c_1 + \cdots + \frac{1}{10^n} c_n,$$

where $c_1, c_2, \ldots, c_n$ are digits and where $c_n > 0$, since otherwise x would have been a point of subdivision at an earlier stage. It follows that I_{n+1} is either the interval $[x, x + 1/10^{n+1}]$ or the interval $[x - 1/10^{n+1}, x]$. In the first case x will be the left-hand end point of all later intervals $I_{n+2}, I_{n+3}, \ldots$, and in the second case, the right-hand end point. We are then led either to the decimal representation

$$x = c_0 + 0.c_1 c_2 \cdots c_n 000 \cdots$$

or the representation

$$x = c_0 + 0.c_1 c_2 \cdots (c_n - 1)99999 \cdots.$$

Hence the only case in which an ambiguity can arise is for a rational number x which can be written as a fraction having a power of ten for its denominator. We can eliminate even this ambiguity by excluding decimal representations in which all digits from a certain point on are nines.

In the decimal representation of real numbers the special role played by the number ten is purely incidental. The only evident reason for the widespread use of the decimal system is the ease of counting by tens on our fingers (digits). Any integer p greater than one can serve equally well. We could use p equal subdivisions at each stage. A real number x would then be represented in the form

$$x = c_0 + 0.c_1c_2c_3\cdots,$$

where c_0 is an integer, and now $c_1, c_2, \ldots$ have one of the values $0, 1, 2, \ldots, p - 1$. This representation again characterizes x by a nested set of intervals, namely

$$c_0 + \frac{1}{p}c_1 + \cdots + \frac{1}{p^n}c_n \leq x \leq c_0 + \frac{1}{p}c_1 + \cdots + \frac{1}{p^n}c_n + \frac{1}{p^n}.$$

If x is positive or zero, the integer c_0 is also positive or zero and c_0 itself has a finite expansion of the form

$$c_0 = d_0 + pd_1 + p^2d_2 + \cdots + p^kd_k,$$

where $d_0, d_1, \ldots, d_k$ take one of the values $0, 1, \ldots, p - 1$. The complete representation of x "to the base p" takes the form

$$x = d_kd_{k-1}\cdots d_1d_0.c_1c_2c_3\cdots.$$

If x is negative, we may use this kind of representation for $-x$.

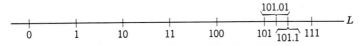

Figure 1.4   The fraction $\frac{21}{4}$ in the binary system.

Bases other than 10 have actually been used extensively. Following the lead of the ancient Babylonians, astronomers for many centuries consistently represented numbers as "sexagesimal" fractions with $p = 60$ as the base.

Binary Representation.   The "binary" system with the base $p = 2$ has special theoretical interest and is useful in the logical design of computing machines. In the binary system the digits have only two possible values, zero and one. The number $\frac{21}{4}$, for example, would be written 101.01 corresponding to the formula

$$\frac{21}{4} = 2^2 \cdot 1 + 2^1 \cdot 0 + 1 \cdot 1 + \frac{1}{2} \cdot 0 + \frac{1}{2^2} \cdot 1 \qquad \text{(see Fig. 1.4)}.$$

Calculating with Real Numbers.   Although the definition of real numbers and their infinite decimal or binary representations, etc., are straightforward, it may not seem obvious that one can operate with the number continuum exactly as with rational numbers, performing the rational operations and retaining the laws of arithmetic, such as the associative, the commutative, and the distributive laws. The proof is simple, although somewhat tedious. Instead of impeding the way to the live substance of analysis by taking up the question here, we shall accept temporarily the possibility of ordinary arithmetic calculation with the real numbers. A deeper understanding of the logical structure underlying the number concept will come when we discover the idea of limit and its implications. (See the Supplement to this chapter, p. 89.)

d. Definition of Neighborhood

Not only the rational operations but also order relations or in-equalities for real numbers obey the same rules as for the rational numbers.

Pairs of real numbers a and b with $a < b$ again give rise to closed intervals $[a, b]$ (given by $a \leq x \leq b$) and open intervals (a, b) (given by $a < x < b$). Frequently we shall be led to associate with a point x_0 the various open intervals that contain that point or specifically have it as center, which we shall call *neighborhoods* of the point. More precisely, for any positive ϵ the ϵ-*neighborhood* of the point x_0 consists of the values x for which $x_0 - \epsilon < x < x_0 + \epsilon$, that is, it is the interval $(x_0 - \epsilon, x_0 + \epsilon)$. Any open interval (a, b) containing a point x_0 always also contains a whole neighborhood of x_0.

Having defined intervals with real end points we can now form *nested sequences* of intervals using the same definition as in the case of rational end points. It is most important for the logical consistency of calculus that for any nested sequence of intervals with real end points there is a real number contained in all of them. (See Supplement, p. 95.)

e. Inequalities

Basic Rules

Inequalities play a far larger role in higher mathematics than in elementary mathematics. Often the precise value of a quantity x is difficult to determine, whereas it may be easy to make an estimate of x, that is, to show that x is greater than some known quantity a and less than some other quantity b. For many purposes, only the information contained in such an estimate of x is significant. We shall therefore briefly recall some of the elementary rules about inequalities.

The basic fact is that the sum and product of two positive real numbers are again positive; that is, if $a > 0$ and $b > 0$, then $a + b > 0$

and $ab > 0$. Moreover, we rely on the fact that the inequality $a > b$ is equivalent to $a - b > 0$. Consequently, two inequalities $a > b$ and $c > d$ can be *added* to yield the inequality $a + c > b + d$ since

$$(a + c) - (b + d) = (a - b) + (c - d)$$

is positive as the sum of two positive numbers. (Subtracting the inequalities to obtain $a - c > b - d$ is not legitimate. Why?) An inequality can be *multiplied by a positive number*; that is, if $a > b$ and $c > 0$, then $ac > bc$. For the proof, we observe that

$$ac - bc = (a - b)c$$

is positive since it is the product of positive numbers. If c is *negative*, we can conclude from $a > b$ that $ac < bc$. More generally, it follows from $a > b > 0$ and $c > d > 0$ that $ac > bd$.

It is geometrically obvious that inequality is *transitive*; that is, if $a > b$ and $b > c$, then $a > c$. Transitivity[1] also follows immediately from the positivity of the sum

$$(a - b) + (b - c) = a - c.$$

The preceding rules also hold if we replace the sign $>$ by $\geq$ everywhere.

Let a and b be *positive* numbers and observe that

$$a^2 - b^2 = (a + b)(a - b).$$

Since $a + b$ is positive, we conclude that $a^2 > b^2$ follows from $a > b$. Thus an inequality between positive numbers can be "squared." Similarly, $a^2 \geq b^2$ whenever $a \geq b \geq 0$. From the equation

$$a - b = \frac{1}{a + b}(a^2 - b^2),$$

valid for all positive a and b, it follows that the converse is also true; that is, for positive a and b, $a^2 > b^2$ implies $a > b$. Applying this result to the numbers $a = \sqrt{x}$, $b = \sqrt{y}$, for arbitrary positive real numbers x, y, we find[2] that $\sqrt{x} > \sqrt{y}$ when $x > y$. More generally, $\sqrt{x} \geq \sqrt{y}$ whenever $x \geq y \geq 0$. Hence it is legitimate to take the

---

[1] Transitivity justifies the use of the compound formula "$a < b < c \ldots$" to express "$a < b$ and $b < c$, etc." Avoid nontransitive arrangements like $x < y > z$; these are confusing and misleading.

[2] Here and hereafter the symbol $\sqrt{z}$ for $z \geq 0$ denotes that nonnegative number whose square is z. With this convention $|c| = \sqrt{c^2}$ for any real c since $|c| \geq 0$ and $|c|^2 = c^2$. From this we obtain the important identity $|xy| = |x| \cdot |y|$ since

$$|xy|^2 = (xy)^2 = x^2 y^2 = (|x| \cdot |y|)^2.$$

square root of both sides of an inequality between nonnegative real numbers.

Suppose that a and b are positive and n is a positive integer. In the factorization

$$a^n - b^n = (a - b)(a^{n-1} + a^{n-2}b + \cdots + b^{n-1})$$

the second factor is positive. Thus $a^n - b^n$ has the same sign as $a - b$; if $a^n > b^n$, then $a > b$ and if $a^n < b^n$, then $a < b$.

Most inequalities we shall encounter occur in the form of estimates for the absolute value of a number. We recall that $|x|$ is defined to be x for $x \geq 0$ and $-x$ for $x < 0$. We may also say that $|x|$ is the larger of the two numbers x and $-x$ when x is not zero and is equal to both of them when x is zero. The inequality $|x| \leq a$ then states that neither x nor $-x$ exceeds a, that is, that $x \leq a$ *and* $-x \leq a$. Since $-x \leq a$ is equivalent to $x \geq -a$, we see that the inequality $|x| \leq a$ means that x

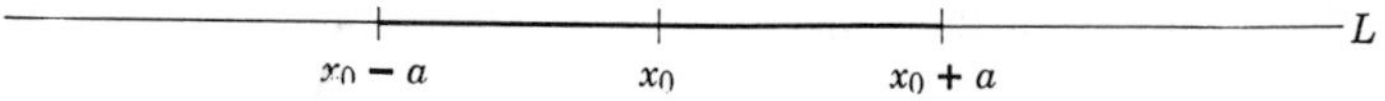

Figure 1.5 The interval $|x - x_0| \leq a$.

lies in the closed interval $-a \leq x \leq a$ with center 0 and length $2a$. The inequality $|x - x_0| \leq a$ then states that $-a \leq x - x_0 \leq a$ or that $x_0 - a \leq x \leq x_0 + a$, thus, that x lies in the closed interval with center x_0 and length $2a$ (see Fig. 1.5). Similarly, the ϵ-neighborhood $(x_0 - \epsilon, x_0 + \epsilon)$ of a point x_0, that is, the open interval $x_0 - \epsilon < x < x_0 + \epsilon$, can be described by the inequality $|x - x_0| < \epsilon$.

Triangle Inequality

One of the most important inequalities involving absolute values is the so-called triangle inequality

$$|a + b| \leq |a| + |b|$$

for any real a, b. The name "triangle inequality" is more appropriate for the equivalent statement

$$|\alpha - \beta| \leq |\alpha - \gamma| + |\gamma - \beta|$$

for which we have set $a = \alpha - \gamma$, $b = \gamma - \beta$. The geometrical interpretation of this statement is that the direct distance from α to β is less than or equal to the sum of the distances via a third point γ; (this also corresponds to the fact that in any triangle the sum of the two sides exceeds the third side).

A formal proof of the triangle inequality is easily given. We distinguish the cases $a + b \geq 0$ and $a + b < 0$. In the first case the inequality states that $a + b \leq |a| + |b|$: but this follows trivially by addition of the inequalities $a \leq |a|$ and $b \leq |b|$. In the second case the triangle inequality reduces to $-(a + b) \leq |a| + |b|$, which again follows by addition from $-a \leq |a|$, $-b \leq |b|$.

We immediately derive an analogous inequality for three quantities:

$$|a + b + c| \leq |a| + |b| + |c|;$$

for, by applying the triangle inequality twice,

$$|a + b + c| = |(a + b) + c| \leq |a + b| + |c| \leq |a| + |b| + |c|.$$

In the same way, the more general inequality

$$|a_1 + a_2 + \cdots + a_n| \leq |a_1| + |a_2| + \cdots + |a_n|$$

is derived.

Occasionally we need estimates for $|a + b|$ from below. We observe that

$$|a| = |(a + b) + (-b)| \leq |a + b| + |-b| = |a + b| + |b|$$

and hence that the inequality

$$|a + b| \geq |a| - |b|$$

holds.

The Cauchy-Schwarz Inequality

Some of the most important inequalities exploit the obvious fact that the square of a real number is never negative and that consequently a sum of squares also cannot be negative. One of the most frequently used results obtained in this way is the Cauchy-Schwarz inequality

$$(a_1 b_1 + a_2 b_2 + \cdots + a_n b_n)^2$$
$$\leq (a_1^2 + a_2^2 + \cdots + a_n^2)(b_1^2 + b_2^2 + \cdots + b_n^2).$$

Putting

$$A = a_1^2 + a_2^2 + \cdots + a_n^2,$$
$$B = a_1 b_1 + a_2 b_2 + \cdots + a_n b_n,$$
$$C = b_1^2 + b_2^2 + \cdots + b_n^2,$$

the inequality becomes $AC \geq B^2$. To prove it we observe that for any real t

$$0 \leq (a_1 + t b_1)^2 + (a_2 + t b_2)^2 + \cdots + (a_n + t b_n)^2$$

since the right-hand side is a sum of squares. Expanding each square and arranging according to powers of t, we find that

$$0 \leq A + 2Bt + Ct^2$$

for all t, where A, B, C have the same meaning as before. Here $C \geq 0$. We may assume that $C > 0$, since certainly $B^2 = AC = 0$ when $C = 0$. Substituting then for t the special value $t = -B/C$ [corresponding to the minimum of the quadratic expression

$$A + 2Bt + Ct^2 = C\left(t + \frac{B}{C}\right)^2 + \left(A - \frac{B^2}{C}\right)\Big],$$

we find

$$0 \leq A - \frac{2B^2}{C} + \frac{B^2}{C} = \frac{AC - B^2}{C}$$

and hence $AC - B^2 \geq 0$.

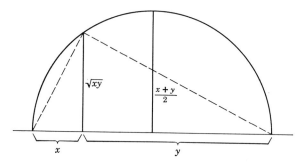

Figure 1.6   Geometric and arithmetic means of x and y.

In the special case $n = 2$ we can choose

$$a_1 = \sqrt{x}, \qquad a_2 = \sqrt{y}, \qquad b_1 = \sqrt{y}, \qquad b_2 = \sqrt{x},$$

where x and y are positive numbers. The inequality then takes the form $(2\sqrt{xy})^2 \leq (x + y)^2$ or

$$\sqrt{xy} \leq \frac{x + y}{2}.$$

This inequality states that the *geometric mean* $\sqrt{xy}$ of two positive numbers x, y never exceeds their *arithmetic mean* $(x + y)/2$. The geometric mean of two numbers x, y can be interpreted as the length of the altitude of a right triangle dividing the hypotenuse into segments of length x and y respectively. The inequality then states that in a right triangle the altitude does not exceed half the hypotenuse (see Fig. 1.6).[1]

1.2 The Concept of Function

From the beginning of modern mathematics in the 17th century the concept of function has been at the very center of mathematical thought. (Leibnitz appears to have been the first to use the word "function".) Although the idea of functional relationships is significant far beyond the mathematical domain, we shall naturally focus our attention on functions in the mathematical sense, that is, on the connection of mathematical quantities by mathematical relations or prescriptions or "operations." A very large part of mathematics and the natural sciences is dominated by functional relationships, for they occur everywhere in analysis, geometry, mechanics, and other fields. For example, the pressure in an ideal gas is a function of density and temperature; the position of a moving molecule is a function of the time; the volume and surface of a cylinder are functions of its radius and height. Whenever the values of certain quantities $a, b, c, \ldots$ are determined by those of certain others $x, y, z, \ldots$, we say that $a, b, c, \ldots$ *depend* on $x, y, z, \ldots$ or are *functions* of $x, y, z, \ldots$.   Examples of functional relations are given by formal expressions such as the following.

(*a*) The formula $A = a^2$ defines A as a function of a. For $a > 0$ we can interpret A as the area of a square of side a.

(*b*) The formula

$$y = \sqrt{1 - x^2}$$

defines y as a function of x for all x for which $-1 \leq x \leq 1$. For $x > 0$ this function expresses the side y of a right triangle with hypotenuse 1 in terms of the other side x.

(*c*) The equations

$$x = t, \qquad y = -t^2$$

assign values of x and y to each t and thus define x and y as functions of t. If we interpret x and y as the rectangular coordinates of a point P in the plane and t as the time, then our equations describe the location of P at the time t; in other words, they describe the *motion* of the point P.

(*d*) The equations

$$a = \frac{x}{x^2 + y^2}, \qquad b = \frac{y}{x^2 + y^2}$$

---

[1] The interested reader will find more material in *An Introduction to Inequalities*, by E. F. Beckenbach and R. Bellman, Random House, 1961, and *Geometric Inequalities*, by N. Kazarinoff, Random House, 1961.

define a and b as functions of x and y for $x^2 + y^2 \neq 0$. Interpreting the pairs of values x, y and a, b as rectangular coordinates of two points, we see that the equations assign to each point (x, y) [with the exception of the origin $(0, 0)$] an "image" (a, b). The reader can verify easily that the image (a, b) always lies on the same ray from the origin as the "original" or "antecedent" (x, y) and has the reciprocal distance from the origin. We speak of "mapping" (x, y) onto (a, b) by means of the equations expressing a, b in terms of x, y.

In the preceding examples the functional law is expressed by simple formulas which determine certain quantities in terms of certain others.[1] The quantities appearing on the left-hand sides, the "dependent variables," are expressed in terms of the "independent variables" on the right. The mathematical law assigning unique values of the dependent variables to given values of the independent variables is called a function. It is unaffected by the names x, y, etc., for these variables. In Example c we have an independent variable t and two dependent variables x, y, whereas in Example d there are two independent variables x, y and two dependent variables a, b.

The dependence of y on x by a functional relation is frequently indicated by the brief expression "y is a function of x."[2]

a. Mapping-Graph

Domain and Range of a Function

We usually interpret the independent variables geometrically as coordinates of a point in one or more dimensions. In Example b this would be a point on the x-axis, in Example d a point in the x,y-plane. Sometimes the independent variables are free to take all values, as in examples a and c. Often, however, there is some restriction, inherent or imposed, and our functions are not defined for all values. The set of values or the points for which a function is defined form the "domain" of the function. In Example a the domain is the whole a-axis, in b the interval $-1 \leq x \leq 1$, in c the whole t-axis, and in d the points of the x,y-plane different from the origin.

To each point P in the domain our functions assign definite values

---

[1] Later we shall gradually realize the need for considering functions not capable of such representation by simple formulas. (See, for example, p. 25.)

[2] This locution is used freely in the sciences, but some of the more pedantic texts avoid it. There is no point in hampering ourselves by an undue concern for hair-splitting "precision" when it has no relation to the substance.

for the dependent variables. These values also can be interpreted as coordinates of a point Q, the *image* of P. We say that P is "mapped" by our functions onto the point Q. Thus in Example d the point $P = (1,2)$ of the x,y-plane is mapped onto the point $Q = (\frac{1}{5}, \frac{2}{5})$ of the a,b-plane. The image points Q form the *range* of the function.[1] Each Q in the range is the image of one (or more) points in the domain of the function.

In Example c points of the t-axis have as their images points in the x,y-plane. The t-axis is mapped *into* the x,y-plane. But not every point of the x,y-plane occurs as image, only those for which $y = -x^2$. Thus the range of the mapping is the parabola $y = -x^2$. We say, the t-axis is mapped *onto* the parabola $y = -x^2$, in the sense that the image points fill this parabola.

In Example d the range consists of the points (a, b) in the a,b-plane whose coordinates can be written in the form $a = x/(x^2 + y^2)$, $b = y/(x^2 + y^2)$ with suitable x, y for which $x^2 + y^2 \neq 0$. In other words, the range consists of those points (a, b) for which the preceding equations have a solution (x, y). As seen immediately the range consists of the points (a, b) for which a and b do not both vanish; each such point (a, b) is image of the point $x = a/(a^2 + b^2)$, $y = b/(a^2 + b^2)$. Every geometrical figure in the x,y-plane is then mapped onto a corresponding figure in the a,b-plane which consists of the images of the points of the first figure. For example, a circle $x^2 + y^2 = r^2$ about the origin is mapped onto the circle $a^2 + b^2 = 1/r^2$ in the a,b-plane.

In this and the following chapters we shall deal almost exclusively with a single independent variable, say x, and a single dependent variable, say y, as indicated in Example b.[2] Ordinarily we represent such a function in the standard way by its *graph* in the x,y-plane, that is, by the curve consisting of those points (x, y) whose ordinate is in the specified functional relationship to the abscissa x (see Fig. 1.7). For Example b the graph is the upper half of a circle of radius one about the origin.

The interpretation of the function as a mapping of a domain on the x-axis onto a range on the y-axis leads to a different visualization of functions. We interpret x and y not as coordinates of the same point in the x,y-plane, but as points on two different, independent number

---

[1] It is often convenient to talk of the point Q as "a function" of P, although in the analytic representation several functions expressing the different coordinates of Q appear.

[2] However, it should be emphasized from the beginning that functions of several variables occur just as naturally in many instances. They will be discussed systematically in Volume II.

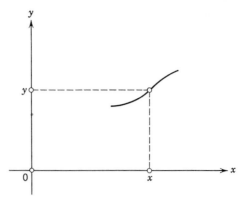

Figure 1.7   Graph of function.

axes.   Then the function maps a point x on the x-axis into a point y on the y-axis.   Such mappings arise frequently in geometry, such as the "affine" mapping which originates by projecting a point x on the x-axis onto a point y on a parallel y-axis from a center 0 located in the plane of the two axes (see Fig. 1.8).   This mapping can be expressed analytically, as easily ascertained, by the linear function $y = ax + b$ with

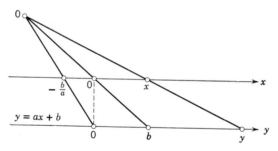

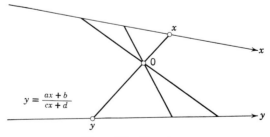

Figure 1.8   Mappings.

constants a and b. Obviously, it is a "one-to-one" mapping in which inversely to the image y, there corresponds a unique original x. Another, more general, example is the "perspective mapping" defined by the same sort of projection, only with the two axes not necessarily parallel. Here the analytical expression is given by a rational linear function of the form $y = (ax + b)/(cx + d)$, with constants a, b, c, d.

Any projection of a surface S in space into another surface S' from some center N can be viewed as a mapping whose domain is S and whose range lies on S'. For example, we can map a sphere onto an equatorial plane by projecting each point P of the sphere onto a point P' of the plane by rays from the North Pole (see Fig. 1.9). This mapping

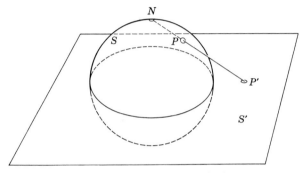

Figure 1.9   Stereographic projection.

is the "stereographic projection" used frequently for maps of the earth. The interpretation of functions as "maps" is suggested by examples of this type.

When more independent or dependent variables are involved, the definition of functions by mapping provides a more flexible and suitable interpretation than that by graphs. This fact will become fully apparent in the second volume.

b. Definition of the Concept of Functions of a Continuous Variable. Domain and Range of a Function

A function of a single independent variable x assigns values y to values x. The *domain* of the function is the totality of values x for which the function is defined. In the cases that concern us most the domain of the function consists of one or several intervals (see Fig. 1.10). We say then that y is a function of a *continuous variable* (in contrast to other cases where, for example, the function might only be defined for rational or for integral values of x). Here the "intervals" forming the domain may or may not contain their end points and may also extend to infinity in one or both directions.[1] Thus the function $y = \sqrt{1 - x^2}$ is defined in the closed interval $-1 \leq x \leq +1$, the function $y = 1/x$ in the two semi-infinite open intervals $x < 0$ and $x > 0$, the function $y = x^2$ in the infinite "interval" $-\infty < x < +\infty$

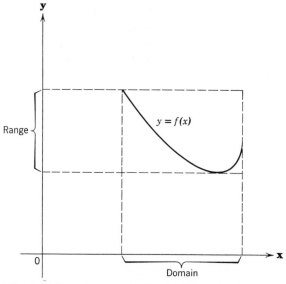

Figure 1.10   Domain and range of a function in graphical representation.

consisting of all x, the function $y = \sqrt{(x^2 - 1)(4 - x^2)}$ in the two separate intervals $1 \leq x \leq 2$ and $-2 \leq x \leq -1$.

Functions are denoted by symbols such as f, F, g, etc. The corresponding relations between x and the associated y-values are written in the form $y = f(x)$ or $y = F(x)$ or $y = g(x)$, etc., or also sometimes $y = y(x)$ to indicate[2] that y depends on x. If, for example, $f(x)$ is defined by the expression $x^2 + 1$ we have $f(3) = 3^2 + 1 = 10, f(-1) = (-1)^2 + 1 = 2$.

---

[1] Ordinarily we will reserve the word "interval" for "bounded," that is, "finite" intervals, that have definite finite end points; then one might indicate the more comprehensive concept as used in the text, by the word "convex sets," meaning sets which when containing two points must contain all intermediate ones.

[2] In this notation we try to emphasize the variables and do not explicitly indicate the functional operation by a symbol such as f. The notation

$$f: \quad x \to y$$

for the function f mapping x into y is also sometimes encountered.

Nature of Functional Relation

In the general definition of a function $f(x)$ nothing is said about the nature of the relation by which the dependent variable is found when the independent variable is given. As said before, often the function is given in "closed form" by a simple expression like $f(x) = x^2 + 1$ or $f(x) = \sqrt{1 + \sin^2 x}$, and in the early days of the calculus such explicit expressions were mostly what mathematicians meant by functions. Often mechanical devices generate geometric curves or

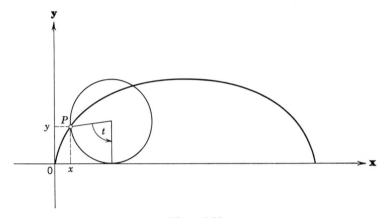

Figure 1.11

graphs which then define functions. A striking example is the cycloid, a curve described by a point fixed on a circle which rolls along the x-axis (see Fig. 1.11). Its functional analytical expression by formulas will be given later (see p. 328).

Logically, we are not restricted to such geometrically or mechanically generated functions. Any rule by which a value of y is assigned to values of x constitutes a function. In some theoretical investigations the wide generality or vagueness of the function concept is, in fact, an advantage. However, for applications, particularly in the calculus, the general concept of function is unnecessarily wide. To make meaningful mathematical developments possible, the "arbitrary" laws of correspondence by which a value of y is assigned to x must be subjected to radical restrictions. During the past century and a half mathematicians have recognized and formulated in precise terms the essential restrictions that have to be imposed on the overly general concept in order to obtain functions that indeed have the useful properties one would expect intuitively.

*Extended or Restricted Domains of Functions

Even for functions given by explicit formulas, it is important to realize that any complete description of a function must include a definition of the *domain* of the function. For us the "function" f described by "$f(x) = x^2$ for $0 < x < 2$" is not strictly the same function as the function g given by "$g(x) = x^2$ in the larger domain $-2 < x < 2$," although $f(x)$ and $g(x)$ have the same values in the interval $0 < x < 2$ where both are defined. Generally, we call a function f a "restriction" of a function g (or g an "extension" of f), if, wherever f is defined, g is also defined and assumes the same values. Of course, the same function f can arise by restriction from many different functions. In our example above f is also a restriction of the function h defined by $h(x) = x^2$ for $0 < x < 2$, $h(x) = -x^2$ for $-2 < x \leq 0$. As a matter of fact this example illustrates the process inverse to that of forming restrictions of a function which might be called "piecing together"; we can generate new functions by simply defining them by different explicit expressions in different portions of the domain.

c. Graphical Representation. Monotonic Functions

The fundamental idea of analytical geometry is to give an analytical representation to a curve originally defined by some geometrical property. This is done usually by regarding one of the rectangular coordinates, say y, as a function $y = f(x)$ of the other coordinate x; for example, a parabola is represented by the function $y = x^2$, the circle with radius 1 about the origin by the two functions $y = \sqrt{1 - x^2}$ and $y = -\sqrt{1 - x^2}$. In the first example we may think of the function as defined in the infinite interval $-\infty < x < \infty$; in the second we must restrict ourselves to the interval $-1 \leq x \leq 1$, since outside this interval the function has no meaning.[1]

Conversely, if instead of starting with a curve defined geometrically we consider a function $y = f(x)$ given analytically, we can represent the functional dependence of y on x graphically, using a rectangular coordinate system in the usual way (cf. Fig. 1.7). If for each abscissa x we take the corresponding ordinate $y = f(x)$, we obtain the geometrical representation of the function. The restrictions to be imposed on the function concept should secure for its geometrical representation the shape of a "reasonable" geometrical curve. This, it is true, expresses an intuitive feeling rather than a strict mathematical condition. However, we shall soon formulate conditions, such as continuity, differentiability, etc., which insure that the graph of a function is a curve capable

---

[1] We do not ordinarily consider imaginary or complex values of x and y.

of being visualized geometrically. This would not be the case if we admitted "pathological" functions such as the following: For every rational value of x, the function y has the value 1; for every irrational value of x, the value of y is 0. This functional prescription assigns a definite value of y to each x; but in every interval of x, no matter how small, the value of y jumps from 0 to 1 and back an infinite number of times. This example demonstrates that the general unrestricted function concept may lead to graphs which we would not consider as curves.

Multivalued Functions

We consider only functions $y = f(x)$ assigning a unique value of y to each value of x in the domain, as, for example, $y = x^2$ or $y = \sin x$. Yet, for a curve described geometrically, it may happen, as for the circle $x^2 + y^2 = 1$, that the whole course of the curve is not given by just one (single-valued) function, but requires several functions—in the case of the circle, the two functions $y = \sqrt{1 - x^2}$ and $y = -\sqrt{1 - x^2}$. The same is true for the hyperbola $y^2 - x^2 = 1$, which is represented by the two functions $y = \sqrt{1 + x^2}$ and $y = -\sqrt{1 + x^2}$. Such curves therefore do not determine unambiguously the corresponding functions. It is sometimes said that the curve is represented by a *multivalued* function; the separate functions representing it are then called the single-valued *branches* of the multivalued function belonging to the curve. For the sake of clarity we shall always use the word "function" to mean a single-valued function. For example, the symbol $\sqrt{x}$ (for $x \geq 0$) will always denote the *nonnegative* number whose square is x.

If a curve is the graph of *one* function, it is intersected by any parallel to the y-axis in at most one point, since to each point x in the interval of definition there corresponds just one value of y. The unit circle represented by the two functions

$$y = \sqrt{1 - x^2} \quad \text{and} \quad y = -\sqrt{1 - x^2},$$

is intersected by such parallels to the y-axis in more than one point. The portions of a curve corresponding to different single-valued branches are sometimes connected with each other so that the complete curve is a single figure which can be drawn with one stroke of the pen, for example, the circle (cf. Fig. 1.12); on the other hand, these portions may be completely separated, as for the hyperbola (cf. Fig. 1.13).

Examples. Let us consider some further examples of the graphical representation of functions.

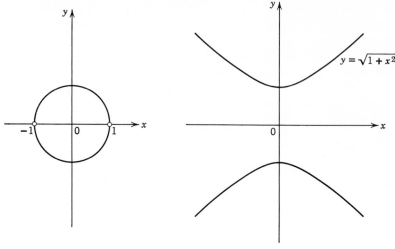

Figure 1.12   Figure 1.13

(*a*) *y* is proportional to *x*,

$$y = ax.$$

The graph (see Fig. 1.14) is a straight line through the origin of the coordinate system.

(*b*) *y* is a "linear function" of *x*,

$$y = ax + b.$$

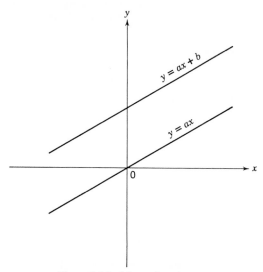

Figure 1.14   Linear functions.

The graph is a straight line through the point $x = 0$, $y = b$, which, if $a \neq 0$, also passes through the point $x = -b/a$, $y = 0$, and if $a = 0$ is horizontal.

(c) y is inversely proportional to x,

$$y = \frac{a}{x}.$$

In particular, for $a = 1$

$$y = \frac{1}{x},$$

so that

$$y = 1 \quad \text{for} \quad x = 1, \qquad y = 2 \quad \text{for} \quad x = \tfrac{1}{2}, \qquad y = \tfrac{1}{2} \quad \text{for} \quad x = 2.$$

The graph (cf. Fig. 1.15) is a *rectangular hyperbola*, a curve symmetrical with respect to the bisectors of the angles between the coordinate axes.

This function is obviously not defined for the value $x = 0$ since division by zero has no meaning. In the neighborhood of the exceptional point $x = 0$, the function has arbitrarily large values, both positive and negative; this is the simplest example of an *infinite discontinuity*, a concept which we shall discuss later (see p. 35).

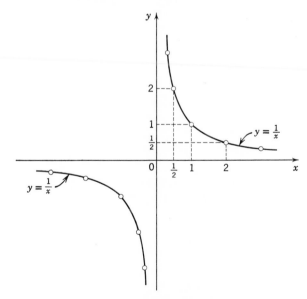

Figure 1.15 Infinite discontinuity.

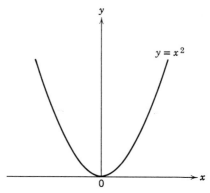

Figure 1.16   Parabola.

(*d*) y is the square of x,

$$y = x^2.$$

As is well known, this function is represented by a parabola (see Fig. 1.16).

Similarly, the function $y = x^3$ is represented by the so-called *cubical parabola* (see Fig. 1.17).

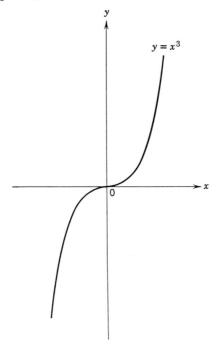

Figure 1.17   Cubical parabola.

Monotone Functions

A function which for all values of x in an interval has the same value $y = a$ is called a *constant*; it is represented graphically by a horizontal straight line. A function $y = f(x)$ for which an increase in the value of x always results in an increase in the value of y that is, for which $f(x) < f(x')$ whenever $x < x'$) is called a *monotonic increasing* function; if, on the other hand, an increase in the value of x always implies a decrease in the value of y, the function is called a *monotonic decreasing* function. Such functions are represented graphically by curves which always rise or always fall as x traverses the interval of definition toward

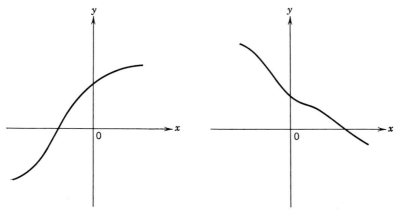

Figure 1.18   Monotone functions.

increasing values (see Fig. 1.18). A monotone function always maps different values of x into different y; that is, the mapping is one-to-one.

Even and Odd Functions

If the curve represented by $y = f(x)$ is symmetrical with respect to the y-axis, that is, if $x = -a$ and $x = a$ yield the same function value

$$f(-x) = f(x)$$

we call the function an *even* function. For example, the function $y = x^2$ is even (see Fig. 1.16). If, on the other hand, the curve is symmetrical with respect to the origin; that is, if

$$f(-x) = -f(x),$$

we say the function is an *odd* function; thus the functions $y = x$, $y = x^3$ (see Fig. 1.17) and $y = 1/x$ (see Fig. 1.15) are odd.

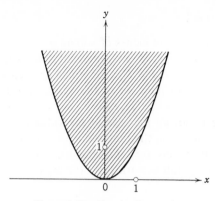

Figure 1.19   Graph of $y > x^2$.

It is frequently helpful to consider the geometrical representation of an inequality. For example, the inequality $y > x^2$ is represented by the domain above the parabola $y = x^2$ (Fig. 1.19). The interior of the unit circle centered at the origin (Fig. 1.20) is described by the inequality $x^2 + y^2 < 1$.

Often several inequalities describe more complicated regions with boundaries consisting of different pieces. Thus the "first" quadrant of the unit circle is described by the system of simultaneous inequalities:

$$x^2 + y^2 < 1, \qquad x > 0, \qquad y > 0.$$

(See Fig. 1.21.)

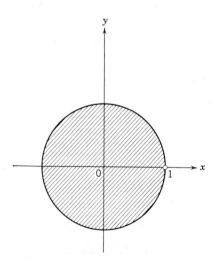

Figure 1.20   Graph of $x^2 + y^2 < 1$.

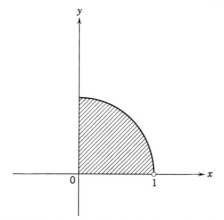

Figure 1.21   Graph of $x^2 + y^2 < 1$, $x > 0$, $y > 0$.

d. Continuity

Intuitive and Precise Explanation

The functions and graphs just considered exhibit a property of greatest importance in the calculus, that of *continuity*. Intuitively, continuity means that a small change in the independent variable x implies only a small change in the dependent variable $y = f(x)$ and excludes a jump in the value of y: thus the graph consists of one piece. In contrast, a graph $y = f(x)$ consisting of pieces separated by a gap at an abscissa x_0 exhibits there a *jump discontinuity*. For example, the function[1] $f(x) = \operatorname{sgn} x$ defined by $f(x) = +1$ for $x > 0$, by $f(x) = -1$ for $x < 0$, and $f(0) = 0$ has a "jump discontinuity"[2] at $x_0 = 0$ (see Fig. 1.22).

The idea of continuity is implicit in the everyday use of elementary mathematics. Whenever a function $y = f(x)$ is described by tables, such as the logarithmic or trigonometric tables, the values of y can be listed only for a "discrete" set of values of the independent variable x, say at intervals of $1/1000$ or $1/100,000$. Yet, unlisted values of the function may be needed for intermediate x. Then we tacitly assume that an unlisted value $f(x_0)$ is approximately the same as that of $f(x)$

---

[1] Pronounced "signum" or "sign" of x.
[2] Technically, the word "jump" refers only to the particular kind of discontinuity in which the function approaches values from the right and left that do not both agree with $f(x_0)$. An "infinite" discontinuity is exhibited by the function $y = 1/x$ for $x \neq 0$ and $y = 0$ for $x = 0$. Still other types of discontinuities will be discussed later.

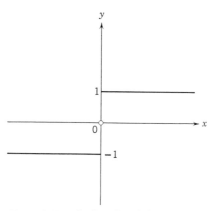

Figure 1.22   The function $f(x) = \text{sgn } x$.

for a neighboring x which appears in the table and that $f(x_0)$ can be approximated as precisely as we want if only the x-values in the table are spaced sufficiently close to each other.

Continuity of the function $f(x)$ for a value x_0 just means that $f(x)$ differs arbitrarily little from the value $f(x_0)$ once x is sufficiently close to x_0. The words "differs arbitrarily little" and "sufficiently close" are somewhat vague and must be explained precisely in quantitative terms.

Prescribe any "margin of precision" or "tolerance," that is, any positive real number ϵ (however small). For continuity of f at x_0 we require that the difference between $f(x)$ and $f(x_0)$ stay within this margin, that is, that $|f(x) - f(x_0)| < \epsilon$, for all values x which are sufficiently close to x_0 (or for all values x lying within some distance δ from x_0).

We can visualize most easily what continuity means if we interpret f as a mapping assigning to points x on the x-axis images on the y-axis. Take any point x_0 on the x-axis and its image $y_0 = f(x_0)$ (see Fig. 1.23).

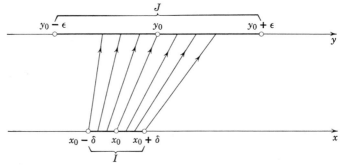

Figure 1.23   Continuity of the mapping $y = f(x)$ at the point x_0.

We mark off an arbitrary open interval J on the y-axis having the point y_0 as center. If 2ϵ is the length of J, then the points y of J are those whose distance from y_0 is less than ϵ or for which $|y - y_0| < \epsilon$. The condition for continuity of $f(x)$ at x_0 is: All points x close enough to x_0 have images lying in J; or: It is possible to mark off an interval I on the x-axis with center x_0, say the interval $x_0 - \delta < x < x_0 + \delta$ such that every point x of I has an image $f(x)$ which lies in J and thus $|f(x) - f(x_0)| < \epsilon$. Continuity of $f(x)$ at the point x_0 means that for an arbitrary ϵ-*neighborhood* J of the point $y_0 = f(x_0)$ on the y-axis a δ-neighborhood I of the point x_0 on the x-axis can be found, all of whose points are mapped into points of J.[1] Of course, this makes sense only for points on the x-axis at which the mapping is defined, that is, which belong to the domain of f. Thus we are led to the following *precise definition of continuity.*

The function $f(x)$ is continuous at a point x_0 of its domain if for every positive ϵ we can find a positive number δ such that

$$|f(x) - f(x_0)| < \epsilon$$

for all values x in the domain of f for which $|x - x_0| < \delta$.

Most useful is the geometric interpretation of continuity when we represent the function f by its graph in the xy-plane (see Fig. 1.24). Let $P_0 = (x_0, y_0)$ be a point on the graph. The points (x, y) with $y_0 - \epsilon < y < y_0 + \epsilon$ now form a horizontal "strip" J containing P_0. Continuity of f at x_0 means that given any such horizontal strip J, however thin, we can find a vertical strip I given by $x_0 - \delta < x < x_0 + \delta$ so thin that every point of the graph lying in I also falls into J.

As an illustration we consider the linear function $f(x) = 5x + 3$; we have

$$|f(x) - f(x_0)| = |(5x + 3) - (5x_0 + 3)| = 5\,|x - x_0|,$$

which expresses that the mapping $y = 5x + 3$ *magnifies* distances by the factor 5. Here obviously $|f(x) - f(x_0)| < \epsilon$ for all x for which

---

[1] In this definition of continuity I and J are intervals having their *centers* respectively at the points x_0 and y_0. This is convenient for the analytic definition of continuity at x_0 which refers to the distances $|x - x_0|$ and $|y - y_0|$, but it is somewhat artificial if we interpret f geometrically as a mapping. We could instead define continuity of $y = f(x)$ at a point x_0 just as well by the requirement that for every *open* interval J on the y-axis which contains the point $y_0 = f(x_0)$ we can find an open interval I on the x-axis containing the point x_0 such that the y-image of any point x in I for which the mapping is defined lies in J. The proof of the equivalence of the two definitions is left to the reader as a simple exercise.

$|x - x_0| < \epsilon/5$. Consequently, the condition for continuity of $f(x)$ at the point x_0 is satisfied if we choose $\delta = \epsilon/5$ (but, of course, any positive number $\delta < \epsilon/5$ is also a possible choice); the image of any point of the interval $x_0 - \delta < x < x_0 + \delta$ will then lie in the interval $y_0 - \epsilon < y < y_0 + \epsilon$. In this example the statement that the distance $|y - y_0|$ is "arbitrarily small" for "sufficiently small" $|x - x_0|$ can be given a quite specific meaning; indeed $|x - x_0|$ is *sufficiently small* if it does not exceed one-fifth of the value of $|y - y_0|$.

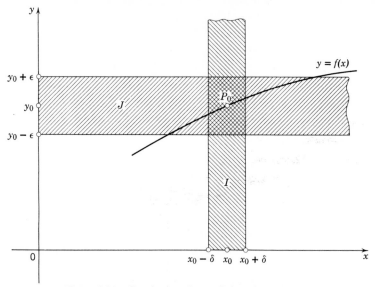

Figure 1.24   Continuity of $y = f(x)$ at the point x_0.

Another example is furnished by the function $f(x) = x^2$. Here we have for $|x - x_0| < \delta$

$$|f(x) - f(x_0)| = |x^2 - x_0^2| = |x - x_0| \, |2x_0 + (x - x_0)|$$
$$\leq |x - x_0| \, (2 \, |x_0| + |x - x_0|) \leq \delta(2 \, |x_0| + \delta).$$

We verify immediately that the condition $|f(x) - f(x_0)| < \epsilon$ is satisfied if we choose $\delta = - |x_0| + \sqrt{\epsilon + |x_0|^2}$.

Intuitively, the idea of continuity seems obvious without explanation, but the precise formulation may initially be somewhat difficult to grasp because of the permissiveness of words such as "one can find" or "arbitrarily chosen." Yet the reader who may at first be well satisfied with some intuitive notion of continuity will gradually learn to appreciate the logical precision and generality of the analytic definition, the outcome of a long and persistent struggle for reconciliation of the need for intuitive understanding with that of logical clarity. In the long run a precise meaning for the word "continuity" is indispensable; the analytic definition given here is the compelling formulation of an important property of functions.

For the beginner it should be emphasized again that "small" is not an absolute designation of a number; rather the term "arbitrarily small" refers to a number that is not fixed at the outset but for which then any positive value may be chosen, and which is subject to a subsequent smaller choice for a refined approximation of $f(x_0)$. "Sufficiently small" refers to a number δ that must be adjusted to suit a margin of tolerance set previously by another number ϵ.

Continuity and Discontinuity Explained by Examples. We can illuminate the definition of continuity by contrast with examples of discontinuity, examples which do not fit the definition above. Recall the simple example of the function $f(x) = \operatorname{sgn} x$ on p. 31. Obviously, for any $x_0 \neq 0$ this function is continuous according to the ϵ, δ-definition above, in fact, with a constant $\delta = |x_0|$ no matter how small ϵ is chosen. But for $x_0 = 0$ no δ at all can be found if ϵ is less than 1 since $|f(x) - f(0)| = |f(x)| = 1 > \epsilon$ for every x unequal to zero, however close x might be to zero.

The function $\operatorname{sgn} x$ illustrates the simple type of discontinuity at a point ξ known as *jump-discontinuity*, in which $f(x)$ approaches limiting values from the right and left as x approaches ξ—limiting values, however, that differ either from each other or from the value of f at the point ξ.[1] The graph at $x = \xi$ then has a gap. Other curves with jump discontinuities are sketched in Fig. 1.25a and b; the definition of these functions should be clear from the figures.[2]

In discontinuities of this kind the limits from the right and the left both exist. We turn to discontinuities in which this is not the case. The most important of these are the *infinite discontinuities* or infinities.

---

[1] The precise definition of *limit* will be given in Section 1.7; an intuitive idea is sufficient for the descriptive remarks made here.

[2] In all these examples of jump discontinuities the limits of the function at the point of discontinuity from the right and left have different values. The trivial example of the function $f(x)$ defined by

$$f(x) = 0 \quad \text{for} \quad x \neq 0, \qquad f(x) = 1 \quad \text{for} \quad x = 0$$

illustrates a jump discontinuity in which the limits from both sides are equal to each other but differ from the value of f at the point of discontinuity ξ itself. We have then a *removable* singularity. Here f can be made continuous by merely changing the value of f at ξ so as to agree with the limits from both sides.

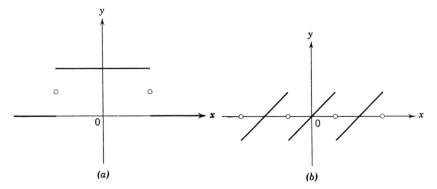

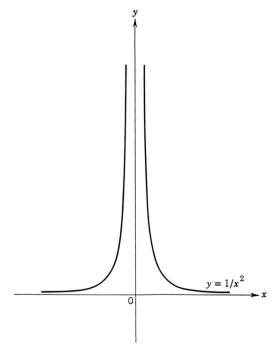

(a)   **(b)**

Figure 1.25

Figure 1.26   Graph of function with infinite
discontinuity.

These are discontinuities like those exhibited by the functions $1/x$ or $1/x^2$ at the point $x = 0$; as $x \to 0$ the absolute value $|f(x)|$ of the function increases beyond all bounds. The function $1/x$ increases numerically beyond all bounds through positive and negative values, respectively, as x approaches the origin from the right and from the left. On the other hand, the function $1/x^2$ has for $x = 0$ an infinite discontinuity at which the value of the function increases beyond any positive bound as x approaches the origin from both sides (cf. Fig. 1.26

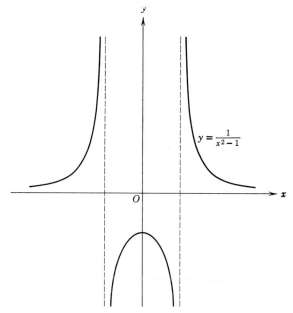

Figure 1.27   Function with infinite discontinuities.

and Fig. 1.27). The function $1/(x^2 - 1)$ shown in Fig. 1.27 has infinite discontinuities both at $x = 1$ and at $x = -1$.

An example of another type of discontinuity in which no limit from the right or from the left exists is the "piecewise linear" even function $y = f(x)$ illustrated in Fig. 1.28, which is defined as follows for all nonzero values of x. This function alternately takes the values $+1$ and -1 for the x-values of the form $\pm 1/2^n$, where n is any integer: $f(\pm 1/2^n) = (-1)^n$. In every interval $1/2^{n+1} < x < 1/2^n$ or $-1/2^n < x < -1/2^{n+1}$ the function $f(x)$ is linear and ranges over all values between -1 and $+1$. Therefore the function swings backward and forward more and more rapidly between the values -1 and $+1$ as x approaches nearer and nearer to the point $x = 0$, and in the immediate neighborhood of that

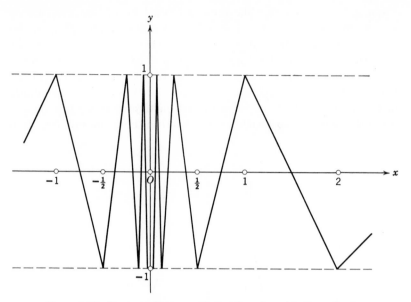

Figure 1.28 Piecewise linear oscillating function with discontinuity.

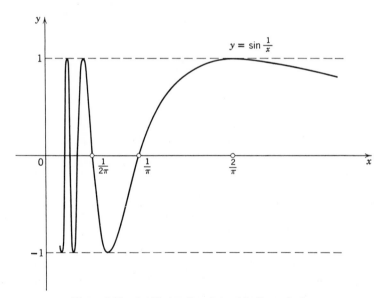

Figure 1.29 Oscillating function with discontinuity.

point an infinite number of such oscillations occur. A similar behavior is exhibited by the smooth curve (Fig. 1.29). [Here $f(x)$ actually is given by an expression in closed form, namely, $f(x) = \sin(1/x)$, with the sine-function defined appropriately as on p. 51].

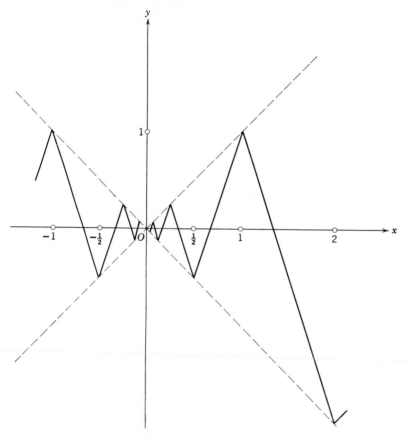

Figure 1.30   Continuous oscillating function.

A contrast to this example is the piecewise linear function $y = f(x)$ that takes the values $f(\pm 1/2^n) = (-\tfrac{1}{2})^n$ for all integers n (see Fig. 1.30) and is linear for intermediate values of x. Here $f(x)$ remains continuous at the point $x = 0$ if we assign to it the value 0 at that point. In the neighborhood of the origin the function oscillates backward and forward an infinite number of times, but the magnitude of these oscillations becomes arbitrarily small as the origin is approached. The situation is the same for the function $y = x \sin(1/x)$ (see Fig. 1.31).

These examples show that continuity permits all sorts of remarkable possibilities foreign to our naive intuition.

Removable Discontinuities

As noted it may happen that at a certain point say $x = 0$, a function is not defined by the original law, as, for example, in the last examples discussed. We are then free to extend the definition of the function by

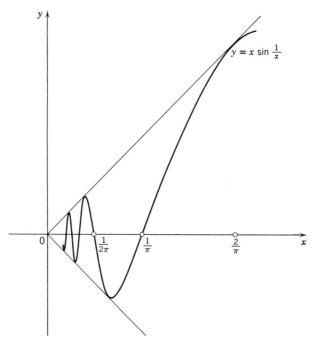

Figure 1.31   Continuous oscillating function.

assigning to it any desired value at such a point. In the last example we can choose the definition in such a way that the function becomes *continuous* at that point also, namely, by choosing $y = 0$ at $x = 0$. A similar continuous extension can be defined whenever the limits from the left and from the right both exist and are equal to one another; then we need only make the value of the function at the point in question equal to these limits in order to make the function continuous there. Whatever discontinuity may be imposed by definition at $x = 0$, this discontinuity is "removable" by assigning a suitable value $f(0)$. For the function $y = \sin 1/x$ or for the function in Fig. 1.28, this is, however, not possible: whatever value we assign to the function at $x = 0$, the extended function is discontinuous.

Modulus of Continuity. Uniform Continuity. Our definition of continuity of the function $f(x)$ at x_0 requires that for every degree of precision $\epsilon > 0$ there *exist* quantities $\delta > 0$ (so-called moduli of continuity) such that $|f(x) - f(x_0)| < \epsilon$ for all x in the domain of f for which $|x - x_0| < \delta$. A modulus of continuity expresses information about the sensitivity of f to changes in x. A modulus of continuity δ is never unique; it can always be replaced (for the same x_0 and ϵ) by any smaller positive quantity δ' since $|x - x_0| < \delta'$ implies $|x - x_0| < \delta$ and thereby $|f(x) - f(x_0)| < \epsilon$. For practical purposes, as in numerical computations, we may be interested in a particular choice of δ; for example, in the largest value for δ. On the other hand, if we merely want to establish the fact that f is continuous at x_0, then we need only to exhibit any one modulus of continuity for every positive ϵ.

In general, as our examples show, this $\delta = \delta(\epsilon)$ depends not only on ϵ but also on the value of x_0. Of course, we need not consider all positive values ϵ. We can always restrict considerations to sufficiently small ϵ, say to $\epsilon \leq \epsilon_0$ for an arbitrarily chosen ϵ_0, since for $\epsilon > \epsilon_0$ we can use the same modulus of continuity as for $\epsilon = \epsilon_0$. Similarly, we only have to take into account the points x of the domain of f lying in an arbitrary neighborhood of x_0, say those with $|x - x_0| < \delta_0$, since we can always replace any modulus of continuity δ by a smaller one which does not exceed δ_0. Continuity of f at x_0 is a *local property*, meaning a property which only depends on the values of f in some neighborhood of x_0 however small.

As we have seen, the function f may be continuous for some x_0 and discontinuous for others. A function is called *continuous in an interval* if it is continuous at each point of the interval. For each x_0 of the interval we have then a modulus of continuity $\delta = \delta(\epsilon)$ which can be expected to vary with x_0 reflecting the different rates at which y changes with changing x near different points x_0.

We call f *uniformly continuous* in an interval if we can find a *uniform modulus of continuity* $\delta = \delta(\epsilon)$ for that interval, that is, one not dependent on the particular point x_0 of the interval. Thus $f(x)$ is uniformly continuous in an interval[1] if for each positive ϵ there exists a positive number δ such that $|f(x) - f(x_0)| < \epsilon$ for *any two* points x and x_0 of the interval for which $|x - x_0| < \delta$.

For a uniformly continuous function $y = f(x)$ the values of y differ "arbitrarily little" from each other for any values of x that are "sufficiently close" *regardless of their location in the interval.* In some respects

---

[1] In this definition the word interval can refer either to closed or open or infinite intervals.

uniform continuity comes closer to intuitive notions than the mere local property of continuity.

For example, the function $f(x) = 5x + 3$ is uniformly continuous for all values of the independent variable since here $|f(x) - f(x_0)| = 5|x - x_0| < \epsilon$ for $|x - x_0| < \epsilon/5$, and thus $\delta(\epsilon) = \epsilon/5$ represents a uniform modulus of continuity.

The function $f(x) = x^2$ for an infinite x-interval is definitely not uniformly continuous. It is clear that small changes in x can produce arbitrarily large changes in x^2 if only x is large enough. A glance at a table of squares of integers x shows how successive squares are spaced further and further apart as x increases. If, however, we only consider pairs of values x and x_0 belonging to a fixed finite closed interval $[a, b]$, we can find a uniform modulus of continuity. Indeed, for $|x - x_0| < \delta$ we have

$$|f(x) - f(x_0)| = |x^2 - x_0^2| = |x - x_0|\,|x + x_0| \leq 2\,|x - x_0|(|b| + |a|)$$
$$< 2\delta(|b| + |a|) = \epsilon$$

if we take $\delta = \epsilon/2(|b| + |a|)$.

A similar situation prevails for the function $f(x) = 1/x$ for $x \neq 0$, $f(0) = 0$. Consider a closed bounded interval $a \leq x \leq b$ throughout which the function is continuous. Such an interval cannot include the origin, which is a point of discontinuity, so that a and b must have the same sign. Suppose a and b are both positive. Then for x and x_0 belonging to the interval and for $|x - x_0| < \delta$ we have

$$\left| \frac{1}{x} - \frac{1}{x_0} \right| = |x_0 - x| \frac{1}{|x_0|\,|x|} < \frac{\delta}{a^2} = \epsilon$$

for $\delta = a^2\epsilon$. Thus the function is uniformly continuous in the interval $[a, b]$. Of course, this proves also that the function $f(x) = 1/x$ is continuous at every point $x_0 > 0$. For every such value x_0 can be enclosed in some interval $a < x_0 < b$ with positive a, b. The expression $\delta = a^2\epsilon$ is then a modulus of continuity for the function at x_0, if we restrict x to a neighborhood of x_0 lying completely in the interval.

The continuous functions of the preceding examples turn out to be uniformly continuous in any closed bounded interval belonging to their domain. They illustrate a general fact which will be proved in the Supplement, p. 100.

Any function, continuous in a closed and bounded interval, automatically is uniformly continuous in that interval.

The restriction to *bounded* intervals is essential as the example of the function x^2 shows. Similarly, we must stipulate that the interval be *closed;* for example, the function $y = 1/x$ is continuous in the open interval $0 < x < 1$ but is not uniformly continuous there; arbitrarily large changes in y can be produced by arbitrarily small changes in x if only x is sufficiently close to the origin. If there existed a uniform modulus of continuity $\delta(\epsilon)$ for the interval $(0, 1)$, we could take, for example, $x_0 < \delta, x = \frac{1}{2}x_0$; obviously then $|1/x - 1/x_0| = 1/x_0$ is greater than any preassigned ϵ whenever x_0 is sufficiently small, so that the assumption of a uniform $\delta(\epsilon)$ leads to a contradiction.

Lipschitz-Continuity—Hölder-Continuity. In the preceding examples of functions uniformly continuous in an interval $[a, b]$ we found a particularly simple modulus of continuity, namely $\delta(\epsilon)$ proportional to ϵ. This most common situation is presented by the so-called *Lipschitz-continuous* functions, that is, by the functions $f(x)$ which satisfy an inequality of the form

$$|f(x_2) - f(x_1)| \leq L\,|x_2 - x_1|$$

(a so-called Lipschitz condition) for all x_1, x_2 in the interval with a fixed value L. Lipschitz-continuity means that the "difference quotient"

$$\frac{f(x_2) - f(x_1)}{x_2 - x_1}$$

formed for any two distinct points of the interval never exceeds a fixed finite value L in absolute value or that the mapping $y = f(x)$ magnifies distances of points on the x-axis at most by the factor L. Clearly, for a Lipschitz-continuous function the expression $\delta(\epsilon) = \epsilon/L$ is a modulus of continuity since $|f(x_2) - f(x_1)| < \epsilon$ for $|x_2 - x_1| < \epsilon/L$. Conversely, any function with a modulus of continuity proportional to ϵ, say $\delta(\epsilon) = c\epsilon$, is Lipschitz-continuous, with $L = 1/c$.

As we shall see in Chapter 2 most of the functions encountered are Lipschitz-continuous except at isolated points, as a consequence of the fact that their derivatives are bounded in any closed interval which excludes these points. However, Lipschitz-continuity is only sufficient but not necessary for uniform continuity. The simplest example of a function which is continuous without being Lipschitz-continuous is given by $f(x) = \sqrt{x}$ for $x \geq 0$ and $x_0 = 0$. Here the difference quotient

$$\frac{f(x) - f(0)}{x - 0} = \frac{1}{\sqrt{x}}$$

becomes arbitrarily large for sufficiently small x and hence cannot be bounded by a fixed constant L. Thus it is not possible to choose $\delta(\epsilon)$ proportional to ϵ; but there exist other, nonlinear, moduli of continuity for this function, for example, $\delta(\epsilon) = \epsilon^2$.

The function $\sqrt{x}$ belongs to the general class of functions called "Höldercontinuous," satisfying a "Hölder-condition"

$$|f(x_2) - f(x_1)| \leq L\,|x_2 - x_1|^\alpha$$

for all x_1, x_2 of an interval, where L and α are fixed constants, the "Hölderexponent" α being restricted to values $0 < \alpha \leq 1$. The Lipschitz-continuous functions arise for the special Hölder-exponent $\alpha = 1$.

Obviously, $\delta = L^{-1/\alpha}\epsilon^{1/\alpha}$ is a possible modulus of continuity for a Höldercontinuous function f; here δ is proportional to $\epsilon^{1/\alpha}$, and not to ϵ itself. The function $f(x) = \sqrt{x}$ is Hölder-continuous with exponent $\alpha = \frac{1}{2}$. This follows from the inequality

$$|\sqrt{x_2} - \sqrt{x_1}| \leq |x_2 - x_1|^{1/2},$$

which we obtain by observing

$$|\sqrt{x_2} - \sqrt{x_1}| \leq |\sqrt{x_2} + \sqrt{x_1}|$$

and multiplying by $|\sqrt{x_2} - \sqrt{x_1}|$. This yields the modulus of continuity $\delta(\epsilon) = \epsilon^2$ for $\sqrt{x}$ as mentioned before.

More generally, the fractional powers $f(x) = x^\alpha$ for $0 < \alpha \leq 1$ are Höldercontinuous with Hölder-exponent α.

The Hölder-continuous functions still do not exhaust the class of all uniformly continuous functions. It is not difficult to construct examples of continuous functions for which powers of ϵ do not suffice as moduli of continuity. (See Problem 13, p. 118.)

e. The Intermediate Value Theorem. Inverse Functions

Intuitively there is no doubt that a function which is continuous, and hence has no "jumps," cannot vary from one value to another without passing through all intermediate values. This fact is expressed by the so-called intermediate value theorem (its precise proof is given in the Supplement, p. 100).

INTERMEDIATE VALUE THEOREM.   *Consider a function $f(x)$ continuous at every point of an interval. Let a and b be any two points of the interval and let η be any number between $f(a)$ and $f(b)$. Then there exists a value ξ between a and b for which $f(\xi) = \eta$.*

Interpreted geometrically, the theorem states that if two points $(a, f(a))$ and $(b, f(b))$ of the graph of a continuous function f lie on

different sides of a parallel $y = \eta$ to the x-axis, then the parallel
intersects the graph at some intermediate point (see Fig. 1.32). There
may, of course, be several intersections. In the important case where the
function $f(x)$ is monotonic increasing or monotonic decreasing throughout the interval, there can be only one intersection for then f cannot
have the same value η for two different values of ξ.

As an example we take the function $f(x) = x^2$ which is monotonic
increasing and continuous in the interval $1 \leq x \leq 2$. Here $f(1) = 1$,
$f(2) = 4$. Taking for η the value 2 intermediate between 1 and 4 we find

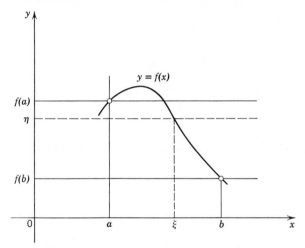

Figure 1.32 The intermediate value theorem.

that there exists a unique ξ between 1 and 2 for which $\xi^2 = 2$. This is,
of course, the number denoted by $\sqrt{2}$.

Continuity of the Inverse Function

For any monotonic increasing continuous function $f(x)$ defined in an
interval $a \leq x \leq b$, we found that for every η with $f(a) \leq \eta \leq f(b)$
there is exactly one ξ with $a \leq \xi \leq b$ for which $f(\xi) = \eta$.[1] Let
$\alpha = f(a)$, $\beta = f(b)$. Since ξ is determined uniquely by η, it represents
a *function* $\xi = g(\eta)$ defined for arguments η in the closed interval
$[\alpha, \beta]$. We call this function g the *inverse* of f. Since larger ξ correspond
to larger $\eta = f(\xi)$, the function g is again monotonic increasing. It is
easy to show that the inverse function g is also continuous.

---

[1] The intermediate value theorem as stated assigns ξ for η in the *open* interval
$f(x) < \eta < f(b)$. However, of course, for $\eta = f(a)$ or $\eta = f(b)$ we have only to
take $\xi = a$ or $\xi = b$.

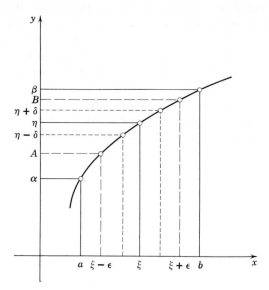

Figure 1.33   Continuity of the inverse of a monotonic continuous function.

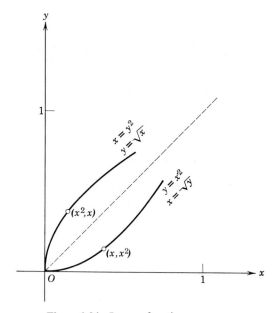

Figure 1.34   Inverse functions.

Indeed, let η be any value between α and β (see Fig. 1.33). Then $\xi = g(\eta)$ must lie between $a = g(\alpha)$ and $b = g(\beta)$. Let ϵ be a given positive number which we can assume to be so small that $a < \xi - \epsilon < \xi + \epsilon < b$. We must show that $|g(y) - g(\eta)| < \epsilon$ for all y sufficiently close to η. Since f is increasing, $\eta = f(\xi)$ lies between the values $f(\xi - \epsilon) = A$ and $f(\xi + \epsilon) = B$ and we can find a δ so small that

$$A < \eta - \delta < \eta + \delta < B.$$

If y is any value with $\eta - \delta < y < \eta + \delta$ and $x = g(y)$, we have $A < y < B$ and hence $g(A) < g(y) < g(B)$, that is, $\xi - \epsilon < g(y) < \xi + \epsilon$ or $|g(y) - g(\eta)| < \epsilon$. The same proof, modified slightly, applies when η is one of the end points α or β of the interval of definition of g.

The relations $y = f(x)$ and $x = g(y)$ are equivalent and are represented by the same graph in the x,y-plane; the points (x, y) in the plane for which $y = f(x)$ are the same as those points for which $x = g(y)$. If we represent the function g in the customary way by $y = g(x)$, we must interchange x and y; then the graph of $y = g(x)$ is obtained from the graph of $y = f(x)$ by taking the mirror image with respect to the line $y = x$. An example is given by the graphs of the function $f(x) = x^2$ for $x \geq 0$ and of the inverse function $g(x) = \sqrt{x}$ for $x \geq 0$ (see Fig. 1.34).

1.3 The Elementary Functions

a. Rational Functions

We turn to a brief review of the familiar elementary functions. The simplest types of function are constructed by repeated application of the elementary operations, addition and multiplication. If we apply these operations to an independent variable x and to a set of real numbers $a_1, \ldots, a_n$ we obtain the *polynomials*

$$y = a_0 + a_1 x + \cdots + a_n x^n.$$

Polynomials are the simplest functions of analysis and in a sense the basic ones.

Quotients of such polynomials, of the form

$$y = \frac{a_0 + a_1 x + \cdots + a_n x^n}{b_0 + b_1 x + \cdots + b_m x^m},$$

are the general *rational functions;* these are defined at all points where the denominator differs from zero.

The simplest polynomial, the *linear function*
$$y = ax + b,$$
is represented graphically by a straight line. Every *quadratic function*
$$y = ax^2 + bx + c$$
is represented by a parabola. The graphs of polynomials of the third degree
$$y = ax^3 + bx^2 + cx + d,$$
are occasionally called parabolas of the third order, etc.

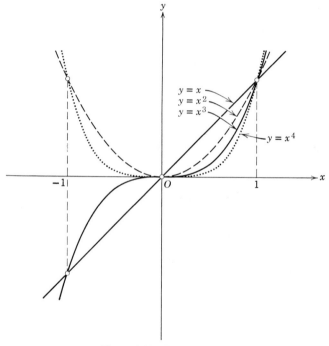

Figure 1.35   Powers of x.

The graphs of the function $y = x^n$ for the exponents $n = 1, 2, 3, 4$ are given in Fig. 1.35. For even values of n the function $y = x^n$ satisfies the equation $f(-x) = f(x)$, and is therefore an even function, whereas for odd values of n the function satisfies the condition $f(-x) = -f(x)$, and is therefore odd.

The simplest example of a rational function which is not a polynomial is the function $y = 1/x$ mentioned on p. 27; its graph is a rectangular hyperbola. Another example is the function $y = 1/x^2$ (cf. Fig. 1.26, p. 36).

b. Algebraic Functions

We are at once forced out of the set of rational functions by the problem of forming their inverses. The typical example of this is the function $\sqrt[n]{x}$, the inverse of x^n. The function $y = x^n$ for $x \geq 0$ is easily seen to be monotonic increasing and continuous. It therefore has a single-valued inverse, which we denote by the symbol $x = \sqrt[n]{y}$, or, interchanging the letters used for the dependent and independent variables,

$$y = \sqrt[n]{x} = x^{1/n}.$$

By definition this root is always nonnegative. For odd values of n the function x^n is monotonic for all values of x, including negative values. Consequently, for odd values of n we may extend the definition of $\sqrt[n]{x}$ uniquely to all values of x; in this case $\sqrt[n]{x}$ is negative for negative values of x.

More generally, we may consider

$$y = \sqrt[n]{R(x)},$$

where $R(x)$ is a rational function. Further functions of similar type are formed by applying rational operations to one or more of these special functions. Thus, for example, we may form the functions

$$y = \sqrt[m]{x} + \sqrt[n]{x^2 + 1}, \qquad y = x + \sqrt{x^2 + 1}.$$

These functions are special cases of *algebraic functions*. (The general concept of an algebraic function will be defined in Volume II.)

c. Trigonometric Functions

The rational functions and the algebraic functions are defined directly by the elementary operations of calculation, but geometry is the source from which we first draw examples of other functions, the so-called *transcendental functions*.[1] Of these we consider here the *elementary transcendental functions*, namely, the trigonometric functions, the exponential function, and the logarithm.

In analytical investigations angles are not measured in degrees, minutes, and seconds, but in *radians*. We place the angle to be measured

---

[1] The word "transcendental" does not mean anything particularly deep or mysterious; it merely suggests that the definition of these functions transcends the elementary operations of calculations, "*quod algebrae vires transcendit.*"

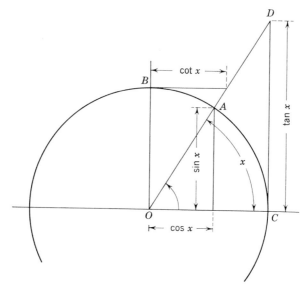

Figure 1.36 The trigonometric functions.

with its vertex at the center of a circle of radius 1, and measure the size of the angle by the length of the arc of the circumference cut out by the angle.[1] Thus an angle of 180° is the same as an angle of π radians (has radian measure π), an angle of 90° has radian measure $\pi/2$, an angle of 45° has radian measure $\pi/4$, an angle of 360° has radian measure 2π. Conversely, an angle of 1 radian expressed in degrees is

$$\frac{180°}{\pi}, \quad \text{or approximately } 57° \ 17' \ 45''.$$

Henceforth, whenever we speak of an angle x, we shall mean an angle whose radian measure is x.

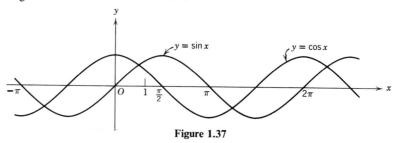

Figure 1.37

---

[1] The radian measure of an angle can also be defined as twice the *area* of the corresponding sector of the circle of radius one.

We briefly recall the meaning of the trigonometric functions sin x, cos x, tan x, cot x.[1] They are shown in Fig. 1.36, in which the angle x is measured from the segment OC (of length 1), angles being reckoned positive in the counterclockwise direction. The functions cos x and sin x are the rectangular coordinates of the point A. The graphs of the functions sin x, cos x, tan x, cot x are given in Figs. 1.37 and 1.38.

Later (see p. 215) we will be able to replace the geometrical definitions by analytical ones.

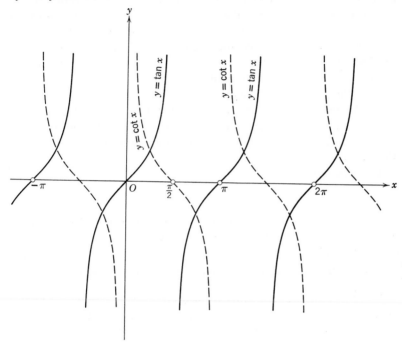

Figure 1.38

d. The Exponential Function and the Logarithm

In addition to the trigonometric functions, the exponential function with the positive base a,

$$y = a^x,$$

and its inverse, the logarithm to the base a,

$$x = \log_a y,$$

---

[1] It is also sometimes convenient to introduce the functions sec $x = 1/\cos x$, cosec $x = 1/\sin x$.

are also included among the elementary transcendental functions. In elementary mathematics it is customary to pass over certain inherent difficulties in their definition, and we too shall postpone the detailed discussion of them until we have better methods at our disposal (cf. Section 2.5, p. 145). We can, however, at least indicate here one "elementary" way of defining these functions. If $x = p/q$ is a rational number (where p and q are positive integers), then—the number a being assumed positive—we define a^x as $\sqrt[q]{a^p} = a^{p/q}$, where the root, according to convention, is to be taken as positive. Since the rational values of x are everywhere dense, it is natural to extend this function a^x to a continuous function defined for irrational values of x as well, giving to a^x when x is irrational, values which are continuous with the values already defined when x is rational. This defines a continuous function $y = a^x$, the "exponential function," which for all *rational* values of x gives the value of a^x found above. That this extension is actually possible and can be carried out in only one way we take for granted at the moment; but it must be borne in mind that we still have to *prove* that this is so.[1]

The function

$$x = \log_a y$$

can then be defined for $y > 0$ as the inverse of the exponential function: $x = \log_a y$ is that number for which $y = a^x$.

e. Compound Functions, Symbolic Products, Inverse Functions

New functions are frequently formed not only by combining known functions by rational operations but by the more general and basic process of forming *functions of functions* or *compound* functions.

Let $u = \phi(x)$ be a function whose domain is in the interval $a \leq x \leq b$ and whose range lies in the interval $\alpha \leq u \leq \beta$. Moreover, let $y = g(u)$ be a function defined for $\alpha \leq u \leq \beta$. Then $g(\phi(x)) = f(x)$ defines a function f for $a \leq x \leq b$ which is "compounded" or "composed" from g and ϕ. For example, $f(x) = 1/(1 + x^{2n})$ is composed of the functions $\phi(x) = 1 + x^{2n}$ and $g(u) = 1/u$. Similarly, the function $f(x) = \sin(1/x)$ is composed of $\phi(x) = 1/x$ and $g(u) = \sin u$.

It is useful to interpret the compound functions in terms of *mappings*. The mapping ϕ takes every point x of the interval $[a, b]$ into a point u in the interval $[\alpha, \beta]$; the mapping g takes any value u in $[\alpha, \beta]$ into a point y. The mapping f is the "symbolic product" $g\phi$ of the mappings

---

[1] This is done on p. 152.

g and ϕ, that is, the mapping carrying out ϕ and g successively, in that order; for any x in $[a, b]$ we form its map u under the mapping ϕ, and then apply g to the image $u = \phi(x)$, obtaining $g(\phi(x)) = f(x) = y$ (see Fig. 1.39). Such a symbolic product $g\phi$ is natural and meaningful for any type of operation; it signifies that we first perform ϕ, and then, on the result, perform g.[1] We must not confuse the symbolic product $g\phi = g(\phi)$ of two functions with the ordinary *algebraic* product $g(x) \cdot \phi(x)$ of the functions, in which both $g(x)$ and $\phi(x)$ are formed for the same argument x (the mappings applied to the same point) and the *product of the values* of the functions is formed.

Naturally, symbolic products cannot be expected to be commutative. In general, $g(\phi)$ and $\phi(g)$ are not the same, even where both are defined;

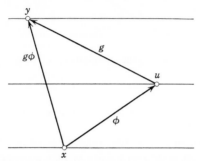

Figure 1.39 Symbolic product $g\phi = f$ of two mappings.

the order in which operations are performed matters very much. If, for example, ϕ stands for the operation of "adding 1 to a number" and g for the operation of "multiplying a number by 2," then

$$g(\phi(x)) = 2(x + 1) = 2x + 2, \qquad \phi(g(x)) = (2x) + 1 = 2x + 1.$$

(See Fig. 1.40.)

In order to be able to form the symbolic product $g\phi$ of two mappings, the "factors" g and ϕ must fit together in the sense that the domain of g must include the range of ϕ; thus we cannot form $g\phi$ when

$$g(u) = \sqrt{u}, \qquad \text{and} \quad \phi(x) = -1 - x^2.$$

---

[1] That the product $g\phi$ corresponds to *first* carrying out ϕ *and then* g (in *that* order) seems unnatural at first glance, but actually corresponds to the convention always adopted in mathematics of writing the argument x of a function $f(x)$ *to the right* of the symbol f for the function. Thus, for example, in sin (log x) it is always understood that we first form the logarithm of x and then take the sine of that, and not the other way around.

It is useful to consider functions which are compounded more than once. Such a function is

$$f(x) = \sqrt{1 + \tan{(x^2)}},$$

which can be built up by successive compositions

$$\phi(x) = x^2, \qquad \psi(\phi) = 1 + \tan{\phi}, \qquad g(\psi) = \sqrt{\psi} = f(x).$$

We would write symbolically $f = g\psi\phi$.

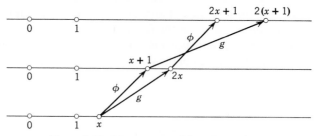

Figure 1.40 Noncommutativity of mappings.

Inverse Functions

The notion of "inverse function" becomes clearer in the context of product of mappings. Consider the mapping ϕ associating with a point x of the domain of ϕ the image $u = \phi(x)$. Assume that our mapping ϕ is such that different x are always mapped into different u. The mapping is then called "one to one." Then a value u is the image of at most one value x. We can associate with every u in the range of ϕ the value $x = g(u)$ of which u is the image under the mapping ϕ. In this way we have defined a mapping g whose domain is the range of ϕ and which when applied to an image $u = \phi(x)$ of the ϕ-mapping reproduces the original value x, that is, $g(\phi(x)) = x$. We call g the *inverse* of ϕ. It is characterized by the symbolic equation $g\phi x = x$.

The Identity Mapping

We define the *identity mapping* I as the one that maps every x into itself; for the inverse g of ϕ then, $g\phi = I$.[1] The mapping I plays the same role for symbolic multiplication as the number 1 in ordinary multiplication; multiplication by I does not change a mapping. Accordingly, the equation $g\phi = I$ suggests the notation $g = \phi^{-1}$ for the inverse of ϕ. For example, the inverse $x = \text{arc sin } u$ of the function $u = \sin x$ is often denoted by $x = \sin^{-1} u$.[2]

---

[1] More precisely $g\phi$ agrees with I, in the domain of ϕ.

[2] This must not be confused with the algebraic *reciprocal* $1/(\sin u)$.

From the definition of the inverse g of ϕ it follows immediately that also ϕ is the inverse of g so that not only $g(\phi) = x$ but also $\phi(g(u)) = u$.

* A monotone function $u = \phi(x)$ defined in an interval $a \leq x \leq b$ clearly defines a 1-1 mapping of that interval. If, in addition, ϕ is continuous, then as we saw earlier as a consequence of the intermediate value theorem (p. 44), the range of ϕ is the interval with end points $\phi(a)$ and $\phi(b)$. In that case the inverse g of ϕ exists and is again monotone and continuous in that latter interval. As a matter of fact *the monotone continuous functions are the only continuous functions that have inverses or define one-to-one mappings.* Indeed, let $u = \phi(x)$ be a continuous function in the closed interval $[a, b]$ mapping different x of the interval into different u. Then in particular the values $\phi(a) = \alpha$ and $\phi(b) = \beta$ are distinct. We assume, say, that $\alpha < \beta$. Then we can show that $\phi(x)$ is monotonic increasing throughout the interval. For if that were not the case we could find two values c and d with $a \leq c < d \leq b$ for which $\phi(d) < \phi(c)$. If here also $\phi(d) \geq \phi(a)$ it would follow from the intermediate value theorem that there exists a ξ in the interval $[a, c]$ for which $\phi(\xi) = \phi(d)$. This ξ would be different from d and our mapping could not be 1-1. If, on the other hand, $\phi(d) < \phi(a) = \alpha$ it would follow that $\phi(a)$ is intermediate between $\phi(d)$ and $\phi(b)$; there would then be a ξ intermediate between d and b for which $\phi(\xi) = \phi(a)$, and this also contradicts the 1-1 nature of ϕ.

An important, almost obvious property of compound functions, is that $g(\phi(x)) = f(x)$ is continuous (where defined) if g and ϕ are. Indeed, for given positive ϵ we have

$$|f(x) - f(x_0)| = |g(\phi(x)) - g(\phi(x_0))| < \epsilon \qquad \text{for} \quad |\phi(x) - \phi(x_0)| < \delta$$

as a consequence of the continuity of the function g. Since, however, ϕ is also continuous, we certainly have $|\phi(x) - \phi(x_0)| < \delta$ for all x satisfying $|x - x_0| < \delta'$ with some suitable positive δ'. Hence

$$|f(x) - f(x_0)| < \epsilon \qquad \text{for} \quad |x - x_0| < \delta'$$

which shows the continuity of f.

It is much easier to appeal to this general theorem in proving continuity of compound functions like $\sqrt{1 - x^2}$ than to try to construct directly a modulus of continuity for the function.

1.4   Sequences

Hitherto we have considered functions of a continuous variable, or functions whose domains consist of one or more intervals. However, numerous cases occur in mathematics in which a quantity a

depends on a positive integer n. Such a function $a(n)$ associates a value with every natural number n. The function $a(n)$ is called a *sequence*, specifically, an *infinite sequence*, if n ranges over *all* positive integers. Usually, we write $a_n{}^1$ instead of $a(n)$ for the "nth element" of the sequence, and think of the elements forming a sequence arranged in order of increasing subscripts n:

$$a_1, a_2, a_3, \ldots .$$

Here the dependence of the numbers a_n on n may be defined by any law whatsoever, and, in particular, the values a_n need not all be distinct from each other. The idea of a sequence will most easily be grasped by examples.

1. The sum of the first n integers

$$S(n) = 1 + 2 + 3 + \cdots + n = \tfrac{1}{2}n(n + 1)$$

is a function of n, giving rise to the sequence

$$1, 3, 6, 10, 15, \ldots .$$

2. Another simple function of n is the expression "n-factorial," the product of the first n integers.

$$n! = 1 \cdot 2 \cdot 3 \cdots \cdot n.$$

3. Every integer $n > 1$ which is not a prime number is divisible by more than two positive integers, whereas the prime numbers are divisible only by themselves and by 1. We can obviously consider the number $T(n)$ of divisors of n as a function of n itself. For the first few numbers it is given by the table:

$$n = 1 \quad 2 \quad 3 \quad 4 \quad 5 \quad 6 \quad 7 \quad 8 \quad 9 \quad 10 \quad 11 \quad 12$$
$$T(n) = 1 \quad 2 \quad 2 \quad 3 \quad 2 \quad 4 \quad 2 \quad 4 \quad 3 \quad 4 \quad 2 \quad 6$$

4. A sequence of great importance in the Theory of Numbers is $\pi(n)$, the number of primes less than the number n. Its detailed investigation is one of the most fascinating problems. The principal result is: The number $\pi(n)$ is given *asymptotically*,[2] for large values of n, by the function $n/\log n$, where by $\log n$ we mean the logarithm to the "natural base" e, to be defined later (p. 77).

---

[1] Pronounced "a-sub-n."
[2] That is, the quotient of the number $\pi(n)$ by the number $n/\log n$ differs arbitrarily little from one, provided only that n is large enough.

1.5   Mathematical Induction

We insert here a discussion of a very important type of reasoning which permeates much of mathematical thought.

The fact that the whole sequence of natural numbers is generated by starting with the number 1 and passing from n to $n + 1$ leads to the fundamental "principle of mathematical induction." In the natural sciences we derive by "empirical induction" from a large number of samples, a law which is expected to hold generally. The degree of certainty of the law depends then on the number of times a sample or an "event" has been observed and the law confirmed. This type of induction can be overwhelmingly convincing, although it does not carry with it the logical certainty of a mathematical proof.

Mathematical induction is used to establish with logical certainty the correctness of a theorem for an infinite sequence of cases. Let A denote a statement referring to an arbitrary natural number n. For example, A might be the statement "The sum of the interior angles in a simple polygon of $n + 2$ sides is n times $180°$" or $n\pi$. To prove a statement of this type it is not sufficient to prove it for the first 10 or the first 100 or even the first 1000 values of n. Instead, we have to apply a mathematical method which we explain first for this example. For $n = 1$ the polygon reduces to a triangle, for which the sum of the angles is known to be $180°$. For a quadrangle corresponding to $n = 2$ we draw a diagonal dividing the quadrangle into two triangles. This shows that the sum of the angles of the quadrangle is equal to the combined sum of the angles of the two triangles, that is, $180° + 180° = 2 \cdot 180°$. Proceeding to the example of a pentagon we can divide this into a quadrangle and a triangle by drawing a suitable diagonal. This yields for the sum of the angles of the pentagon the value $2 \cdot 180° + 1 \cdot 180° = 3 \cdot 180°$. We can go on in this manner and prove the general theorem successively for $n = 4, 5$, etc. The correctness of the statement A for any n follows from its correctness for the preceding n; in this way its general validity is established for all n.

General Formulation

What is essential in the proof of statement A in our example is that A is proved successively for the special cases $A_1, A_2, \ldots A_n, \ldots$. The possibility of doing this depends on two factors: (1) a general proof has to be given showing that the statement A_{r+1} is correct whenever A_r is correct and (2) the statement A_1 must be proved. That these two conditions are sufficient to prove the correctness of all $A_1, A_2, A_3, \ldots$

constitutes the *principle of mathematical induction*. In what follows we accept the validity of this principle as a basic fact of logic.

The principle can be formulated in a more general abstract form. "Let S be any set consisting of natural numbers which has the following two properties: (1) whenever S contains a number r, then it also contains the number $r + 1$ and (2) S contains the number 1. Then it is true that S is the set of all natural numbers." The previous formulation of the principle of mathematical induction follows if we take for S the set of all natural numbers for which statement A is correct.

Often the principle is applied without specific mention or its use is indicated only by the expression, "etc." This happens particularly often in elementary mathematics. However, in more complicated situations an explicit appeal to the principle is preferable.

Examples. Two applications follow as illustrations.

First we prove a formula for the sum of the first n squares. By some trial we find for small n, (say $n < 5$), that the following formula,[1] denoted by A_n, holds:

$$1^2 + 2^2 + 3^2 + \cdots + n^2 = \frac{n(n + 1)(2n + 1)}{6}.$$

We conjecture that this formula is correct for all n. For the proof we assume that r is any number for which the formula A_r is correct, that is, that

$$1^2 + 2^2 + 3^2 + \cdots + r^2 = \frac{r(r + 1)(2r + 1)}{6};$$

adding $(r + 1)^2$ to both sides, we obtain

$$1^2 + 2^2 + \cdots + r^2 + (r + 1)^2 = \frac{r(r + 1)(2r + 1)}{6} + (r + 1)^2$$

$$= \frac{(r + 1)(r + 2)[2(r + 1) + 1]}{6}.$$

This, however, is just the statement A_{r+1} obtained by substituting $r + 1$ for n in A_n. Thus the truth of A_r implies that of A_{r+1}. To complete the proof of A_n for general n we need only to verify the correctness of A_1, that is, of

$$1^2 = \frac{1 \cdot 2 \cdot 3}{6}.$$

---

[1] Incidentally, this result was used by the Greek mathematician Archimedes in his work on spirals.

Since this is obviously correct, the formula A_n is established for all natural n.

The reader should prove by a similar argument that

$$1^3 + 2^3 + 3^3 + \cdots + n^3 = \left[\frac{n(n + 1)}{2}\right]^2.$$

As a further illustration for the principle of induction we prove

THE BINOMIAL THEOREM. *The statement A_n of the theorem is represented by the formula*

$$(a + b)^n = a^n + \frac{n}{1}a^{n-1}b + \frac{n(n - 1)}{1 \cdot 2}a^{n-2}b^2$$

$$+ \frac{n(n - 1)(n - 2)}{1 \cdot 2 \cdot 3}a^{n-3}b^3 + \cdots + \frac{n(n - 1)(n - 2) \cdot \cdots \cdot 2 \cdot 1}{1 \cdot 2 \cdot 3 \cdot \cdots \cdot (n - 1) \cdot n}b^n.$$

It is customary to write the formula in the form

$$(a + b)^n = \binom{n}{0}a^n + \binom{n}{1}a^{n-1}b + \binom{n}{2}a^{n-2}b^2 + \cdots + \binom{n}{n}b^n$$

where the *binomial coefficient* $\binom{n}{k}$ is defined by

$$\binom{n}{k} = \frac{n(n - 1)(n - 2) \cdots (n - k + 1)}{k!} = \frac{n!}{k!(n - k)!}$$

for $k = 1, 2, \ldots, n - 1$ and

$$\binom{n}{0} = \binom{n}{n} = 1.$$

(If we define $0! = 1$, the general formula for $\binom{n}{k}$ applies also to the cases $k = 0$ and $k = n$.)

If A_n holds for a certain n, we find by multiplying both sides with $(a + b)$ that

$$(a + b)^{n+1} = (a + b)\left[\binom{n}{0}a^n + \binom{n}{1}a^{n-1}b + \cdots + \binom{n}{n}b^n\right]$$

$$= \binom{n}{0}a^{n+1} + \left[\binom{n}{0} + \binom{n}{1}\right]a^nb + \left[\binom{n}{1} + \binom{n}{2}\right]a^{n-1}b^2$$

$$+ \cdots + \left[\binom{n}{n - 1} + \binom{n}{n}\right]ab^n + \binom{n}{n}b^{n+1}.$$

Now

$$\binom{n}{k} + \binom{n}{k+1}$$

$$= \frac{n(n-1)\cdots(n-k+1)}{k!} + \frac{n(n-1)\cdots(n-k+1)(n-k)}{(k+1)!}$$

$$= \frac{n(n-1)(n-2)\cdots(n-k+1)}{k!}\left(1 + \frac{n-k}{k+1}\right)$$

$$= \frac{(n+1)n(n-1)\cdots(n-k+1)}{(k+1)!} = \binom{n+1}{k+1}.$$

Since $\binom{n}{0} = \binom{n+1}{0} = 1$ and $\binom{n}{n} = \binom{n+1}{n+1} = 1$, we have

$$(a+b)^{n+1} = \binom{n+1}{0}a^{n+1} + \binom{n+1}{1}a^n b + \binom{n+1}{2}a^{n-1}b^2 + \cdots$$

$$+ \binom{n+1}{n}ab^n + \binom{n+1}{n+1}b^{n+1},$$

which is the formula A_{n+1}. Since also for $n = 1$

$$(a+b)^1 = \binom{1}{0}a + \binom{1}{1}b = a + b,$$

the binomial theorem holds for all natural numbers n.

1.6   The Limit of a Sequence

The fundamental concept on which the whole of mathematical analysis ultimately rests is that of the *limit* of an infinite sequence a_n. A number a is often described by an infinite sequence a_n of approximations; that is, the value a is given by the value a_n with any desired degree of precision if we choose the index n sufficiently large. We have already encountered such *representations of numbers a as "limits"* of sequences in their representations as infinite decimal fractions; the real numbers then appeared as limits for increasing n of the sequences of ordinary decimal fractions with n digits. In Section 1.7 we shall give a precise general discussion of the limit concept; at this point we illustrate the idea of limit by some significant examples.

Sequences $a_1, a_2, \ldots$ can be depicted conveniently by a succession of "blocks," the element a_n corresponding to the rectangle in the xy-plane bounded by the lines $x = n-1$, $x = n$, $y = a_n$, $y = 0$, having $|a_n|$

as area,[1] or equivalently, by the graph of a piecewise constant function $a(x)$ of a continuous variable x with jump discontinuities at the points $x = n$.

a. $a_n = \dfrac{1}{n}$

We consider the sequence

$$1, \frac{1}{2}, \frac{1}{3}, \ldots, \frac{1}{n}, \ldots .$$

(See Fig. 1.41.) No number of this sequence is zero; but as the number n grows larger, a_n approaches zero. Furthermore, if we take any

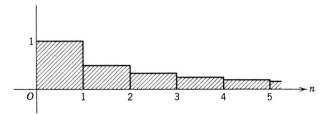

Figure 1.41 The sequence $a_n = \dfrac{1}{n}$.

interval centered at the origin, no matter how small, then from a definite index onward all numbers a_n will be in this interval. This situation is expressed by saying that as n increases the numbers a_n *tend to* zero or that they possess the *limit* zero or that the sequence $a_1, a_2, a_3, \ldots$ *converges* to zero.

If the numbers are represented as points on a line, this means that the points $1/n$ crowd closer and closer to the point zero as n increases.

The situation is similar for the sequence

$$1, -\frac{1}{2}, \frac{1}{3}, -\frac{1}{4}, \ldots, \frac{(-1)^{n-1}}{n}, \ldots .$$

(See Fig. 1.42.) Here too, the numbers a_n tend to zero as n increases; the only difference is that the numbers a_n are sometimes greater and sometimes less than the limit zero; as we say, the sequence *oscillates* about the limit.

---

[1] We might just as well have chosen the rectangle bounded by the lines $x = n$, $x = n + 1$, $y = a_n$, $y = 0$ to represent a_n.

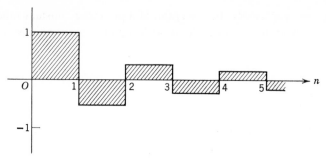

Figure 1.42   The sequence $a_n = \dfrac{(-1)^{n-1}}{n}$.

The convergence of the sequence to zero is usually expressed symbolically by the equation

$$\lim_{n \to \infty} a_n = 0,$$

or occasionally by the abbreviation

$$a_n \to 0.$$

b. $a_{2m} = \dfrac{1}{m}$; $a_{2m-1} = \dfrac{1}{2m}$

In the preceding examples, the absolute value of the difference between a_n and the limit steadily becomes smaller as n increases. This is not necessarily the case, as is shown by the sequence

$$\frac{1}{2}, 1, \frac{1}{4}, \frac{1}{2}, \frac{1}{6}, \frac{1}{3}, \dots, \frac{1}{2m}, \frac{1}{m}, \dots ;$$

(see Fig. 1.43) given for even values $n = 2m$ by $a_n = a_{2m} = 1/m$; for odd values $n = 2m - 1$ by $a_n = a_{2m-1} = 1/2m$.   This sequence

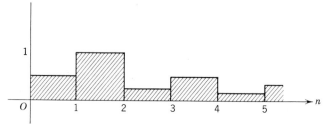

Figure 1.43   The sequence $a_{2n} = \dfrac{1}{n}$, $a_{2n-1} = \dfrac{1}{2n}$.

also has the limit zero; for every interval about the origin, no matter how small, contains all the numbers a_n from a certain value of n onward; but it is not true that every number lies nearer to the limit zero than the preceding one.

c. $a_n = \dfrac{n}{n+1}$

We consider the sequence

$$a_1 = \frac{1}{2}, a_2 = \frac{2}{3}, \ldots, a_n = \frac{n}{n+1}, \ldots .$$

Writing $a_n = 1 - 1/(n+1)$, we see that as n increases the number a_n will approach the number 1, in the sense that if we mark off any interval about the point 1 all the numbers a_n following a certain a_N must fall in that interval. We write

$$\lim_{n \to \infty} a_n = 1.$$

The sequence

$$a_n = \frac{n^2 - 1}{n^2 + n + 1}$$

behaves in a similar way. This sequence also tends to a limit as n increases, in fact to the limit one; $\lim_{n \to \infty} a_n = 1$. We see this most readily if we write

$$a_n = 1 - \frac{n+2}{n^2 + n + 1} = 1 - r_n;$$

we need only show that the numbers r_n tend to zero as n increases. For all values of n greater than 2 we have $n + 2 < 2n$ and $n^2 + n + 1 > n^2$. Hence for the remainder r_n, we have

$$0 < r_n < \frac{2n}{n^2} = \frac{2}{n} \qquad (n > 2),$$

from which we see that r_n tends to zero as n increases. Our discussion at the same time gives an estimate of the largest amount by which the number a_n (for $n > 2$) can differ from the limit one; this difference cannot exceed $2/n$.

This example illustrates the fact, that for large values of n the terms with the highest exponents in the numerator and denominator of the fraction for a_n predominate and determine the limit.

d. $a_n = \sqrt[n]{p}$

Let p be any fixed positive number. We consider the sequence $a_1, a_2, a_3, \ldots, a_n, \ldots,$ where

$$a_n = \sqrt[n]{p}.$$

We assert that

$$\lim_{n \to \infty} a_n = \lim_{n \to \infty} \sqrt[n]{p} = 1.$$

We shall prove this by using a lemma that we shall also find useful for other purposes.

LEMMA. *If h is a positive number and n a positive integer, then*

$$(1) \qquad\qquad (1 + h)^n \geq 1 + nh.$$

This inequality is a trivial consequence of the binomial theorem (see p. 59) according to which

$$(1 + h)^n = 1 + nh + \frac{n(n - 1)}{2} h^2 + \cdots + h^n,$$

if we observe that all terms in the expansion of $(1 + h)^n$ are non-negative. The same argument yields the stronger inequality

$$(1 + h)^n \geq 1 + nh + \frac{n(n - 1)}{2} h^2.$$

Returning to our sequence, we distinguish between the cases $p > 1$ and $p < 1$ (if $p = 1$, then $\sqrt[n]{p}$ is equal to 1 for every n, and our statement is certainly true).

If $p > 1$, then $\sqrt[n]{p}$ also is greater than 1; we set $\sqrt[n]{p} = 1 + h_n$, where h_n is a positive quantity depending on n; by the inequality (1) we have

$$p = (1 + h_n)^n \geq 1 + nh_n,$$

implying

$$0 < h_n \leq \frac{p - 1}{n}.$$

As n increases the number h_n must tend to 0, which proves that a_n converges to the limit one, as stated. At the same time we have a means for estimating how close any a_n is to the limit one, since the difference h_n between a_n and one is not greater than $(p - 1)/n$.

If $p < 1$, then $1/p > 1$ and $\sqrt[n]{1/p}$ converges to the limit one. However,

$$\sqrt[n]{p} = \frac{1}{\sqrt[n]{1/p}}.$$

As the reciprocal of a quantity tending to one $\sqrt[n]{p}$ itself tends to one.

e. $a_n = \alpha^n$

We consider the sequence $a_n = \alpha^n$, where α is fixed and n runs through the sequence of positive integers.

First, let α be a positive number less than one. We then put $\alpha = 1/(1 + h)$, where h is positive, and the inequality (1) gives

$$a_n = \frac{1}{(1 + h)^n} \leq \frac{1}{1 + nh} < \frac{1}{nh}.$$

Since h, and consequently $1/h$, depends only on α and does not change as n increases, we see that α^n tends to zero as n increases:

$$\lim_{n \to \infty} \alpha^n = 0 \qquad (0 < \alpha < 1).$$

The same relationship holds when α is zero, or negative but greater than -1. This is immediately obvious, since in any case $\lim_{n \to \infty} |\alpha|^n = 0$.

If $\alpha = 1$, then α^n always is equal to one and we shall have to regard the number one as the limit of α^n.

If $\alpha > 1$, we put $\alpha = 1 + h$, where h is positive, and at once see from our inequality that as n increases α^n does not tend to any definite limit, but increases beyond all bounds. We say that α^n *tends to infinity* as n increases or that α^n *becomes infinite*; in symbols,

$$\lim_{n \to \infty} \alpha^n = \infty \qquad (\alpha > 1).$$

We explicitly emphasize that *the symbol ∞ does not denote a number and that we cannot calculate with it according to the usual rules*; statements asserting that a quantity is or becomes infinite never have the same sense as an assertion involving definite quantities. In spite of this, such modes of expression and the use of the symbol ∞ are extremely convenient, as we shall often see in the following pages.

If $\alpha = -1$, the value of α^n does not tend to any limit, but as n runs through the sequence of positive integers α^n takes the values $+1$ and -1 alternately. Similarly, if $\alpha < -1$ the value of α^n increases numerically beyond all bounds, but its sign is alternately positive and negative.

f. Geometrical Illustration of the Limits of α^n and $\sqrt[n]{p}$

If we consider the graphs of the functions $y = x^n$ and $y = x^{1/n} = \sqrt[n]{x}$ and restrict ourselves for the sake of convenience to nonnegative values of x, the preceding limits are illustrated by Figs. 1.44 and 1.45 respectively. We see that in the interval from 0 to 1 the curves $y = x^n$ come

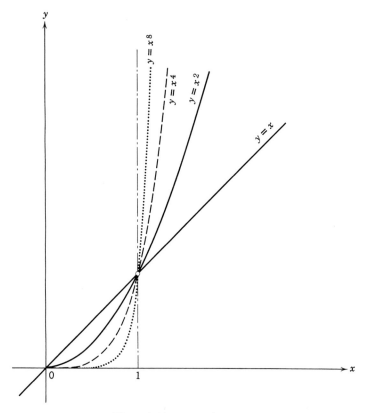

Figure 1.44 x^n as n increases.

closer and closer to the x-axis as n increases, whereas outside that interval they climb more and more steeply and approach a line parallel to the y-axis. All the curves pass through the point with coordinates $x = 1$, $y = 1$ and the origin.

The graphs of the functions $y = x^{1/n} = +\sqrt[n]{x}$, come closer and closer to the line parallel to the x-axis and at a distance 1 above it; again all the curves must pass through the origin and the point $(1, 1)$. Hence in the limit the curves approach the broken line consisting of the part of the y-axis between the points $y = 0$ and $y = 1$ and of the parallel to the x-axis $y = 1$. Moreover, it is clear that the two figures are closely related, as one would expect from the fact that the functions $y = \sqrt[n]{x}$ are the inverse functions of the nth powers, from which we infer that for each n the graph of $y = x^n$ is transformed into that of $y = \sqrt[n]{x}$ by reflection in the line $y = x$.

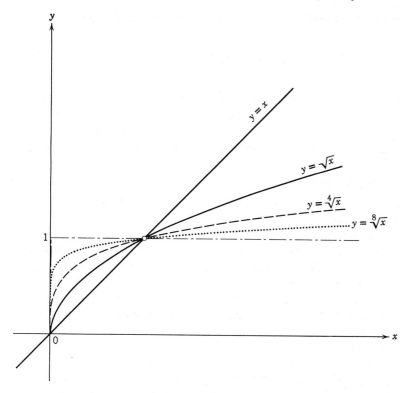

Figure 1.45 $x^{1/n}$ as n increases.

g. The Geometric Series

An example of a limit familiar from elementary mathematics is furnished by the *geometric series*

$$1 + q + q^2 + \cdots + q^{n-1} = S_n;$$

the number q is called the *common ratio* or *quotient* of the series. The value of this sum may, as is well known, be expressed in the form

$$S_n = \frac{1 - q^n}{1 - q}$$

provided that $q \neq 1$; we can derive this expression by multiplying the sum S_n by q and subtracting the equation thus obtained from the original equation or we may verify the formula by division.

What becomes of the sum S_n when n increases indefinitely? The answer is: The sequence of sums S_n has a definite limit S if q lies between -1 and $+1$, these end values being excluded, and

$$S = \lim_{n \to \infty} S_n = \frac{1}{1 - q}.$$

In order to verify this statement we write S_n as $(1 - q^n)/(1 - q)$ $= 1/(1 - q) - q^n/(1 - q)$. We have already shown that provided $|q| < 1$ the quantity q^n tends to zero as n increases; hence under this assumption $q^n/(1 - q)$ also tends to zero and S_n tends to the limit $1/(1 - q)$ as n increases.

The passage to the limit $\lim_{n \to \infty} (1 + q + q^2 + \cdots + q^{n-1}) = 1/(1 - q)$ is usually expressed by saying that when $|q| < 1$ *the sum of the infinite geometric series is the expression* $1/(1 - q)$.

The sums S_n of the finite geometric series are also called the *partial sums* of the infinite geometric series $1 + q + q^2 + \ldots$. (We must draw a distinction between the *sequence* of numbers q^n and the partial sums of the geometric *series*.)

The fact that the partial sums S_n of the geometric series tend to the limit $S = 1/(1 - q)$ as n increases is also expressed by saying that the infinite geometric series $1 + q + q^2 + \cdots$ *converges* to the sum $S = 1/(1 - q)$ when $|q| < 1$.

In passing it should be noted if q is rational, for example, $q = \frac{1}{2}$ or $q = \frac{1}{3}$, then the sum of the infinite geometric series has a rational value (in the cases mentioned the values are 2 and $\frac{3}{2}$, respectively). This observation is behind the well-known fact that *periodic decimal fractions always represent rational numbers*.[1] The general proof of this fact will be clear from the example of the number

$$x = 0.343434 \cdots$$

which can be evaluated by writing

$$x = \frac{34}{10^2} + \frac{34}{10^4} + \frac{34}{10^6} + \cdots$$

$$= \frac{34}{10^2}\left(1 + \frac{1}{10^2} + \frac{1}{10^4} + \cdots\right)$$

$$= \frac{34}{100} \frac{1}{1 - 1/100} = \frac{34}{99}.$$

---

[1] See Courant and Robbins, *What Is Mathematics?*, p. 66.

h. $a_n = \sqrt[n]{n}$

We show that the sequence of numbers

$$a_1 = 1, \qquad a_2 = \sqrt{2}, \qquad a_3 = \sqrt[3]{3}, \ldots, \qquad a_n = \sqrt[n]{n}, \ldots$$

tends to 1 as n increases:

$$\lim_{n \to \infty} \sqrt[n]{n} = 1.$$

Since a_n exceeds the value 1, we set $a_n = 1 + h_n$, with h_n positive. Then (see p. 64)

$$n = (a_n)^n = (1 + h_n)^n$$

$$\geq 1 + nh_n + \frac{n(n-1)}{2} h_n{}^2 \geq \frac{n(n-1)}{2} h_n{}^2.$$

It follows for $n > 1$ that

$$h_n{}^2 \leq \frac{2}{n-1} \,;$$

hence

$$h_n \leq \frac{\sqrt{2}}{\sqrt{n-1}} \,.$$

We now have

$$1 \leq a_n = 1 + h_n \leq 1 + \frac{\sqrt{2}}{\sqrt{n-1}}.$$

The right-hand side of this inequality obviously tends to one, and therefore so does a_n.

i. $a_n = \sqrt{n+1} - \sqrt{n}$

In this example the a_n are differences of two terms, each of which increases beyond all bounds. Attempting to pass to the limit separately with each of the two terms, we obtain the meaningless symbolic expression $\infty - \infty$. In such a case the existence of a limit and what its value may be depends completely on the special case. We assert that in our example

$$\lim_{n \to \infty} (\sqrt{n+1} - \sqrt{n}) = 0.$$

For the proof we need only write the expression in the form

$$\sqrt{n+1} - \sqrt{n} = \frac{(\sqrt{n+1} - \sqrt{n})(\sqrt{n+1} + \sqrt{n})}{\sqrt{n+1} + \sqrt{n}} = \frac{1}{\sqrt{n+1} + \sqrt{n}} \,;$$

and see at once that it tends to zero as n increases.

j. $a_n = \dfrac{n}{\alpha^n}$, **for** $\alpha > 1$

Formally, the limit of the a_n is of the indeterminate type ∞/∞ already encountered in Example *c*. We assert that in this example the sequence of numbers $a_n = n/\alpha^n$ tends to the limit zero.

For the proof we put $\alpha = 1 + h$, where $h > 0$, and again make use of the inequality

$$(1 + h)^n \geq 1 + nh + \frac{n(n - 1)}{2} h^2 > \frac{n(n - 1)}{2} h^2.$$

Hence for $n > 1$

$$a_n = \frac{n}{(1 + h)^n} < \frac{2}{(n - 1)h^2}.$$

Since a_n is positive and the right-hand side of this inequality tends to zero, a_n must also tend to zero.

1.7   Discussion of the Concept of Limit

a. Definition of Convergence and Divergence

From the examples discussed in Section 1.6 we abstract the following general concept of limit:

Suppose that for a given infinite sequence of points $a_1, a_2, a_3, \ldots$ there is a number l such that every open interval, no matter how small, marked off about the point l, contains all the points a_n except for at most a finite number. The number l is then called the limit of the sequence $a_1, a_2, \ldots$, or we say that the sequence $a_1, a_2, \ldots$ is convergent and converges to l; in symbols, $\lim_{n \to \infty} a_n = l$.

The following definition of limit is equivalent:

To any positive number ϵ, no matter how small, we can assign a sufficiently large integer $N = N(\epsilon)$ such that from the index N onward [that is, for $n > N(\epsilon)$] we always have $|a_n - l| < \epsilon$.

Of course, it is true as a rule that $N(\epsilon)$ will have to be chosen larger and larger for smaller and smaller values of the tolerance ϵ; in other words, $N(\epsilon)$ will usually increase beyond all bounds as ϵ tends to zero. The vague intuitive notion of limit suggests a picture of the a_n *moving closer and closer* to l. This picture is replaced here by the precise "static" definition: Any neighborhood of l contains all a_n with at most a finite number of exceptions.[1]

Obviously, a sequence $a_1, a_2, \ldots$ cannot have more than one limit l. If on the contrary two distinct numbers l and l' were limits of the same sequence $a_1, a_2, \ldots$, we could mark off open intervals about each of the points l and l' which do not overlap. Since each interval contains all but a finite number of the a_n, the sequence could not be infinite. The limit of a convergent sequence is therefore uniquely determined.

Another obvious but useful remark is: If from a convergent sequence we omit any number of terms the resulting sequence converges to the same limit as the original sequence.

A sequence which does not converge is said to be divergent. If as n increases the numbers a_n increase beyond all positive bounds, we say that the sequence diverges to $+\infty$; as we have already done occasionally, we write then $\lim_{n \to \infty} a_n = \infty$. Similarly, we write $\lim_{n \to \infty} a_n = -\infty$ if, as n increases, the numbers $-a_n$ increase beyond all bounds in the positive direction. But divergence may manifest itself in other ways, as for the sequence $a_1 = -1, a_2 = +1, a_3 = -1, a_4 = +1, \ldots$, whose terms swing back and forth between two different values.

Clearly, neither divergence nor convergence of a sequence is affected by removing finitely many terms.

A sequence $a_1, a_2, \ldots$ is *bounded* if there is a finite interval containing *all* points of the sequence. Any finite interval is contained in some finite interval that has the origin as center. Hence the requirement that the sequence is bounded means that there exists a number M such that $|a_n| \leq M$ for all n.

A convergent sequence $a_1, a_2, \ldots$ *necessarily is also bounded.* For let l be the limit of the sequence. Taking $\epsilon = 1$ we find from the definition of convergence that all a_n from a certain N onward lie in the interval of length 2 centered at l. The only terms a_n of the sequence that may lie outside that interval are $a_1, \ldots, a_{N-1}$. We can then, however, find a larger finite interval that also includes $a_1, \ldots, a_{N-1}$.

b. Rational Operations with Limits

From the definition of limit it follows at once that we can perform the elementary operations of addition, multiplication, subtraction, and division of limits according to the following rules.

---

[1] The reader will notice the analogy with the definition of continuity of a function $f(x)$ at a point x_0. The role played there by the sufficiently small quantity $\delta(\epsilon)$ is played here by the sufficiently large integer $N(\epsilon)$. We shall see indeed on p. 82 that continuity of a function at a point can be formulated in terms of limits of sequences.

If $a_1, a_2, \ldots$ is a sequence with the limit a and $b_1, b_2, \ldots$ is a sequence with the limit b, then the sequence of numbers $c_n = a_n + b_n$ also has a limit c, and

$$c = \lim_{n \to \infty} c_n = a + b.$$

The sequence of numbers $c_n = a_n b_n$ likewise converges and

$$\lim_{n \to \infty} c_n = ab.$$

Similarly, the sequence $c_n = a_n - b_n$ converges and

$$\lim_{n \to \infty} c_n = a - b.$$

Provided the limit b differs from zero, the numbers $c_n = a_n/b_n$ likewise converge and have the limit

$$\lim_{n \to \infty} c_n = \frac{a}{b}.$$

In words: We can *interchange* the rational operations of calculation with the process of forming the limit; we obtain the same result whether we first perform a passage to the limit and then a rational operation or vice versa.

The proofs of all these rules become clear if one of them is carried out. We consider the multiplication of limits. If the relations $a_n \to a$ and $b_n \to b$ hold, then for any positive number ϵ, we can insure both

$$|a - a_n| < \epsilon \qquad \text{and} \quad |b - b_n| < \epsilon$$

by choosing n sufficiently large, say $n > N(\epsilon)$. If we write

$$ab - a_n b_n = b(a - a_n) + a_n(b - b_n)$$

and recall that there is a positive bound M, independent of n, such that $|a_n| < M$, we obtain

$$|ab - a_n b_n| \leq |b|\,|a - a_n| + |a_n|\,|b - b_n| < (|b| + M)\epsilon.$$

Since the quantity $(|b| + M)\epsilon$ can be made arbitrarily small by choosing ϵ small enough, the difference between ab and $a_n b_n$ actually becomes as small as we please for all sufficiently large values of n; this is precisely the statement made in the equation

$$ab = \lim_{n \to \infty} a_n b_n.$$

Using this example as a model, the reader can prove the rules for the remaining rational operations.

By means of these rules many limits can be evaluated easily; thus, we have

$$\lim_{n \to \infty} \frac{n^2 - 1}{n^2 + n + 1} = \lim_{n \to \infty} \frac{1 - \dfrac{1}{n^2}}{1 + \dfrac{1}{n} + \dfrac{1}{n^2}} = 1,$$

since in the second expression we can pass directly to the limit in the numerator and denominator.

The following simple rule is frequently useful: *If* $\lim a_n = a$ *and* $\lim b_n = b$, *and if in addition* $a_n > b_n$ *for every n, then* $a \geq b$. We are, however, by no means entitled to expect that a will always be *greater* than b, as is shown by the sequences $a_n = 1/n$, $b_n = 1/2n$, for which $a = b = 0$.

c. Intrinsic Convergence Tests. Monotone Sequences

In all the examples given the limit of the sequence considered was a known number. In fact, to apply the above definition of limit of a sequence we must know the limit before we can verify convergence. If the concept of limit of a sequence yielded nothing more than the recognition that some known numbers can be approximated by certain sequences of other known numbers, we should have gained very little from it. The advantage of the concept of limit in analysis lies essentially on the fact that important problems often have numerical solutions which may not otherwise be directly known or expressible, but can be described as limits. The whole of higher analysis consists of a succession of examples of this fact which will become steadily clearer in the following chapters. The representation of the irrational numbers as limits of rational numbers may be regarded as the first and typical example.

Any *convergent* sequence of known numbers $a_1, a_2, \ldots$ *defines* a number l, its limit. However, the only test for convergence that arises from the definition of convergence consists in estimating the differences $|a_n - l|$, and this is applicable only if the number l is known already. It is essential to have "*intrinsic*" tests for convergence that do not require an *a priori* knowledge of the value of the limit but only involve the terms of the sequence themselves. The simplest such test applies to a special class of sequences, the monotone sequences, and includes most of the important examples.

Limits of Monotone Sequences

A sequence $a_1, a_2, \ldots$ is called *monotonically increasing* if each term a_n is larger, or at least not smaller than the preceding one; that is,

$$a_n \geq a_{n-1}.$$

Similarly, the sequence is *monotonically decreasing* if $a_n \leq a_{n-1}$ for all n. A *monotone* sequence is one that is either monotonically increasing or decreasing. With this definition we have the basic principle:

A sequence that is both monotone and bounded converges.[1]

This principle is convincingly suggested, but not proved, by intuition; it is intimately related to the properties of real numbers and in fact is equivalent to the continuity axiom for real numbers.

The axiom (see Section 1b) that every nested sequence of intervals contains a point is easily seen to be a consequence of the convergence of bounded monotone sequences. For let $[a_1, b_1], [a_2, b_2], \ldots$ be a sequence of nested intervals. By the definition of nested sequences we have

$$a_1 \leq a_2 \leq \cdots \leq a_n < b_n \leq b_{n-1} \leq \cdots \leq b_1.$$

Obviously, the infinite sequence $a_1, a_2, \ldots$ is monotonically increasing. It is also bounded since $a_1 \leq a_n \leq b_1$ for all n. Hence $l = \lim_{n \to \infty} a_n$ exists. Moreover, for any m and for any number $n > m$ we have

$$a_m \leq a_n \leq b_m.$$

Hence also

$$a_m \leq \lim_{n \to \infty} a_n = l \leq b_m.$$

Thus all intervals of the nested sequence contain one and the same point l. (That they have no other point in common follows from the further property $\lim (b_n - a_n) = 0$ of nested sequences of intervals.)

Cauchy's Criteria for Convergence

A convergent sequence is automatically bounded but need not be monotone (see Example b, p. 62). Hence, in dealing with general sequences, it is desirable to have a test for convergence that is also

---

[1] The assumption of boundedness is essential since no unbounded sequence can converge. Oberve that a monotonically increasing sequence $a_1, a_2, \ldots$ is always "bounded from below": $a_n \geq a_1$ for all n. In order to prove that a monotonically increasing sequence converges it is sufficient then to find a number M such that $a_n \leq M$ for all n.

applicable to nonmonotone sequences. This need is satisfied by a simple condition, the *Cauchy test* for *convergence*; this criterion characterizes sequences of real numbers which have a limit; most importantly it does not require *a priori* knowledge of the value of the limit: *Necessary and sufficient for convergence of a sequence* $a_1, a_2, \ldots$ *is that the elements* a_n *of the sequence with sufficiently large index n differ arbitrarily little from each other.* Formulated precisely: a sequence a_n is convergent if for every $\epsilon > 0$ there exists a natural number $N = N(\epsilon)$ such that $|a_n - a_m| < \epsilon$ whenever $n > N$ and $m > N$. Geometrically, the Cauchy condition states that a sequence converges if there exist arbitrarily small intervals outside of which there lie only a finite number of points of the sequence. The correctness of Cauchy's test for convergence will be proved and its significance discussed in the Supplement.

d. Infinite Series and the Summation Symbol

A sequence is just an ordered infinite array of numbers $a_1, a_2, \ldots$. *An infinite series*

$$a_1 + a_2 + a_3 + \cdots$$

requires the terms to be added in the order in which they appear. To arrive at a precise meaning of the *sum* of an infinite series we consider the *n*th *partial sum* that is, the sum of the first *n* terms of the series

$$s_n = a_1 + a_2 + \cdots + a_n.$$

The partial sums s_n for different *n* form a sequence

$$s_1 = a_1, \qquad s_2 = a_1 + a_2, \qquad s_3 = a_1 + a_2 + a_3,$$

and so on. The sum *s* of the infinite series is then defined as

$$s = \lim_{n \to \infty} s_n,$$

provided this limit exists. In that case we call the infinite series *convergent*. If the sequence s_n diverges, the infinite series is called *divergent;* For example, the sequence $1, q, q^2, q^3, \ldots$ gives rise to the infinite geometric series

$$1 + q + q^2 + q^3 + \cdots$$

whose partial sums are

$$s_n = 1 + q + q^2 + \cdots + q^{n-1}.$$

For $|q| < 1$ the sequence s_n converges toward the limit

$$s = \frac{1}{1-q},$$

which then represents the sum of the infinite series. For $|q| \geq 1$ the partial sums s_n have no limit and the series diverges (see p. 67).

It is customary to use for $a_1 + a_2 + \cdots + a_n$ the symbol

$$\sum_{k=1}^{n} a_k$$

which indicates that the sum of the a_k is to be taken with k running through the integers from $k = 1$ to $k = n$. For example,

$$\sum_{k=1}^{4} \frac{1}{k!} \text{ stands for } \frac{1}{1!} + \frac{1}{2!} + \frac{1}{3!} + \frac{1}{4!},$$

whereas

$$\sum_{k=1}^{n} a^k b^{2k} \text{ stands for } a^1 b^2 + a^2 b^4 + a^3 b^6 + \cdots + a^n b^{2n}.$$

More generally, $\sum_{k=m}^{n} a_k$ means the sum of all a_k obtained by giving k the values $m, \ m+1, \ m+2, \ldots, n$. Thus

$$\sum_{k=3}^{5} \frac{1}{k!} = \frac{1}{3!} + \frac{1}{4!} + \frac{1}{5!}.$$

In these examples we have used the letter k for the index of summation. Of course, the sum is independent of the letter denoting this index. Thus

$$s_n = \sum_{k=1}^{n} a_k = \sum_{i=1}^{n} a_i.$$

We use the symbol

$$\sum_{k=1}^{\infty} a_k$$

to denote the sum of the whole infinite series. Similarly, $\sum_{k=0}^{\infty} a_k$ would stand for the sum of the infinite series $a_0 + a_1 + a_2 + \ldots$, whose nth partial sum is $s_n = a_0 + a_1 + a_2 + \cdots + a_{n-1}$.

Many of our earlier results can be written more concisely in this summation notation. The formula of p. 58, for the sum of the first n squares becomes

$$\sum_{k=1}^{n} k^2 = \frac{n(n+1)(2n+1)}{6}.$$

The formula for the sum of a geometric series is

$$\sum_{k=0}^{\infty} q^k = \frac{1}{1-q} \qquad \text{for } |q| < 1.$$

Finally, the binomial theorem is expressed by

$$(a + b)^n = \sum_{k=0}^{n} \binom{n}{k} a^{n-k} b^k.$$

Since an infinite series is merely the limit of a sequence s_n, convergence can be decided on the basis of the convergence tests for sequences. For example, the convergence of the series

$$\sum_{k=1}^{\infty} \frac{1}{k^k} = \frac{1}{1^1} + \frac{1}{2^2} + \frac{1}{3^3} + \cdots$$

follows immediately from the fact that the partial sums

$$s_n = \sum_{k=1}^{n} \frac{1}{k^k} = \frac{1}{1^1} + \frac{1}{2^2} + \frac{1}{3^3} + \cdots + \frac{1}{n^n}$$

increase monotonically with n and are bounded since

$$1 \le s_n \le 1 + \frac{1}{2^2} + \frac{1}{2^3} + \frac{1}{2^4} + \cdots + \frac{1}{2^n}$$

$$= 1 + \frac{1}{4} \frac{1 - 1/2^{n-1}}{1 - \frac{1}{2}} = 1 + \frac{1}{2} - \frac{1}{2^n} < \frac{3}{2}.$$

Later, in Chapter 7, we shall study infinite series more systematically.

e. The Number e

As a first example of a number which is generated as the limit of a sequence, we consider

$$e = 1 + \frac{1}{1!} + \frac{1}{2!} + \frac{1}{3!} + \cdots.$$

Thus e stands for $\lim_{n \to \infty} S_n$, where

$$S_n = 1 + \frac{1}{1!} + \frac{1}{2!} + \cdots + \frac{1}{n!}.^1$$

---

[1] Remembering the convention defining 0! as 1, we can write the first term of the series as 1/0! in agreement with the law of formation of the following terms. Notice that in our notation S_n is really the $(n + 1)$st partial sum of the infinite series, instead of the nth. This is, however, of no significance.

The numbers e and π are the most widely used transcendental constants in mathematical analysis. In order to prove the existence of the limit e we need only prove that the sequence S_n is bounded since the numbers S_n increase monotonically. For all values of n we have

$$S_n = 1 + 1 + \frac{1}{2} + \frac{1}{2 \cdot 3} + \frac{1}{2 \cdot 3 \cdot 4} + \cdots + \frac{1}{2 \cdot 3 \cdot 4 \cdot \cdots \cdot n}$$

$$\leq 1 + 1 + \frac{1}{2} + \frac{1}{2^2} + \frac{1}{2^3} + \cdots + \frac{1}{2^{n-1}}$$

$$= 1 + \frac{1 - 1/2^n}{1 - \frac{1}{2}} < 3.$$

The numbers S_n therefore have the upper bound 3, and since they form a monotonic increasing sequence, they possess a limit which we denote by e.

The expression for e as a series permits us to compute e rapidly with great accuracy. The error committed in approximating e by a partial sum S_m can be estimated by the same method of comparison with a geometric series that furnished the upper bound 3 for e. We have for any $n > m$

$$S_n = S_m + \frac{1}{(m+1)!} + \frac{1}{(m+2)!} + \cdots + \frac{1}{n!}$$

$$\leq S_m + \frac{1}{(m+1)!}\left[1 + \frac{1}{m+2} + \frac{1}{(m+2)(m+3)} + \cdots\right]$$

$$\leq S_m + \frac{1}{(m+1)!}\left[1 + \frac{1}{m+1} + \frac{1}{(m+1)^2} + \cdots\right]$$

$$= S_m + \frac{1}{(m+1)!}\frac{1}{1 - \dfrac{1}{m+1}} = S_m + \frac{1}{m}\frac{1}{m!}.$$

Hence for $n > m$

$$S_m < S_n \leq S_m + \frac{1}{m}\frac{1}{m!}.$$

Letting n increase beyond all bounds while holding m fixed we find also that

$$S_m < e \leq S_m + \frac{1}{m}\frac{1}{m!}.$$

Hence e differs from S_m by at most $(1/m)(1/m!)$. Since $m!$ increases extremely rapidly with m, the number S_m is a good approximation for e already for fairly small m; for example, S_{10} differs from e by less than 10^{-7}. In this way we find that $e = 2.718281 \cdots$.

e is an *irrational number*. The estimate for e in terms of S_m can also be used to establish this fact. Indeed, if e were rational, we could write e in the form p/m with positive integers p, m; here, $m \geq 2$, since e, lying between 2 and 3, cannot be an integer. Comparing e with the partial sum S_m, we would have

$$S_m < \frac{p}{m} \leq S_m + \frac{1}{m}\frac{1}{m!}.$$

If we here multiply both sides by $m!$, we find that

$$m!\, S_m < p(m-1)! \leq m!\, S_m + \frac{1}{m} < m!\, S_m + 1.$$

But

$$m!\, S_m = m! + m! + \frac{m!}{2!} + \frac{m!}{3!} + \cdots + \frac{m!}{m!}$$

is an integer since each term in the sum is. Thus, if e were rational, the integer $p(m-1)!$ would lie between two successive integers, which is impossible.[1]

e As Limit of $(1 + 1/n)^n$. The number e that was defined here as the sum of an infinite series can also be obtained as the limit of the sequence

$$T_n = \left(1 + \frac{1}{n}\right)^n.$$

The proof is simple and at the same time an instructive example of operations with limits. According to the binomial theorem,

$$T_n = \left(1 + \frac{1}{n}\right)^n$$

$$= 1 + n\frac{1}{n} + \frac{n(n-1)}{2!}\frac{1}{n^2} + \cdots + \frac{n(n-1)(n-2)\cdots 1}{n!}\frac{1}{n^n}$$

$$= 1 + 1 + \frac{1}{2!}\left(1 - \frac{1}{n}\right) + \cdots$$

$$+ \frac{1}{n!}\left(1 - \frac{1}{n}\right)\left(1 - \frac{2}{n}\right)\cdots\left(1 - \frac{n-1}{n}\right)$$

---

[1] The irrationality of the number e means that there is no linear equation $ax + b = 0$ with rational coefficients a, b and $a \neq 0$ having e as a solution. A much stronger statement has been proved (by Hermite), that there exists no polynomial equation $a_0 x^n + a_1 x^{n-1} + \cdots + a_{n-1}x + a_n = 0$ of any degree n whatsoever and with rational coefficients $a_0, a_1, \ldots, a_n$ (with $a_0 \neq 0$) with $x = e$ as a root. One says that e is a *transcendental* number in contrast to "algebraic" numbers like $\sqrt{2}$ or $\sqrt[3]{10}$ that are roots of certain polynomial equations with rational coefficients.

From this we see at once that $T_n \leq S_n < 3$. Furthermore, since we obtain T_{n+1} from T_n by replacing the factors $1 - 1/n$, $1 - 2/n$, ... by the larger factors $1 - 1/(n + 1)$, $1 - 2/(n + 1)$, ... and finally adding a positive term we see that the T_n's also form a monotonic increasing sequence, from which the existence of the limit $\lim\limits_{n \to \infty} T_n = T$ follows. To prove that $T = e$, we observe that for $m > n$

$$T_m > 1 + 1 + \frac{1}{2!}\left(1 - \frac{1}{m}\right) + \cdots + \frac{1}{n!}\left(1 - \frac{1}{m}\right) \cdots \left(1 - \frac{n - 1}{m}\right).$$

If we now keep n fixed and let m increase beyond all bounds, we obtain on the left the number T and on the right the expression S_n, so that $T \geq S_n$. Thus $T \geq S_n \geq T_n$ for every value of n. We now let n increase, so that T_n tends to T; from the double inequality it follows that $T = \lim\limits_{n \to \infty} S_n = e$. This was the statement to be proved.

We shall later (Section 2.6, p. 149) be led to this number e again from still another point of view.

f. The Number π as a Limit

A limiting process which in essence goes back to classical antiquity (Archimedes) is that by which the number π is defined. Geometrically, π means the area of the circle of radius one. We regard it as obvious that this area can be expressed by a (rational or irrational) number, denoted by π. However, this definition is not of much help to us if we wish to calculate the number with any accuracy. We then have no choice but to represent the number by means of a limiting process, namely, as the limit of a sequence of known and easily calculated numbers. Archimedes already used this process in his method of exhaustion, which consists of approximating the circle by means of regular polygons with an increasing number of sides fitting it more and more closely. If we let f_m denote the area of the regular m-gon (polygon of m sides) inscribed in the circle, the area of the inscribed $2m$-gon is given by the formula [proved by elementary geometry or from the expression $f_n = (n/2) \sin (2\pi/n)$ (see Fig. 1.46)]

$$f_{2m} = \frac{m}{2}\sqrt{2 - 2\sqrt{1 - (2f_m/m)^2}}.$$

We now let m range, not through the sequence of all positive integers but through the sequence of powers of 2, that is, $m = 2^n$; in other words, we form those regular polygons whose vertices are obtained by repeated bisection of the circumference. It is clear from the geometric interpretation that the f_{2^n} form an increasing and bounded sequence and thus *have* a limit which is the area of the circle:

$$\pi = \lim_{n \to \infty} f_{2^n}.$$

This representation of π as a limit serves actually as a basis for numerical computations; for, starting with the value $f_4 = 2$, we can calculate in order the terms of our sequence tending to π. An estimate of the accuracy with which any term f_{2^n} represents π can be obtained by constructing the lines touching the circle and parallel to the sides of the inscribed 2^n-gon. These lines form a circumscribed polygon similar to the inscribed 2^n-gon and having larger dimensions in the ratio $1 : \cos (\pi/2^n)$. Hence the area F_{2^n} of the circumscribed polygon may be found from the ratio given by

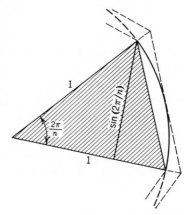

$$\frac{f_{2^n}}{F_{2^n}} = \left(\cos \frac{\pi}{2^n} \right)^2.$$

Figure 1.46

Since the area of the circumscribed polygon is greater than that of the circle, we have

$$f_{2^n} < \pi < F_{2^n} = \frac{f_{2^n}}{\left(\cos \dfrac{\pi}{2^n} \right)^2} = \frac{2f_{2^n}}{1 + \sqrt{1 - (f_{2^n}/2^{n-1})^2}} \, .$$

For example, $f_8 = 2\sqrt{2}$, so that we have the estimate

$$2\sqrt{2} < \pi < \frac{4\sqrt{2}}{1 + \frac{1}{2}\sqrt{2}} \, .$$

These are matters with which the reader will be more or less familiar. What we wish to point out, however, is that the calculation of areas by means of exhaustion by rectilinear figures whose areas can be calculated easily forms the basis for the concept of integral, to be introduced in Chapter 2. For the actual numerical computation of π much more efficient methods are available, as we shall see in Section 6.26.

1.8   The Concept of Limit for Functions of a Continuous Variable

Hitherto we have considered limits of sequences, that is, of functions of an integral variable n. The notion of limit, however, frequently occurs in connection with a function $f(x)$ that is defined for all x in some interval.

We say that the value of the function $f(x)$ tends to a limit η as x tends to ξ, or in symbols,

$$\lim_{x \to \xi} f(x) = \eta$$

if $f(x)$ differs arbitrarily little from η for all x for which $f(x)$ is defined and which lie sufficiently near to ξ.[1] Expressed more precisely the definition of $\lim f(x)$ is as follows.

Whenever an arbitrary positive quantity ϵ is assigned, we can mark off an interval $|x - \xi| < \delta$ so small that for any x which belongs both to the domain of f and to that interval the inequality $|f(x) - \eta| < \epsilon$ holds, then $\lim_{x \to \xi} f(x) = \eta$.

There is a close connection between the concepts of limit of a function and continuity. If ξ belongs to the domain of f, that is, if $f(\xi)$ is defined, then $\lim_{x \to \xi} f(x)$, if it exists at all, must have the value $f(\xi)$. Indeed, the definition of $\eta = \lim_{x \to \xi} f(x)$ implies in particular $|f(\xi) - \eta| < \epsilon$ for every positive ϵ and hence $\eta = f(\xi)$. Now, comparing the definitions of limit and of continuity, we see that the relation $\lim_{x \to \xi} f(x) = f(\xi)$ just expresses the continuity of the function f at the point ξ. Hence for ξ in the domain of f the existence of $\lim_{x \to \xi} f(x)$ just signifies that f is continuous at ξ. More generally, if $f(x)$ is not defined at ξ but $\lim_{x \to \xi} f(x)$ exists and has the value η, we can assign to f at the point ξ the value η and the function f, thus completed, will be continuous at ξ. (Removable Singularity. See p. 35.)

The *limit of a function* can also be described completely in terms of *limits of sequences*. The statement

$$\lim_{x \to \xi} f(x) = \eta$$

means that

$$\lim_{n \to \infty} f(x_n) = \eta$$

for every sequence x_n with limit ξ (where it is assumed, of course, that the x_n belong to the domain of f). For if $\lim_{x \to \xi} f(x) = \eta$ and if $\lim_{n \to \infty} x_n = \xi$, then $f(x)$

---

[1] It is assumed here that arbitrarily close to ξ there are points where f is defined.

is arbitrarily close to η for x sufficiently close to ξ; but x_n is sufficiently close to ξ if only n is large enough, and consequently, $\lim_{n \to \infty} f(x_n) = \eta$. If, on the other hand, $\lim_{n \to \infty} f(x_n) = \eta$ whenever $x_n \to \xi$, we must have $\lim_{x \to \xi} f(x) = \eta$. Otherwise there would exist a positive ϵ such that $|f(x) - \eta| \geq \epsilon$ for some x arbitrarily close to ξ; there would then also exist a sequence x_n converging to ξ for which $|f(x_n) - \eta| \geq \epsilon$, but then $\lim_{n \to \infty} f(x_n)$ could not be η.

Continuity of the function $f(x)$ at the point ξ implies then: $\lim_{n \to \infty} f(x_n) = f(\xi)$, for every sequence x_n in the domain of f that converges to ξ. More generally, for a function continuous in an *interval* the relation

$$\lim_{n \to \infty} f(x_n) = f(\lim_{n \to \infty} x_n)$$

is valid for any sequence in the domain of f which converges to a point of the interval. We see that for a continuous function the limit symbol can be interchanged (or, as one says, "commutes") with the symbols for the function.

Limits of sums, products, and quotients of functions are found by the same rules as for sequences (see p. 71): If $\lim_{x \to \xi} f(x) = \eta$ and $\lim_{x \to \xi} g(x) = \zeta$, exist, then

$$\lim_{x \to \xi} (f(x) + g(x)) = \eta + \zeta, \qquad \lim_{x \to \xi} (f(x)g(x)) = \eta\zeta$$

and for $\zeta \neq 0$ also

$$\lim_{x \to \xi} \frac{f(x)}{g(x)} = \frac{\eta}{\zeta} .$$

The proofs are the same as for sequences. (The rules would also follow from those for sequences by writing limits of functions as limits of sequences.) Consequently, when ξ belongs to the domain of f and g, *the sum, product, and quotient of two functions $f(x)$ and $g(x)$ which are continuous at a point ξ are again continuous* (where for quotients we have to assume that $g(\xi) \neq 0$).

The cases where ξ does not belong to the domain of f will turn out to be of particular importance for differential calculus. As a first example we consider the relation

$$\lim_{x \to \xi} \frac{x^n - \xi^n}{x - \xi} = n\xi^{n-1},$$

for n a positive integer. Of course, $f(x) = (x^n - \xi^n)/(x - \xi)$ is a function defined only for $x \neq \xi$. But for $x \neq \xi$ the algebraic identity

$$\frac{x^n - \xi^n}{x - \xi} = x^{n-1} + x^{n-2}\xi + x^{n-3}\xi^2 + \cdots + \xi^{n-1},$$

is valid as a consequence of the summation formula for the geometric series. To find the limit we only have to let x tend to ξ and to evaluate the limit of the right-hand side by the rules for limits of sums and quotients.

Less obvious is the formula

$$\lim_{x \to 0} \frac{\sin x}{x} = 1$$

(where, of course, the angle x is measured in "radians," as explained on p. 50). Again the quotient $(\sin x)/x$ is defined only for $x \neq 0$.

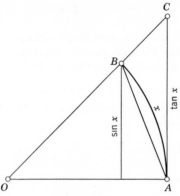

Figure 1.47

But, if we *define* $(\sin x)/x = 1$ for $x = 0$ we complete the quotient as a function which is continuous also at $x = 0$. For the proof of the limit formula we appeal here to a geometric argument.

From Fig. 1.47 we find by comparing the areas of the triangles OAB and OAC and the sector OAB[1] of the unit circle that if $0 < x < \pi/2$

$$\tfrac{1}{2} \sin x < \tfrac{1}{2} x < \tfrac{1}{2} \tan x.$$

From this it follows that if $0 < |x| < \pi/2$,

$$1 < \frac{x}{\sin x} < \frac{1}{\cos x}.$$

Hence the quotient $(\sin x)/x$ lies between the numbers 1 and $\cos x$. We know that $\cos x$ tends to 1 as $x \to 0$, and from this it follows that the quotient $(\sin x)/x$ can differ only arbitrarily little from 1, provided that

---

[1] Of course, we could have defined the angle x in the first place as twice the area of sector OAB.

x is near enough to 0. This is exactly what is meant by the equation which was to be proved.

From the result just proved it follows that

$$\lim_{x \to 0} \frac{\tan x}{x} = \lim_{x \to 0} \frac{\sin x}{x} \lim_{x \to 0} \frac{1}{\cos x} = 1,$$

and also

$$\lim_{x \to 0} \frac{1 - \cos x}{x} = 0.$$

This last follows from the formula, valid for $0 < |x| < \pi/2$,

$$\frac{1 - \cos x}{x} = \frac{(1 - \cos x)(1 + \cos x)}{x(1 + \cos x)} = \frac{1 - \cos^2 x}{x(1 + \cos x)}$$

$$= \frac{\sin x}{x} \cdot \frac{1}{1 + \cos x} \cdot \sin x.$$

For $x \to 0$ the first factor on the right tends to 1, the second to $\frac{1}{2}$, and the third to 0; the product therefore tends to 0, as was stated.

Dividing the same formula by x, we obtain

$$\frac{1 - \cos x}{x^2} = \left(\frac{\sin x}{x}\right)^2 \frac{1}{1 + \cos x},$$

from which

$$\lim_{x \to 0} \frac{1 - \cos x}{x^2} = \frac{1}{2}.$$

Limits for $x \to \infty$.   Finally we remark that it is just as well possible to consider limiting processes in which the continuous variable x increases beyond all bounds. For example, the meaning of the equation

$$\lim_{x \to \infty} \frac{x^2 + 1}{x^2 - 1} = \lim_{x \to \infty} \frac{1 + 1/x^2}{1 - 1/x^2} = 1$$

is clear. It signifies that the function on the left differs arbitrarily little from one, provided only that x is sufficiently large. The rules for forming the limits of this kind for sums, products, and quotients are the same as before.

* There is one further result which is frequently useful in the calculation of limits, the rule for obtaining the limit of a compound function. The *compound* function $f(g(z))$ is defined for those values of z for which $x = g(z)$ lies in the domain of $f(x)$. The function $g(z)$ may be a function of a continuous variable or an integer variable, but $f(x)$ must be a function of a continuous variable.

If $\lim_{z \to \zeta} g(z) = \xi$ where ξ lies within an open interval of the domain of f and if $\lim_{x \to \xi} f(x) = \eta$, then $\lim_{z \to \zeta} f(g(z)) = \eta$. As a corollary we observe that a continuous function of a continuous function is itself continuous (as already mentioned on p. 55).

The result is obvious from the fact that we can make $f(x)$ arbitrarily close to η by taking x sufficiently close to ξ and to make $x = g(z)$ close enough to ξ we have only to take z sufficiently close to ζ. With slight modifications, the same statements apply when any of the variables is allowed to increase beyond all bounds.

a. Some Remarks about the Elementary Functions

So far we tacitly assumed that the elementary functions are continuous. The proof of this fact is very simple. First, the function $f(x) = x$ is continuous; therefore $x^2 = x \cdot x$ is continuous, as the product of two continuous functions, and every power of x is likewise continuous. Thus every polynomial is continuous, being the sum of continuous functions. Every rational function, as a quotient of continuous functions, is likewise continuous in every interval in which the denominator does not vanish.

The function x^n is continuous and monotonic for $x \geq 0$. Hence the nth root, being the inverse function of the nth power, is continuous. From this fact it is easy to conclude that the nth root of a rational function is continuous (except where the denominator vanishes).

The continuity of the trigonometric functions could now be proved, using the concepts already developed. However, we omit the discussion here, since in Chapter 2 (p. 166), the continuity of all these functions will be seen to follow simply as a consequence of their differentiability.

We merely make a few remarks about the definition and continuity of the exponential function a^x, the general power function x^α, and the logarithm. We assume, as in Section 1.3 (p. 51), that a is a positive number, say greater than one, and $r = p/q$ is a positive rational number (p and q being integers); then $a^r = a^{p/q}$ is the positive number whose qth power is a^p. If α is any irrational number and $r_1, r_2 \cdots r_m, \ldots$ is a sequence of rational numbers approaching α, we assert that $\lim_{m \to \infty} a^{r_m}$ exists; we then call this limit a^α.

In order to prove the existence of this limit by Cauchy's test, we need show only that $|a^{r_n} - a^{r_m}|$ is arbitrary small, provided that n and m are sufficiently large. We suppose, for example, that $r_n > r_m$, or that $r_n - r_m = \delta$, where $\delta > 0$. Then

$$a^{r_n} - a^{r_m} = a^{r_m}(a^\delta - 1).$$

Since the r_m converge to α, they are bounded and so are the a^{r_m}; thus it suffices to show that

$$|a^\delta - 1| = a^\delta - 1$$

is arbitrarily small when the values of n and m are sufficiently large. However, the rational number δ certainly may be made as small as we please provided the values of n and m are sufficiently large. Hence if l is an arbitrarily large positive integer, $\delta < 1/l$ if n and m are large enough. Now the relations $\delta < 1/l$ and $a > 1$ give[1]

$$1 < a^\delta < a^{1/l},$$

and since $a^{1/l}$ tends to one as l increases to infinity (cf. p. 64), our assertion follows immediately.

It can be shown that the function a^x extended to irrational values in this way is also continuous everywhere, and, moreover, that it is monotonic. For negative values of x this function is naturally defined by the equation

$$a^x = \frac{1}{a^{-x}}.$$

As x runs from $-\infty$ to $+\infty$, a^x takes all values between zero and $+\infty$. Consequently, it possesses a continuous and monotonic inverse function, which we call the *logarithm to the base a*. In like manner we can prove that the general power x^α is a continuous function of x, where α is any fixed rational or irrational number and x varies over the interval $0 < x < \infty$, and is monotonic if $\alpha \neq 0$.

The "elementary" discussion of the exponential function, the logarithm, and the power x^α outlined here will later (p. 149) be replaced by another discussion which in principle is much simpler.

Supplement

One of the great achievements of Greek mathematics was the reduction of mathematical statements and theorems in a logically coherent way to a small number of very simple postulates or axioms, the well-known axioms of geometry or the rules of arithmetic governing relations among a few basic objects, such as integers or geometrical points. The basic objects originate as abstractions or idealizations from physical reality. The axioms, whether considered as "evident" from a philosophical point of view or merely as overwhelmingly plausible, are accepted without proof; on them the crystalized structure of mathematics rests. For many centuries the axiomatic Euclidean

---

[1] This statement follows from the fact that for $a > 1$ the power $a^{m/n}$ is greater than one if m/n is positive. For $a = (a^{1/n})^m$ is the product of m factors all greater than one, and so is greater than one.

mathematics was accepted as a model for mathematical style and even imitated for other intellectual endeavors. (For example, philosophers, such as Descartes and Spinoza, tried to make their speculations more convincing by presenting them axiomatically or, as they said, "more geometrico.")

The axiomatic method was discarded when after the stagnation during the Middle Ages mathematics in union with natural science started an explosively vigorous development based on the new calculus. Ingenious pioneers vastly extending the scope of mathematics could not be hampered by having to subject the new discoveries to consistent logical analysis and thus in the seventeenth century an invocation of intuitive evidence became a widely used substitute for deductive proof. Mathematicians of first rank operated with the new concepts guided by an unerring feeling for the correctness of the results, sometimes even with mystical associations as in references to "infinitesimals" or "infinitely small quantities." Faith in the sweeping power of the new manipulations of calculus carried the investigators far along paths impossible to travel if subjected to the limitations of complete rigor. Only the sure instinct of great masters could guard against gross errors.

The uncritical but enormously fruitful enthusiasm of the early period gradually met with countercurrents which rose to full strength in the nineteenth century but did not impede the development of constructive analysis initiated earlier. Many of the great mathematicians of the nineteenth century, in particular Cauchy and Weierstrass, played a role in the effort toward critical reappraisal. The result was not only a new and firm foundation of analysis, but also increased lucidity and simplicity as a basis for further remarkable progress.

An important goal was to replace indiscriminate reliance on imprecise "intuition" by precise reasoning based on operations with numbers; for naive geometric thinking leaves an undesirable margin of vagueness as we shall see time and again in the following chapters. For example, the general concept of a continuous curve eludes geometrical intuition. A continuous curve, representing a continuous function, as defined earlier, need not have a definite direction at every point; we can even construct continuous functions whose graphs nowhere have a direction, or to which no length can be assigned.

Yet one must never forget that abstract deductive reasoning is merely one aspect of mathematics while the driving motivation and the great universal scope of analysis stem from physical reality and intuitive geometry.

This supplement will provide a rigorous buttressing (with some repetitions) for basic concepts treated intuitively earlier in this chapter.

S.1 Limits and the Number Concept

We start with the ideas of Section 1.1, analyzing fully the concept of real number and its connection with that of limit. We define the number continuum by a constructive procedure based on the natural numbers. We then prove that the extended number concept satisfies the rules of arithmetic and the other requirements, making it the adequate tool for measurement.

Since a complete exposition would require a separate book,[1] we shall indicate only the main steps. In struggling through the somewhat tedious material the student will marvel at the fact that on the basis of the natural numbers the human mind could erect a logically consistent number system superbly suited to the task of scientific measurement.[2]

a. The Rational Numbers

Limits Defined by Rational Intervals. We begin by accepting the system of rational numbers with all its usual properties, derived from the basic properties of natural numbers. Thus the rational numbers are *ordered* by magnitude, permitting us to define "rational" intervals as sets of rational numbers lying between two given rational numbers (intervals including the end points are called *closed*). The length of the interval with end points a, b is $|b - a|$. As observed in Section 1a the rational numbers are *dense* and every rational interval contains infinitely many rational numbers. For the time being, all quantities occurring are assumed to be rational numbers.

Within the domain of rational numbers we define sequences and limits (see p. 70). Given an infinite sequence of rational numbers $a_1, a_2, \ldots$ and a *rational* number r we say that

$$\lim_{n \to \infty} a_n = r$$

---

[1] See for example, E. Landau, *Foundations of Analysis*, 2nd Ed., Chelsea, New York, 1960.

[2] Real numbers can also be introduced purely axiomatically, with all their basic properties accepted as *axioms*. In the approach we shall take here we accept, in principle, only the axioms for natural numbers (including the principle of mathematical induction). The rational numbers and real numbers are then *constructed* on that basis. The "axioms" for real numbers are then, in principle, merely *theorems* about natural numbers for which proofs are required. Actually, we shall start already with the rational numbers as known elements, since the construction of the rational from the natural numbers and the derivation of the basic properties of rational numbers present no difficulties at all.

if every rational interval containing r in its interior also contains "*almost all*" a_n, that is, all a_n with at most a *finite* number of exceptions. It follows immediately that a sequence of rational numbers cannot have more than one rational limit and that the usual rules for limits of sums, differences, products, and quotients (see p. 71) are valid for sequences of rational numbers with rational limits.

An entirely obvious consequence of this definition is that passing to the limit preserves order: if $\lim a_n = a$, $\lim b_n = b$ and for every n, $a_n \leq b_n$, then $a \leq b$. Note that even assuming $a_n < b_n$ strictly, we cannot say more than $a \leq b$, or exclude possible equality of the limits (for example, both sequences $a_n = 1 - 2/n$ and $b_n = 1 - 1/n > a_n$ have the limit 1).

Statements about limits can be expressed in terms of rational *null-sequences*, that is, sequences $a_1, a_2, \ldots$ of rational numbers for which

$$\lim_{n \to \infty} a_n = 0.$$

One says a_n "becomes arbitrarily small as n tends to infinity," meaning that for any positive rational ϵ, no matter how small, the inequality $|a_n| < \epsilon$ holds for almost all n. Obviously the sequence $a_n = 1/n$ is a null-sequence.

Thus a sequence of rational numbers a_n has the rational limit r if and only if the numbers $r - a_n$ form a null-sequence.

b. Real Numbers Determined by Nested Sequences of Rational Intervals

We observed on p. 5 that intuitively the rational points are dense on the real axis and that there are always rational numbers between any two real numbers. This suggests the possibility of rigorously *defining* a real number entirely in terms of order relations with respect to the rationals, a procedure we shall now follow.

A nested sequence of rational intervals (see p. 8) is a sequence of *closed* intervals J_n with *rational* end points a_n, b_n, with each interval contained in the preceding one, whose lengths form a null-sequence

$$a_{n-1} \leq a_n \leq b_n \leq b_{n-1}$$

and

$$\lim_{n \to \infty} (b_n - a_n) = 0.$$

Since each interval $J_n = [a_n, b_n]$ of a nested sequence contains all succeeding intervals, a rational number r lying outside any J_n also lies outside and on the same side of all succeeding intervals. Thus a nested sequence of rational intervals gives rise to a separation of all rational numbers into three classes.[1] The first class consists of the rational numbers r lying to the left of the intervals J_n for sufficiently large n, or for which $r < a_n$ for almost all n. The second class consists of the rational numbers r contained in all intervals J_n. This class contains at most one number, since the length of the interval J_n shrinks to zero with increasing n. The third class consists of the rational numbers r for which $r > b_n$ for almost all n. It is clear that any number of the first class is less than any of the second class, and any number of the second class is less than any of the third class. The points a_n themselves are either in the first or second class, and the numbers b_n either in the second or third class.

If the second class is not empty, it consists of a single rational number r. In this case the first class consists of the rational numbers less than r, the third class of the rational numbers greater than r. We say then that the nested sequence of intervals J_n *represents* the rational number r. For example, the nested sequence of intervals $[r - 1/n, r + 1/n]$ represents the number r.

If the second class is empty, then the nested sequence does not represent a rational number; these nested sequences then serve to represent irrational numbers. The individual intervals $[a_n, b_n]$ of the sequence are for this purpose unimportant; only the separation of the rational numbers into three classes generated by this sequence is essential, telling us where the irrational number fits in among the rational ones.

Thus we call two nested sequences of rational intervals $[a_n, b_n]$ and $[a_n', b_n']$ *equivalent* if they give rise to the same separation of the rational numbers into three classes. The reader should prove as an exercise that necessary and sufficient for equivalence is: $a_n' - a_n$ is a null-sequence, or also: the inequalities

$$a_n \leq b_n', \qquad a_n' \leq b_n$$

hold for all n.

We assign a *real number* to *a nested sequence of rational intervals* $[a_n, b_n]$. *The real numbers determined by two different nested sequences will be considered to be equal if the sequences are equivalent.* A real number then is represented by the separation of the rational numbers into three classes generated by equivalent nested sequences of rational intervals. If the second class consists of a rational number r, we consider the real number represented by this separation into classes as identical with the rational number r.

---

[1] A so-called "Dedekind Cut."

*c. Order, Limits, and Arithmetic Operations for Real Numbers

Having defined real numbers, we can now define the notions of order, sum, difference, product, limit, etc., for them and prove that they have the usual properties. To be consistent any definition concerning real numbers must: (1) have the ordinary meaning in case the real numbers are rational and (2) be independent of the individual nested sequences intervals used to represent the real numbers.

*Intervals with Real End Points

Although so far, even for the definition of irrational numbers, the end points of nested intervals were assumed to be rational, we must now remove such restrictions and show that we can operate with real numbers exactly as we do with rational numbers. In carrying out this program we have to be careful at each step to avoid reliance on facts not yet proved by logical deduction from our basis of departure, the rational numbers.

We shall denote real numbers by letters $x, y, \ldots$. If the real number x is given by the nested sequence of rational intervals $[a_n, b_n]$, we write $x \sim \{[a_n, b_n]\}$. From our definition of real number we draw a natural definition of order for a real number $x \sim \{[a_n, b_n]\}$ relative to a rational number r. We say that $r < x, r = x, r > x$ according as r belongs to the first, second, or third class of the separation of the rational numbers generated by the sequence of nested intervals. This definition is obviously independent of the special nested sequence $\{[a_n, b_n]\}$ defining x and has the ordinary meaning when x is rational. Equivalently, we say that $r < x$ if $r < a_n$ for almost all n, $r = x$ if $a_n \leq r \leq b_n$ for all n, and $r > x$ if $r > b_n$ for almost all n.

By comparing real numbers with rational numbers we can compare real numbers with each other. Let $x \sim \{[a_n, b_n]\}$, $y \sim \{[\alpha_n, \beta_n]\}$. We say $x < y$ if there is a rational number r such that $x < r < y$. Clearly, this definition does not depend on the particular representations of x and y by nested sequences since comparisons with rational r are independent of such representations. Thus we say that $x < y$ if there exists a rational r such that $b_n < r < \alpha_n$ for almost all n, or simply if $b_n < \alpha_n$, for almost all n. The relation $x < y$ precludes the possibility that $y < x$ or $x = y$. Obviously $x < y$ and $y < z$ implies $x < z$.

For any two real numbers x and y, one of the relations $x < y$, $x = y$, $y < x$ must hold. For if $x \neq y$ and either number, say y, is rational, then y must be in the first or third class of the separation generated by x, that is, either $y < x$ or $x < y$. If neither x nor y is rational, the second classes of the corresponding subdivisions are empty, and there must be a rational number r in the first class with respect to one of the numbers and in the third class with respect to the other. Thus either $x < y$ or $y < x$.

Density. An immediate consequence of these definitions is the density of the rational numbers in the sense that between any two real numbers x, y there is always a rational number r. We also observe that if a real number x is represented by a nested sequence of rational intervals $[a_n, b_n]$, then $a_n \leq x \leq b_n$ for all n. For if $x < a_m$ for some m, then $b_n < a_m$ for almost all n, contradicting the inequality $a_m < b_n$ which holds for all n. Hence every real number can be confined to a rational interval $[a_n, b_n]$ of arbitrarily small length.

Once the real numbers are ordered we can talk of intervals with *real end points*. The density of the rational numbers guarantees that every such interval includes rational numbers.

Limits. A real number x is called the *limit* of a sequence $x_1, x_2, \ldots$ of real numbers if every open interval with real end points containing x also contains x_n for almost all n. This definition is consistent with the definition in terms of rational intervals given earlier, in the sense that a rational limit of a rational sequence is a limit of the same sequence of numbers in the more general sense of a real limit. As a consequence of the definition of limit we find that for a real number x represented by a nested sequence of rational intervals $[a_n, b_n]$

$$x = \lim_{n \to \infty} a_n = \lim_{n \to \infty} b_n.$$

*Arithmetic. We next define the *arithmetic operations* for real numbers $x \sim \{[a_n, b_n]\}$ and $y \sim \{[\alpha_n, \beta_n]\}$: This is achieved most easily for the operations of addition and subtraction. We define

$$x + y \sim \{[a_n + \alpha_n, b_n + \beta_n]\}, \qquad x - y \sim \{[a_n - \beta_n, b_n - \alpha_n]\}.$$

To prove these definitions meaningful is a simple exercise whose details are left to the reader (see Problem 3, p. 116). For example, for $x - y$ it is necessary only to verify the intervals $[a_n - \beta_n, b_n - \alpha]$ form a nested sequence with lengths tending to zero, and hence that they represent a real number z. The fact that z does not depend on the special representations of x and y is proved by characterizing the separation of rational numbers into three classes generated by z directly in terms of x and y; for instance, the first class consists of the rational numbers $r < z$, or of the r which are exceeded by $a_n - \beta_n$ for some n; these r are easily seen to be the rational numbers of the form $s - t$, where s and t are rational numbers for which $s < x$ and $t > y$.

The product of the two real numbers x, y is for $y > 0$ defined by

$$x \cdot y \sim \{[a_n \alpha_n, b_n \beta_n]\},$$

where we have assumed that all $\alpha_n > 0$; it is obvious what nested sequences are proper to use for xy in the case $y < 0$ and $y = 0$. Whenever y is a positive *rational* number, the product $x \cdot y$ also is representable in the form

$$x \cdot y \sim \{[a_n y, b_n y]\}.$$

For a natural number $y = m$, the product $x \cdot y = mx$ also can be obtained by repeated addition of x, that is, $mx = x + (m - 1)x = x + x + \cdots + x$.

The arithmetic operations obey the usual laws. In particular, the relation $x < y$ is equivalent to $0 < y - x$. We can introduce the *absolute value* of a real number and prove the triangle inequality $|x + y| \leq |x| + |y|$. The notion of limit of a sequence of real numbers defined above in terms of order relations can then be given the equivalent formulation: $x = \lim_{n \to \infty} x_n$ if for every real positive ϵ the relation $|x - x_n| < \epsilon$ holds for almost all n.

We now verify the so-called

AXIOM OF ARCHIMEDES. *If x and y are real numbers and x is positive, then there exists a natural number m such that $mx > y$.*

In essence this means a real number cannot be "infinitely small" or "infinitely large" compared with another (except if one of them is zero). To prove the Axiom of Archimedes (which in our context is really a *theorem*) we observe that for rational numbers it is a consequence of the common properties of integers. If now $x \sim \{[a_n, b_n]\}$ and $y \sim \{[\alpha_n, \beta_n]\}$ are real numbers and x is positive, then $a_n > 0$ for almost all n. Since a_n and β_n are rational numbers, we can then find an m so large that $ma_n > \beta_n$, whence $mx > \beta_n \geq y$.

d. Completeness of the Number Continuum. Compactness of Closed Intervals. Convergence Criteria

Real numbers make possible limit operations with rational numbers, but they would be of little value if the corresponding limit operations carried out with them necessitated the introduction of some further kind of "unreal" numbers which would have to be fitted in between the real ones, and so on ad infinitum. Fortunately, the definition of real number is so comprehensive that no further extension of the number system is possible without discarding one of its essential properties (as "order" must be discarded for complex numbers).

Principle of Continuity

This completeness of the real number continuum is expressed by the basic continuity principle (cf. p. 8): Every nested sequence of intervals with real end points contains a real number. To prove this, consider closed intervals $[x_n, y_n]$, each interval contained in the preceding one, whose lengths $y_n - x_n$ form a null-sequence. We claim there is a real x contained in all $[x_n, y_n]$: The sequences x_n and y_n will then have x as limit. To prove this we replace the nested sequence $[x_n, y_n]$ by a nested sequence of rational intervals $[a_n, b_n]$, containing the $[x_n, y_n]$. This rational sequence will then define the desired real number x. For each n let a_n be the largest rational number of the form $p/2^n$ less than x_n, and b_n the smallest rational number of the form $q/2^n$ greater than y_n, where p and q are integers. Clearly, the intervals $[a_n, b_n]$ form a nested sequence representing a real number x. If x lay outside one of the intervals $[x_m, y_m]$, say $x < x_m$, there would exist a rational r with $x < r < x_m$, whence for all sufficiently large n we would have

$$y_n \leq b_n < r < x_m \leq x_n,$$

which is impossible. Hence all intervals $[x_m, y_m]$ contain the point x.

Weierstrass' Principle—Compactness

Several other versions of this principle of *continuity* are important. The first is the Weierstrass principle of *existence of limit points or accumulation points of bounded sequences.* A point x is a *limit point* of a sequence $x_1, x_2, \ldots$ if every open interval containing x also contains points x_n for *infinitely many* n. Notice the difference between this definition and the definition of *limit*, where the x_n for *almost all n* must lie in the open interval, or for all n with at most a finite number of exceptions or for all sufficiently large n. If a sequence has a limit, then this limit is also a limit point of the sequence and is in fact the only one. There may be no limit point (as in the example of the sequence 1, 2, 3, 4, ...) or a single limit point (as in a convergent sequence) or several limit points (for example, the sequence 1, −1, 1, −1, ... has the two limit points +1 and −1). The Weierstrass principle asserts: *Every bounded sequence has at least one limit point.*

To prove this we observe that since the sequence $x_1, x_2, \ldots$ is *bounded*, there exists an interval $[y_1, z_1]$ containing all x_n. Starting with $[y_1, z_1]$ we construct by induction over n a nested sequence of intervals $[y_n, z_n]$ each containing points x_m for infinitely many m. If $[y_n, z_n]$ contains infinitely many x_m, we divide $[y_n, z_n]$ into two equal parts by its mid-point. At least one of the two resulting closed intervals must again contain infinitely many x_m and can be taken as the interval $[y_{n+1}, z_{n+1}]$. It is clear that the $[y_n, z_n]$ form a nested sequence representing a real number x. Every open interval containing x will contain the intervals $[y_n, z_n]$ for sufficiently large n and hence must contain infinitely many x_m.

Limit points can also be defined as *limits of subsequences* of the given infinite sequence $x_1, x_2, \ldots$. A subsequence is any infinite sequence extracted from the given sequence, or of the form $x_{n_1}, x_{n_2}, x_{n_3}, \ldots$, where $n_1 < n_2 < n_3 < \cdots$. Obviously, a point x is a limit point of the sequence $x_1, x_2, \ldots$ if it is limit of some subsequence. Conversely, for any limit point x we can, by induction, construct a subsequence $x_{n_1}, x_{n_2}, \ldots$ converging to x. If $x_{n_1}, \ldots, x_{n_{k-1}}$ are defined already we take for n_k one of the infinitely many integers n for which $n > n_{k-1}$ and $|x_n - x| < 2^{-k}$.

We restate the Weierstrass principle in the form:

THEOREM. *Every bounded infinite sequence of real numbers has a convergent subsequence.*

A set is called *compact* if every sequence of its elements contains a subsequence converging to an element of the set. Rephrasing our theorem we say that *closed intervals of real numbers are compact sets.*

Monotone Sequences

A special consequence of this theorem is that *every bounded monotone sequence converges.* Indeed, let the sequence $x_1, x_2, \ldots$ be monotone, say monotonic increasing. If the sequence is also bounded, it has a limit point x. Arbitrarily close to x there must be points x_n of the sequence, none exceeding x, since the subsequent terms increase, and if $x_n > x$ then $x_m \geq x_n > x$ for $m > n$. It follows that every interval containing x contains almost all x_n, or x is the limit of the sequence.

Cauchy's Convergence Criterion

The condition that a sequence is bounded and monotone is *sufficient* for convergence. The significance of this statement is that it often permits us to prove existence of the limit of a sequence without requiring a priori knowledge of the value of the limit; in addition, boundedness and monotonicity of a sequence are properties usually easy to check in concrete applications. However, not every convergent sequence need be monotone (although it has to be bounded) and it is important to have a more generally applicable criterion for convergence.

Such is the *intrinsic convergence test of Cauchy* which is a necessary and sufficient condition for the existence of the limit of a sequence.

The sequence $x_1, x_2, x_3, \ldots$ converges if and only if for every positive ϵ there exists an N such that $|x_n - x_m| < \epsilon$ for all n and m exceeding N.

In other words, a sequence converges if any two of its elements with sufficiently large indices differ by less than ϵ from each other.

We claim that the condition is necessary for convergence. If $x = \lim x_n$ then every x_n with sufficiently large n differs from x by less than $\epsilon/2$, and hence by the triangle inequality every two such values x_n and x_m will differ from each other by less than ϵ. Conversely, consider a sequence for which $|x_n - x_m| < \epsilon$ for any $\epsilon > 0$, for all sufficiently large n and m. Then there exists a value N such that almost all x_n differ from x_N by less than 1. This means that almost all x_n can be enclosed in an interval of length 2. We can then find an interval so large that it includes also the finite number of x_n which may lie outside the interval about x_N. Thus the sequence is bounded and hence has a limit point x. Every open interval containing x will also contain some points x_m with arbitrarily large m. Since points x_n differ arbitrarily little from each other for sufficiently large n, it follows that the open interval about x must contain almost all x_n, and so x is the limit of the sequence.

e. Least Upper Bound and Greatest Lower Bound

It is of great importance that a bounded set of real numbers has "best possible" upper and lower bounds. A set S of real numbers x is *bounded*, if all numbers of S can be enclosed in one and the same finite interval. There are then *upper bounds* of S, numbers B which are not exceeded by any number x of S:

$$x \leq B \qquad \text{for all } x \text{ in } S.$$

Similarly, there are *lower bounds* A of S:

$$A \leq x \qquad \text{for all } x \text{ in } S.$$

Thus for the set of reciprocals of natural numbers $1, \frac{1}{2}, \frac{1}{3}, \frac{1}{4}, \ldots$, any number $B \geq 1$ is an upper bound, any number $A \leq 0$, a lower bound; here the number 1, a member of the set is the least upper bound, and the number 0, a limit point of the elements of the set although not a member, is the greatest lower bound. The least upper bound of a set of real numbers is often called its *supremum*, the greatest lower bound its *infimum*. In general the supremum and infimum of a set are either members of the set or at least limits of sequences of members of the set. For, if the least upper bound b of S does not belong to S, there must be some members of S lying arbitrarily close to b, since otherwise we could find upper bounds of S smaller than b; thus we can select successively a sequence of numbers x_1, x_2, ... from S which lie closer and closer to b and converge to b.

The *existence of a least upper bound* of a bounded set S follows immediately from the convergence of monotone bounded sequences. For any n we define B_n as the smallest rational upper bound of S with denominator 2^n. Clearly, for any x in S and any n

$$x \leq B_{n+1} \leq B_n \leq B_1.$$

Thus the B_n form a monotonically decreasing and bounded sequence which must have a limit b. It is easy to see that b is an upper bound of S and that there exists no smaller upper bound. The existence of the greatest lower bound is proved in the same way.

f. Denumerability of the Rational Numbers

A surprising discovery concerning the rational numbers was made late in the nineteenth century and stimulated the creation by Georg Cantor of the Theory of Sets after 1872. Although the rational numbers are dense and cannot be ordered by size, they can be arranged nevertheless as an infinite sequence r_1, r_2, ..., r_n, ... in which every rational number appears once. In this way the rational numbers can be enumerated, or counted off, as a first, second, ..., nth, ... rational number, where, of course, the order of the numbers in the sequence does not correspond at all to their order by magnitude. This result, which holds just as well for the rational numbers in any interval, is expressed by the statement: *The rational numbers are denumerable*, or *they form a denumerable set*.

To prove this result we simply give a prescription for arranging the positive rational numbers as a sequence. Every such number can be written in the form p/q, where p and q are natural numbers. For each positive integer k there are exactly $k - 1$ fractions p/q for which $p + q = k$. These are arranged in order of increasing p. Writing the different arrays of numbers for $k = 2, 3, 4, \ldots$ successively, we obtain (see Fig. 1.S.1) a sequence which contains all positive rational numbers. Omitting fractions, in which numerator and denominator have a common factor greater than 1, and thus represent the same rational number as a previous fraction, we obtain the sequence

$$\frac{1}{1}, \frac{1}{2}, \frac{2}{1}, \frac{1}{3}, \frac{3}{1}, \frac{1}{4}, \frac{2}{3}, \frac{3}{2}, \frac{4}{1}, \frac{1}{5}, \frac{5}{1}, \frac{1}{6}, \frac{2}{5}, \ldots$$

in which every positive rational number occurs exactly once. A similar sequence containing all rational numbers or all rational numbers in some particular interval is easily constructed.

This result is seen in proper perspective only in the light of another basic fact: that *the set of all real* numbers is not denumerable.[1] This

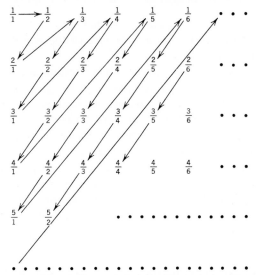

Figure 1.S.1 Denumerability of the positive rationals.

is an indication that the set of real numbers contains "many more" elements than that of the rational numbers, although both sets are infinite; thus denumerability is indeed a highly restrictive property of a set.

The Theory of Sets plays an important clarifying role in mathematics, although its use in unrestricted generality has led to paradoxical results and controversies. Such paradoxes, however, do not affect the substance of constructive mathematics and are absent from the theory of sets of real numbers.

S.2 Theorems on Continuous Functions

Important properties of continuous functions are established on the basis of the completeness property for real numbers. We recall the definition of continuity: the function $f(x)$ is continuous at the point ξ if for any given positive ϵ the inequality $|f(x) - f(\xi)| < \epsilon$ holds for

---

[1] For proof and a brief general discussion of the basic facts of set theory see *What is Mathematics?* by Courant and Robbins, p. 81.

all x sufficiently close to ξ, or, for all x differing from ξ by less than a suitable quantity δ, which generally depends on the choice of ϵ and ξ. It is understood in this definition that only values of x and ξ for which f is defined are considered.

A more concise definition of continuity in terms of convergence of sequences is: $f(x)$ *is continuous at the point* ξ *if* $\lim_{n \to \infty} f(x_n) = f(\xi)$ *for every sequence* $x_1, x_2, \ldots$ *with limit* ξ (where again the values x_n and ξ are in the domain of f). The equivalence of the two definitions was proved in Section 1.8, p. 82.

We call f *continuous in an interval* if f is continuous at each point of the interval. $f(x)$ is *uniformly continuous* if for given $\epsilon > 0$ we have $|f(x) - f(\xi)| < \epsilon$ whenever x and ξ are sufficiently close regardless of their location in the interval; thus f is uniformly continuous if the quantity δ appearing in the definition of continuity can be chosen independently of ξ: For every $\epsilon > 0$ there exists a $\delta = \delta(\epsilon) > 0$ such that $|f(x) - f(\xi)| < \epsilon$ whenever $|x - \xi| < \delta$. For practical purposes this means that if we subdivide the interval in which f is defined into a sufficiently large number of equal subintervals, then f will vary by less than a prescribed amount ϵ in each subinterval: At any point, f will then differ by less than ϵ from its value at any other point of the same subinterval.

We now prove: *Every function continuous in a closed interval* $[a, b]$ *is uniformly continuous in that interval.*

If f were not uniformly continuous in $[a, b]$, there would exist a fixed $\epsilon > 0$ and points x, ξ in $[a, b]$ arbitrarily close to each other for which $|f(x) - f(\xi)| \geq \epsilon$. It would then be possible for every n to choose points x_n, ξ_n in $[a, b]$ for which $|f(x_n) - f(\xi_n)| \geq \epsilon$ and $|x_n - \xi_n| < 1/n$. Since the x_n form a bounded sequence of numbers we could find a subsequence converging to a point η of the interval (using the *compactness* of closed intervals). The corresponding values ξ_n would then also converge to η: since f is continuous at η, we would find that $\eta = \lim f(x_n) = \lim f(\xi_n)$ for n tending to infinity in the subsequence, which is impossible if $|f(x_n) - f(\xi_n)| \geq \epsilon$ for all n.

The *intermediate value theorem* asserts: If for a function $f(x)$ continuous in an interval $a \leq x \leq b$, γ is any value between $f(a)$ and $f(b)$, then $f(\xi) = \gamma$ for some suitable ξ between a and b. Thus the *existence of a solution* ξ of the equation $f(\xi) = \gamma$ is certain if one exhibits two values a and b for which $f(a) < \gamma$ and $f(b) > \gamma$ respectively. This immediately implies the existence of a uniquely determined *inverse function* if f is continuous and monotonic, as we have seen (p. 44).

To prove the intermediate value theorem let $a < b$, $f(a) = \alpha$, $f(b) = \beta$, and $\alpha < \gamma < \beta$. Let S be the set of points x of the interval $[a, b]$ for which $f(x) < \gamma$. S is bounded and has a least upper bound ξ also belonging to the closed interval $[a, b]$. Then $f(x) \geq \gamma$ for $\xi < x \leq b$. The point ξ either belongs to S or is the limit of a sequence of points x_n of S. In the first case $f(\xi) < \gamma$; hence $\xi < b$, since $f(b) > \gamma$, and there are points x between ξ and b, arbitrarily close to ξ for which $f(x) \geq \gamma$. This is impossible if f is continuous at ξ and $f(\xi) < \gamma$. In the second case, $f(\xi) \geq \gamma$, we find from $f(x_n) < \gamma$ and $\lim_{n \to \infty} x_n = \xi$ that $f(\xi) \leq \gamma$; since we saw already that $f(\xi) < \gamma$ is impossible, we must have $f(\xi) = \gamma$.

A third basic property of a continuous function $f(x)$ in a closed interval $[a, b]$ is the *existence of a largest value* (*maximum*), meaning that there exists a point ξ in the interval $[a, b]$ such that $f(x) \leq f(\xi)$ for all x in the interval. Similarly, f will assume its *least value* (*minimum*) at some point η of the interval: $f(x) \geq f(\eta)$ for all x in the interval. It is essential to have the interval *closed:* for example, the functions $f(x) = x$ or $f(x) = 1/x$ are continuous, but they do not have a largest value in the open interval $0 < x < 1$; the maximum may just occur at one of the end points or not exist at all if f is not continuous at the end points.

To prove this principle we observe that a function f continuous in $[a, b]$ is necessarily *bounded:* that is, the values $f(x)$ forming the "range" S of f lie in some finite interval. Indeed by the uniform continuity of f we can find a finite number of points $x_1, x_2, \ldots, x_n$ in the interval such that $f(x)$ at any x of the interval differs by less than one from one of the numbers $f(x_1), f(x_2), \ldots, f(x_n)$ which can all be fitted into a finite interval. Since then the set S of values $f(x)$ is bounded, it has a least upper bound M. This M is the smallest number such that $f(x) \leq M$ for all x in $[a, b]$. Either M belongs to S or is the limit of a sequence of points of S. In the first case, there exists a ξ in $[a, b]$ with $f(\xi) = M$. In the second case, there exists a sequence of points x_n in $[a, b]$ with $\lim_{n \to \infty} f(x_n) = M$; thus we can find a subsequence of the x_n which converges to a point ξ of $[a, b]$ and again $f(\xi) = M$ by continuity of f at ξ. Clearly, $f(\xi)$ is the maximum of f.

S.3 Polar Coordinates

In Chapter 1 we have represented functions geometrically by curves. Analytical geometry follows the reverse procedure, beginning with a curve and representing it by a function, for example, by a function expressing one of the coordinates of a point of the curve in terms of the other. This point of view naturally leads us to consider, in addition to the rectangular coordinates to which we restricted ourselves, other systems of coordinates possibly better suited for the representation of curves given geometrically. The most important example is that of *polar coordinates* r, θ connected with the rectangular coordinates x, y of a point P by the equations

$$x = r \cos \theta, \quad y = r \sin \theta, \quad r^2 = x^2 + y^2, \quad \tan \theta = \frac{y}{x},$$

whose geometrical interpretation is made clear in Fig. 1.S.2.[1]

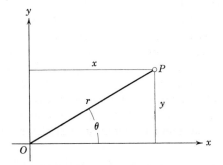

Figure 1.S.2   Polar coordinates.

We consider, for example, the *lemniscate*. This is geometrically defined as the locus of all points P for which the product of the distances r_1 and r_2 from the fixed points F_1 and F_2 with the rectangular coordinates $x = a$, $y = 0$ and $x = -a$, $y = 0$ respectively, has the constant value a^2 (cf. Fig. 1.S.3). Since

$$r_1{}^2 = (x - a)^2 + y^2, \quad r_2{}^2 = (x + a)^2 + y^2,$$

a simple calculation gives us the equation of the lemniscate in the form

$$(x^2 + y^2)^2 - 2a^2(x^2 - y^2) = 0.$$

Introducing polar coordinates, we obtain

$$r^4 - 2a^2r^2(\cos^2 \theta - \sin^2 \theta) = 0;$$

---

[1] The polar coordinates are not completely determined by the point P. In addition to θ, any of the angles $\theta \pm 2\pi$, $\theta \pm 4\pi$, ... can be considered a polar angle of P.

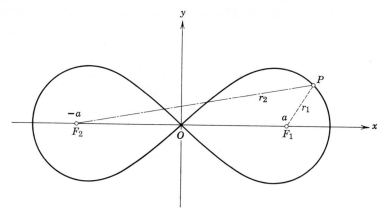

Figure 1.S.3 Lemniscate.

dividing by r^2 and using a simple trigonometrical formula this becomes

$$r^2 = 2a^2 \cos 2\theta.$$

Thus the equation of the lemniscate is simpler in polar coordinates than in rectangular.

S.4 Remarks on Complex Numbers

Our studies will be based chiefly on the continuum of real numbers. Nevertheless, with a view to discussions in Chapters 7, 8, and 9, we remind the reader that the problems of algebra have led to a still wider extension of the concept of number, the *complex numbers*. The advance from the natural numbers to the real numbers arose from the desire to eliminate exceptional phenomena and to make certain operations, such as subtraction, division, and correspondence between points and numbers, always possible. Similarly, we are compelled by the requirement that every quadratic equation and in fact every algebraic equation shall have a solution, to introduce the complex numbers. If, for example, we wish the equation

$$x^2 + 1 = 0$$

to have roots, we are obliged to introduce new symbols i and $-i$ as the roots. (As is shown in the theory of functions of a complex variable, this is sufficient to insure that *every* algebraic equation shall have a solution.[1])

---

[1] An algebraic equation is of the form $P(x) = 0$, where P is a polynomial with complex coefficients.

If *a* and *b* are ordinary real numbers, the *complex number* $c = a + ib$ denotes a pair of numbers (a, b) with which calculations are performed according to the following general rule: We add, multiply, and divide complex numbers (among which the real numbers are included as the special case $b = 0$), treating the symbol *i* as an undetermined quantity, and simplify all expressions using the equation $i^2 = -1$ to remove all powers of *i* higher than the first, leaving only an expression of the form $a + ib$.

We assume that the reader already has a certain degree of familiarity with the complex numbers. We nevertheless emphasize a particularly

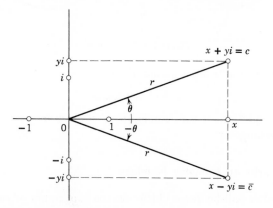

Figure 1.S.4   Geometric representation of a complex number $x + yi$ and of its conjugate.

important relationship which we shall explain in connection with the geometrical or trigonometrical representation of the complex numbers. If $c = x + iy$ is such a number, we represent it in a rectangular co-ordinate system by the point *P* with coordinates *x* and *y*. By means of the equations $x = r \cos \theta$, $y = r \sin \theta$, we introduce the polar coordinates *r* and θ (cf. p. 101) instead of the rectangular coordinates *x* and *y*. Then $r = \sqrt{x^2 + y^2}$ is the distance of the point *P* from the origin, and θ the angle between the positive *x*-axis and the segment *OP*. The complex number *c* is represented in the form

$$c = r(\cos \theta + i \sin \theta).$$

The angle θ is called the *amplitude* of the complex number *c*, the quantity *r* its *absolute value* or *modulus*, for which we also write $|c|$. To the "conjugate" complex number $\bar{c} = x - iy$ there obviously corresponds the same absolute value, but the amplitude $-\theta$ (Fig. 1.S.4).

Clearly,
$$r^2 = |c|^2 = c\bar{c} = x^2 + y^2.$$

If we use this trigonometrical representation, the multiplication of complex numbers takes a particularly simple form, for then

$$c \cdot c' = r(\cos \theta + i \sin \theta) \cdot r'(\cos \theta' + i \sin \theta')$$
$$= rr'[(\cos \theta \cos \theta' - \sin \theta \sin \theta') + i(\cos \theta \sin \theta' + \sin \theta \cos \theta')].$$

If we use the addition theorems for the trigonometric functions, this becomes

$$c \cdot c' = rr'(\cos (\theta + \theta') + i \sin (\theta + \theta')).$$

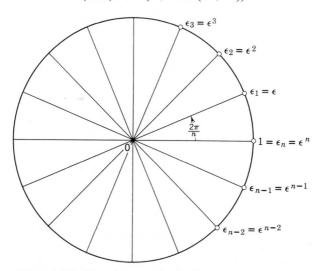

Figure 1.S.5   The nth roots of unity (for $n = 16$).

We therefore multiply complex numbers by multiplying their absolute values and adding their amplitudes. The remarkable formula

$$(\cos \theta + i \sin \theta)(\cos \theta' + i \sin \theta') = \cos (\theta + \theta') + i \sin (\theta + \theta')$$

is usually called *De Moivre's theorem*. It leads us to the relation

$$(\cos \theta + i \sin \theta)^n = \cos n\theta + i \sin n\theta,$$

which, for example, at once enables us to solve the equation $x^n = 1$ for positive integers n; the roots (the so-called roots of unity) are

$$\epsilon_1 = \epsilon = \cos \frac{2\pi}{n} + i \sin \frac{2\pi}{n}, \qquad \epsilon_2 = \epsilon^2 = \cos \frac{4\pi}{n} + i \sin \frac{4\pi}{n}, \ldots,$$

$$\epsilon_{n-1} = \epsilon^{n-1} = \cos \frac{2(n-1)\pi}{n} + i \sin \frac{2(n-1)\pi}{n}, \qquad \epsilon_n = \epsilon^n = 1$$

(Fig. 1.S.5).

Geometrically, the points corresponding to the roots of unity form the vertices of a regular n-gon inscribed in the circle of radius 1 about the origin.

Finally, if we imagine the expression on the left-hand side of the equation $(\cos \theta + i \sin \theta)^n = \cos n\theta + i \sin n\theta$ expanded by the binomial theorem, we need only separate real and imaginary parts in order to obtain expressions for $\cos n\theta$ and $\sin n\theta$ in terms of powers and products of powers of $\sin \theta$ and $\cos \theta$:

$$\cos n\theta = \cos^n \theta - \binom{n}{2} \cos^{n-2} \theta \sin^2 \theta$$

$$+ \binom{n}{4} \cos^{n-4} \theta \sin^4 \theta + - \cdots,$$

$$\sin n\theta = \binom{n}{1} \cos^{n-1} \theta \sin \theta - \binom{n}{3} \cos^{n-3} \theta \sin^3 \theta$$

$$+ \binom{n}{5} \cos^{n-5} \theta \sin^5 \theta + - \cdots.$$

PROBLEMS

SECTION 1.1a, page 2

1. (*a*) If a is rational and if x is irrational, prove that $a + x$ is irrational, and if $a \neq 0$, that ax is irrational.

(*b*) Show that between any two rational numbers there exists at least one irrational number and, consequently, infinitely many.

2. Prove that the following numbers are not rational: (*a*) $\sqrt{3}$. (*b*) $\sqrt{n}$, where the integer n is not a perfect square, that is, not the square of an integer. (*c*) $\sqrt[3]{2}$. (*d*) $\sqrt[p]{n}$, where n is not a perfect pth power.

*****3.** (*a*) Prove for any rational root of a polynomial with integer coefficients,

$$a_n x^n + a_{n-1} x^{n-1} + \cdots + a_1 x + a_0, \qquad (a_n \neq 0),$$

if written in lowest terms as p/q, that the numerator p is a factor of a_0 and the denominator q is a factor of a_n. (This criterion permits us to obtain all rational real roots and hence to demonstrate the irrationality of any other real roots.)

(*b*) Prove the irrationality of $\sqrt{2} + \sqrt[3]{2}$ and $\sqrt{3} + \sqrt[3]{2}$.

SECTION 1.1c, page 9

1. Let $[x]$ denote the integer part of x; that is, $[x]$ is the integer satisfying

$$x - 1 < [x] \leq x.$$

Set $c_0 = [x]$, and $c_n = [10^n(x - c_0) - 10^{n-1}c_1 - 10^{n-2}c_2 - \cdots - 10c_{n-1}]$ for $n = 1, 2, 3, \ldots$. Verify that the decimal representation if x is

$$x = c_0 + 0 \cdot c_1 c_2 c_3 \cdots$$

and that this construction excludes the possibility of an infinite string of 9's.

2. Define inequality $x > y$ for two real numbers in terms of their decimal representations (see Supplement, p. 92).

*3. Prove if p and q are integers, $q > 0$, that the expansion of p/q as a decimal either terminates (all the digits following the last place are zeros) or is periodic; that is, from a certain point on the decimal expansion consists of the sequential repetition of a given string of digits. For example, $\frac{1}{4} = 0.25$ is terminating, $\frac{1}{11} = 0.090909 \cdots$ is periodic. The length of the repeated string is called the *period* of the decimal; for $\frac{1}{11}$ the period is 2. In general, how large may the period of p/q be?

SECTION 1.1e, page 12

1. Using signs of inequality alone (not using signs of absolute value) specify the values of x which satisfy the following relations. Discuss all cases.
 (*a*) $|x - a| < |x - b|$.
 (*b*) $|x - a| < x - b$.
 (*c*) $|x^2 - a| < b$.

2. An interval (see definitions in text) may be defined as any connected part of the real continuum. A subset S of the real continuum is said to be *connected* if with every pair of points a, b in S, the set S contains the entire closed interval $[a, b]$. Aside from the open and closed intervals already mentioned, there are the "half-open" intervals $a \leq x < b$ and $a < x \leq b$ (sometimes denoted by $[a, b)$ and $(a, b]$, respectively) and the unbounded intervals that may be either the whole real line or a ray, that is, a "half-line" $x \leq a, x < a, x > a, x \geq a$ (sometimes denoted by $(-\infty, \infty)]$ and $(-\infty, a]$, $(-\infty, a), (a, \infty), [a, \infty)$, respectively) (see also footnote, p. 22).
 *(*a*) Prove that the cases of intervals specified above exhaust all possibilities for connected subsets of the number axis.
 (*b*) Determine the intervals in which the following inequalities are satisfied.
  (i) $x^2 - 3x + 2 < 0$.
  (ii) $(x - a)(x - b)(x - c) > 0$, for $a < b < c$.
  (iii) $|1 - x| - x \geq 0$.
  (iv) $\dfrac{x - a}{x + a} \geq 0$.
  (v) $\left| x + \dfrac{1}{x} \right| \geq 6$.
  (vi) $[x] \leq x/2$. See Problem 1 of this page.
  (vii) $\sin x \geq \sqrt{2}/2$.
 (*c*) Prove if $a \leq x \leq b$, then $|x| \leq |a| + |b|$.

3. Derive the inequalities

(a) $x + \dfrac{1}{x} \geq 2$, for $x > 0$,

(b) $x + \dfrac{1}{x} \leq -2$, for $x < 0$,

(c) $\left| + \dfrac{1}{x} \right| \geq 2$, for $x \neq 0$.

4. The harmonic mean ξ of two positive numbers a, b is defined by

$$\frac{1}{\xi} = \frac{1}{2}\left(\frac{1}{a} + \frac{1}{b} \right).$$

Prove that the harmonic mean does not exceed the geometric mean; that is, that $\xi \leq \sqrt{ab}$. When are the two means equal?

5. Derive the following inequalities:

(a) $x^2 + xy + y^2 \geq 0$,

*(b) $x^{2n} + x^{2n-1}y + x^{2n-2}y^2 + \cdots + y^{2n} \geq 0$,

*(c) $x^4 - 3x^3 + 4x^2 - 3x + 1 \geq 0$.

When does equality hold?

*6.** What is the geometrical interpretation of Cauchy's inequality for $n = 2, 3$?

7. Show that the equality sign holds in Cauchy's inequality if and only if the a_ν are proportional to the b_ν: that is, $ca_\nu + db_\nu = 0$ for all ν where c and d do not depend on ν and are not both zero.

8. (a) $|x - a_1| + |x - a_2| + |x - a_3| \geq a_3 - a_1$, for $a_1 < a_2 < a_3$. For what value of x does equality hold?

*(b) Find the largest value of y for which for all x

$$|x - a_1| + |x - a_2| + \cdots + |x - a_n| \geq y,$$

where $a_1 < a_2 < \cdots < a_n$. Under what conditions does equality hold?

9. Show that the following inequalities hold for positive a, b, c.

(a) $a^2 + b^2 + c^2 \geq ab + bc + ca$.

(b) $(a + b)(b + c)(c + a) \geq 8abc$.

(c) $a^2b^2 + b^2c^2 + c^2a^2 \geq abc(a + b + c)$.

10. Assume that the numbers x_1, x_2, x_3 and a_{ik} $(i, k = 1, 2, 3)$ are all positive, and in addition, $a_{ik} \leq M$ and $x_1^2 + x_2^2 + x_3^2 \leq 1$. Prove that

$$a_{11}x_1^2 + a_{12}x_1x_2 + \cdots + a_{33}x_3^2 \leq 3M.$$

*11.** Prove the following inequality and give its geometrical interpretation for $n \leq 3$,

$$\sqrt{(a_1 - b_1)^2 + \cdots + (a_n - b_n)^2} \leq \sqrt{(a_1^2 + \cdots + a_n^2)} + \sqrt{(b_1^2 + \cdots + b_n^2)}.$$

12. Prove, and interpret geometrically for $n \leq 3$,

$$\sqrt{(a_1 + b_1 + \cdots + z_1)^2 + \cdots + (a_n + b_n + \cdots + z_n)^2}$$
$$\leq \sqrt{a_1^2 + \cdots + a_n^2} + \sqrt{b_1^2 + \cdots + b_n^2} + \cdots + \sqrt{z_1^2 + \cdots + z_n^2}.$$

13. Show that the geometric mean of n positive numbers is not greater than the arithmetic mean; that is, if $a_i > 0$ $(i = 1, \ldots, n)$, then

$$\sqrt[n]{a_1 a_2 \cdots a_n} \le \frac{1}{n}(a_1 + a_2 + \cdots + a_n).$$

(*Hint:* Suppose $a_1 \le a_2 \le \cdots \le a_n$. For the first step replace a_n by the geometric mean and adjust a_1 so that the geometric mean is left unchanged.)

SECTION 1.2d, page 31

1. If $f(x)$ is continuous at $x = a$ and $f(a) > 0$, show that the domain of f contains an open interval about a where $f(x) > 0$.

2. In the definition of continuity show that the centered intervals

$$|f(x) - f(x_0)| < \epsilon \quad \text{and} \quad |x - x_0| < \delta$$

may be replaced by an arbitrary open interval containing $f(x_0)$ and a sufficiently small open interval containing x_0, as indicated on p. 33.

3. Let $f(x)$ be continuous for $0 \le x \le 1$. Suppose further that $f(x)$ assumes rational values only and that $f(x) = \frac{1}{2}$ when $x = \frac{1}{2}$. Prove that $f(x) = \frac{1}{2}$ everywhere.

4. (*a*) Let $f(x)$ be defined for all values of x in the following manner:

$$f(x) = \begin{cases} 0, & x \text{ irrational} \\ 1, & x \text{ rational.} \end{cases}$$

Prove that $f(x)$ is everywhere discontinuous.

(*b*) On the other hand, consider

$$g(x) = \begin{cases} 0, & x \text{ irrational} \\ \dfrac{1}{q}, & x = \dfrac{p}{q} \text{ rational in lowest terms.} \end{cases}$$

(The rational number p/q is said to be in lowest terms if the integers p and q have no common factor larger than 1, and $q > 0$. Thus $f(16/29) = 1/29$.) Prove that $g(x)$ is continuous for all irrational values and discontinuous for all rational values.

*5. If $f(x)$ satisfies the functional equation

$$f(x + y) = f(x) + f(y)$$

for all values of x and y, find the values of $f(x)$ for rational values of x and prove if $f(x)$ is continuous that $f(x) = cx$ where c is a constant.

6. (*a*) If $f(x) = x^n$, find a δ which may depend on ξ such that

$$|f(x) - f(\xi)| < \epsilon$$

whenever

$$|x - \xi| < \delta.$$

*(b) Do the same if $f(x)$ is any polynomial

$$f(x) = a_n x^n + a_{n-1} x^{n-1} + \cdots + a_1 x + a_0,$$

where $a_n \ne 0$.

SECTION 1.2e, page 44

1. Prove that if $f(x)$ is monotonic on $[a, b]$ and satisfies the intermediate value property, then $f(x)$ is continuous. Can you draw the same conclusion if f is not monotonic?

2. (a) Show that x^n is monotonic for $x > 0$. As a consequence, show for $a > 0$ that $x^n = a$ has a unique positive solution $\sqrt[n]{a}$.

(b) Let $f(x)$ be a polynomial

$$f(x) = a_n x^n + a_{n-1} x^{n-1} + \cdots + a_1 x + a_0, \qquad (a_n \neq 0).$$

Show (i) if n is odd, then $f(x)$ has at least one real root, (ii) if a_n and a_0 have opposite signs, then $f(x)$ has at least one positive root, and, in addition, if n is even, $n \neq 0$, then $f(x)$ has a negative root as well.

*3. (a) Prove that there exists a line in each direction which bisects any given triangle, that is, divides the triangle into two parts of equal area.

(b) For any pair of triangles prove that there exists a line which bisects them simultaneously.

SECTION 1.3b, page 49

1. (a) Prove that $\sqrt{x}$ is not a rational function. (*Hint:* Examine the possibility of representing $\sqrt{x}$ as a rational function for $x = y^2$. Use the fact that a nonzero polynomial can have at most finitely many roots.)

(b) Prove $\sqrt[n]{x}$ is not a rational function.

SECTION 1.3c, page 49

1. (a) Show that a straight line may intersect the graph of a polynomial higher than first degree in at most finitely many points.

(b) Obtain the same result for general rational functions.

(c) Verify that the trigonometric functions are not rational.

SECTION 1.5, page 57

1. Prove the following properties of the binomial coefficients.

(a) $1 + \binom{n}{1} + \binom{n}{2} + \cdots + \binom{n}{n-1} + \binom{n}{n} = 2^n.$

(b) $1 - \binom{n}{1} + \binom{n}{2} - \binom{n}{3} + \cdots + (-1)^n \binom{n}{n} = 0.$

(c) $\binom{n}{1} + 2\binom{n}{2} + 3\binom{n}{3} + \cdots + n\binom{n}{n} = n(2^{n-1}).$ (*Hint:* Represent the binomial coefficients in terms of factorials.)

(d) $1 \cdot 2\binom{n}{2} + 2 \cdot 3\binom{n}{3} + \cdots + (n-1)n\binom{n}{n} = n(n-1)2^{n-2}.$

(e) $1 + \frac{1}{2}\binom{n}{1} + \frac{1}{3}\binom{n}{2} + \cdots + \frac{1}{n+1}\binom{n}{n} = \frac{2^{n+1} - 1}{n + 1}.$

(f) $\binom{n}{0}^2 + \binom{n}{1}^2 + \cdots + \binom{n}{n}^2 = \binom{2n}{n}$. (*Hint:* Consider the coefficient

of x^n in $(1 + x)^{2n}$.)

(g) $S_n = -\binom{n}{0} - \frac{1}{3}\binom{n}{1} + \frac{1}{5}\binom{n}{2} - \frac{1}{7}\binom{n}{3} + \cdots + \frac{(-1)^n}{2n+1}\binom{n}{n}$

$$= \frac{4^n(n!)^2}{(2n+1)!}.$$

$\left(\text{Hint: Prove } \dfrac{2n+2}{2n+3} S_n = S_{n+1}.\right)$

2. Prove $(1 + x)^n \geq 1 + nx$, for $x > -1$.

3. Prove by induction that $1 + 2 + \cdots + n = \frac{1}{2}n(n + 1)$.

*4. Prove by induction the following:

(a) $1 + 2q + 3q^2 + \cdots + nq^{n-1} = \dfrac{1 - (n+1)q^n + nq^{n+1}}{(1-q)^2}$.

(b) $(1 + q)(1 + q^2) \cdots (1 + q^{2^n}) = \dfrac{1 - q^{2^{n+1}}}{1 - q}$.

5. Prove for all natural numbers n greater than 1 that n is either a prime or can be expressed as a product of primes. (*Hint:* Let A_{n-1} be the assertion for all integers k with $k \leq n$ that k is either prime or a product of primes.)

*6. Consider the sequence of fractions

$$\frac{1}{1}, \frac{3}{2}, \frac{7}{5}, \ldots, \frac{p_n}{q_n}, \ldots,$$

where $p_{n+1} = p_n + 2q_n$ and $q_{n+1} = p_n + q_n$.

(a) Prove for all n that p_n/q_n is in lowest terms.

(b) Show that the absolute difference between p_n/q_n and $\sqrt{2}$ can be made arbitrarily small. Prove also that the error of approximation to $\sqrt{2}$ alternates in sign.

7. Let a, b, a_n and b_n be integers such that

$$(a + b\sqrt{2})^n = a_n + b_n\sqrt{2},$$

where a is the integer closest to $b\sqrt{2}$. Prove that a_n is the integer closest to $b_n\sqrt{2}$.

*8. Let a_n and b_n be defined by

$$a_1 = 3, \qquad a_{n+1} = 3^{a_n}, \qquad \text{and} \quad b_1 = 9, \qquad b_{n+1} = 9^{b_n}.$$

For each value of n, determine the minimum value m such that $a_m \geq b_n$.

9. If n is a natural number, show that

$$\frac{(1 + \sqrt{5})^n - (1 - \sqrt{5})^n}{2^n\sqrt{5}}$$

is a natural number.

10. Determine the maximum number of pieces into which a plane may be cut by n straight lines. Show that the maximum occurs when no two of the lines are parallel and no three meet in a common point, and determine the number of pieces when concurrences and parallelisms are permitted.

11. Prove for each natural number n that there exists a natural number k such that

$$(\sqrt{2} - 1)^n = \sqrt{k} - \sqrt{k - 1}.$$

12. Prove Cauchy's inequality inductively.

SECTION 1.6, page 60

1. Prove that $\lim\limits_{n \to \infty} (\sqrt{n + 1} - \sqrt{n})(\sqrt{n + \frac{1}{2}}) = \frac{1}{2}$.

2. Prove that $\lim\limits_{n \to \infty} (\sqrt[3]{n + 1} - \sqrt[3]{n}) = 0$.

3. Let $a_n = 10^n/n!$. (a) To what limit does a_n converge? (b) Is the sequence monotonic? (c) Is it monotonic from a certain n onward? (d) Give an estimate of the difference between a_n and the limit. (e) From what value of n onward is this difference less than $1/100$?

4. Prove that $\lim\limits_{n \to \infty} \dfrac{n!}{n^n} = 0$.

5. (a) Prove that $\lim\limits_{n \to \infty} \left(\dfrac{1}{n^2} + \dfrac{2}{n^2} + \cdots + \dfrac{n}{n^2} \right) = \frac{1}{2}$.

(b) Prove that $\lim\limits_{n \to \infty} \left(\dfrac{1}{n^2} + \dfrac{1}{(n + 1)^2} + \cdots + \dfrac{1}{(2n)^2} \right) = 0$. (*Hint:* Compare

the sum with its largest term.)

(c) Prove that $\lim\limits_{n \to \infty} \left(\dfrac{1}{\sqrt{n}} + \dfrac{1}{\sqrt{n + 1}} + \cdots + \dfrac{1}{\sqrt{2n}} \right) = \infty$.

*(d) Prove that $\lim\limits_{n \to \infty} \left(\dfrac{1}{\sqrt{n^2 + 1}} + \dfrac{1}{\sqrt{n^2 + 2}} + \cdots + \dfrac{1}{\sqrt{n^2 + n}} \right) = 1$.

6. Prove that every periodic decimal represents a rational number. (Compare Section 1.1c, Problem 3.)

7. Prove that $\lim\limits_{n \to \infty} \dfrac{n^{100}}{1.01^n}$ exists and determine its value.

8. Prove that if a and $b \le a$ are positive, the sequence $\sqrt[n]{a^n + b^n}$ converges to a. Similarly, for any k fixed positive numbers $a_1, a_2, \ldots, a_k$ prove that $\sqrt[n]{a_1^n + a_2^n + \cdots + a_k^n}$ converges and find its limit.

9. Prove that the sequence $\sqrt{2}, \sqrt{2\sqrt{2}}, \sqrt{2\sqrt{2\sqrt{2}}}, \ldots$, converges. Find its limit.

10. If $\nu(n)$ is the number of prime factors of n, prove that

$$\lim_{n \to \infty} \frac{\nu(n)}{n} = 0.$$

11. Prove that if $\lim_{n\to\infty} a_n = \xi$, then $\lim_{n\to\infty} \sigma_n = \xi$, where σ_n is the arithmetic mean $(a_1 + a_2 + \cdots + a_n)/n$.

12. Find

(a) $\lim_{n\to\infty} \left(\dfrac{1}{1 \cdot 2} + \dfrac{1}{2 \cdot 3} + \cdots + \dfrac{1}{n(n+1)} \right).$

$\left(Hint: \dfrac{1}{k(k+1)} = \dfrac{1}{k} - \dfrac{1}{k+1}. \right)$

(b) $\lim_{n\to\infty} \left(\dfrac{1}{1 \cdot 2 \cdot 3} + \dfrac{1}{2 \cdot 3 \cdot 4} + \cdots + \dfrac{1}{n(n+1)(n+2)} \right).$

13. If $a_0 + a_1 + \cdots + a_p = 0$, prove that
$$\lim_{n\to\infty} (a_0 \sqrt{n} + a_1 \sqrt{n+1} + \cdots + a_n \sqrt{n+p}) = 0.$$

(*Hint:* Take $\sqrt{n}$ out as a factor.)

14. Prove that $\lim_{n\to\infty} {}^{(2n+1)}\sqrt{(n^2 + n)} = 1$.

15. Let a_n be a given sequence such that the sequence $b_n = pa_n + qa_{n+1}$, where $|p| < q$, is convergent. Prove that a_n converges. If $|p| \geqslant q > 0$, show that a_n need not converge.

16. Prove the relation
$$\lim_{n\to\infty} \frac{1}{n^{k+1}} \sum_{i=1}^{n} i^k = \frac{1}{k+1}$$

for any nonnegative integer k. (*Hint:* Use induction with respect to k and use the relation
$$\sum_{i=1}^{n} [i^{k+1} - (i-1)^{k+1}] = n^{k+1},$$

expanding $(i-1)^{k+1}$ in powers of i.)

SECTION 1.7, page 70

1. Let a_1 and b_1 be any two positive numbers, and let $a_1 < b_1$. Let a_2 and b_2 be defined by the equations
$$a_2 = \sqrt{a_1 b_1}, \qquad b_2 = \frac{a_1 + b_1}{2}.$$

Similarly, let
$$a_3 = \sqrt{a_2 b_2}, \qquad b_3 = \frac{a_2 + b_2}{2},$$

and, in general,
$$a_n = \sqrt{a_{n-1} b_{n-1}}, \qquad b_n = \frac{a_{n-1} + b_{n-1}}{2}.$$

Prove (a) that the sequence $a_1, a_2, \ldots$, converges, (b) that the sequence $b_1, b_2, \ldots$, converges, and (c) that the two sequences have the same limit. (This limit is called the *arithmetic-geometric mean* of a_1 and b_1.)

*2. Prove that the limit of the sequence

$$\sqrt{2}, \sqrt{2 + \sqrt{2}}, \sqrt{2 + \sqrt{2 + \sqrt{2}}}, \ldots$$

(*a*) exists and (*b*) it is equal to 2.

*3. Prove that the limit of the sequence

$$a_n = \frac{1}{n} + \frac{1}{n+1} + \cdots + \frac{1}{2n}$$

exists. Show that the limit is less than 1 but not less than $\frac{1}{2}$.

4. Prove that the limit of the sequence

$$b_n = \frac{1}{n+1} + \cdots + \frac{1}{2n}$$

exists, is equal to the limit of the previous example.

5. Obtain the following bounds for the limit L in the two previous examples: $37/60 < L < 57/60$.

*6. Let a_1, b_1 be any two positive numbers, and let $a_1 \leqslant b_1$. Let

$$a_2 = \frac{2a_1 b_1}{a_1 + b_1}, \qquad b_2 = \sqrt{a_1 b_1},$$

and in general

$$a_n = \frac{2a_{n-1} b_{n-1}}{a_{n-1} + b_{n-1}}, \qquad b_n = \sqrt{a_{n-1} b_{n-1}}.$$

Prove that the sequences $a_1, a_2, \ldots$ and $b_1, b_2, \ldots$ converge and have the same limit.

*7. Show that $1/e = 1 - 1 + \dfrac{1}{2!} - \dfrac{1}{3!} + \cdots + \dfrac{(-1)^n}{n!} + \cdots$. (*Hint:* Consider the product of the nth partial sums of the expansions for e and $1/e$.)

8. (*a*) Without reference to the binomial theorem show that $a_n = (1 + 1/n)^n$ is monotone increasing and $b_n = (1 + 1/n)^{n+1}$ is monotone decreasing. (*Hint:* Consider a_{n+1}/a_n and b_n/b_{n+1}. Use the result of Section 1.5, Problem 2.)

(*b*) Which is the larger number $(1,000,000)^{1,000,000}$ or $(1,000,001)^{999,999}$?

9. (*a*) From the results of Problem 8*a* show that

$$\left(\frac{n}{e}\right)^n < n! < e(n+1)\left(\frac{n}{e}\right)^n.$$

(*b*) For $n > 6$ derive the sharper inequality

$$n! < n\left(\frac{n}{e}\right)^n.$$

*10. If $a_n > 0$, and $\lim\limits_{n \to \infty} \dfrac{a_{n+1}}{a_n} = L$, then $\lim\limits_{n \to \infty} \sqrt[n]{a_n} = L$.

11. Use Problem 10 to evaluate the limits of the following sequences:

(a) $\sqrt[n]{n}$,   (b) $\sqrt[n]{n^5 + n^4}$,   (c) $\sqrt[n]{\dfrac{n!}{n^n}}$

12. Use Problem 11c to show

$$n! = n^n e^{-n} a_n,$$

where a_n is a number whose nth root tends to 1. (See Appendix, Chapter 7.)

13. (a) Evaluate

$$\frac{1}{1 \cdot 3} + \frac{1}{2 \cdot 4} + \cdots + \frac{1}{n(n + 2)}.$$

(*Hint:* Compare Section 1.6, Problem 12a.)

(b) From the result above, prove that $\sum\limits_{k=1}^{\infty} \dfrac{1}{n^2}$ converges.

14. Let p and q be arbitrary natural numbers. Evaluate

(a) $\sum\limits_{k=1}^{n} \dfrac{1}{(k + p)(k + p + q)}$.

(b) $\sum\limits_{k=1}^{n} \dfrac{1}{(k + p)(k + p + q)}$.

15. Evaluate

(a) $\dfrac{1}{1 \cdot 2 \cdot 3} + \dfrac{1}{2 \cdot 3 \cdot 4} + \cdots + \dfrac{1}{n(n + 1)(n + 2)}$.

(b) $\sum\limits_{k=1}^{n} \dfrac{1}{k(k + 1)(k + 3)}$.

(c) Evaluate the limit on each of the above expressions as $n \to \infty$.

*(d) Let $a_1, a_2, \ldots, a_m$ be nonnegative integers with $a_1 < a_2 < \cdots < a_m$. Show how to obtain a formula for

$$S_n = \sum\limits_{k=1}^{n} \frac{1}{(k + a_1)(k + a_2) \cdots (k + a_m)}$$

and how to find $\lim\limits_{n \to \infty} S_n$.

16. If a_k is monotone and $\sum\limits_{k=1}^{\infty} a_k$ converges, show that $\lim\limits_{k \to \infty} k a_k = 0$.

17. If a_k is monotone decreasing with limit 0 and $b_k = a_k - 2a_{k+1} + a_{k+2} \geq 0$ for all k, then show $\sum\limits_{k=1}^{\infty} k b_k = a_1$.

SECTION 1.8, page 82

1. Prove that $\lim\limits_{m \to \infty} (\cos \pi x)^{2m}$ exists for each value of x and is equal to 1 or 0 according to whether x is an integer or not.

2. (a) Prove that $\lim\limits_{n \to \infty} [\lim\limits_{m \to \infty} (\cos n! \, \pi x)^{2m}]$ exists for each value of x and is equal to 1 or 0 according to whether x is rational or irrational.

(b) Discuss the continuity of these limit functions.

3. Let $f(x)$ be continuous for $0 \leq x \leq 1$. Suppose further that $f(x)$ assumes rational values only, and that $f(x) = \frac{1}{2}$ when $x = \frac{1}{2}$. Prove that $f(x) = \frac{1}{2}$ everywhere.

SECTION 1.S.1, page 89

1. Let $r = p/q$, $s = m/n$ be arbitrary rational numbers where p, q, m, n are integers and q, n are positive. In terms of the integers p, q, m, n, define

(a) $r + s$,   (b) $r - s$,   (c) rs,   (d) $\dfrac{r}{s}$,   (e) $r < s$.

2. Prove for nested sequences of rational numbers $[a_n, b_n]$ and $[a_n', b_n']$ that each of the following conditions is necessary and sufficient for equivalence:
(a) $a_n' - a_n$ is a null sequence,
(b) $a_n \leq b_n'$ and $a_n' \leq b_n$.

3. Given $x \sim \{[a_n, b_n]\}$, $y \sim \{[\alpha_n, \beta_n]\}$, (a) verify that the definitions of addition and subtraction,

$$x + y = \{[a_n + \alpha_n, b_n + \beta_n]\}, \qquad x - y = \{[a_n - \beta_n, b_n - \alpha_n]\},$$

are meaningful. Specifically, verify that
(i) the given representations are, in fact, nested sets for $x + y$ and $x - y$ when x and y are rational;
(ii) if $x < y$, then $x + z < y + z$, where z is an arbitrary real number.
(b) Define the product xy and verify specifically that your definition of product is meaningful.
(i) that the given nested set is, in fact, a nested set for xy when x and y are rational.
(ii) that if $x < y$ and $z > 0$, then $xz < yz$.

4. Prove that the following principles are equivalent in the sense that any one can be derived as a consequence of any other.
(a) Every nested sequence of intervals with real end points contains a real number.
(b) Every bounded monotone sequence converges.
(c) Every bounded infinite sequence has at least one accumulation or limit point.
(d) Every Cauchy sequence converges.
(e) Every bounded set of real numbers has an infimum and a supremum.

Miscellaneous Problems

1. If $w_1, w_2, \ldots, w_n > 0$, prove that the weighted average

$$\frac{w_1 x_1 + w_2 x_2 + \cdots + w_n x_n}{w_1 + w_2 + \cdots w_2}$$

lies between the greatest and the least of the x's.

2. Prove

$$2(\sqrt{n+1} - 1) < 1 + \frac{1}{\sqrt{2}} + \frac{1}{\sqrt{3}} + \cdots + \frac{1}{\sqrt{n}} < 2\sqrt{n}.$$

3. Prove for $x, y > 0$

$$\frac{x^n + y^n}{2} \geq \left(\frac{x + y}{2}\right)^n.$$

Interpret this result geometrically in terms of the graph of x^n.

4. If $a_1 \geq a_2 \geq \cdots \geq a_n$ and $b_1 \geq b_2 \geq \cdots \geq b_n$, prove

$$n \sum_1^n a_i b_i \geq \left(\sum_{i=1}^n a_i\right)\left(\sum_{i=1}^n b_i\right).$$

5. (*a*) Show that the sequence $a_1, a_2, a_3, \ldots$ can be written as the sequence of partial sums of the series $u_1, u_2, u_3, \ldots$ where $u_n = a_n - a_{n-1}$ for $n \geq 1$ and $u_1 = a_1$.

(*b*) Write the sequence $a_n = n^3$ as the sequence of partial sums of a series.

(*c*) From the result obtain a formula for the nth partial sum of the series

$$1 + 4 + 9 + \cdots + n^2 + \cdots.$$

(*d*) From the formula for $1^2 + 2^2 + \cdots + n^2$, find a formula for

$$1^2 + 3^2 + 5^2 + \cdots + (2n + 1)^2.$$

6. A sequence is called an arithmetic progression of the first order if the differences of successive terms are constant. It is called an arithmetic progression of the second order if the differences of successive terms form an arithmetic progression of the first order; and, in general, it is called an arithmetic progression of order k if the differences of successive terms form an arithmetic progression of order $(k - 1)$.

The numbers 4, 6, 13, 27, 50, 84 are the first six terms of an arithmetic progression. What is its least possible order? What is the eighth term of the progression of smallest order with these initial terms?

7. Prove that the nth term of an arithmetic progression of the second order can be written in the form $an^2 + bn + c$, where a, b, c are independent of n.

*8. Prove that the nth term of an arithmetic progression of order k can be written in the form $an^k + bn^{k-1} + \cdots + pn + q$, where $a, b, \ldots, p, q$ are independent of n.

Find the nth term of the progression of smallest order in Problem 6.

9. Find a formula for the nth term of the arithmetic progressions of smallest order for which the following are the initial terms:

(*a*) 1, 2, 4, 7, 11, 16, ...

(*b*) $-7, -10, -9, 1, 25, 68, \ldots$.

*10. Show that the sum of the first n terms of an arithmetic progression of order k is

$$a_k S_k + a_{k-1} S_{k-1} + \cdots + a_1 S_1 + a_0 n,$$

where S_v represents the sum of the first n vth powers and the a_i are independent of n. Use this result to evaluate the sums for the arithmetic progressions of Problem 9.

11. By summing

$$v(v + 1)(v + 2) \cdots (v + k + 1) - (v - 1)v(v + 1) \cdots (v + k)$$

from $\nu = 1$ to $\nu = n$, show that

$$\sum_{\nu=1}^{n} \nu(\nu + 1)(\nu + 2) \cdots (\nu + k) = \frac{n(n + 1) \cdots (n + k + 1)}{k + 2}.$$

12. Evaluate $1^3 + 2^3 + \cdots + n^3$ by using the relation

$$\nu^3 = \nu(\nu + 1)(\nu + 2) - 3\nu(\nu + 1) + \nu.$$

13. Show that the function

$$f(x) = \begin{cases} \dfrac{1}{\log_2 |x|}, & x \neq 0 \\[2mm] 0, & x = 0 \end{cases}$$

is continuous but not Hölder-continuous. (*Hint:* Show Hölder continuity with exponent α fails at the origin by considering the values $x = 1/2^{n/\alpha}$.)

14. Let a_n be a monotone decreasing sequence of nonnegative numbers. Show that $\sum_{n=1}^{\infty} a_n$ converges if and only if $\sum_{\nu=0}^{\infty} 2^\nu a_{2^\nu}$ does.

15. Investigate for convergence and determine the limit when possible,
(a) $n!\, e - [n!\, e]$
(b) a_n/a_{n+1}, where $a_1 = 0$, $a_2 = 1$, and $a_{k+2} = a_{k+1} + a_k$.

2

The Fundamental Ideas of the
Integral and Differential Calculus

The fundamental limiting processes of calculus are integration and differentiation. Isolated instances of these processes of calculus were considered even in antiquity (culminating in the work of Archimedes), and with increasing frequency in the sixteenth and seventeenth centuries. However, the systematic development of calculus, started only in the seventeenth century, is usually credited to the two great pioneers of science, Newton and Leibnitz. The key to this systematic development is the insight that the two processes of differentiation and integration, which had been treated separately, are intimately related by being reciprocal to each other.[1]

A fair historical assessment of the merits cannot attribute the invention of calculus to sudden unexplainable flashes of genius on the part of one or two individuals. Many people, such as Fermat, Galileo, and Kepler, stimulated by the revolutionary new ideas in science, contributed to the foundations of calculus. In fact, Newton's teacher, Barrow, was almost in full possession of the basic insight into the reciprocity between differentiation and integration, the cornerstone of the systematic calculus of Newton and Leibnitz. Newton has stated the concepts somewhat more clearly; on the other hand, Leibnitz's ingenious notation and methods of calculation are highly suggestive and remain indispensable. The work of these two men immediately stimulated the higher branches of analysis including the calculus of variations and the theory of differential equations, and led to innumerable applications in science. Curiously enough, although Newton,

---

[1] This fact constitutes the "fundamental theorem of calculus."

Leibnitz, and their immediate successors made such varied uses of the powerful tool put into their hands, none succeeded in completely clarifying the basic concepts involved in their work. Their arguments employed "infinitely small quantities" in ways which are logically indefensible and unconvincing. Clarification came at last in the nineteenth century with the careful formulation of the concept of limit and with the analysis of the number continuum as explained in Chapter 1.[1]

We begin with a discussion of the fundamental concepts. They can be fully appreciated only through concrete illustrations and examples; it is therefore recommended here, as at many places in this book, that theoretical and general sections be carefully studied again after the reader has absorbed more specific and concrete material in subsequent sections.

2.1   The Integral

a. Introduction

Only after a lengthly development the systematic procedures of integration and differentiation met the need for precise mathematical descriptions of intuitive notions arising in geometry and natural science. Differentiation is the concept needed for describing the notions of *tangents* to curves and of velocity of moving particles, or more generally, the concept of *rate of change*. The intuitive concept of *area* of a region with curved boundaries, finds its precise mathematical formulation in the process of integration. Many other related concepts in geometry and physics also require integration, as we shall see later. In this section we introduce the concept of integral, in connection with the problem of measuring the area of a plane region bounded by curves.

Areas.   We have an intuitive feeling that a region contained in a closed curve has an "area" which measures the number of square units inside the curve. Yet, the question, of *how* this measure for the area can be described in precise terms, necessitates a chain of mathematical steps. The basic properties of area which intuition suggests are: area is a (positive) number (depending on the choice of the unit of length); this number is the same for congruent figures; for all

---

[1] The emergence of calculus extending over more than 2000 years represents one of the most fascinating chapters in the history of scientific discovery. Interested readers are referred to Carl B. Boyer, *Concept of the Calculus*, Hafner Publishing Company, 1949. See also O. Toeplitz, *Calculus, A Genetic Approach*, University of Chicago, 1963.

rectangles it is the product of the lengths of two adjacent sides; and finally, for a region decomposed into parts, the area of the whole is equal to the sum of the areas of the parts.

An immediate consequence is the fact: for a region A which is part of a region B, the area of A cannot exceed the area of B.

These properties permit the direct computation of the area of any figure that can be decomposed into a finite number of rectangles. More generally, to assign a value F to the area of a region R we consider two other regions R' (inscribed) and R'' (circumscribed) decomposable

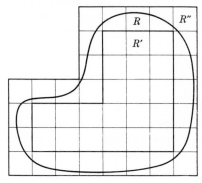

Figure 2.1   Approximation of an area.

into rectangles, where R'' contains R and R' is contained in R (cf. Fig. 2.1). We know then at least that F has to lie between the areas of R' and R''. The value of F is completely determined if we find sequences of circumscribed regions R_n'' and inscribed regions R_n' which are both decomposable into rectangles and such that the areas of R_n'' and R_n' have the same limit as n tends to infinity. This is the method of "exhaustion", going back to antiquity which is used in elementary geometry to describe the area of a circle.[1] The precise formulation of this intuitive idea now leads to the notion of integration.

b. The Integral as an Area

Area under a Curve

The analytic notion of integral arises when we associate areas with *functions*: We consider the area of a region bounded on the left and

---

[1] Of course, we may use any kind of inscribed and circumscribed *polygon*, since a polygon can be decomposed into right triangles and the area of a right triangle clearly is half that of a rectangle with the same sides.

right by vertical lines $x = a$ and $x = b$, below by the x-axis and above by the graph of a positive continuous function $f(x)$ (Fig. 2.2). This is referred to in brief as the area "under the curve." For the moment we accept as intuitive the idea that the area of such a region is a definite number. We call this area F_a^b the *integral of the function f* between the

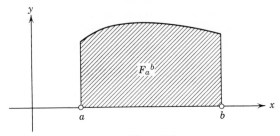

Figure 2.2

limits[1] a and b. In seeking the numerical value of F_a^b we make use of approximations by sums of areas of rectangles. For that purpose we divide the interval (a, b) of the x-axis into n (small) parts, not necessarily of the same size, which we shall call *cells*. At each point of division we draw the line perpendicular to the x-axis up to the curve. The region with area F_a^b is thus divided into n strips, each bounded by a portion of

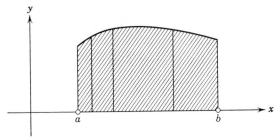

Figure 2.3

the graph of the function $f(x)$ and by three straight line segments (Fig. 2.3).

Area or Integral as Limit of a Sum. Calculating the area of such strips precisely is not easier than calculating that of the original region. It is a step forward, however, to approximate the area of each strip from above and from below by the areas of the circumscribed and

---

[1] No confusion should arise from the use of the word "limit" for boundary points of the interval of integration.

inscribed rectangles with the same base, where the curved boundary of the strip is replaced by a horizontal line at a distance from the x-axis which is either the greatest or the smallest value of $f(x)$ in the cell (Fig. 2.4). More generally, we obtain an intermediate approximation if we replace the strip by a rectangle of the same base and bounded on

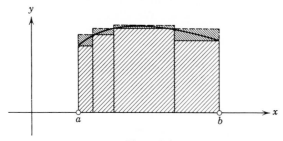

Figure 2.4

top by any horizontal line which intersects the curved boundary of the strip (see Fig. 2.5). Analytically, this amounts to replacing the function $f(x)$ in each of the cells by some intermediate constant value. We denote by F_n the sum of the n rectangular areas. Intuition tells us that the values F_n tend to F_a^b if we make the subdivision finer and finer, that is, if we let n increase without limit while the largest length of the

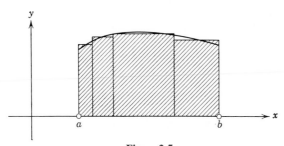

Figure 2.5

individual cells tends to zero. In this way F_a^b is represented as a limit of areas consisting of rectangles.

c. Analytic Definition of the Integral. Notations

Definition and Existence of Integrals

In the last paragraph we accepted the area under a curve as a quantity given intuitively and subsequently we represented it as a limiting value. Now we shall reverse the procedure. We no longer invoke intuition to assign an area to the region under a continuous curve; on the contrary, we shall begin in a purely analytic way with the sums F_n defined previously, and we shall *prove* that these sums tend to a definite limit. This limit is then the precise *definition* of the integral and of the area.

Let the function $f(x)$ be continuous (but not necessarily positive) in the closed interval $a \leq x \leq b$. We divide the interval by $(n-1)$

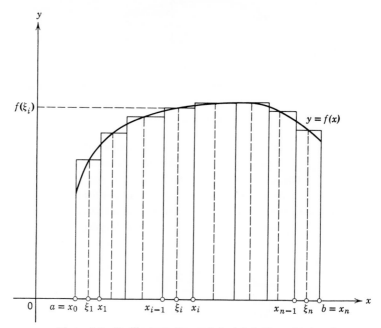

Figure 2.6   To illustrate the analytical definition of integral.

points $x_1, x_2, \ldots, x_{n-1}$ into n equal or unequal cells with the lengths

$$x_i - x_{i-1} = \Delta x_i, \qquad (i = 1, 2, \ldots, n)^1,$$

where in addition we put $x_0 = a$, $x_n = b$ (cf. Fig. 2.6). In each closed subinterval $[x_{i-1}, x_i]$ or cell we choose any point ξ_i whatever. We form the sum

$$F_n = f(\xi_1)(x_1 - x_0) + f(\xi_2)(x_2 - x_1) + \cdots + f(\xi_n)(x_n - x_{n-1})$$
$$= f(\xi_1)\, \Delta x_1 + f(\xi_2)\, \Delta x_2 + \cdots + f(\xi_n)\, \Delta x_n.$$

---

[1] The symbol Δ must not be interpreted as a factor but only as indicating a difference in values of the variable which follows. Thus the symbol Δx_i means the difference $x_i - x_{i-1}$ of consecutive values of x.

Using the summation symbol we write more concisely

$$F_n = \sum_{i=1}^{n} f(\xi_i)(x_i - x_{i-1})$$

or

$$F_n = \sum_{i=1}^{n} f(\xi_i)\,\Delta x_i.$$

If $f(x)$ is positive, the value F_n represents the area under the curve obtained by replacing f in each subinterval by the constant value $f(\xi_i)$. Of course, the sums F_n can be formed without assuming f to be positive. It appears intuitively plausible that the sums F_n must tend to a limit F_a^b as the number n of intervals increases indefinitely and at the same time the length of the largest subinterval tends to zero. This would imply that the value of the limit F_a^b is independent of the particular manner in which the points of division $x_1, x_2, \ldots, x_{n-1}$ and the intermediate points $\xi_1, \xi_2, \ldots, \xi_n$ are chosen. We call F_a^b the integral of $f(x)$ between the limits a and b.

Geometric intuition, no matter how convincing, can only serve as a guide to our analytical limiting process; therefore an analytic justification is needed, and we must furnish a proof for the existence of the integral as the limit described above. Furthermore, as already said, we need not at all insist on the assumption that the function f is positive in the interval.

Thus we assert

THEOREM OF EXISTENCE. *For any continuous function $f(x)$ in a closed interval $[a, b]$ the integral over this interval exists as the limit of the sums F_n described above (independently of the choice of the points of subdivision $x_1, \ldots, x_{n-1}$ and of the intermediate points $\xi_1, \ldots, \xi_n$ as long as the largest of the lengths Δx_i tends to zero).*

We shall first gain some experience and insight before considering the existence proof for the integral in the Supplement (p. 192).

Leibnitz's Notation for the Integral

The definition of the integral as the limit of a sum led Leibnitz to express the integral by the following symbol:

$$\int_a^b f(x)\,dx.$$

The integral sign is a modification of the summation sign in the shape of a long S used at Leibnitz's time. The passage to the limit from a finite subdivision into portions Δx_i is indicated by the use of the letter d in place of Δ. In using this notation, however, we must not tolerate the eighteenth century mysticism of considering dx as an "infinitely small" or "infinitesimal quantity," or considering the integral as a "sum of an infinite number of infinitely small quantities." Such a conception is devoid of clear meaning and obscures what we have previously formulated with precision. From our present viewpoint the individual symbol dx has not been defined at all. The suggestive combination of symbols $\int_a^b f(x)\,dx$ is defined for a function $f(x)$ in the interval $[a, b]$ by forming the ordinary sums F_n and passing to the limit as $n \to \infty$.

The particular symbol we use for the variable of integration is a matter of complete indifference (just as in the notation for sums it

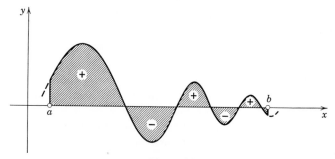

Figure 2.7

did not matter what we called the index of summation); instead of $\int_a^b f(x)\,dx$ we can equally well write $\int_a^b f(t)\,dt$ or $\int_a^b f(u)\,du$. The *integrand* denoted by f is a function of an independent variable over the interval $[a, b]$ and the name of the variable is irrelevant. Only the end points of the interval of integration a and b affect the value of the integral for given f. Expressions like $\int_a^x f(x)\,dx$ or $\int_a^b f(a)\,da$ in which the same letter is used for the variable of integration and an end‑point of the interval are misleading under our definition and should, at first, be avoided.

If the *integrand* $f(x)$ is positive in the interval $[a, b]$, we can immediately identify $\int_a^b f(x)\,dx$ with the area bounded by the graph of f and the lines $x = a$, $x = b$, and $y = 0$. The integral of f, however, is defined analytically as the limit of sums F_n independent of any assumption on the sign of f. If $f(x)$ is negative in all or part of our interval, the only effect is to make the corresponding factors $f(\xi_i)$ in our sum

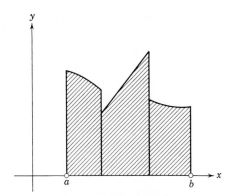

Figure 2.8

negative instead of positive. To the region bounded by the part of the curve below the x-axis we shall then naturally assign a negative area. The integral will thus be the sum of positive and negative terms, corresponding respectively to portions of the curve above and below the x-axis[1] (see Fig. 2.7).

It is intuitively convincing that our limit process converges even if the function $f(x)$ is not everywhere continuous, but has jump discontinuities at one or several points like the function indicated by the curve in Fig. 2.8, where clearly an area under the curve exists.[2]

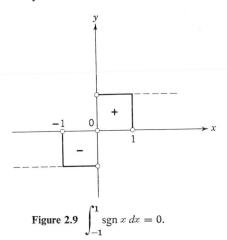

Figure 2.9 $\displaystyle\int_{-1}^{1} \operatorname{sgn} x \, dx = 0.$

---

[1] Areas of regions bounded by arbitrary closed curves will be considered in Chapter 4.

[2] As another example consider $f(x) = \operatorname{sgn} x$ on $[-1, 1]$. We have $f(x) = -1$ for $x < 0$ and $f(x) = +1$ for $x > 0$ (see Fig. 2.9). Then $\displaystyle\int_{-1}^{+1} f(x) \, dx = 0.$

Thus the preceding limit process may well result in a definite limit of the sum F_n for functions having some discontinuities; we indicate this possibility by calling such functions *integrable*. In the middle of the nineteenth century, the great Bernhard Riemann first analyzed the applicability of the process of integration to general functions. More recently, various extensions of the concept of integration itself have been introduced. Yet such refinements have less immediate importance for the calculus aimed at intuitively accessible phenomena, and it will not be necessary for us always to emphasize the integrability of our functions as a reminder that nonintegrable functions can be defined.

In advanced courses the integral we have defined here is called the Riemann integral to distinguish it from various generalized concepts of integral; the approximating sums F_n are called *Riemann* sums.

2.2   Elementary Examples of Integration

In a number of significant cases we are now able to calculate the integral of a function by carrying out the prescribed limiting process. This we shall do by an explicit evaluation of the sums F_n for a suitable choice of intermediate points ξ_i (usually the left or right end point of the cells). The theorem on the existence of the integral of a continuous function assures that the limit of the F_n is the same for any other choice of the intermediate points ξ_i and for any method of subdivision.

a. Integration of a Linear Function

First we verify that the integral indeed gives the correct value of the area for some simple figures we know from geometry.

Let $f(x) = $ constant $= \gamma$. To calculate the integral of $f(x)$ between the limits of a and b we form the sums F_n (see Fig. 2.10). Since here $f(\xi_i) = \gamma$, we find

$$F_n = \sum_{i=1}^{n} \gamma \, \Delta x_i = \gamma \sum_{i=1}^{n} \Delta x_i = \gamma(b - a).$$

Hence, likewise

$$\lim_{n \to \infty} F_n = \int_{a}^{b} \gamma \, dx = \gamma(b - a).$$

This is just the formula for the area of a rectangle of height γ and base $b - a$.

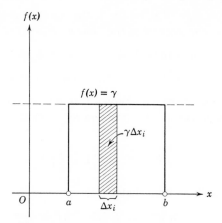

Figure 2.10   Integral of a constant.

The integral of the function $f(x) = x$,

$$\int_a^b x \, dx,$$

(Fig. 2.11), as we know from elementary geometry, has the value

$$\tfrac{1}{2}(b - a)(b + a) = \tfrac{1}{2}(b^2 - a^2).$$

To confirm that our limiting process leads analytically to the same result, we subdivide the interval from a to b into n equal parts by means of the points of division

$$a + h, a + 2h, \ldots, a + (n - 1)h,$$

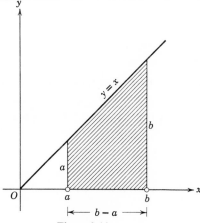

Figure 2.11

where $h = (b - a)/n$. Taking for ξ_i the right-hand end point of each interval we find the integral as the limit as $n \to \infty$ of the sum

$$F_n = (a + h)h + (a + 2h)h + \cdots + (a + nh)h$$

$$= nah + (1 + 2 + 3 + \cdots + n)h^2 = nah + \tfrac{1}{2}n(n + 1)h^2,$$

where we have used the well-known formula for the sum of an arithmetic progression (see p. 111, Problem 3). Substituting $h = (b - a)/n$, we see that

$$F_n = a(b - a) + \frac{1}{2}\left(1 + \frac{1}{n}\right)(b - a)^2,$$

from which it follows immediately that

$$\lim_{n \to \infty} F_n = a(b - a) + \tfrac{1}{2}(b - a)^2 = \tfrac{1}{2}(b^2 - a^2).$$

b. Integration of x^2

Elementary geometry does not so easily lead to the integration of the function $f(x) = x^2$, that is, to the determination of the area of the region[1] bounded by a segment of a parabola, a segment of the x-axis, and two coordinates. A genuine limit process is needed. Assuming $a < b$ we choose the same points of division and the same intermediate points as in the previous example (see Fig. 2.12). It follows then that the integral of x^2 between the limits a and b is the limit of the sums

$$F_n = (a + h)^2 h + (a + 2h)^2 h + \cdots + (a + nh)^2 h$$

$$= na^2 h + 2ah^2(1 + 2 + 3 + \cdots + n)$$

$$+ h^3(1^2 + 2^2 + 3^2 + \cdots + n^2);$$

by using the known values of the sums enclosed in parentheses we find (see p. 58)

$$F_n = na^2 h + n(n + 1)ah^2 + \frac{1}{6}[n(n + 1)(2n + 1)]h^3$$

$$= a^2(b - a) + \left(1 + \frac{1}{n}\right)a(b - a)^2 + \frac{1}{6}\left(1 + \frac{1}{n}\right)\left(2 + \frac{1}{n}\right)(b - a)^3.$$

Since $\lim\limits_{n \to \infty} \dfrac{1}{n} = 0$, we have

$$\lim_{n \to \infty} F_n = a^2(b - a) + a(b - a)^2 + \frac{1}{3}(b - a)^3 = \frac{1}{3}(b^3 - a^3).$$

---

[1] Sometimes referred to as "squaring" the region.

Figure 2.12   Area under a parabolic arc by arithmetic subdivision.

Thus, for $a < b$,

$$\int_a^b x^2 \, dx = \frac{1}{3}(b^3 - a^3).$$

*c. Integration of x^α for Integers $\alpha \neq -1$

The next examples of this section are instructive illustrations showing that in some cases the integration can be carried out by special elementary devices. Later in Section 2.9d (p. 191) we shall achieve the same results more simply by using general methods.

The same kind of argument as used for x and x^2, applied to the functions $x^3, x^4, \ldots$, results in the relation

(1) $$\int_a^b x^\alpha \, dx = \frac{1}{\alpha + 1}(b^{\alpha+1} - a^{\alpha+1}),$$

where α is any positive integer; this can be proved by finding appropriate formulas for the sums $1^\alpha + 2^\alpha + \cdots + n^\alpha$, such as the relation

$$\lim_{n \to \infty} \left[(1^\alpha + 2^\alpha + \cdots + n^\alpha) \frac{1}{n^{\alpha+1}} \right] = \frac{1}{\alpha + 1}$$

which can be proved by induction over α (see Problem 16, p. 113). In the following section, formula (1) will be proved in a different way, with greater generality and simplicity, indicating the power of the methods that we will develop. Its validity will be extended to all real values of α except $\alpha = -1$.

Fortunately, the definition of the integral leaves us a great deal of latitude in the choice of subdivisions and furnishes a much simpler way to evaluate the integral. We do not have to use sums based on equidistant points of division. Instead, with the "quotient" $\sqrt[n]{b/a} = q$ we

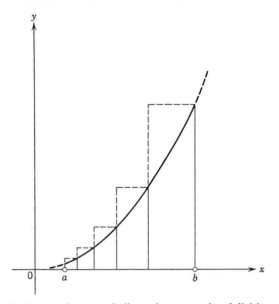

Figure 2.13   Area under a parabolic arc by geometric subdivision.

subdivide the interval $[a, b]$ by the points of a geometric progression (Fig. 2.13),

$$a, aq, aq^2, \ldots, aq^{n-1}, aq^n = b;$$

we then need only to evaluate the sum of a geometric series. Given the points of division $x_i = aq^i$ the length of the ith cell is given by

$$\Delta x_i = aq^i - aq^{i-1} = \frac{aq^i(q-1)}{q}.$$

The largest Δx_i is the last:

$$\Delta x_n = \frac{b(q-1)}{q}.$$

For $n \to \infty$ the number q tends toward the value one (see Example d, p. 64), and hence the length Δx_n of the largest cell, and then also the lengths of all cells tend to zero. For the intermediate points ξ_i we choose again the right-hand end points x_i of each cell. The sum

$$
(2) \qquad F_n = \sum_{i=1}^{n} (\xi_i)^\alpha \, \Delta x_i = \sum_{i=1}^{n} (aq^i)^\alpha aq^i \frac{q-1}{q}
$$

$$
= a^{\alpha+1} \frac{q-1}{q} \sum_{i=1}^{n} (q^{1+\alpha})^i
$$

is known explicitly from the sum of the geometric progression with ratio $q^{1+\alpha}$. Applying the well-known formula (p. 67), we find

$$
F_n = a^{\alpha+1} \frac{q-1}{q} q^{\alpha+1} \frac{q^{n(\alpha+1)}-1}{q^{\alpha+1}-1}
$$

$$
= a^{\alpha+1}(q-1)q^\alpha \frac{(b/a)^{\alpha+1}-1}{q^{\alpha+1}-1} = (b^{\alpha+1} - a^{\alpha+1})q^\alpha \frac{q-1}{q^{\alpha+1}-1}.
$$

Since $q \neq 1$, we can use once more the formula for the sum of a geometric progression and write

$$
\frac{q-1}{q^{1+\alpha}-1} = \frac{1}{q^\alpha + q^{\alpha-1} + \cdots + 1}.
$$

For $n \to \infty$ all powers of q tend to one and it follows that

$$
\lim_{n \to \infty} F_n = \frac{1}{1+\alpha} (b^{1+\alpha} - a^{1+\alpha}).
$$

In this way we have verified the formula (1) for the integral of x^α for $0 < a < b$ and any positive integer α.

The same method applies also for negative integers α, provided that $\alpha \neq -1$. For the sum F_n we obtain as before

$$
F_n = (b^{\alpha+1} - a^{\alpha+1})q^\alpha \frac{q-1}{q^{\alpha+1}-1}
$$

$$
= (b^{\alpha+1} - a^{\alpha+1}) \frac{q-1}{q(1-q^{-\alpha-1})},
$$

where we recall that $-\alpha$ is positive and greater than one. Applying the formula for a geometric progression, we obtain

$$
\frac{1}{q}\left(\frac{q-1}{q^{-\alpha-1}-1} \right) = \frac{1}{q^{-\alpha-1} + q^{-\alpha-2} + \cdots + q}
$$

which tends to $1/(-\alpha-1)$ as $n \to \infty$. Consequently, as before,

$$
\lim_{n \to \infty} F_n = \frac{1}{\alpha+1} (b^{\alpha+1} - a^{\alpha+1}),
$$

The integral formula is meaningless for $\alpha = -1$, since both numerator and denominator on the right-hand side would then be zero. We find instead from our original expression (2) for F_n for the case $\alpha = -1$ that $F_n = n(q - 1)/q$. Consequently, observing that $q = \sqrt[n]{b/a}$ tends to one as $n \to \infty$, we find

$$(3) \qquad \int_a^b \frac{1}{x}\, dx = \lim_{n \to \infty} n(\sqrt[n]{b/a} - 1).$$

Here the limit on the right-hand side cannot be expressed in terms of powers of a and b but can be expressed in terms of logarithms of those quantities as we shall see later (p. 145).

d. Integration of x^α for Rational α Other Than -1

The result obtained previously may be generalized considerably without essentially complicating the proof. Let $\alpha = r/s$ be a positive rational number, r and s being positive integers: then in the evaluation of the integral given above nothing is changed except the evaluation of the limit $(q - 1)/(q^{\alpha+1} - 1)$ as q approaches one. This expression is now simply $(q - 1)/(q^{(r+s)/s} - 1)$. Let us put $q^{1/s} = \tau \; (\tau \neq 1)$: Then as q tends to one, τ also tends to one. We have therefore to find the limiting value of $(\tau^s - 1)/(\tau^{r+s} - 1)$ as τ approaches one. If we divide both numerator and denominator by $\tau - 1$ and transform them as before by the formula for geometric progressions, the limit simply becomes

$$\lim_{\tau \to 1} \frac{\tau^{s-1} + \tau^{s-2} + \cdots + 1}{\tau^{r+s-1} + \tau^{r+s-2} + \cdots + 1}.$$

Since both numerator and denominator are continuous in τ, this limit is at once obtained by substituting $\tau = 1$, and thus equals $s/(r + s) = 1/(\alpha + 1)$; hence for every positive rational value of α we obtain the integral formula

$$\int_a^b x^\alpha\, dx = \frac{1}{\alpha + 1}\, (b^{\alpha+1} - a^{\alpha+1}),$$

just as with positive integers.

This formula remains valid for negative rational values of $\alpha = -r/s$ as well, provided we exclude the value $\alpha = -1$ (for which the formula used above for the sum of the geometric progression loses its meaning).

For negative α we again evaluate the limit of $(q - 1)/(q^{\alpha+1} - 1)$ by putting $q^{-1/s} = \tau$ for $\alpha = -r/s$; this is left as an exercise for the reader.

It is natural to guess that the range of validity of our last formula extends also to irrational values of α. We shall actually establish our integral formula for all real values of α (except $\alpha = -1$) in Section 2.7 (p. 154) in a quite simple way as a consequence of the general theory.

*e. Integration of $\sin x$ and $\cos x$

The last elementary example to be treated here by means of a special device is the integral of $f(x) = \sin x$. The integral

$$\int_a^b \sin x \, dx$$

clearly is the limit of the sum

$$S_h = h[\sin (a + h) + \sin (a + 2h) + \cdots + \sin (a + nh)],$$

arising from division of the interval of integration into cells of size $h = (b - a)/n$. We multiply the right-hand expression by $2 \sin h/2$ and recall the well-known trigonometrical formula

$$2 \sin u \sin v = \cos (u - v) - \cos (u + v).$$

Provided h is not a multiple of 2π, we obtain the formula

$$S_h = \frac{h}{2 \sin \frac{h}{2}} \left[\cos \left(a + \frac{h}{2} \right) - \cos \left(a + \frac{3}{2} h \right) + \cos \left(a + \frac{3}{2} h \right) \right.$$

$$\left. - \cos \left(a + \frac{5}{2} h \right) + \cdots + \cos \left(a + \frac{2n - 1}{2} h \right) - \cos \left(a + \frac{2n + 1}{2} h \right) \right]$$

$$= \frac{h}{2 \sin \frac{h}{2}} \left[\cos \left(a + \frac{h}{2} \right) - \cos \left(a + \frac{2n + 1}{2} h \right) \right].$$

Since $a + nh = b$, the integral becomes the limit of

$$\frac{h}{2 \sin \frac{h}{2}} \left[\cos \left(a + \frac{h}{2} \right) - \cos \left(b + \frac{h}{2} \right) \right] \qquad \text{as } h \to 0.$$

Now we know from Chapter 1 (p. 84) that for $h \to 0$, the expression $(h/2)/(\sin h/2)$ approaches the limit one. The desired limit is then simply $\cos a - \cos b$, and we arrive at the integral

$$\int_a^b \sin x \, dx = -(\cos b - \cos a).$$

Similarly,

$$\int_a^b \cos x \, dx = \sin b - \sin a \qquad \text{(see Problem 3, p. 196).}$$

Each of the preceding examples was treated with a special device. Yet the essential point of the systematic integral and differential calculus is the very fact that, instead of such special devices, we use general considerations which lead directly to the result. We shall arrive at these methods by first discussing some general rules concerning integrals and then introducing the concept of the derivative, and finally establishing the connection between integral and derivative.

2.3   Fundamental Rules of Integration

The basic properties of the integral follow directly from its definition as the limit of a sum:

$$\int_a^b f(x)\,dx = \lim_{n\to\infty} \sum_{i=1}^n f(\xi_i)\,\Delta x_i,$$

where the interval $[a, b]$ is broken up into subintervals or cells of length Δx_i, the number ξ_i stands for any value in the ith subinterval, and the largest Δx_i is required to tend to zero for $n \to \infty$.

a. Additivity

Let c be any value between a and b. If we interpret integrals as areas and remember that the area of a region consisting of several parts is the sum of the areas of the parts (Fig. 2.14), we are led to the rule

(4) $$\int_a^b f(x)\,dx = \int_a^c f(x)\,dx + \int_c^b f(x)\,dx.$$

For an analytical proof we choose our subdivisions in such a manner that the point c appears as a point of division, say $c = x_m$ (where m varies with n). Then

$$\sum_{i=1}^n f(\xi_i)\,\Delta x_i = \sum_{i=1}^m f(\xi_i)\,\Delta x_i + \sum_{i=m+1}^n f(\xi_i)\,\Delta x_i,$$

where the first sum on the right-hand side corresponds to a subdivision of the interval $[a, c]$ in m cells and the second sum to a subdivision of the interval $[c, b]$. Now for $n \to \infty$ we obtain our rule for integrals.

So far we have only defined $\int_a^b f(x)\,dx$ when $a < b$. For $a = b$ or $a > b$ we define the integral in such a way that the rule of additivity is preserved. Therefore for $c = a$ we must define

(5) $$\int_a^a f(x)\,dx = 0,$$

and then for $b = a$ it follows that

$$\int_a^c f(x)\, dx + \int_c^a f(x)\, dx = \int_a^a f(x)\, dx = 0.$$

This leads us to *define* $\int_a^c f(x)\, dx$ for $c < a$ by the formula

(6) $$\int_a^c f(x)\, dx = -\int_c^a f(x)\, dx,$$

where the right side has the meaning originally established. Its *geometric meaning is that the area* under the curve $y = f(x)$ is to be counted

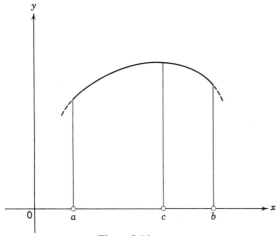

Figure 2.14

as *negative* if the direction of moving from the lower limit of integration to the upper limit is that of decreasing x. A glance at the previous examples of integrals confirms that indeed an interchange in the limits of integration a and b results in changing the sign in the value of the integral.

b. Integral of a Sum and of a Product with a Constant

If $f(x)$ and $g(x)$ are any two (integrable) functions, the basic laws of operating with limits imply

$$\int_a^b f(x)\, dx + \int_a^b g(x)\, dx = \lim_{n \to \infty}\left[\sum_{i=1}^n f(\xi_i)\,\Delta x_i\right] + \lim_{n \to \infty}\left[\sum_{i=1}^n g(\xi_i)\,\Delta x_i\right]$$

$$= \lim_{n \to \infty}\left[\sum_{i=1}^n f(\xi_i)\,\Delta x_i + \sum_{i=1}^n g(\xi_i)\,\Delta x_i\right]$$

$$= \lim_{n \to \infty}\left\{\sum_{i=1}^n [f(\xi_i) + g(\xi_i)]\,\Delta x_i\right\};$$

and hence the important rule for the sum of two functions

(7) $$\int_a^b f(x)\, dx + \int_a^b g(x)\, dx = \int_a^b [f(x) + g(x)]\, dx;$$

similarly for the difference

$$\int_a^b f(x)\, dx - \int_a^b g(x)\, dx = \int_a^b [f(x) - g(x)]\, dx.$$

Furthermore, with any constant α

$$\int_a^b \alpha f(x)\, dx = \lim_{n \to \infty} \sum_{i=1}^n \alpha f(\xi_i)\, \Delta x_i$$

$$= \alpha \lim_{n \to \infty} \sum_{i=1}^n f(\xi_i)\, \Delta x_i,$$

and so

(8) $$\int_a^b \alpha f(x)\, dx = \alpha \int_a^b f(x)\, dx.$$

The last two rules enable us to integrate "*linear combinations*" of two or more functions that can be integrated individually. Thus for any quadratic function $y = Ax^2 + Bx + C$ with any constants A, B, C, we have

$$\int_a^b (Ax^2 + Bx + C)\, dx = \int_a^b Ax^2\, dx + \int_a^b Bx\, dx + \int_a^b C\, dx$$

$$= A \int_a^b x^2\, dx + B \int_a^b x\, dx + C \int_a^b 1\, dx$$

$$= \frac{A}{3}(b^3 - a^3) + \frac{B}{2}(b^2 - a^2) + C(b - a).$$

In the same way we integrate the general *polynomial*

$$y = A_0 x^n + A_1 x^{n-1} + \cdots + A_{n-1} x + A_n:$$

$$\int_a^b y\, dx = \frac{1}{n+1} A_0(b^{n+1} - a^{n+1}) + \frac{1}{n} A_1(b^n - a^n) + \cdots$$

$$+ \tfrac{1}{2} A_{n-1}(b^2 - a^2) + A_n(b - a).$$

c. Estimating Integrals

Another obvious observation concerning integrals is basic. Consider for $a < b$ a function $f(x)$ which is positive or zero at each point of the interval $[a, b]$. Then

(9) $$\int_a^b f(x)\, dx \geq 0.$$

is follows immediately if we write the integral as limit of a sum and
:ice that the sums contain only nonnegative terms.

More generally, if we have two functions f and g with the property
.t $f(x) \geq g(x)$ for all x in the interval $[a, b]$, then

)
$$\int_a^b f(x)\, dx \geq \int_a^b g(x)\, dx.$$

r we have

$$\int_a^b f(x)\, dx - \int_a^b g(x)\, dx = \int_a^b [f(x) - g(x)]\, dx \geq 0,$$

ce $f(x) - g(x)$ is never negative.

We apply this result to a function $f(x)$ which is continuous in the
:rval $[a, b]$. Let M be the greatest value and m the least value of f
that interval. Since

$$m \leq f(x) \leq M$$

all x in $[a, b]$, we have

$$\int_a^b m\, dx \leq \int_a^b f(x)\, dx \leq \int_a^b M\, dx.$$

calling that for any constant C

$$\int_a^b C\, dx = C \int_a^b 1\, dx = C(b - a),$$

obtain the inequality

)
$$m(b - a) \leq \int_a^b f(x)\, dx \leq M(b - a),$$

ich gives simple upper and lower bounds for the definite integral of
y continuous function.

Again this estimate is intuitively obvious. If we think of the integral
:rpreted as an area, the quantities $M(b - a)$ and $m(b - a)$ represent
:as of a circumscribed and an inscribed rectangle on the common
:e of length $b - a$ (see Fig. 2.15).

1. The Mean Value Theorem for Integrals

:egral as a Mean Value

Significant is a slightly different interpretation of our inequalities
terms of the *average of the function f in an interval* $[a, b]$. For a
ite number of quantities $f_1, f_2, \ldots, f_n$ the *average* or *arithmetic mean*
:he number

$$\frac{f_1 + f_2 + \cdots + f_n}{n}.$$

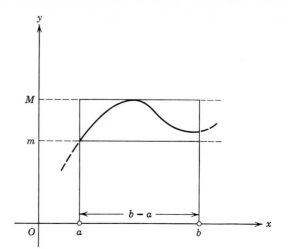

Figure 2.15

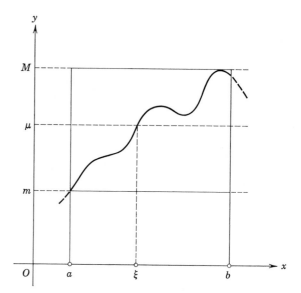

Figure 2.16   The mean value μ of a function.

ve want to assign a meaning to the average value of the infinitely
ny quantities $f(x)$ corresponding to arbitrary x in the interval
$b]$, it is natural to pick out first a finite number n of values of f,
$f(x_1), f(x_2), \ldots, f(x_n)$, to form their average

$$\frac{f(x_1) + \cdots + f(x_n)}{n},$$

1 then to take the limit as n increases beyond all bounds. The value
this limit, if it exists at all, will depend very much on how the points
are spaced in the interval $[a, b]$. A definite value for the average of f
attained if we take for the x_i the points obtained when we divide the
erval $[a, b]$ into n *equal* parts of length $\Delta x_i = (b - a)/n$. We have
n

$$\frac{f(x_1) + \cdots + f(x_n)}{n} = \frac{1}{b - a} \sum_{i=1}^{n} f(x_i) \, \Delta x_i,$$

1 it is clear that in the limit for $n \to \infty$ the nth averages converge
vards the value

$$\mu = \frac{1}{b - a} \int_a^b f(x) \, dx = \frac{\displaystyle\int_a^b f(x) \, dx}{\displaystyle\int_a^b dx}.$$

: shall call μ the "arithmetic average" or the *mean value* of f in the
erval $[a, b]$. Our inequalities then simply state that the mean value
a continuous function cannot be larger than the greatest value or
s than the least value of the function (Fig. 2.16).
Since the function $f(x)$ is continuous in the interval $[a, b]$, there
ist be points in the interval where f has the value M or the value m.
the intermediate value theorem for continuous functions there must
en also be a point ξ in the interval where f actually assumes the
ermediate value μ. We have proved then:

MEAN VALUE THEOREM. *For a continuous function $f(x)$ in the
erval $[a, b]$ there exists a value ξ in the interval such that*

2) $$\int_a^b f(x) \, dx = f(\xi)(b - a).$$

is is the simple but very important *mean value theorem of integral
lculus.* In words, it states that the mean value of a continuous
iction in an interval belongs to the range of the function.
The theorem asserts only the *existence* of at least one ξ in the interval
: which $f(\xi)$ is equal to the average value of f but gives no further
ormation about the location of ξ.

Note that the formula expressing the mean value theorem st. valid if the limits a and b are interchanged; hence the mean va theorem is correct also when $a > b$.

The Generalized Mean Value Theorem. Instead of the simple arithme average we often have to consider "weighted averages" of n quanti $f_1, \ldots, f_n$ given by

$$\frac{p_1 f_1 + p_2 f_2 + \cdots + p_n f_n}{p_1 + p_2 + \cdots + p_n} = \mu,$$

where the "weight factors" p_i are any positive quantities. If, for exam $p_1, p_2, \ldots, p_n$ are actually the weights of particles located respectively at points $f_1, f_2, \ldots, f_n$ of the x-axis, then μ will represent the location the center of gravity. If all weights p_i are equal, the quantity μ is just *arithmetic* average defined above.

For a function $f(x)$ we can form analogously the weighted average

(13)
$$\mu = \frac{\displaystyle\int_a^b f(x)p(x)\,dx}{\displaystyle\int_a^b p(x)\,dx}$$

over the interval $[a, b]$ where $p(x)$, the *weight function*, is any positive functi in the interval. The assumption that p is positive guarantees that denominator does not vanish.

The weighted average μ also lies between the largest value M and the small value m of the function f in the interval.

For multiplying the inequality

$$m \leq f(x) \leq M,$$

by the *positive* number $p(x)$, we find that

$$mp(x) \leq f(x)p(x) \leq Mp(x).$$

Integration then yields

$$m \int_a^b p(x)\,dx \leq \int_a^b f(x)p(x)\,dx \leq M \int_a^b p(x)\,dx.$$

Dividing by the positive quantity $\displaystyle\int_a^b p(x)\,dx$, we indeed obtain the result

$$m \leq \mu \leq M.$$

If here $f(x)$ is continuous, we conclude from the intermediate val theorem (p. 44) that $\mu = f(\xi)$, where ξ is a suitable value in the inter $a \leq \xi \leq b$. This leads to the following *generalized mean value theorem integral calculus*:

If $f(x)$ and $p(x)$ are continuous in the interval $[a, b]$ and moreover $p(x)$ is positive in that interval, then there exists a value ξ in the interval such that

(14) $$\int_a^b f(x)p(x)\, dx = f(\xi) \int_a^b p(x)\, dx.$$

The special case $p(x) = 1$ leads to our earlier mean value theorem.

2.4   The Integral as a Function of the Upper Limit (Indefinite Integral)

Definition and Basic Formula

The value of the integral of a function $f(x)$ depends on the limits of integration a and b: The integral is a function of the two limits a and b. In order to study this dependence on the limits more closely we imagine the lower limit to be a fixed number, say α, denote the variable of integration no longer by x but by u (see p. 126), and denote the upper limit by x instead of by b in order to indicate that we shall consider the upper limit as the variable and that we wish to investigate the value of the integral as a function of this upper limit. Accordingly, we write

$$\phi(x) = \int_\alpha^x f(u)\, du.$$

We call the function $\phi(x)$ an *indefinite integral* of the function $f(x)$. When we speak of *an* and not of *the* indefinite integral, we suggest that instead of the lower limit α any other could be chosen, in which case we should ordinarily obtain a different value for the integral. Geometrically, the indefinite integral $\phi(x)$ is given by the area (shown by shading in Fig. 2.17) under the curve $y = f(u)$ and bounded by the u-axis, the ordinate $u = \alpha$ and the variable ordinate $u = x$, the sign being determined by the rules discussed earlier (p. 126).

Any particular *definite* integral is found from the indefinite integral $\phi(x)$. Indeed, by our basic rules for integrals,

$$\int_a^b f(u)\, du = \int_a^\alpha f(u)\, du + \int_\alpha^b f(u)\, du$$
$$= -\int_\alpha^a f(u)\, du + \int_\alpha^b f(u)\, du = \phi(b) - \phi(a).$$

In particular, we can express any other indefinite integral with a lower limit α' in terms of $\phi(x)$:

$$\int_{\alpha'}^x f(u)\, du = \phi(x) - \phi(\alpha').$$

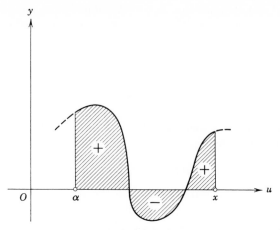

Figure 2.17   The indefinite integral as an area.

As we see, *any indefinite integral differs from the special indefinite integral $\phi(x)$ only by a constant.*

Continuity of the Indefinite Integral

If the function $f(x)$ is continuous in the interval $[a, b]$ and α is a point of that interval, then the indefinite integral

$$\phi(x) = \int_\alpha^x f(u) \, du$$

represents a function of x which is again defined in the same interval. As easily seen: *The indefinite integral $\phi(x)$ of a continuous function $f(x)$ is likewise continuous.* For if x and y are any two values in the interval we have by the mean value theorem that

(15) $$\phi(y) - \phi(x) = \int_x^y f(u) \, du = f(\xi)(y - x)$$

where ξ is some value in the interval with end points x and y. From the continuity of f we have then

$$\lim_{y \to x} \phi(y) = \lim_{y \to x} [\phi(x) + f(\xi)(y - x)] = \phi(x) + f(x) \cdot 0 = \phi(x),$$

which shows that ϕ is continuous. More specifically, in any closed interval we have $|\phi(y) - \phi(x)| \leq M \, |y - x|$, where M is the maximum of $|f|$ in the interval, so that ϕ is even Lipschitz-continuous.

Formula (15) for $\phi(y) - \phi(x)$ shows: that $\phi(x)$ is an increasing function of x in case f is positive throughout the interval, namely, for $y > x$

$$\phi(y) = \phi(x) + f(\xi)(y - x) > \phi(x).$$

Forming the indefinite integral of a function is an important way of *generating new types of functions*. In Section 2.5 we shall apply this method to introduce the logarithm function. This will also give us a first glimpse of the fact that general theorems of mathematical analysis lead to the most remarkable specific formulas.

As we shall see in Section 3.14a (p. 298), the definition of new functions by means of integrals of already defined functions is a satisfactory procedure if we wish to put definitions (for example, of the trigonometric functions) on a purely analytical basis instead of relying on intuitive geometrical explanations.

2.5   Logarithm Defined by an Integral

a. Definition of the Logarithm Function

In Section 2.2 we had succeeded in expressing $\int_a^b x^\alpha \, dx$ for any rational $\alpha \neq -1$ in terms of powers of a and b. For $\alpha = -1$ we were only able to represent the integral as limit of a sequence

$$\int_a^b \frac{1}{u} \, du = \lim_{n \to \infty} n(\sqrt[n]{b/a} - 1).$$

Independently of the discussions of Section 2.2 we now *introduce* the function represented by the indefinite integral

$$\int_1^x \frac{1}{u} \, du,^1$$

or, geometrically, by the area under a hyperbola as indicated in Fig. 2.18. We call it the *logarithm* of x, or more accurately the *natural logarithm* of x, and write

(16) $$\log x = \int_1^x \frac{1}{u} \, du.$$

Since $y = 1/u$ is a continuous and positive function for all $u > 0$, the function $\log x$ is defined for all $x > 0$, is moreover continuous, and also is monotonically increasing. The choice of 1 as the lower limit in

---

[1] In this section we again freely use the fact that the integral of a continuous function (here the function $1/u$) exists; the general proof is given in the Supplement.

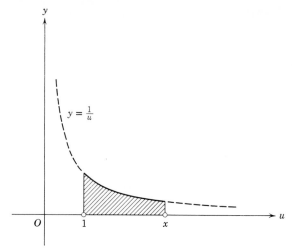

Figure 2.18   Log x represented by an area.

the indefinite integral for log x is a matter of convenience. It implies that

(17) $\log 1 = 0,$

and that log x is positive for $x > 1$ and negative for x between zero and 1 (Fig. 2.19). Any definite integral of $1/u$ between positive limits a and b can be expressed in terms of logarithms by the formula (see p. 143)

(18) $$\int_a^b \frac{1}{u}\, du = \log b - \log a.$$

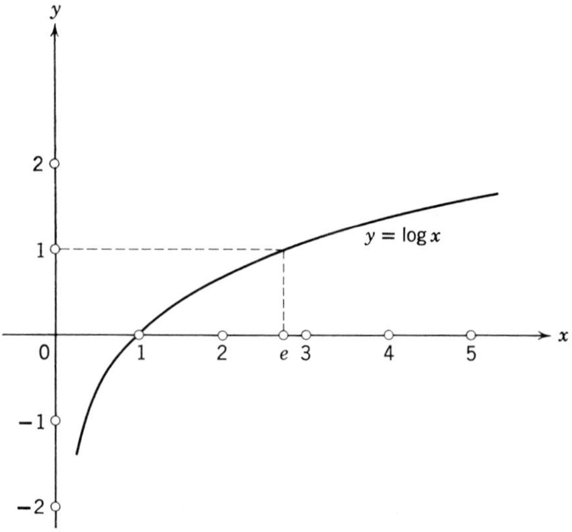

Figure 2.19   The natural logarithm.

ometrically, this integral represents the area under the hyperbola
$= 1/x$ between the ordinates $x = a$ and $x = b$.

. The Addition Theorem for Logarithms

The fundamental property which justifies the traditional name for
x is expressed by the

ADDITION THEOREM.   *For any positive x and y*

$$\log (xy) = \log x + \log y.$$

PROOF.   We write the addition theorem in the form

$$\log (xy) - \log y = \log x$$

$$\int_y^{xy} \frac{1}{v}\, dv = \int_1^x \frac{1}{u}\, du,$$

ere we have deliberately chosen different letters for the variables of
egration in the two integrals. The equality of the two integrals will
low from the fact that the approximating sums have the same value
suitable choices of subdivisions and of intermediate points. Assume
first $x > 1$. Then

$$\int_1^x \frac{1}{u}\, du = \lim_{n \to \infty} \sum_{i=1}^n \frac{1}{\xi_i} \Delta u_i,$$

ere $u_0 = 1, u_1, u_2, \ldots, u_n = x$ represent the points arising in a sub-
ision of the interval $[1, x]$ and ξ_i lies in the ith cell. Putting $v_i = yu_i$,
$= y\xi_i$ we see that the points $v_0, v_1, \ldots, v_n$ correspond to a sub-
ision of the interval $[y, xy]$ with intermediate points $\eta_i = \xi_i y$.
viously,

$$\Delta v_i = y\, \Delta u_i,$$

that

$$\sum_{i=1}^n \frac{1}{\eta_i} \Delta v_i = \sum_{i=1}^n \frac{1}{\xi_i} \Delta u_i.$$

r n tending to infinity we obtain the desired identity between integrals
the case $x > 1$.
For $x = 1$ the addition theorem holds trivially, since $\log 1 = 0$.
prove the theorem also for the case $0 < x < 1$, we observe that then

$1/x > 1$, and hence

$$\log x + \log y = \log x + \log \left(\frac{1}{x} xy\right)$$

$$= \log x + \log \frac{1}{x} + \log (xy)$$

$$= \log \frac{1}{x} + \log x + \log (xy)$$

$$= \log \left(\frac{1}{x} x\right) + \log (xy)$$

$$= \log 1 + \log (xy) = \log (xy).$$

This completes the proof of the addition theorem.

A proof of the addition theorem can also be based on formula (p. 134), according to which

$$\log x = \lim_{n \to \infty} n(\sqrt[n]{x} - 1).$$

Then

$$\log (xy) = \lim_{n \to \infty} n(\sqrt[n]{xy} - 1)$$

$$= \lim_{n \to \infty} [n(\sqrt[n]{x} - 1)\sqrt[n]{y} + n(\sqrt[n]{y} - 1)]$$

$$= [\lim_{n \to \infty} n(\sqrt[n]{x} - 1)](\lim_{n \to \infty} \sqrt[n]{y}) + \lim_{n \to \infty} n(\sqrt[n]{y} - 1)$$

$$= \log x + \log y,$$

since $\lim_{n \to \infty} \sqrt[n]{y} = 1$ (see p. 64).

Applying the addition theorem to the special case $y = 1/x$ leads

$$\log 1 = \log x + \log \frac{1}{x}$$

or

(20)
$$\log \frac{1}{x} = -\log x.$$

More generally then

(21)
$$\log \frac{y}{x} = \log y + \log \frac{1}{x} = \log y - \log x.$$

Repeated application of the addition theorem to a product of factors yields

$$\log (x_1 x_2 \cdots x_n) = \log x_1 + \log x_2 + \cdots + \log x_n.$$

particular, we find that for any positive integer n

$$\log (x^n) = n \log x.$$

is identity also holds for $n = 0$, since $x^0 = 1$, and can be extended negative integers n by observing that

$$og (x^n) = \log \left(\frac{1}{x^{-n}} \right) = -\log (x^{-n}) = -(-n) \log x = n \log x.$$

For any rational $\alpha = m/n$ and any positive a we can form $a^\alpha = {}^{/n} = x$. We have then

$$\log x = \frac{1}{n} \log x^n = \frac{1}{n} \log a^m = \frac{m}{n} \log a = \alpha \log a.$$

us the identity

$$\log (a^\alpha) = \alpha \log a$$

lds for any positive real a and any rational α.

Exponential Function and Powers

a. The Logarithm of the Number e

The constant e obtained on p. 79 as the limit of $(1 + 1/n)^n$ plays a tinguished role for the function $\log x$. Indeed, the number e is aracterized by the equation[1]

$$\log e = 1.$$

r the proof we observe that the continuity of the function $\log x$ plies

$$\log e = \log \left[\lim_{n \to \infty} \left(1 + \frac{1}{n} \right)^n \right] = \lim_{n \to \infty} \log \left[\left(1 + \frac{1}{n} \right)^n \right]$$

$$= \lim_{n \to \infty} n \log \left(1 + \frac{1}{n} \right).$$

w by the mean value theorem of integral calculus

$$\log \left(1 + \frac{1}{n} \right) = \int_1^{1+1/n} \frac{1}{u} \, du = \frac{1}{\xi} \frac{1}{n},$$

---

his means geometrically that the area bounded by the hyperbola $y = 1/x$ and the es $y = 0$, $x = 1$, and $x = e$ has the value one (see Fig. 2.18).

where ξ is some number between 1 and $1 + 1/n$ which depends on choice of n. Obviously, $\lim_{n \to \infty} \xi = 1$ so that

(24) $$\log e = \lim_{n \to \infty} \frac{1}{\xi} = 1.$$

b. The Inverse Function of the Logarithm. The Exponential Funct

From the relation $\log e = 1$ it follows that for any rational α

$$\log (e^{\alpha}) = \alpha \log e = \alpha.$$

This shows that every rational number α occurs as a value of $\log$ for some positive x. Since $\log x$ is continuous, it assumes then any va intermediate between two rational values; this means all real valu It follows that for x varying over all positive values the values $y = \log x$ range over all numbers y. Since $\log x$ is monotonica increasing, there exists for any real y exactly one positive x such t $\log x = y$. The solution x of the equation $y = \log x$ is given by inverse function of the logarithm which we shall denote by $x = E$ We know then that $E(y)$ (Fig. 2.20) is defined and positive for all and again continuous and increasing (see p. 45)

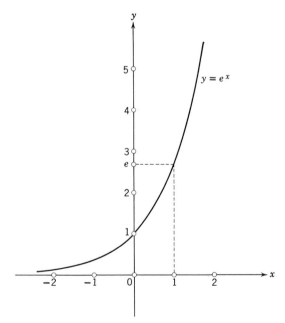

Figure 2.20 The exponential function.

Since the equations $y = \log x$ and $x = E(y)$ stand for the same relation between x and y, we can write the equation $\alpha = \log(e^{\alpha})$, which is valid for rational α, also in the form

$$E(\alpha) = e^{\alpha}.$$

We see: for any rational α the value of $E(\alpha)$ is the αth power of the number e. For rational $\alpha = m/n$ the power e^{α} is defined directly as $\sqrt[n]{e^m}$. For irrational α the expression e^{α} is defined most naturally by representing α as the limit of a sequence of rational numbers α_n and putting $e^{\alpha} = \lim\limits_{n \to \infty}(e^{\alpha_n})$. Since $e^{\alpha_n} = E(\alpha_n)$ and since the function $E(y)$ depends continuously on y, we can be sure that the limit of the e^{α_n} exists and that it has the value $E(\alpha)$ independently of the special sequence used to approximate α. This proves that the equation $E(\alpha) = e^{\alpha}$ holds for irrational α as well. For all real α we can now write e^{α} instead of $E(\alpha)$. We call e^{x} the *exponential function*. This function is defined and continuous for all x, is increasing, and positive everywhere.

Since the equations $y = \log x$ and $x = e^{y}$ are two ways of expressing the same relation between the numbers x and y, we see that $\log x$, the "natural logarithm" of x (as defined here by an integral) stands for the *logarithm to the base e*, as that term would be used in elementary mathematics; that is, $\log x$ is the exponent of that power of e which is equal to x or

(25) $$e^{\log x} = x.$$

We can write[1] $\log x = \log_e x$.

Similarly, $x = e^{y}$ is that number whose logarithm is y, or

(26) $$\log e^{y} = y.$$

From the point of view of calculus it is really easier to introduce natural logarithms first as integrals of the simple function $y = 1/x$, as we did here, and to define powers of e by taking the inverse of the logarithm function. In this way the continuity and monotonicity of the functions $\log x$ and e^{x} arise just as consequences of general theorems and require no special arguments.

---

[1] The reader may feel that the name "natural logarithm" should have been reserved rather for logarithms to the base 10. However, historically the first table of logarithms published by Napier in 1614 essentially gave logarithms to the base e. Logarithms to the base 10 were introduced only subsequently by Briggs because of their obvious computational advantages.

c. The Exponential Function as Limit of Powers

Originally we obtained the number e as the limit

$$e = \lim_{n \to \infty} \left(1 + \frac{1}{n}\right)^n.$$

A more general formula represents e^x for any x as a limit

(27) $$e^x = \lim_{n \to \infty} \left(1 + \frac{x}{n}\right)^n.$$

For the proof it is sufficient to show that the sequence

$$s_n = \log\left(1 + \frac{x}{n}\right)^n$$

has the limit x. For then the sequence of values

$$e^{s_n} = \left(1 + \frac{x}{n}\right)^n$$

must tend to e^x since the exponential function is continuous. Now

$$s_n = n \log\left(1 + \frac{x}{n}\right) = n \int_1^{1 + x/n} \frac{1}{\xi}\, d\xi.$$

By the mean value theorem of integral calculus we have

$$s_n = n \frac{1}{\xi_n}\left[\left(1 + \frac{x}{n}\right) - 1\right] = \frac{x}{\xi_n},$$

where ξ_n is some value between one and $1 + x/n$. Since obviously ξ_n tends to one for n tending to ∞, we have indeed $\lim_{n \to \infty} s_n = x$.

d. Definition of Arbitrary Powers of Positive Numbers

Arbitrary powers of any positive numbers can now be expressed in terms of the exponential and logarithmic functions.[1]

We found for rational α and any positive x that the relation

$$\log(x^\alpha) = \alpha \log x$$

holds. We write this equation in the form

$$x^\alpha = e^{\alpha \log x}.$$

---

[1] This obviates the more clumsy "elementary" definition and justification of these processes by passage to the limit from rational exponents indicated on p. 86.

For irrational α we again represent α as limit of a sequence of rational numbers α_n and define

$$x^\alpha = \lim_{n \to \infty} x^{\alpha_n} = \lim_{n \to \infty} e^{\alpha_n \log x}.$$

The continuity of the exponential function implies again that the limit exists and that it has the value $e^{\alpha \log x}$, since

$$e^{\alpha \log x} = e^{\lim (\alpha_n \log x)} = \lim e^{\alpha_n \log x}.$$

Hence the equation

$$(28) \qquad\qquad x^\alpha = e^{\alpha \log x}$$

holds quite generally for any α and any positive x. Putting $\log x = \beta$ or, what is the same, $x = e^\beta$ we infer

$$(29) \qquad\qquad (e^\beta)^\alpha = e^{\alpha\beta},$$

and more generally then for any positive x

$$(x^\alpha)^\beta = (e^{\alpha \log x})^\beta = e^{\alpha\beta \log x} = x^{\alpha\beta}.$$

Another rule for working with powers which is easily established in complete generality, is the *multiplication law*

$$x^\alpha x^\beta = x^{\alpha+\beta},$$

where x is a positive number and α and β are arbitrary. It is sufficient to prove the corresponding formula obtained by taking the logarithms of both sides:

$$\log (x^\alpha x^\beta) = \log (x^{\alpha+\beta}).$$

Now by the rules (19), (26), and (28) already established it follows that

$$\begin{aligned}
\log (x^\alpha x^\beta) &= \log x^\alpha + \log x^\beta = \log (e^{\alpha \log x}) + \log (e^{\beta \log x}) \\
&= \alpha \log x + \beta \log x = (\alpha + \beta) \log x \\
&= \log (e^{(\alpha+\beta) \log x}) = \log (x^{\alpha+\beta}).
\end{aligned}$$

e. Logarithms to Any Base

It is easy to express logarithms to a base other than e in terms of natural logarithms. If for a positive number a the equation $x = a^y$ is satisfied, we write

$$y = \log_a x.$$

Now $a^y = e^{y \log a}$, so that $x = e^{y \log a}$ or $y \log a = \log x$. It follows that

$$(30) \qquad\qquad \log_a x = \frac{\log x}{\log a},$$

where $\log x$ is the natural logarithm to the base e. In particular, the common logarithms to the base 10 are given by

$$\log_{10} x = \frac{\log x}{\log 10}.$$

Since logarithms to any base a are proportional to natural logarithms, they satisfy the same addition theorem:

$$\log_a x + \log_a y = \log_a (xy).$$

2.7   The Integral of an Arbitrary Power of x

In Section 2.2 we obtained the formula

$$\int_a^b u^\alpha \, du = \frac{b^{\alpha+1} - a^{\alpha+1}}{\alpha + 1},$$

for any rational $\alpha \neq -1$. (The case $\alpha = -1$ was seen to lead to the logarithm.) To evaluate the integral when α is an irrational number, it is sufficient to discuss the indefinite integral

$$\phi(x) = \int_1^x u^\alpha \, du$$

from which all definite integrals with positive limits a and b can be obtained. Assume $x > 1$ (the case $x < 1$ can be handled in the same fashion after interchanging the limits). We have then by (28)

$$u^\alpha = e^{\alpha \log u},$$

where $\log u \geq 0$ for u in the interval of integration. Let β and γ be any two rational numbers different from -1 for which

$$\beta \leq \alpha \leq \gamma.$$

Then also

$$\beta \log u \leq \alpha \log u \leq \gamma \log u.$$

Since the exponential function is increasing, this implies

$$e^{\beta \log u} \leq e^{\alpha \log u} \leq e^{\gamma \log u};$$

that is,

$$u^\beta \leq u^\alpha \leq u^\gamma.$$

We have then

$$\int_1^x u^\beta \, du \leq \phi(x) \leq \int_1^x u^\gamma \, du.$$

The integrals of u^β and u^γ were evaluated before, leading to

$$\frac{1}{\beta + 1}(x^{\beta+1} - 1) \le \phi(x) \le \frac{1}{\gamma + 1}(x^{\gamma+1} - 1).$$

If we now let the rational numbers β and γ converge to α, we obtain in the limit

$$\phi(x) = \frac{1}{\alpha + 1}(x^{\alpha+1} - 1),$$

since $x^{\beta+1} = e^{(\beta+1)\log x}$ and $x^{\gamma+1} = e^{(\gamma+1)\log x}$ tend to $e^{(\alpha+1)\log x} = x^{\alpha+1}$ because of the continuity of the exponential function. The same result follows for x between zero and one. Thus generally for positive a, b

$$\int_a^b u^\alpha \, du = \phi(b) - \phi(a) = \frac{1}{\alpha + 1}(b^{\alpha+1} - a^{\alpha+1})$$

just as for rational α.

When α is a positive integer, the formula remains valid even when the limits a or b become zero or negative; it is easy to extend the formula directly to those cases.

2.8 The Derivative

The concept of the derivative, like that of the integral, has an immediate intuitive origin and is easy to grasp. Yet it opens the door to an enormous wealth of mathematical facts and insights; the student will only gradually become aware of the variety of significant applications and of the power of the techniques which we shall develop in this book.

The concept of derivative is first suggested by the intuitive notion of the *tangent to a smooth curve* $y = f(x)$ at a point P with the coordinates x and y. This tangent is characterized by the angle α between its direction and the positive x-axis. But how does one obtain this angle from the analytical description of the function $f(x)$? The knowledge of the values of x and y at the point P does not suffice to determine the angle α since there are infinitely many different lines besides the tangent passing through P. On the other hand, to determine α one does not need to know the function $f(x)$ in its total over-all behavior; the knowledge of the function in an arbitrary neighborhood of the point P must be sufficient to determine the direction α, no matter how tiny a neighborhood is chosen. This indicates that we should define the direction of the tangent to a curve $y = f(x)$ by a limiting process, as we shall presently do.

The problem of calculating the direction of tangents, or of "differentiation," was impressed on mathematicians as early as the sixteenth century by optimization problems, that is, questions of maxima and minima arising in geometry, mechanics and optics. (See the discussion in Section 3.6.)

Another problem of paramount importance which leads to differentiation is that of giving a precise mathematical meaning to the intuitive notion of *velocity* in an arbitrary nonuniform motion (see p. 162).

We shall start with the problem of describing the tangent to a curve analytically by a limit process.

a. The Derivative and the Tangent

Geometric Definition.   In conformity with naive intuition, we define the tangent to the given curve $y = f(x)$ at one of its points P by means

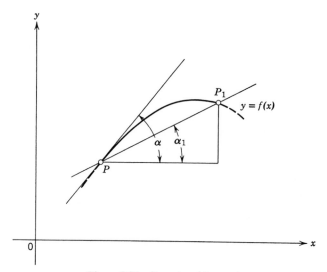

Figure 2.21   Secant and tangent.

of the following geometrical limiting process (Fig. 2.21). We consider a second point P_1 near P on the curve. Through the two points P, P_1 we draw a straight line, a secant of the curve. If now the point P_1 moves along the curve towards the point P, then the secant is expected to approach a limiting position which is independent of the side from which P_1 tends to P. This *limiting position of the secant is the tangent*; the statement that such a limiting position of the secant exists is equivalent to the assumption that the curve has a definite tangent or a definite direction at the point *P*. (We have used the word "assumption" because we have actually made one. The hypothesis that the tangent exists at every point is by no means true for all curves representing simple functions. For example, any curve with a corner or vertex at a point *P* does not have a uniquely determined direction there, such as the curve defined by $y = |x|$ at $(0, 0)$. (See the discussion on p. 166.)

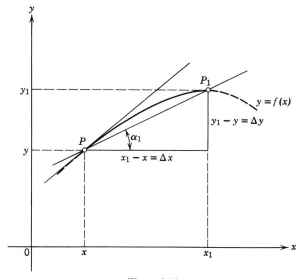

Figure 2.22

Since our curve is represented by means of a function $y = f(x)$, we must formulate the geometric limiting process analytically, with reference to $f(x)$. This analytical limit process is called *differentiation* of $f(x)$.

Consider the angle which a straight line makes with the *x*-axis as the one through which the positive *x*-axis must be turned in the positive direction or counterclockwise[1] in order to become for the first time parallel to the line. (This would be an angle α in the interval $0 \leq \alpha < \pi$.) Let α_1 be the angle which the secant PP_1 forms with the positive *x*-axis (cf. Fig. 2.22) and α the angle which the tangent forms with the positive *x*-axis. Then

$$\lim_{P_1 \to P} \alpha_1 = \alpha,$$

---

[1] That is, in such a direction that a rotation of $\pi/2$ brings it into coincidence with the positive *y*-axis.

where the meaning of the symbols is obvious. Let x, y and x_1, y_1 be the coordinates of the points P and P_1 respectively. Then we immediately have[1]

$$\tan \alpha_1 = \frac{y_1 - y}{x_1 - x} = \frac{f(x_1) - f(x)}{x_1 - x};$$

thus our limiting process (disregarding the case $\alpha = \pi/2$ of a perpendicular tangent) is represented by the equation

$$\lim_{x_1 \to x} \frac{f(x_1) - f(x)}{x_1 - x} = \lim_{x_1 \to x} \tan \alpha_1 = \tan \alpha.$$

Notation.   The expression

$$\frac{f(x_1) - f(x)}{x_1 - x} = \frac{y_1 - y}{x_1 - x} = \frac{\Delta y}{\Delta x}$$

we call the *difference quotient* of the function $y = f(x)$ where the symbols Δy and Δx denote the differences of the function $y = f(x)$ and of the independent variable x. (Here, as on p. 124, the symbol Δ is an abbreviation for *difference*, and is *not* a factor.) The trigonometric tangent of α, the "slope" of the curve,[2] is therefore equal to the limit to which the difference quotient of our function tends when x_1 tends to x.

We call this limit of the difference quotient the *derivative*[3] of the function $y = f(x)$ at the point x. We shall generally use either the notation of Lagrange, $y' = f'(x)$, to denote the derivative, or, as Leibnitz did, the symbol[4] dy/dx or $df(x)/dx$ or $(d/dx)f(x)$. On p. 171 we shall discuss the meaning of Leibnitz's notation in more detail; here we point out: The notation $f'(x)$ indicates the fact that the *derivative is itself a function of x* since a value of $f'(x)$ corresponds to each value of x in the interval considered. This fact is sometimes emphasized by the use of the terms *derived function, derived curve*. The definition of the derivative appears in several different forms:

$$f'(x) = \lim_{x_1 \to x} \frac{f(x_1) - f(x)}{x_1 - x} = \lim_{h \to 0} \frac{f(x + h) - f(x)}{h},$$

---

[1] In order that this equation may have a meaning, we must assume that both x and x_1 lie in the domain of f. In what follows, corresponding assumptions will often be made tacitly in the steps leading up to limiting processes.

[2] The word *gradient* or *direction coefficient* is used occasionally.

[3] The term *differential coefficient* is also used in older textbooks.

[4] Cauchy's notation $Df(x)$ and Newton's notation $\dot{y}$ are also used.

where in the second expression x_1 is replaced by $x + h$, or in Leibnitz's notation,

$$\frac{dy}{dx} = \frac{df(x)}{dx} = f'(x) = \lim_{x_1 \to x} \frac{f(x_1) - f(x)}{x_1 - x} = \lim_{\Delta x \to 0} \frac{\Delta y}{\Delta x}.$$

If f is defined in a neighborhood of the point x, then the quotient $[f(x + h) - f(x)]/h$ is defined as a function of h for all values $h \neq 0$ for which $|h|$ is sufficiently small to ensure that $x + h$ is in the interval under consideration. The definition of $f'(x)$ as a limit requires that $\left| \dfrac{f(x + h) - f(x)}{h} - f'(x) \right|$ is arbitrarily small for all $h \neq 0$ (positive *or* negative) for which $|h|$ is sufficiently small.

Analytic Calculation of Derivatives.   The intuitive concept and the general analytic notion of derivative are simple and straightforward. Less obvious is the procedure of actually carrying out such limiting processes.

It is impossible to find the derivative merely by putting $x_1 = x$ in the expression for the difference quotient, for then the numerator and denominator would both be equal to zero and we would be led to the meaningless expression $0/0$. Thus the passage to the limit in each case depends on certain preliminary steps (transformation of the difference quotient).

For example, for the function $f(x) = x^2$ we have

$$\frac{f(x_1) - f(x)}{x_1 - x} = \frac{x_1{}^2 - x^2}{x_1 - x} = x_1 + x \quad \text{whenever } x \neq x_1.$$

This function $x_1 + x$ does not have exactly the same domain as $(x_1{}^2 - x^2)/(x_1 - x)$: The function $x_1 + x$ is defined at the one point $x_1 = x$, where the quotient $(x_1{}^2 - x^2)/(x_1 - x)$ is undefined. For all other values of x_1 the two functions are equal to one another; hence in the passage to the limit, for which we specifically require that $x_1 \neq x$, we obtain the same value for $\lim_{x_1 \to x} (x_1{}^2 - x^2)/(x_1 - x)$ as for $\lim_{x_1 \to x} (x_1 + x)$. However, since the function $x_1 + x$ is defined and continuous at the point $x_1 = x$, we can do with it what we could not do with the quotient, namely, pass to the limit by simply putting $x_1 = x$. For the derivative we then obtain

$$f'(x) = \frac{d(x^2)}{dx} = 2x.$$

As another example we differentiate, that is, calculate the derivative of the function $y = \sqrt{x}$ for $x > 0$. We have for $x_1 \neq x$

$$\frac{f(x_1) - f(x)}{x_1 - x} = \frac{\sqrt{x_1} - \sqrt{x}}{x_1 - x} = \frac{(\sqrt{x_1} - \sqrt{x})(\sqrt{x_1} + \sqrt{x})}{(x_1 - x)(\sqrt{x_1} + \sqrt{x})}$$

$$= \frac{x_1 - x}{(x_1 - x)(\sqrt{x_1} + \sqrt{x})} = \frac{1}{\sqrt{x_1} + \sqrt{x}}.$$

Hence (for $x > 0$)

$$\frac{d\sqrt{x}}{dx} = \lim_{x_1 \to x} \frac{1}{\sqrt{x_1} + \sqrt{x}} = \frac{1}{2\sqrt{x}}.$$

For $x = 0$ we have a singularity: The *derivative is infinite*, since $(\sqrt{x_1} - 0)/(x_1 - 0) = 1/\sqrt{x_1} \to \infty$ for $x_1 \to 0$.

Analytic Definition

It is extremely significant that the process of differentiating a function has a definite analytic meaning quite apart from the geometric intuitive conception of the tangent. The analytic definition of the integral, freed from the geometric visualization of area, allowed us to base the notion of area on that of integral. In a similar spirit, independently of the geometrical representation of a function $y = f(x)$ by means of a curve, we *define* the derivative of the function $y = f(x)$ as the new function $y' = f'(x)$ given by the limit of the difference quotient $\Delta y/\Delta x$ provided that the limit exists.

Here the differences $\Delta y = y_1 - y = f(x_1) - f(x)$ and $\Delta x = x_1 - x$ are "corresponding changes" in the variables y and x. The ratio $\Delta y/\Delta x$ can be called the "average rate of change" of y with respect to x in the interval $(x, x + \Delta x)$. The limit $f'(x) = dy/dx$ represents then the "instantaneous rate of change" or simply the "rate of change" of y with respect to x.

If this limit exists, we say that the function $f(x)$ is *differentiable*. We shall always assume that every function dealt with is differentiable unless specific mention is made to the contrary.[1] We emphasize that if the function $f(x)$ is to be differentiable at the point x the limit as $h \to 0$ of the quotient $[f(x + h) - f(x)]/h$ must exist, where h can have any value $\neq 0$ for which $x + h$ belongs to the domain of f. If, in particular, f is defined in a whole interval containing the point x in its interior, then the limit must exist *independently of the manner in*

---

[1] Examples in which this assumption is not satisfied will be given later (see p. 167). Such examples justify mentioning differentiability as an assumption if the context warrants it.

which h tends to zero, whether it be through positive values or through negative values, without restriction upon sign.

Having now an analytic definition for the derivative $f'(x)$, we take the direction angle α to the positive x-axis given by the equation $\tan \alpha = f'(x)$ as the direction of the tangent to the curve at the point (x, y).[1] By thus basing the geometric definition on the analytic one we avoid the difficulties which might arise from the vagueness of the geometric visualization. In fact, we have now *defined* precisely what we mean by a tangent to the graph of $y = f(x)$ at a point (x, y), and we have an analytic criterion for deciding whether or not a curve has a tangent at a given point (x, y).

Monotone Functions

Nevertheless, the visual interpretation of the derivative as the slope of the tangent to the curve is a highly useful aid to understanding, even in purely analytic discussions. A case in question is the following statement based on geometric intuition:

The function $f(x)$ is monotonically increasing when $f'(x) > 0$ and monotonically decreasing when $f'(x) < 0$.

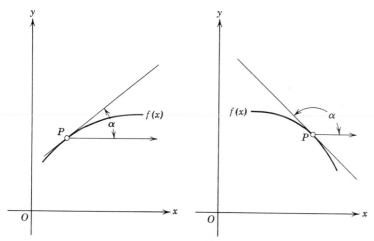

Figure 2.23   Tangents to graphs of increasing and decreasing functions.

Indeed, if $f'(x)$ is positive and the curve is traversed in the direction of increasing x, then the tangent slants upwards, that is, toward increasing y (α is an "acute angle"); therefore at the point in question the curve

---

[1] The angle α is not determined quite uniquely but can be replaced by $\alpha \pm \pi$, $\alpha \pm 2\pi$, etc., unless we specify as above that $0 \leq \alpha < \pi$.

rises as x increases; if, on the other hand, $f'(x)$ is negative, the tangent slants downwards (α is an "obtuse angle") and the curve falls as x increases (see Fig. 2.23). Analytically this will be proved on p. 177.

b. The Derivative as a Velocity

The need to replace the intuitive concept of velocity or speed by a precise definition leads once again to exactly the same limiting process we have already called differentiation.

Consider the example of a point moving on a straight line, the directed y-axis, the position of the point being determined by a single coordinate y. This coordinate y is the distance, with its proper sign, of our moving point from a fixed initial point on the line. The motion is given if we know y as a function of the time t: $y = f(t)$. If this function is a linear function $f(t) = ct + b$, we speak of a *uniform motion with the velocity c*, and for every pair of distinct values t and t_1 we can obtain the velocity by dividing the distance traversed in a time interval by the length of that time interval:

$$c = \frac{f(t_1) - f(t)}{t_1 - t}.$$

The velocity is therefore the difference quotient of the function $ct + b$, and this difference quotient is independent of the particular pair of instants which we fix upon. But what are we to understand by the velocity of motion at an instant t if the motion is no longer uniform?

To answer this question we consider the difference quotient

$$[f(t_1) - f(t)]/(t_1 - t),$$

which we shall call the *average velocity* in the time interval between t_1 and t. Now if this average velocity tends to a definite limit when we let t_1 tend to t, we shall define this limit as the velocity at the time t. In other words: *the velocity*, that is, the instantaneous rate of change of distance with respect to time *at the time t, is the derivative*

$$f'(t) = \lim_{t_1 \to t} \frac{f(t_1) - f(t)}{t_1 - t}.$$

Newton emphasized the interpretation of derivatives[1] as velocity, and wrote $\dot{y}$ or $f(x)$ instead of $f'(t)$, a notation which we shall occasionally use. Again, the differentiability of the function is a necessary assumption if the notion of velocity is to have a meaning.

---

[1] Called by him "fluxions."

A simple example is the motion of freely falling bodies. We start from the experimentally established law that the distance traversed in time t by a freely falling body starting from rest at $t = 0$ is proportional to t^2; it is therefore represented by a function of the form

$$y = f(t) = at^2$$

with constant a. As on p. 159, the velocity then is given by the expression $f'(t) = 2at$; thus: *the velocity of a freely falling body increases in proportion to the time.*

c. Examples of Differentiation

We now illustrate the technique of differentiation by a number of typical examples.

Linear Functions

For the function $y = f(x) = c$ with constant c we see for all x, $f(x + h) - f(x) = c - c = 0$, so that $\lim_{h \to 0} [f(x + h) - f(x)]/h = 0$; that is, *the derivative of a constant function is zero.*

For a linear function $y = f(x) = cx + b$, we find

$$f'(x) = \lim_{h \to 0} \frac{f(x + h) - f(x)}{h} = \lim_{h \to 0} \frac{ch}{h} = c.$$

The derivative of a linear function is constant.

Powers of x

Next, we differentiate the power function

$$y = f(x) = x^\alpha,$$

at first assuming that α is a positive integer. Provided $x_1 \neq x$, we have

$$\frac{f(x_1) - f(x)}{x_1 - x} = \frac{x_1^\alpha - x^\alpha}{x_1 - x} = x_1^{\alpha-1} + x_1^{\alpha-2}x + \cdots + x^{\alpha-1},$$

where we divide directly or use the formula for the sum of a geometric progression. This simple algebraic manipulation is the key to the passage to the limit; for the last expression on the right-hand side of the equation is a continuous function of x_1, in particular for $x_1 = x$, and so we can carry out the passage to the limit $x_1 \to x$ for this expression simply by replacing x_1 everywhere by x. Each term then takes the value $x^{\alpha-1}$, and since the number of terms is exactly α, we obtain

$$y' = f'(x) = \frac{d(x^\alpha)}{dx} = \alpha x^{\alpha-1}.$$

We arrive at the same result if α is a negative integer $-\beta$; we must, however, assume that x is not zero. We then find

$$\frac{f(x_1) - f(x)}{x_1 - x} = \frac{\dfrac{1}{x_1{}^\beta} - \dfrac{1}{x^\beta}}{x_1 - x} = -\frac{x^\beta - x_1{}^\beta}{x - x_1} \cdot \frac{1}{x^\beta x_1{}^\beta}$$

$$= -\frac{x^{\beta-1} + x^{\beta-2}x_1 + \cdots + x_1{}^{\beta-1}}{x_1{}^\beta x^\beta}.$$

Once again we can carry out the passage to the limit simply by substituting x for x_1. Then just as before we obtain for the limit

$$y' = -\beta\,\frac{x^{\beta-1}}{x^{2\beta}} = -\beta x^{-\beta-1}.$$

Hence for *negative* integral values $\alpha = -\beta$ the derivative is again given by the formula

$$y' = \alpha x^{\alpha-1}.$$

Finally, we shall prove the same formula where x is positive and α any rational number. We suppose that $\alpha = p/q$, where p and q are both integers and, moreover, positive. (If one of them were negative, no essential changes in the proof would be needed; for $\alpha = 0$ the result is already known, since x^α is then constant.) We now have

$$\frac{f(x_1) - f(x)}{x_1 - x} = \frac{x_1{}^{p/q} - x^{p/q}}{x_1 - x}.$$

If we now put $x^{1/q} = \xi$ and $x_1{}^{1/q} = \xi_1$, we obtain

$$\frac{f(x_1) - f(x)}{x_1 - x} = \frac{\xi_1{}^p - \xi^p}{\xi_1{}^q - \xi^q} = \frac{\xi_1{}^{p-1} + \xi_1{}^{p-2}\xi + \cdots + \xi^{p-1}}{\xi_1{}^{q-1} + \xi_1{}^{q-2}\xi + \cdots + \xi^{q-1}}.$$

After this last transformation we can immediately perform the passage to the limit $x_1 \to x$ (or what amounts to the same thing, $\xi_1 \to \xi$), and thus obtain for the limiting value the expression

$$y = \frac{p}{q}\,\frac{\xi^{p-1}}{\xi^{q-1}} = \frac{p}{q}\,\xi^{p-q} = \frac{p}{q}\,x^{(p-q)/q} = \frac{p}{q}\,x^{(p/q)-1},$$

or finally,

$$f'(x) = y' = \alpha x^{\alpha-1},$$

which is formally the same result as before. We leave it for the reader to prove for himself that the same differentiation formula holds also for negative rational exponents.

We shall come back (p. 186) to the differentiation of powers and prove the general validity of the preceding formula for arbitrary exponents α.

Trigonometric Functions

As a last example we consider the *differentiation of the trigonometric functions* sin x and cos x. We use the elementary trigonometric addition formula to transform the difference quotient

$$\frac{\sin(x+h) - \sin x}{h} = \frac{\sin x \cos h + \cos x \sin h - \sin x}{h}$$

$$= \sin x \frac{\cos h - 1}{h} + \cos x \frac{\sin h}{h}.$$

Recalling the relations of Section 1.8, pp. 84–85,

$$\lim_{h \to 0} \frac{\sin h}{h} = 1, \qquad \lim_{h \to 0} \frac{\cos h - 1}{h} = 0,$$

we immediately obtain

$$y' = \frac{d(\sin x)}{dx} = \cos x.$$

The function $y = \cos x$ can be differentiated in exactly the same way. Starting with

$$\frac{\cos(x+h) - \cos x}{h} = \cos x \frac{\cos h - 1}{h} - \sin x \frac{\sin h}{h}$$

and taking the limit as $h \to 0$, we obtain the derivative[1]

$$y' = \frac{d(\cos x)}{dx} = -\sin x.$$

d. Some Fundamental Rules for Differentiation

Just as in the case of the integral, there exist certain basic rules for differentiation that follow immediately from the definition and suffice for forming the derivative for many functions.

1. If $\phi(x) = f(x) + g(x)$, then $\phi'(x) = f'(x) + g'(x)$.
2. If $\psi(x) = cf(x)$ (where c is a constant), then $\psi'(x) = cf'(x)$.

---

[1] If x is interpreted as an angle, then these simple formulas for the derivatives of sin x and cos x presupposes, of course, that the angle x is measured in radians.

We have

$$\frac{\phi(x + h) - \phi(x)}{h} = \frac{f(x + h) - f(x)}{h} + \frac{g(x + h) - g(x)}{h}$$

and

$$\frac{\psi(x + h) - \psi(x)}{h} = c\frac{f(x + h) - f(x)}{h},$$

and our statements follow directly by passage to the limit.

Thus, for example, the derivative of the function $\phi(x) = f(x) + ax + b$ (where a and b are constants) is given by the equation

$$\phi'(x) = f'(x) + a.$$

With the help of these rules and of the formula for the derivative of a power we can immediately differentiate any polynomial $y = a_0x^n + a_1x^{n-1} + \cdots + a_n$ and find

$$y' = na_0x^{n-1} + (n - 1)a_1x^{n-2} + \cdots + 2a_{n-2}x + a_{n-1}.$$

e. Differentiability and Continuity of Functions

It is useful to know that differentiability is a stronger condition than continuity:

If a function is differentiable it is automatically continuous.

For if the difference quotient $[f(x + h) - f(x)]/h$ approaches a definite limit as h tends to zero, the numerator of the fraction, that is, $f(x + h) - f(x)$ must[1] tend to zero with h; this just expresses the continuity of the function $f(x)$ at the point x. Hence, separate cumbersome continuity proofs are unnecessary for functions that can be shown to be differentiable (that is, for most functions we shall encounter).

Discontinuities of the Derivative-Corners

The converse, however, is false; it is *not* true that every continuous function has a derivative at every point. The simplest counter-example is the function $f(x) = |x|$, that is, $f(x) = -x$ for $x \leq 0$ and $f(x) = x$ for $x \geq 0$; its graph is shown in Fig. 2.24. At the point $x = 0$ this function is continuous, but has no derivative. The limit of $[f(x + h) - f(x)]/h$ is equal to 1 if h tends to zero through positive

---

[1] Since then

$$\lim_{h \to 0} [f(x + h) - f(x)] = \left[\lim_{h \to 0}\frac{f(x + h) - f(x)}{h}\right](\lim_{h \to 0} h) = f'(x) \cdot 0 = 0.$$

values, and is equal to -1 if h tends to zero through negative values; if we do not restrict the sign of h, no limit exists. We say that our function has different *forward* and *backward derivatives* at the point $x = 0$, where by forward derivative and backward derivative we mean respectively the limiting values of $[f(x + h) - f(x)]/h$ as h approaches zero through positive values only and negative values only. The *differentiability* of a function defined in an interval about the point

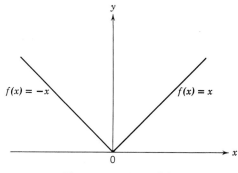

Figure 2.24 $f(x) = |x|$.

considered thus requires not merely that the forward and backward derivatives exist, but that they are equal. Geometrically the inequality of the two derivatives means that the curve has a *corner*. Differentiability expresses in a precise way what intuitively would be called *smoothness* of the graph of the function.

Infinite Discontinuities

As further examples of points where a continuous function is not differentiable we consider the points where the derivative becomes infinite, that is, the points at which there exists neither a forward nor a backward derivative, the difference quotient $[f(x + h) - f(x)]/h$ increasing beyond all bounds as $h \to 0$. For example, the function $y = f(x) = \sqrt[3]{x} = x^{1/3}$ is defined and continuous for all values of x. For all nonzero values of x its derivative is given (p. 164) by the formula $y' = \frac{1}{3}x^{-2/3}$. At the point $x = 0$ we have $[f(x + h) - f(x)]/h = h^{1/3}/h = h^{-2/3}$, and we see at once that as $h \to 0$ the expression has no limiting value, but, on the contrary, tends to infinity. This state of affairs is often briefly described by saying that the function possesses an infinite derivative, or the derivative infinity, at the point in question; as we should remember, however, this merely means that as h tends to zero the difference quotient increases beyond all bounds, and that the

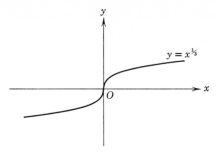

Figure 2.25

derivative in the sense in which we have defined it really does not exist. The geometrical meaning of an *infinite derivative* is that the tangent to the curve is vertical (cf. Fig. 2.25).

The function $y = f(x) = \sqrt{x}$, which is defined and continuous for $x \geq 0$, is also not differentiable at the point $x = 0$. Since y is not defined for negative values of x, we here consider the right-hand derivative only. The equation $[f(h) - f(0)]/h = 1/\sqrt{h}$ shows that this derivative is infinite; the curve touches the y-axis at the origin (Fig. 2.26).

Finally, in the function $y = \sqrt[3]{x^2} = x^{\frac{2}{3}}$ we have a case in which the right-hand derivative at the point $x = 0$ is positive and infinite, whereas the left-hand derivative is negative and infinite, as follows from the relation

$$\frac{f(h) - f(0)}{h} = \frac{1}{\sqrt[3]{h}}.$$

As a matter of fact, the continuous curve $y = x^{\frac{2}{3}}$, the so-called *semicubical parabola* or *Neil's parabola*, has at the origin a *cusp* with a tangent perpendicular to the x-axis (cf. Fig. 2.27).

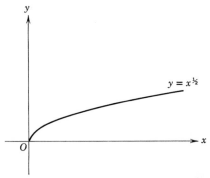

Figure 2.26

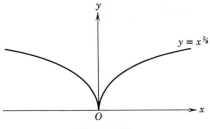

Figure 2.27

f. Higher Derivatives and Their Significance

The graph of the derivative $f'(x)$ of a function is called the *derived curve* of the graph of $f(x)$. For example, the derived curve of the parabola $y = x^2$ is a straight line, represented by the function $y = 2x$. The derived curve of the sine curve $y = \sin x$ is the cosine curve $y = \cos x$; similarly, the derived curve of the curve $y = \cos x$ is the curve $y = -\sin x$. (These latter curves can be obtained from each other by translation in the direction of the x-axis, as is shown in Fig. 2.28.)

It is quite natural to form the derived curves of the derived curves, that is, to form the derivative of the function $f'(x) = \phi(x)$. This derivative

$$\phi'(x) = \lim_{h \to 0} \frac{f'(x + h) - f'(x)}{h},$$

provided that it exists, is called the *second derivative* of the function $f(x)$; we shall denote it by $f''(x)$.

Similarly, we may attempt to form the derivative of $f''(x)$, the so-called *third derivative* of $f(x)$, which we then denote by $f'''(x)$. For most functions that concern us there is nothing to hinder us from repeating the process of differentiation as many times as we like, thus defining an nth *derivative* $f^{(n)}(x)$.[1] Occasionally, it will be convenient to call the function $f(x)$ its own 0th derivative.

If the independent variable is interpreted as the time t and the motion of a point is represented as previously by the function $f(t)$, the *physical* meaning of the second derivative is the rate of change of the velocity $f'(t)$ with respect to time, or, as it is usually called, the *acceleration*. In the example of the freely falling body the distance traveled in the time t was given by the function $y = f(t) = at^2$. We found $f'(t) = 2at$ for the velocity at the time t. The acceleration has then the constant

---

[1] The terms *second, third, . . . , nth differential coefficient* are also used, or $D^2 f, \ldots,$ $D^n f$ (cf. footnote 3, p. 158).

value $f''(t) = 2a$ (which is usually identified with the gravitational constant g). Later (p. 236), we shall discuss the geometrical interpretation of the second derivative in detail. Here, however, we take note of the following facts: At a point where $f''(x)$ is positive, $f'(x)$ increases

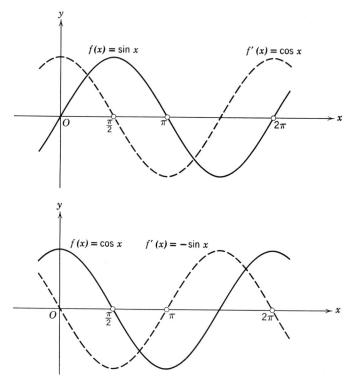

Figure 2.28   Derived curves of $\sin x$ and $\cos x$.

as x increases; if here $f'(x)$ is positive, the curve becomes steeper for increasing x. If, on the other hand, $f''(x)$ is negative, $f'(x)$ decreases as x increases, and if $f'(x)$ is positive, the curve becomes less steep as x increases.

Finally, we observe that the higher derivatives may be used to define a function. Thus one can characterize the trigonometric functions by a so-called *differential equation* involving the function and its second derivative. From the formulas $(d \cos x)/dx = -\sin x$, $(d \sin x)/dx = \cos x$ we obtain immediately by differentiating again,

$$\frac{d^2}{dx^2} \cos x = -\cos x, \qquad \frac{d^2}{dx^2} \sin x = -\sin x.$$

Hence if the symbol u stands for either of the functions $\sin x$ or $\cos x$, we have the relation (*differential equation*)

$$u'' = -u.$$

This differential equation is also clearly satisfied by any linear combination $u = a \cos x + b \sin x$ with constant coefficients a, b. We shall see on p. 312 that such linear combinations, with arbitrary constants a and b, are the only functions u for which $u'' = -u$.

In all types of applications involving oscillations or wave phenomena, such as motions of springs or waves on the surface of water, we are led directly from physical considerations to a differential equation of the type $u'' = -u$ for the physically significant variable u (usually the independent variable is *time*). It is therefore important to recognize that u can be represented simply in terms of trigonometric functions (see Chapter 9).

g. Derivative and Difference Quotient. Leibnitz's Notation

In Leibnitz's notation the passage to the limit in the process of differentiation is symbolically expressed by replacing the symbol Δ by the symbol d, motivating Leibnitz's symbol for the derivative *defined* by the equation

$$\frac{dy}{dx} = \lim_{\Delta x \to 0} \frac{\Delta y}{\Delta x}.$$

If we wish to obtain a clear grasp of the meaning of the differential calculus, we must beware of the old fallacy of imagining the derivative as the quotient of two "quantities" dy and dx which are actually "infinitely small." The *difference quotient* $\Delta y / \Delta x$ has a meaning only for differences Δx which are not equal to zero. *After* forming this genuine difference quotient we must perform the passage to the limit by means of a transformation or some other device which also in the limit avoids division by zero. It does not make sense to suppose that *first* Δx and Δy go through something like a limiting process and reach values which are infinitesimally small but still not zero, so that Δx and Δy are replaced by "infinitely small quantities" or "infinitesimals" dx and dy, and that the quotient of these quantities is then formed. Such a conception of the derivative is incompatible with mathematical clarity; in fact, it is entirely meaningless. For many people it undoubtedly has a certain charm of mystery, always associated with the word "infinite"; in the early days of the differential calculus even Leibnitz himself was capable of combining these vague mystical ideas with a thoroughly clear handling of the limiting process. But today the mysticism of infinitely small quantities has no place in the calculus.

The notation of Leibnitz, however, is not merely suggestive in itself, but it is actually extremely flexible and useful. The reason is that in many calculations and formal transformations we can deal with the symbols dy and dx exactly *as if* they were ordinary numbers. By treating dx and dy like numbers we can give neater expression to many calculations which can admittedly be carried out without their use. In the following chapters we shall see this fact verified over and over again and shall find ourselves justified in making free and repeated use of it, provided we do not lose sight of the symbolical character of the signs dy and dx.

*For the second and higher derivatives too, Leibnitz devised a suggestive notation. He considered the second derivative as the limit of the "second difference quotient" in the following manner: In addition to the variable x we consider $x_1 = x + h$ and $x_2 = x + 2h$. We then take the second difference quotient, meaning the first difference quotient of the first difference quotient, that is, the expression

$$\frac{1}{h}\left(\frac{y_2 - y_1}{h} - \frac{y_1 - y}{h}\right) = \frac{1}{h^2}(y_2 - 2y_1 + y),$$

where $y = f(x)$, $y_1 = f(x_1)$, and $y_2 = f(x_2)$. Writing $h = \Delta x$, $y_2 - y_1 = \Delta y_1$, and $y_1 - y = \Delta y$, we may appropriately call the expression in the last parentheses the difference of the difference of y or the *second difference* of y and write symbolically[1]

$$y_2 - 2y_1 + y = \Delta y_1 - \Delta y = \Delta(\Delta y) = \Delta^2 y.$$

In this symbolic notation the second difference quotient is then written $\Delta^2 y/(\Delta x)^2$, where the denominator is really the square of Δx, whereas in the numerator the superscript 2 symbolically denotes the repetition of the difference process. The second derivative is then expressed by

$$f''(x) = \lim_{x_1 \to x} \frac{\Delta^2 f}{(\Delta x)^2}.$$

This symbolism for the difference quotient[2] led Leibnitz to introduce

---

[1] Here $\Delta\Delta = \Delta^2$ is merely a symbol for "difference of difference" or "second difference."

[2] As we must emphasize, the statement that the second derivative may be represented as the limit of the second difference quotient requires proof. We previously defined the second derivative, not in this way, but as the limit of the first difference quotient of the first derivative. The two definitions are equivalent, provided the second derivative is continuous; the proof, however, will be given only later (see Chapter 5, Appendix II since we have no particular need of the result.

the notation

$$y'' = f''(x) = \frac{d^2y}{dx^2}, \qquad y''' = f'''(x) = \frac{d^3y}{dx^3}, \quad \text{etc.},$$

for the second and higher derivatives, and we shall find that this notation also stands the test of usefulness.[1]

h. The Mean Value Theorem of Differential Calculus

The difference quotient involves the values of a function for distinct values of x, whereas the derivative at a point tells us nothing about the function at any other point; the difference quotient reflects properties of the function "in-the-large," while the derivative reflects a local property or a property "in-the-small." We shall often need to derive over-all or "global" properties of a function from the local properties given by its derivative. For this purpose we utilize a fundamental relation between the difference quotient and derivative known as "the mean value theorem of differential calculus."

The mean value theorem is easily appreciated intuitively. We consider the difference quotient

$$\frac{f(x_1) - f(x_2)}{x_1 - x_2} = \frac{\Delta f}{\Delta x}$$

of a function $f(x)$, and assume that the derivative exists everywhere in the closed interval $x_1 \leq x \leq x_2$, so that the graph of the curve has a tangent everywhere. The difference quotient is the tangent of the angle α of inclination of the secant, shown in Fig. 2.29. Imagine this secant shifted parallel to itself. At least once it will reach a position in which it is a tangent to the curve at a point between x_1 and x_2, certainly at that point $x = \xi$ of the curve which is at the greatest distance from the secant say at $x = \xi$. Hence there exists an intermediate value ξ in the interval such that

$$\frac{f(x_1) - f(x_2)}{x_1 - x_2} = f'(\xi).$$

This statement is called *the mean value theorem of the differential calculus*.[2] We can also express it somewhat differently by noticing that

---

[1] This is the customary notation. Writing $y'' = d^2y/(dx)^2$, $y''' = d^3y/(dx)^3$ with parentheses, would be somewhat clearer, but is not done ordinarily.

[2] A more appropriate name would be the *intermediate value theorem* of differential calculus.

the number ξ may be written in the form

$$\xi = x_1 + \Theta(x_2 - x_1),$$

where all we know about Θ is that it lies between 0 and 1. Although Θ (or ξ) generally cannot be specified more exactly, the theorem is extremely powerful in application.

Consider, for example, the case where x is the time and $y = f(x)$ the distance of a car from its starting point along a certain road. Then

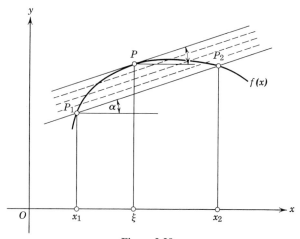

Figure 2.29

$f'(x)$ is the velocity of the car at the time x. If, say, during the first two hours ($\Delta x = 2$) the driver has covered a distance $\Delta f = 120$ miles, we can conclude from the mean value theorem that at least at one moment ξ during those two hours the driver had a speed of exactly 60 miles per hour (provided the velocity exists at every moment). The driver cannot claim, for instance, to have traveled all the time at less than 50 miles per hour. On the other hand, there is nothing to indicate what the time ξ was at which the precise speed of 60 miles per hour was attained; it might have been at some time during the first hour or during the second hour or on several occasions.

A precise statement of the mean value theorem is the following:

If $f(x)$ is continuous in the closed interval $x_1 \leq x \leq x_2$ and differentiable at every point of the open interval $x_1 < x < x_2$, then there exists at least one value θ, where $0 < \theta < 1$, such that

$$\frac{f(x_2) - f(x_1)}{x_2 - x_1} = f'[x_1 + \theta(x_2 - x_1)].$$

If we replace x_1 by x and x_2 by $x + h$, we can express the mean value theorem by the formula

$$\frac{f(x + h) - f(x)}{h} = f'(\xi) = f'(x + \theta h), \qquad x < \xi < x + h.$$

Although it is essential that $f(x)$ should be continuous for all points of the interval, including the end points, we need not assume that the derivative exists at the end points.

If at any point in the interior of the interval the derivative fails to exist, the mean value theorem is not necessarily true. It is easy to see this from the example of $f(x) = |x|$.

i. Proof of the Theorem

The mean value theorem is usually derived by reduction to a special case which we establish first.

ROLLE'S THEOREM. *If a function $\phi(x)$ is continuous in the closed interval $x_1 \leq x \leq x_2$ and differentiable in the open interval $x_1 < x < x_2$, and if in addition $\phi(x_1) = 0$ and $\phi(x_2) = 0$, then there exists at least one point ξ in the interior of the interval at which $\phi'(\xi) = 0$.*

Interpreted geometrically, this means that if a curve reaches the x-axis at two points, then it must have a horizontal tangent at some intermediate point (Fig. 2.30).

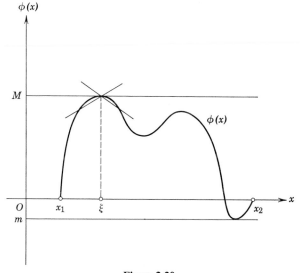

Figure 2.30

Indeed, since $\phi(x)$ is continuous in the closed interval $[x_1, x_2]$ there exists a greatest value M of $\phi(x)$ and a smallest value m in that interval (see p. 101). Since ϕ vanishes in the end points, we must have $m \leq 0 \leq M$. If these greatest and least values should be equal, then necessarily $m = M = 0$ and $\phi(x) = 0$ at all points of the interval; then also $\phi'(x) = 0$ in the interval, and hence $\phi'(\xi) = 0$ for every ξ in the interval. Thus we only have to consider the case where m and M are not both zero. If, in particular, M is not zero, then M must be positive. There exists a point ξ of the interval $[x_1, x_2]$ where $\phi(\xi) = M$. Since ϕ vanishes in the end points of the interval, the point ξ must be an interior point. Furthermore, $\phi(x) \leq \phi(\xi) = M$ for all x in $[x_1, x_2]$. Consequently, for every number h whose absolute value $|h|$ is small enough, the inequality $\phi(\xi + h) - \phi(\xi) \leq 0$ holds. This implies that the quotient

$$\frac{\phi(\xi + h) - \phi(\xi)}{h}$$

is negative or zero for $h > 0$ and positive or zero for $h < 0$. If we let h tend to zero through positive values, we find that $\phi'(\xi) \leq 0$, whereas for h tending to zero through negative values it follows that $\phi'(\xi) \geq 0$. Hence $\phi'(\xi) = 0$ and we have proved Rolle's theorem in the case $M \neq 0$. The same argument holds for $m \neq 0$.

To prove the mean value theorem we apply Rolle's theorem to a function which represents the vertical distance between the point $(x, f(x))$ of the graph and its secant:

$$\phi(x) = f(x) - f(x_1) - \frac{f(x_2) - f(x_1)}{x_2 - x_1}(x - x_1).$$

This function[1] obviously satisfies the condition $\phi(x_1) = \phi(x_2) = 0$, and is of the form $\phi(x) = f(x) + ax + b$ with constant coefficients $a = -[f(x_2) - f(x_1)]/(x_2 - x_1)$ and b. From p. 166 we know that

$$\phi'(x) = f'(x) + a,$$

and thus by Rolle's theorem

$$0 = \phi'(\xi) = f'(\xi) + a$$

---

[1] This function also is proportional to the distance of the point $(x, f(x))$ of the curve from the secant; the reader can easily verify this for himself, for example, by using the fact from elementary analytical geometry that the expression $(y - mx - b)/\sqrt{1 + m^2}$ represents the (signed) distance of the point (x, y) from the line with the equation $y - mx - b = 0$. In this way we find that indeed at the points of the curve having greatest distance from the secant the tangent is parallel to the secant.

for a suitably chosen intermediate value ξ; hence

$$f'(\xi) = -a = \frac{f(x_2) - f(x_1)}{x_2 - x_1} ;$$

thus the mean value theorem is proved.

Significance of the Theorem

The derivative of a function had been defined as the limit of difference quotients for an interval as the end points approach each other. The mean value theorem establishes a connection between difference quotients and derivatives of a differentiable function which does not involve the shrinking of the interval. Each difference quotient is equal to the derivative at a suitable intermediate point ξ.

Examples. Just as in the mean value theorem of integral calculus there is nothing specific asserted in the intermediate value theorem about the location of ξ beyond the fact that ξ lies in the interior of the interval. For the example of the quadratic function $y = f(x) = x^2$ with derivative $f'(x) = 2x$ we find

$$\frac{f(x_2) - f(x_1)}{x_2 - x_1} = x_1 + x_2 = f'(\xi),$$

where $\xi = (x_1 + x_2)/2$ is the midpoint of the interval $[x_1, x_2]$. In general, however, ξ might lie anywhere else between x_1 and x_2. For example, if $f(x) = x^3$, we have $[f(1) - f(0)]/(1 - 0) = 1 = f'(\xi) = 3\xi^2$, where $\xi = 1/\sqrt{3}$.

Monotonic Functions. As one of many applications of the mean value theorem of differential calculus we prove that if the derivative of $f(x)$ has a constant sign, then f is monotonic. Specifically, we assume $f(x)$ to be continuous in the closed interval $[a, b]$ and differentiable at each point of the open interval (a, b). If then $f'(x) > 0$ *for x in (a, b), then the function $f(x)$ is monotonic increasing; similarly, if $f'(x) < 0$, the function is monotonic decreasing.* The proof is obvious: Let x_1 and x_2 be any two values in the closed interval $[a, b]$. Then there exists a ξ between x_1 and x_2, and hence also between a and b, such that

$$f(x_2) - f(x_1) = f'(\xi)(x_2 - x_1).$$

If $f'(x) > 0$ everywhere in (a, b) we have in particular $f'(\xi) > 0$. Hence $f(x_2) - f(x_1)$ is positive for $x_2 > x_1$; that is, $f(x)$ is increasing. Similarly, f is decreasing if $f'(x) < 0$ in (a, b).

In the same way we show that *a function $f(x)$ continuous in $[a, b]$*

and differentiable in the open interval (a, b) *must be a constant if* $f'(x) = 0$ *everywhere in* (a, b). For then

$$f(x_2) - f(x_1) = f'(\xi)(x_2 - x_1) = 0.$$

This important statement corresponds to the intuitively obvious fact that a curve whose tangent at every point is parallel to the x-axis must be a straight line which is parallel to the x-axis.

Lipschitz-Continuity of Differentiable Functions.   It was mentioned earlier that a function $f(x)$ having a derivative is necessarily continuous. The mean value theorem of differential calculus furnishes much more precise quantitative information, namely, a *modulus of continuity*. We consider a function $f(x)$ which is defined in the closed interval $[a, b]$ and has a derivative $f'(x)$ at each point of that interval. Assume that $f'(x)$ is *bounded* in the interval (this is certainly the case provided $f'(x)$ is defined and continuous in the closed interval $[a, b]$); there exists then a number M such that $|f'(x)| \leq M$. For any two values x_1, x_2 in (a, b) we infer from the mean value theorem

$$|f(x_2) - f(x_1)| = |f'(\xi)(x_2 - x_1)| \leq M |x_2 - x_1|.$$

For given $\epsilon > 0$ we have thus produced a simple modulus of continuity $\delta = \epsilon/M$ such that

$$|f(x_2) - f(x_1)| \leq \epsilon \qquad \text{for} \quad |x_2 - x_1| \leq \delta.$$

Take, for example, the function $f(x) = x^2$ in the interval $-a \leq x \leq +a$. Since

$$|f'(x)| = |2x| \leq 2a$$

we see that here

$$|f(x_2) - f(x_1)| \leq \epsilon \qquad \text{for} \quad |x_2 - x_1| \leq \epsilon/2a.$$

We said that a function $f(x)$ "satisfies a Lipschitz-condition" or is "Lipschitz-continuous" if there is a constant M such that

$$|f(x_2) - f(x_1)| \leq M |x_2 - x_1|$$

for all x_1, x_2 in question. This means that all *difference quotients*

$$\frac{f(x_2) - f(x_1)}{x_2 - x_1}$$

have the same upper bound M for their absolute value. We see that every function f with continuous derivative f' on a closed interval is Lipschitz-continuous. However, even functions that do not have a derivative at every point can be Lipschitz-continuous, as the example

$f(x) = |x|$ shows. The reader can verify for himself that for this function always $|f(x_2) - f(x_1)| \leq |x_2 - x_1|$.

On the other hand, not every continuous function is Lipschitz-continuous. This is shown by the example of $f(x) = x^{\frac{1}{3}}$; here

$$\frac{f(x) - f(0)}{x - 0} = x^{-\frac{2}{3}}$$

is not bounded for small x; hence $f(x)$ is not Lipschitz-continuous at $x = 0$. This is consistent with the fact that the derivative $f'(x) = 1/3x^{\frac{2}{3}}$ does not remain bounded as x tends to zero. The functions which are Lipschitz-continuous form an important class intermediate between those that are merely continuous and those that have a continuous derivative.

j. The Approximation of Functions by Linear Functions. Definition of Differentials

Definition.   The derivative of a function $y = f(x)$ was defined by

$$f'(x) = \lim_{h \to 0} \frac{f(x + h) - f(x)}{h} = \lim_{\Delta x \to 0} \frac{\Delta y}{\Delta x},$$

where $\Delta x = h$. If for a fixed x and a variable h, we define a quantity ϵ by

$$\epsilon(h) = \frac{f(x + h) - f(x)}{h} - f'(x) = \frac{\Delta y}{\Delta x} - f'(x),$$

then the fact that $f'(x)$ is the derivative of f at the point x amounts to the equation

$$\lim_{h \to 0} \epsilon(h) = 0.$$

The quantity $\Delta y = f(x + h) - f(x)$ represents the change or *increment* in the value of the dependent variable y that results when the value x of the independent variable is changed by the amount $\Delta x = h$. Since

$$\Delta y = f'(x) \, \Delta x + \epsilon \, \Delta x,$$

the quantity Δy appears as the sum of two parts, namely, a part $f'(x) \, \Delta x$ which is proportional to Δx and a part $\epsilon \, \Delta x$ which can be made as small as we please compared to Δx by making Δx itself small enough. The dominant, linear part in the expression for Δy we shall call the *differential dy* of y and write for it

$$dy = df(x) = f'(x) \, \Delta x.$$

For any differentiable function f and for a fixed x this differential is a well-defined linear function of $h = \Delta x$. For example, for the function $y = x^2$ we have $dy = d(x^2) = 2x \, \Delta x = 2xh$. For the particular function $y = x$ whose derivative has the constant value one, we simply have $dx = \Delta x$. It is then consistent with our definition to write dx for Δx when x is the independent variable; hence the differential of any function $y = f(x)$ can also be written as

$$dy = df(x) = f'(x) \, dx.$$

The increment of the dependent variable

$$\Delta y = f'(x) \, dx + \epsilon \, dx = dy + \epsilon \, dx$$

differs from the differential dy by the amount $\epsilon \, dx$, which in general is not zero. In the example of the function $y = x^2$ we have $dy = 2x \, dx$, whereas

$$\Delta y = (x + dx)^2 - x^2 = 2x \, dx + (dx)^2 = dy + \epsilon \, dx,$$

where $\epsilon = dx$.

Earlier we used the symbol dy/dx purely symbolically to denote the limit of the quotient $\Delta y/\Delta x$ for Δx tending to zero. With our present definition of the differentials dy and dx the derivative dy/dx can actually be considered as the ordinary quotient of dy and dx. Here, however, dy and dx are now not in any sense "infinitely small" quantities or "infinitesimals;" such an interpretation would be devoid of meaning. Instead dy and dx are well-defined linear functions of $h = \Delta x$ which for large Δx may have large numerical values. There is nothing remarkable in the fact that the quotient dy/dx of those quantities has the same value as the derivative $f'(x)$. This is merely a tautology restating the definition of dy as $f'(x) \, dx$.[1]

Rewriting the relation between increment and differential of f in the form

$$f(x + h) = f(x) + hf'(x) + \epsilon h,$$

we see that the expression $f(x + h)$ considered as a function of h is represented by the linear function $f(x) + hf'(x)$ wth an error ϵh which is arbitrarily small compared to h if h is sufficiently small. This approximate representation of $f(x + h)$ by the linear function $f(x) + hf'(x)$ means geometrically that we replace the curve by its tangent at the point x (see Fig. 2.31).

---

[1] Similarly, higher-order differentials could be defined by $d^2y = f''(x)h^2 = f''(x)(dx)^2$, $d^3y = f'''(x)(dx)^3$, etc., in agreement with Leibnitz' notation for the higher derivatives.

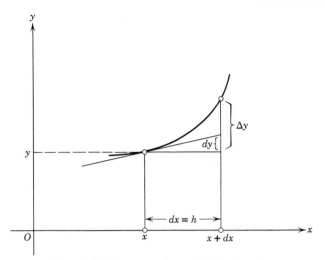

Figure 2.31   Increment Δy and differential dy.

Linear Approximation

A more precise estimate of the magnitude of the "*error*," that is, of the deviation of the function $f(x)$ from the linear function representing the tangent, is given by the mean value theorem of differential calculus. We have for a suitable ξ between x and $x + h$

$$f(x + h) - f(x) = hf'(\xi),$$

so that

$$\epsilon = \frac{f(x + h) - f(x)}{h} - f'(x) = f'(\xi) - f'(x).$$

If, as usually in applications, the function $f'(x)$ itself has a derivative $f''(x)$, we find by applying the mean value theorem a second time that

$$f'(\xi) - f'(x) = (\xi - x)f''(\eta),$$

where η is a value intermediate between x and ξ and hence also between x and $x + h$. It follows that

$$|\epsilon| = |(\xi - x)f''(\eta)| = |\xi - x|\,|f''(\eta)| \leq hM,$$

where M is any upper bound for the absolute value of the second derivative of f in the interval $[x, x + h]$. Then $|\epsilon h|$, which measures the deviation of $f(x + h)$ from the linear function $f(x) + hf'(x)$, is at most Mh^2. For sufficiently small h the expression Mh^2 is, of course, much smaller than $f'(x)h$, unless $f'(x)$ happens to have the value zero. This *approximation of a function in a small interval by a linear function* is of greatest significance both for practical applications and for advanced mathematical analysis. We shall return to this topic in later chapters, and incidentally derive then the better estimate $|\epsilon h| \leq \frac{1}{2}Mh^2$.

Interpolation

*When a function $f(x)$ is described numerically by a table of values, f is ordinarily determined by *linear interpolation* for arguments x intermediate between those for which f is listed. This procedure also corresponds to replacing the function f by a linear function in an interval. In this case the graph of the linear function is given by a

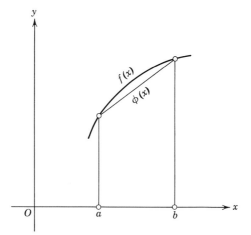

Figure 2.32   Linear interpolation.

secant rather than by a *tangent* to the curve representing f. If, say, the values of f are known at two points a and b, we replace $f(x)$ for intermediate x by the expression

$$\phi(x) = f(a) + (x - a)\frac{f(b) - f(a)}{b - a},$$

which is linear in x and gives the correct values of f at the end points $x = a$ and $x = b$ of the interval (see Fig. 2.32). Here again by use of the mean value theorem we can estimate the error in this approximation. We have

$$f(x) - \phi(x) = (x - a)\left[\frac{f(x) - f(a)}{x - a} - \frac{f(b) - f(a)}{b - a}\right]$$

$$= (x - a)[f'(\xi_1) - f'(\xi_2)].$$

Since ξ_1 lies between a and x, it also lies between a and b, as does ξ_2. A second application of the mean value theorem of differential calculus then yields

$$f(x) - \phi(x) = (x - a)f''(\eta)(\xi_1 - \xi_2),$$

where η is between ξ_1 and ξ_2 and hence also between a and b. Consequently, denoting by M an upper bound for $|f''|$ in the interval $[a, b]$, we find that

$$|f(x) - \phi(x)| \leq |x - a| \, |\xi_1 - \xi_2| \, |f''(\eta)| \leq M(b - a)^2.$$

Once again the deviation of f from its linear approximation can be estimated by the square of the length of the interval.

As a numerical example we take from a table of trigonometric functions in radian measure the values

$$\sin 0.75 = 0.6816, \qquad \sin 0.76 = 0.6889,$$

where the errors do not exceed 0.00005. If we want to deduce the value of the sine function for the intermediate argument 0.754, we find by linear interpolation that

$$\sin 0.754 \approx 0.6816 + \tfrac{4}{10}(0.6889 - 0.6816) \approx 0.6845.$$

For the function $f(x) = \sin x$ the first derivative is $f'(x) = \cos x$, the second derivative $f''(x) = -\sin x$. Obviously, $|f''(x)| \leq 1$, so that the error in the value found for $\sin 0.754$ as a result of the linear interpolation procedure does not exceed $1 \times (0.01)^2 = 0.0001$. To this error estimate we must add possible errors due to round-off in the tabulated values and in the interpolation.

We can compare this value obtained by linear interpolation with the value we would obtain by replacing the sine curve by its tangent at the point $x = 0.75$. Taking $f'(0.75) = \cos 0.75 = 0.7317$ from the table, we find

$$\sin 0.754 \approx f(0.75) + f'(0.75)(0.004) = \sin 0.75 + 0.004 \cos 0.75$$

$$\approx 0.6845.$$

Incidentally, the true value of $\sin 0.754$ correct to six significant digits is 0.684560.

k. Remarks on Applications to the Natural Sciences

In applying mathematics to natural phenomena we never deal with *precisely known* quantities. Whether a length is *exactly* a meter is a question which cannot be decided by any experiment and which consequently has no physical meaning. Moreover, there is no immediate physical meaning in saying that the length of a material rod is rational or irrational; we can always measure it with any desired degree of accuracy by rational numbers, and the only meaningful question is whether we can manage to perform such a measurement using rational numbers with relatively small denominators. Just as the question of rationality or irrationality in the rigorous sense of "exact mathematics" has no physical meaning, carrying out limiting processes in applications is usually not more than a mathematical idealization.

The practical—and overwhelming—significance of such idealizations lies in the fact that through the idealizations analytical expressions become essentially simpler and more manageable. For example, it is vastly simpler and more convenient to work with the notion of instantaneous velocity, which is a function of only *one* definite instant of time, than with the notion of average velocity between two different instants. Without such idealization every scientific investigation of nature would be condemned to hopeless complications and would bog down at the outset.

We do not intend to enter into a philosophical discussion of the relationship of mathematics to reality. For the sake of better understanding of the theory, it should be emphasized that in applications we have the right to replace a derivative by a difference quotient and vice versa, provided only that the differences are small enough to guarantee a sufficiently close approximation. The physicist, the biologist, the engineer, or anyone else who has to deal with these ideas in practice will therefore have the right to identify the difference quotient with the derivative within his limits of accuracy. The smaller the increment $h = dx$ of the independent variable, the more accurately can he represent the increment $\Delta y = f(x + h) - f(x)$ by the differential $dy = hf'(x)$. As long as he keeps knowingly within the limits of accuracy required by the problem, he might even be permitted to speak of the quantities $dx = h$ and $dy = hf'(x)$ as "infinitesimals." These "physically infinitesimal" quantities have a precise meaning. They are variables with values which are finite, unequal to zero, and chosen small enough for the given investigation, for example, smaller than a fractional part of a wavelength or smaller than the distance between two electrons in an atom; in general, smaller than the degree of accuracy required.

2.9   The Integral, the Primitive Function, and the Fundamental Theorems of the Calculus

a. The Derivative of the Integral

As already stated, the connection between integration and differentiation is the cornerstone of the differential and integral calculus.

We recall from Section 2.4 that an indefinite integral of a continuous function $f(x)$ is defined as a function $\phi(x)$ of the upper end point of integration by the formula

$$\phi(x) = \int_\alpha^x f(u)\, du,$$

where α was any point in the domain of f. We shall now prove

FUNDAMENTAL THEOREM OF CALCULUS (Part One). *The indefinite integral $\phi(x)$ of a continuous function $f(x)$ always possesses a derivative $\phi'(x)$, and moreover*

$$\phi'(x) = f(x).$$

That is, differentiation of the indefinite integral of a continuous function always reproduces the integrand

$$\frac{d}{dx} \int_\alpha^x f(u)\, du = f(x).$$

This inverse character of the operations of differentiation and integration is the basic fact of calculus. The proof is an immediate consequence of the mean value theorem of integral calculus. According to that theorem we have for any values x and $x + h$ of the domain of f

$$\phi(x + h) - \phi(x) = \int_x^{x+h} f(u)\, du = hf(\xi),$$

where ξ is some value in the interval with end points x and $x + h$. For h tending to zero the value ξ must tend to x so that

$$\lim_{h \to 0} \frac{\phi(x + h) - \phi(x)}{h} = \lim_{h \to 0} f(\xi) = f(x),$$

since f is continuous. Hence $\phi'(x) = f(x)$ as stated by the theorem.

Applications. (*a*) We can use the theorem to find derivatives for some of the functions introduced earlier. The natural logarithm was defined for $x > 0$ by the indefinite integral

$$\log x = \int_1^x \frac{1}{u}\, du.$$

It follows immediately that

$$\frac{d \log x}{dx} = \frac{1}{x}.$$

(*b*) More general logarithms to an arbitrary base *a* were expressible in the form

$$\log_a x = \frac{\log x}{\log a}.$$

Applying the rule for the derivative of the product of a constant and of a function we find that

$$\frac{d}{dx} \log_a x = \frac{1}{x \log a}.$$

(*c*) We found that

$$\frac{d}{dx} x^\alpha = \alpha x^{\alpha-1}$$

in the case where the exponent α is an integer or more generally a rational number. We can now extend this formula to arbitrary α. For that purpose we recall the integration formula

$$\int_a^b u^\beta \, du = \frac{1}{\beta + 1} (b^{\beta+1} - a^{\beta+1}),$$

which we had proved for any positive numbers a, b, and any $\beta \neq -1$. If we replace here the upper limit b by the variable x and differentiate both sides with respect to x, it follows that for $x > 0$

$$x^\beta = \frac{d}{dx} \frac{1}{\beta + 1} (x^{\beta+1} - a^{\beta+1}).$$

Using the rules for the derivative of a sum and of a constant times a function, we can write this result in the form

$$x^\beta = \frac{1}{\beta + 1} \frac{d}{dx} x^{\beta+1}.$$

Substituting α for $\beta + 1$, we obtain the formula

$$\frac{d}{dx} x^\alpha = \alpha x^{\alpha-1}$$

for any $\beta \neq -1$, that is, for $\alpha \neq 0$. However, the formula also holds trivially for $\alpha = 0$ since then $x^\alpha = 1$ and the derivative of a constant is zero.

b. The Primitive Function and Its Relation to the Integral

Inverting Differentiation

The fundamental theorem shows that the indefinite integral $\phi(x)$, that is, the integral with a variable upper limit x, of a function $f(x)$, is a solution of the following problem: *Given f(x), determine a function F(x) such that*

$$F'(x) = f(x).$$

This problem requires us to *reverse the process of differentiation.* It is typical of the inverse problems that occur in many parts of mathematics and that we have already found to be a fruitful mathematical method for generating new concepts. (For example, the first extension of the idea of natural numbers is suggested by the desire to invert certain elementary processes of arithmetic. Again new kinds of functions were obtained from the inverses of known functions.)

Any function $F(x)$ such that $F'(x) = f(x)$ is called a *primitive function of f(x)* or simply a *primitive of f(x)*; this terminology suggests that the function $f(x)$ is derived from $F(x)$.

This problem of the inversion of differentiation or of the finding of a primitive function at first sight is of quite different character from the problem of integration. The first part of the fundamental theorem asserts, however:

Every indefinite integral $\phi(x)$ of the function f(x) is a primitive of f(x).

Yet this result does not completely solve the problem of finding the primitive functions. For we do not yet know if we have found *all* the solutions of the problem. The question about the set of all primitive functions is answered by the following theorem, sometimes referred to as the *second part of the fundamental theorem* of the differential and integral calculus

The difference of two primitive functions $F_1(x)$ and $F_2(x)$ of the same function f(x) is always a constant

$$F_1(x) - F_2(x) = c.$$

Thus from any one primitive function F(x) we can obtain all the others in the form

$$F(x) + c$$

by suitable choice of the constant c. Conversely, for every value of the constant c the expression $F_1(x) = F(x) + c$ represents a primitive function of f(x).

It is clear that for any value of the constant c the function $F(x) + c$ is a primitive function, provided that $F(x)$ itself is. For we have (cf. p. 166)

$$\frac{d}{dx}[F(x) + c] = \frac{d}{dx}F(x) + \frac{d}{dx}c = F'(x) = f(x).$$

Thus to complete the proof of our theorem it remains only to show that the difference of two primitive functions $F_1(x)$ and $F_2(x)$ is always constant. For this purpose we consider the difference

$$F_1(x) - F_2(x) = G(x).$$

Clearly,

$$G'(x) = F_1'(x) - F_2'(x) = f(x) - f(x) = 0.$$

However we had proved on p. 178 from the mean value theorem of differential calculus that a function whose derivative vanishes everywhere in an interval is a constant. Hence $G(x)$ is a constant c, and the theorem follows.

Combining the two parts just proved we can now formulate the

FUNDAMENTAL THEOREM OF CALCULUS. *Every primitive function* $F(x)$ *of a given function* $f(x)$ *continuous on an interval can be represented in the form*

$$F(x) = c + \phi(x) = c + \int_a^x f(u) \, du,$$

where c and a are constants, and conversely, for any constant values of a and c chosen arbitrarily[1] *this expression always represents a primitive function.*

Notations

It may be surmised that the constant c can as a rule be omitted because by changing the lower limit a we change the primitive function by an additive constant; that is, that all primitive functions are indefinite integrals. Frequently, however, we cannot obtain all the primitive functions if we omit the c, as the example $f(x) = 0$ shows. For this function the indefinite integral will always be zero, independently of the lower limit; yet any arbitrary constant is a primitive function of $f(x) = 0$. A second example is the function $f(x) = \sqrt{x}$, which is defined for nonnegative values of x only. The indefinite integral is

$$\phi(x) = \tfrac{2}{3}x^{3/2} - \tfrac{2}{3}a^{3/2},$$

and we see that no matter how we choose the lower limit a the indefinite integral $\phi(x)$ is always obtained from $\tfrac{2}{3}(x)^{3/2}$ by addition of a constant $-\tfrac{2}{3}a^{3/2}$ which is less than or equal to zero; yet such a function as $\tfrac{2}{3}x^{3/2} + 1$ is also a primitive function for $\sqrt{x}$. Thus in the general expression for the primitive function we cannot dispense with the arbitrary additive constant.

---

[1] As long as a lies in the domain of f.

The relationship which we have found suggests extending the notion of the indefinite integral so as to include all primitive functions. We shall henceforth call every expression of the form $c + \phi(x) = c + \int_a^x f(u)\,du$ an indefinite integral of $f(x)$, *and we shall no longer distinguish between the primitive function and the indefinite integral.* Nevertheless, if the reader is to have a proper understanding of the interrelations of these concepts, it is absolutely necessary to bear in mind that in the first instance integration and inversion of differentiation are two different things, and that it is only the knowledge of the relationship between them that gives us the right to apply the term "indefinite integral" to the primitive function also.

It is quite customary to use a notation which is not perfectly clear without comment: we write

$$F(x) = \int f(x)\,dx,$$

when we mean that the function $F(x)$ is of the form

$$F(x) = c + \int_a^x f(u)\,du$$

for suitable constants c and a, that is, we omit the upper limit x, the lower limit a and the additive constant c and use the letter x for the variable of integration. Strictly speaking, of course, there is a slight inconsistency in using the same letter for the variable of integration and the upper limit x which is the independent variable in $F(x)$. In using the notation $\int f(x)\,dx$ we must never lose sight of the indeterminacy connected with it, that is, the fact that the symbol always denotes *one* of the primitive functions of f only. The formula $F(x) = \int f(x)\,dx$ is just a symbolic way of writing the relation

$$\frac{d}{dx}F(x) = f(x).$$

c. The Use of the Primitive Function for Evaluation of Definite Integrals

Suppose that we know any one primitive function $F(x)$ for the function $f(x)$ and that we wish to evaluate the definite integral $\int_a^b f(u)\,du$. We know that the indefinite integral

$$\phi(x) = \int_a^x f(u)\,du,$$

being also a primitive of $f(x)$, can only differ from $F(x)$ by an additive constant. Therefore

$$\phi(x) = F(x) + c,$$

and the additive constant c is determined at once because the indefinite integral $\phi(x) = \int_a^x f(u)\,du$ must take the value zero when $x = a$. We thus obtain $0 = \phi(a) = F(a) + c$, from which $c = -F(a)$ and $\phi(x) = F(x) - F(a)$. In particular, for the value $x = b$ we have the basic formula

$$\int_a^b f(u)\,du = F(b) - F(a),$$

if

$$F'(u) = f(u).$$

Therefore,

If $F(x)$ is any primitive function of the continuous function $f(x)$ whatsoever, the definite integral of $f(x)$ between the limits a and b is equal to the difference $F(b) - F(a)$.

If we use the relation $F'(x) = f(x)$, this consequence of the fundamental theorem may be written in the form

$$(31) \qquad F(b) - F(a) = \int_a^b F'(x)\,dx = \int_a^b \frac{dF(x)}{dx}\,dx = \int_a^b dF(x),$$

where now $F(x)$ can be any function with a continuous derivative $F'(x)$, and where we use the suggestive symbolic notation $dF(x) = F'(x)\,dx$ of Leibnitz.

In applying our rule we often use a vertical bar to denote the difference of values at the end points, writing

$$\int_a^b \frac{dF(x)}{dx}\,dx = F(b) - F(a) = F(x)\,\Big|_a^b.$$

We can write (31) in the form

$$(32) \qquad \frac{F(b) - F(a)}{b - a} = \frac{1}{b - a}\int_a^b F'(x)\,dx.$$

Recalling the definition of the average of a function in an interval from p. 141, the rule states then that *the difference quotient of the function $F(x)$ formed for the points a and b is equal to the arithmetic mean or average of the derivative of $F(x)$ in the interval with end points a and b.* When we considered the motion of a particle on a straight line, we called the change in distance s divided by the change in time t

the "average velocity." We see now that indeed $\Delta s/\Delta t$ is precisely the average of the velocities ds/dt for the given time interval if t is the independent variable used in forming the average.

RELATION BETWEEN THE MEAN VALUE THEOREMS

The formula

$$(33) \qquad F(b) - F(a) = \int_a^b f(x)\,dx$$

which holds for any continuous function f and one of its primitives F also makes evident the relation between the mean value theorems of integral calculus (p. 141) and of differential calculus (p. 173). By the mean value theorem of integral calculus we conclude from (33) that

$$F(b) - F(a) = (b - a)f(\xi).$$

Since F is a primitive of f, we can replace $f(\xi)$ by $F'(\xi)$ and obtain the mean value theorem of differential calculus for the function F. Of course, the requirement that F have a continuous derivative is stronger than the requirement of the mean value theorem of differential calculus, that the derivative merely exist.

d. Examples

In Chapter 3 we shall make extensive use of the fundamental theorem in evaluating integrals. For the moment we illustrate the method that is based on the use of the formula

$$\int_a^b \frac{dF(x)}{dx} = F(b) - F(a)$$

by some examples.

On p. 163 we derived the formula

$$\frac{d}{dx}\,x^n = nx^{n-1}$$

for positive integers n. This formula is really a trivial consequence of the *binomial theorem* since

$$\frac{d}{dx}\,x^n = \lim_{h \to 0} \frac{1}{h}\,[(x + h)^n - x^n]$$

$$= \lim_{h \to 0} \frac{1}{h}\left(x^n + nhx^{n-1} + \frac{n(n-1)}{2}\,h^2x^{n-2} + \cdots + h^n - x^n\right)$$

$$= \lim_{h \to 0}\left(nx^{n-1} + \frac{n(n-1)}{2}\,hx^{n-2} + \cdots + h^{n-1}\right) = nx^{n-1}.$$

Integrating between the limits a and b we find that

$$\int_a^b n x^{n-1}\, dx = b^n - a^n.$$

Writing m for $n-1$ we obtain the formula

$$\int_a^b x^m\, dx = \frac{1}{m+1}(b^{m+1} - a^{m+1})$$

for integers $m \geq 0$. This derivation of the expression for the integral of x^m is much simpler than the one given on p. 131 which was based on a geometric subdivision of the interval $[a, b]$; moreover, the result is now actually more general since we can dispense with the assumption that a and b are positive.

The formulas

$$\frac{d \sin x}{dx} = \cos x, \qquad \frac{d \cos x}{dx} = -\sin x$$

were obtained on p. 165 by applying the addition theorems for trigonometric functions and using $\lim\limits_{h \to 0} \left(\dfrac{\sin h}{h} \right) = 1$. Integrating we immediately obtain

$$\int_a^b \cos x\, dx = \sin b - \sin a, \qquad \int_a^b \sin x\, dx = \cos a - \cos b.$$

Again this derivation of the integration formulas from the fundamental theorem is simpler than the one based on the definition of the definite integral as limit of a sum.

Supplement. The Existence of the Definite Integral of a Continuous Function

We have yet to prove the fact that the integral of a function $f(x)$ between the limits a and b ($a < b$) exists whenever $f(x)$ is continuous in the closed interval $[a, b]$. The proof will be based mainly on the uniform continuity of $f(x)$ (see p. 41): for any given positive ϵ the values of f at any two points ξ and η of the interval differ by less than ϵ if ξ and η are sufficiently close to each other, the degree of closeness dependent solely upon ϵ and independent of ξ, η; in other words, there exists a uniform modulus of continuity $\delta(\epsilon)$ such that $|f(\xi) - f(\eta)| < \epsilon$ for any values ξ, η in $[a, b]$ for which $|\xi - \eta| < \delta$.

The definition of integral as a limit of sums requires that we subdivide the interval $[a, b]$ into n parts by successive points $x_0, x_1, \ldots, x_n$, where $x_0 = a$, $x_n = b$ and $x_0 < x_1 < \cdots < x_n$. Let S_n be a name for a particular subdivision of $[a, b]$ of this type into n cells. The *coarseness* of the subdivision will be measured by the length of the largest of the resulting cells, that is, the largest of the quantities $\Delta x_i = x_i - x_{i-1}$, which we shall call the "span" of S_n. Because of the uniform continuity of f the values of f in any two points of the same cell differ by less than ϵ as soon as the span of S_n is less than $\delta = \delta(\epsilon)$. An approximating sum based on the subdivision S_n is obtained by choosing a value ξ_i in each cell $[x_{i-1}, x_i]$ and forming

$$F_n = \sum_{i=1}^{n} f(\xi_i)\, \Delta x_i.$$

We have to prove that for a sequence of subdivisions S_n with span tending to zero the sums F_n converge toward a limit, which we shall denote by $\int_a^b f(x)\, dx$, and that the value of this limit does not depend on the particular choice of subdivisions and of intermediate points ξ_i. To carry out the proof we first compare the values F_n and F_N belonging to two subdivisions S_n and S_N where the span of S_n is less than δ and where the subdivision S_N is a "refinement" of S_n; that is, all points of subdivision of S_n occur among those of S_N. We have in appropriately modified notation

$$F_N = \sum_{j=1}^{N} f(\eta_j)\, \Delta y_j,$$

where the values y_j are the points of subdivision of S_N, where $\Delta y_j = y_j - y_{j-1}$, and η_j lies in the interval $[y_{j-1}, y_j]$. Two successive subdivision points x_{i-1} and x_i of S_n also occur among the values y_j, say $x_{i-1} = y_{r-1}, x_i = y_s$. In S_N the cell $[x_{i-1}, x_i]$ is broken up into intervals, say $[y_{r-1}, y_r], [y_r, y_{r+1}], \ldots, [y_{s-1}, y_s]$, making the total contribution

$$\sum_{j=r}^{s} f(\eta_j)(y_j - y_{j-1})$$

to F_N. We compare this to the contribution of the cell $[x_{i-1}, x_i]$ to F_n given by $f(\xi_i)(x_i - x_{i-1})$, which can be written as

$$\sum_{j=r}^{s} f(\xi_i)(y_j - y_{j-1})$$

(see Fig. 2.33) and find for the absolute value of the difference of the contributions

$$\left| \sum_{j=r}^{s} [f(\eta_j) - f(\xi_i)](y_j - y_{j-1}) \right| \leq \sum_{j=r}^{s} \epsilon \cdot (y_j - y_{j-1}) = \epsilon(x_i - x_{i-1}).$$

Hence, adding up the differences of the contributions to F_n and F_N for all cells $[x_{i-1}, x_i]$ of S_n, we find the estimate

$$|F_N - F_n| \leq \sum_{i=1}^{n} \epsilon(x_i - x_{i-1}) = \epsilon(b - a),$$

whenever S_n has span less than $\delta(\epsilon)$ and S_N is a refinement of S_n.

If now S_n and S_m are any two subdivisions, we can consider the subdivision S_N formed by all the points of subdivision S_n together with

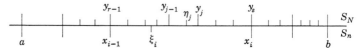

Figure 2.33

all those of S_m. Then S_N will be a refinement of both S_n and S_m. Assume that both S_n and S_m have span less than $\delta(\epsilon)$. Choosing any intermediate points η_j of the cells of S_N to define F_N, we find

$$|F_n - F_m| = |(F_n - F_N) + (F_N - F_m)| \leq |F_n - F_N| + |F_m - F_N|$$
$$\leq 2\epsilon(b - a).$$

We see then that any two approximating sums differ arbitrarily little from each other, if the spans of the corresponding subdivisions are sufficiently small. Consider now any sequence of subdivisions S_n whose spans tend to zero for $n \to \infty$. Let F_n be the corresponding approximating sums. For any $\epsilon > 0$ the span of S_n is less than $\delta(\epsilon)$ for all sufficiently large n. Hence

$$|F_n - F_m| < 2\epsilon(b - a)$$

for both n and m sufficiently large. It follows that the sequence F_n satisfies the Cauchy convergence criterion (see p. 97); consequently,

$$\lim_{n \to \infty} F_n = F$$

exists.

It remains to show that the value of $\lim_{n \to \infty} F_n$ does not depend on the particular subdivisions and intermediate points. If then $S_n{}'$ denotes any other sequence of subdivisions with spans tending to zero, then the corresponding sum $F_n{}'$ has a limit F'. Since

$$|F_n{}' - F_n| < 2\epsilon(b - a)$$

as soon as the spans of S_n and $S_n{}'$ are less than $\delta(\epsilon)$, we find for $n \to \infty$ that also $|F - F'| \leq 2\epsilon(b - a)$. Since here ϵ is an arbitrary positive number, it follows that $F = F'$. Hence the limit F, which we denote by $\int_a^b f(x)\, dx$, is uniquely determined.

The proof of the existence of the definite integral of a continuous function is thus complete.

More General Approximating Sums. Our proof indicates more clearly what is essential in the approximation of an integral by a sum. It makes evident the fact that a somewhat more general limiting process could be formulated, leading also to the integral, and that the following more general form of the theorem is true: f_i need not be a function value in order that the sums $F_n = \Sigma f_i\, \Delta x_i$ converge to the integral; it suffices instead that $|f_i - f(\xi_i)| < \delta(\epsilon)$ for some point ξ_i in the interval $[x_{i-1}, x_i]$, where $\delta(\epsilon) \to 0$ for $\epsilon \to 0$.

This general statement is often useful. If, for example, $f(x) = \phi(x)\psi(x)$, then instead of the sum $\Sigma f(\xi_\nu)\, \Delta x_\nu$ we may consider the more general sum

$$\sum \phi(\xi_\nu')\psi(\xi_\nu'')\, \Delta x_\nu,$$

where ξ_ν' and ξ_ν'' are two not necessarily coincident points of the cell. This sum also tends to the integral

$$\int_a^b f(x)\, dx = \int_a^b \phi(x)\psi(x)\, dx$$

as n increases, provided that the length of the longest cell tends to zero.

A corresponding statement holds for other sums formed in an analogous way; for example, the sum

$$\sum_{\nu=1}^n \sqrt{\phi(\xi_\nu')^2 + \psi(\xi_\nu'')^2}\, \Delta x_\nu$$

tends to the integral

$$\int_a^b \sqrt{\phi(x)^2 + \psi(x)^2}\, dx.$$

To prove these statements we only have to show that the change D in the approximating sums due to the deviation of ξ_ν'' from ξ_ν' tends to zero in the limit. This is obvious in the first example where the change in the approximating sum is

$$D = \sum_{\nu=1}^n \phi(\xi_\nu')[\psi(\xi_\nu'') - \psi(\xi_\nu')]\, \Delta x_\nu.$$

Since ϕ is bounded and ψ uniformly continuous, D can be made arbitrarily small by choosing sufficiently small cells.

The change in the second sum is represented by

$$D = \sum_{v=1}^{n} (\sqrt{\phi(\xi_v')^2 + \psi(\xi_v'')^2} - \sqrt{\phi(\xi_v')^2 + \psi(\xi_v')^2}) \, \Delta x_v.$$

Using the triangle inequality applied to the triangle with vertices $(a,0)$, $(0,b)$, $(0,c)$ in the form $|\sqrt{a^2 + b^2} - \sqrt{a^2 + c^2}| \leq |b - c|$, we find that

$$|\sqrt{\phi(\xi_v')^2 + \psi(\xi_v'')^2} - \sqrt{\phi(\xi_v')^2 + \psi(\xi_v')^2}| \leq |\psi(\xi_v'') - \psi(\xi_v')|$$

from which follows immediately that D tends to zero.

PROBLEMS

SECTION 2.1, Page 120

1. Let f be a positive monotone function defined on $[a, b]$, where $0 < a < b$. Let ϕ be the inverse of f and set $\alpha = f(a)$, $\beta = f(b)$. Using the interpretation of integral as area show that

$$\int_\alpha^\beta \phi(y) \, dy = b\beta - a\alpha - \int_a^b f(x) \, dx \,.$$

SECTION 2.2, Page 128

1. Prove for any natural number p that

$$\int_a^b x^p \, dx = \frac{1}{p+1} (b^{p+1} - a^{p+1})$$

using a subdivision of $[a, b]$ into cells of equal length. Employ the techniques in Chapter 1, miscellaneous Problems 5 to 12, to evaluate the approximating sums F_n.

2. Derive the formula for $\int_a^b x^\alpha \, dx, a, b > 0$, when α is rational and negative, say $\alpha = -r/s$, where r and s are natural numbers. (*Hint:* Set $q^{-1/s} = \tau$, where $q = \sqrt[n]{b/a}$.)

3. By the method used to find the integral of $\sin x$, derive the formula

$$\int_a^b \cos x \, dx = \sin b - \sin a \,.$$

4. Make a general statement about $\int_{-a}^a f(x) \, dx$ when $f(x)$ is (*a*) an odd function and (*b*) an even function.

5. Calculate $\int_0^{\pi/2} \sin x \, dx$ and $\int_0^{\pi/2} \cos x \, dx$. Explain on geometrical grounds why these should be the same. Furthermore, explain why

$$\int_a^{a+2\pi} \sin x \, dx = \int_b^{b+2\pi} \cos x \, dx$$

for all values of a and b.

6. (*a*) Evaluate $I_n = \int_0^a x^{1/n}\, dx$. What is $\lim\limits_{n \to \infty} I_n$? Interpret geometrically.

(*b*) Do the same for $I_n = \int_0^a x^n\, dx$.

7. Evaluate

$$\lim_{n \to \infty} \frac{1}{\sqrt{n}} \left(1 + \frac{1}{\sqrt{2}} + \cdots + \frac{1}{\sqrt{n}} \right).$$

SECTION 2.3, Page 136

*1. *Cauchy's inequality for integrals.* Prove that for all continuous functions $f(x), g(x)$

$$\int_a^b [f(x)]^2\, dx \int_a^b [g(x)]^2\, dx \geq \left(\int_a^b f(x)g(x)\, dx \right)^2.$$

*2. Prove that if $f(x)$ is continuous and

$$f(x) = \int_0^x f(t)\, dt,$$

then $f(x)$ is identically zero.

*3. Let $f(x)$ be Lipschitz-continuous on [0, 1]; that is,

$$|f(x) - f(y)| < M\, |x - y|$$

for all x, y in the interval. Prove that

$$\left| \int_0^1 f(x)\, dx - \frac{1}{n} \sum_{k=1}^n f\left(\frac{k}{n} \right) \right| < \frac{M}{2n}$$

SECTION 2.5, Page 145

1. Prove

$$\log \frac{p}{q} \leq \frac{p - q}{\sqrt{pq}} \qquad (q \leq p).$$

(*Hint:* Apply *Cauchy's* inequality, Problem 1.)

2. (*a*) Verify that $\log (1 + x) = \int_0^x \frac{1}{1 + u}\, du$, where $x > -1$.
(*b*) Show for $x > 0$ that

$$x - \frac{x^2}{2} < \log (1 + x) < x.$$

*(c) More generally, show for $0 < x < 1$ that

$$x - \frac{x^2}{2} + \frac{x^3}{3} - \cdots - \frac{x^{2n}}{2n} < \log (1 + x) < x - \frac{x^2}{2} + \frac{x^3}{3} - \cdots + \frac{x^{2n+1}}{2n + 1}.$$

(*Hint:* Compare $1/(1 + u)$ with a geometric progression.)

SECTION 2.6, Page 149

1. (*a*) Prove

$$\int_a^b e^x \, dx = e^b - e^a$$

using a subdivision of $[a, b]$ into equal cells. [*Hint:* Apply $\log \alpha = \lim_{n \to \infty} n(\sqrt[n]{\alpha} - 1)$].

(*b*) Find $\displaystyle\int_a^b \log x \, dx$. (See Section 2.1, Problem 1.)

(*c*) Show for $x \geq 0$ that

$$1 + x + \frac{x^2}{2!} + \cdots + \frac{x^n}{n!} \leq e^x \leq 1 + x + \frac{x^2}{2!} + \cdots + \frac{x^n}{n!} + \frac{e^x x^{n+1}}{(n+1)!}.$$

(*Hint:* Obtain upper and lower estimates for $\displaystyle\int_0^x e^u \, du$ and integrate repeatedly.)

Obtain estimates of the same type for e^x when $x < 0$.

SECTION 2.8c, Page 163

Calculate the derivatives of the following functions wherever defined directly as the limits of their difference quotients.

1. $\tan x$.

2. $\sec^2 x$.

3. $\sin \sqrt{x}$.

4. $\sqrt{\sin x}$.

5. $\dfrac{1}{\sin x}$.

6. $\sin \dfrac{1}{x}$.

7. x^α, where α is rational and negative.

SECTION 2.8i, Page 175

1. Show $x > \sin x$ for positive x and $x < \tan x$ for x in $\left(0, \dfrac{\pi}{2}\right)$.

2. If $f(x)$ is continuous and differentiable for $a \leq x \leq b$, show that if $f'(x) \leq 0$ for $a \leq x < \xi$ and $f'(x) \geq 0$ for $\xi < x \leq b$, the function is never less than $f(\xi)$.

*3. If the continuous function $f(x)$ has a derivative $f'(x)$ at each point x in the neighborhood of $x = \xi$, and if $f'(x)$ approaches a limit L as $x \to \xi$, then $f'(\xi)$ exists and is equal to L.

*4. Let $f(x)$ be defined and differentiable on the entire x-axis. Show that if $f(0) = 0$ and everywhere $|f'(x)| \leq |f(x)|$, then $f(x) = 0$ identically.

SECTION 2.9, Page 184

*1. If a particle traverses distance 1 in time 1, beginning and ending at rest, then at some point in the interval it must have been subjected to an acceleration equal to 4 or more.

SUPPLEMENT, Page 192. Existence of the Definite Integral

1. Let $f(x)$ be defined and bounded on $[a, b]$. We define the upper sum Σ and lower sum σ for the subdivision

$$a = x_0 < x_1 < x_2 \cdots < x_n = b$$

to be

$$\Sigma = \sum_{i=1}^{n} M_i \, \Delta x_i, \qquad \sigma = \sum_{i=1}^{n} m_i \, \Delta x_i,$$

where M_i is the least upper bound, m_i the greatest lower bound of $f(x)$ in the cell $[x_{i-1}, x_i]$.

(a) Show that in any refinement of a subdivision the upper sum either decreases or remains unchanged and, similarly, the lower sum increases or remains unchanged.

(b) Prove that each upper sum is greater than or equal to every lower sum.

(c) The *upper Darboux integral* F^+ is defined as the greatest lower bound of the upper sums and the *lower Darboux integral* F^- as the least upper bound of the lower sums over all subdivisions. From (b), $F^+ \geq F^-$. If $F^+ = F^-$ we call the common value the *Darboux integral* of f. Prove that the Darboux integral of f is actually the ordinary Riemann integral; furthermore, show that the Riemann integral exists if and only if the upper and lower Darboux integrals exist and are equal.

2. Let $f(x)$ be a monotone function defined on $[a, b]$.

(a) Show that the difference between the upper and lower sums for a subdivision into n equal cells is given exactly by

$$\Sigma - \sigma = |f(b) - f(a)| \, (b - a)/n.$$

and explain this result geometrically.

(b) Use the result of (a) to prove that the Darboux integral exists.

(c) Estimate $\Sigma - \sigma$ in terms of $f(a)$, $f(b)$ and the span of the subdivision if the cells of the subdivision may be unequal.

(d) Mostly $f(x)$, if not monotone, can be written as the sum of monotone functions, $f(x) = \phi(x) + \psi(x)$ where ϕ is nonincreasing and ψ is nondecreasing. Estimate the difference between the upper and lower sums in that case.

3. Show that if $f(x)$ has a continuous derivative in the closed interval $[a, b]$, then $f(x)$ can be written as the sum of monotone functions as in Problem 2d.

MISCELLANEOUS PROBLEMS

1. Prove that

(a) $\displaystyle\int_{-1}^{1} (x^2 - 1)^2 \, dx = \frac{16}{15};$ (b) $\displaystyle(-1)^n \int_{-1}^{1} (x^2 - 1)^n \, dx = \frac{2^{2n+1}(n!)^2}{(2n + 1)!}.$

2. Prove for the binomial coefficient $\dbinom{n}{k}$ that

$$\binom{n}{k} = \left[(n + 1) \int_{0}^{1} x^k (1 - x)^{n-k} \, dx \right]^{-1}.$$

*3. If $f(x)$ possesses a derivative $f'(x)$ (not necessarily continuous) at each point x of $a \leq x \leq b$, and if $f'(x)$ assumes the values m and M it also assumes every value μ between m and M.

4. If $f''(x) \geq 0$ for all values of x in $a \leq x \leq b$, the graph of $y = f(x)$ lies on or above the tangent line at any point $x = \xi$, $y = f(\xi)$ of the graph.

5. If $f''(x) \geq 0$ for all values of x in $u \leq x \leq b$, the graph of $y = f(x)$ in the interval $x_1 \leq x \leq x_2$ lies below the line segment joining the two points of the graph for which $x = x_1$, $x = x_2$.

6. If $f''(x) \geq 0$, then $f\left(\dfrac{x_1 + x_2}{2}\right) \leq \dfrac{f(x_1) + f(x_2)}{2}$.

*7. Let $f(x)$ be a function such that $f''(x) \geq 0$ for all values of x and let $u = u(t)$ be an arbitrary continuous function. Then

$$\frac{1}{a} \int_0^a f[u(t)]\, dt \geq f\left(\frac{1}{a} \int_0^a u(t)\, dt\right).$$

8. (*a*) Differentiate directly and write down the corresponding integration formulas: (i) $x^{1/2}$; (ii) $\tan x$.

(*b*) Evaluate

$$\lim_{n \to \infty} \frac{1}{n}\left(1 + \sec^2 \frac{\pi}{4n} + \sec^2 \frac{2\pi}{4n} + \cdots + \sec^2 \frac{n\pi}{4n}\right).$$

9. Let $f(x)$ have first and second derivatives for all real values of x. Prove that if $f(x)$ is everywhere positive and concave, then $f(x)$ is constant.

3

The Techniques of Calculus

Part A Differentiation and Integration of the Elementary Functions

3.1 The Simplest Rules for Differentiation and Their Applications

Although problems of integration are usually of greater importance than those of differentiation, the latter offer less formal difficulty than the former. Therefore it is a natural procedure first to master the art of differentiating the widest possible classes of functions; then by the fundamental theorem (Section 2.9) the results of differentiation are available for evaluating integrals. In the following sections we shall pursue such applications of the fundamental theorem. To a certain extent we shall make a fresh start and develop techniques of integration systematically on the basis of certain general rules for differentiation.

a. Rules for Differentiation

We assume that in the interval under consideration the functions $f(x)$ and $g(x)$ are differentiable; then the following rules are basic.

Rule 1. Multiplication by a Constant. For any constant c, the function $\phi(x) = cf(x)$ is differentiable, and

(1) $$\phi'(x) = cf'(x).$$

The obvious proof was given in Chapter 2, p. 165.

Rule 2. Derivative of a Sum. If $\phi(x) = f(x) + g(x)$, then $\phi(x)$ is differentiable and

(2) $$\phi'(x) = f'(x) + g'(x);$$

that is, the operations of differentiation and addition are interchangeable. The same holds for the sum of a finite number n of differentiable functions

$$\phi(x) = \sum_{\nu=1}^{n} f_\nu(x),$$

for which we obtain

$$\phi'(x) = \sum_{\nu=1}^{n} f_\nu'(x).$$

The proof is obvious from the definition of derivative.

Rule 3. Derivative of a Product. If $\phi(x) = f(x)g(x)$, then $\phi(x)$ is differentiable and

(3) $$\phi'(x) = f(x)g'(x) + g(x)f'(x).$$

The proof follows from the equation

$$\frac{\phi(x+h) - \phi(x)}{h} = \frac{f(x+h)g(x+h) - f(x)g(x)}{h}$$

$$= f(x+h)\frac{g(x+h) - g(x)}{h} + g(x)\frac{f(x+h) - f(x)}{h}.$$

Taking the limit in this expression as $h \to 0$ yields Eq. (3).

This formula becomes more elegant if we divide[1] by $\phi(x) = f(x)g(x)$. We then obtain

$$\frac{\phi'(x)}{\phi(x)} = \frac{f'(x)}{f(x)} + \frac{g'(x)}{g(x)}.$$

Using the notation of differentials (Chapter 2, p. 179) we may also rewrite Eq. (3) as

$$d(fg) = f\,dg + g\,df.$$

By induction we obtain for the derivative of a product of n factors an expression consisting of n terms, each of which consists of the derivative of one factor multiplied by all the other factors of the

---

[1] We must, of course, assume that $\phi(x)$ is nowhere equal to zero.

original product:

$$\phi'(x) = \frac{d}{dx}\left[f_1(x)f_2(x)\cdots f_n(x)\right]$$

$$= f_1'(x)f_2(x)\cdots f_n(x) + f_1(x)f_2'(x)f_3(x)\cdots f_n(x)$$
$$+ \cdots + f_1(x)f_2(x)\cdots f_n'(x)$$

$$= \sum_{\nu=1}^{n} f_\nu'(x)\,\frac{\phi(x)}{f_\nu(x)},$$

or on division by $\phi(x) = f_1(x)f_2(x)\cdots f_n(x)$

$$\frac{\phi'(x)}{\phi(x)} = \frac{f_1'(x)}{f_1(x)} + \frac{f_2'(x)}{f_2(x)} + \cdots + \frac{f_n'(x)}{f_n(x)} = \sum_{\nu=1}^{n}\frac{f_\nu'(x)}{f_\nu(x)},$$

which is valid where $\phi(x) \neq 0$.

By repeated application of the rule for the derivative of a product we can obtain *formulas for the second and higher derivatives* as well. We have for the second derivative

$$\frac{d^2 fg}{dx^2} = \frac{d}{dx}\left(\frac{dfg}{dx}\right) = \frac{d}{dx}\left(f\frac{dg}{dx} + \frac{df}{dx}g\right)$$

$$= \frac{d}{dx}\left(f\frac{dg}{dx}\right) + \frac{d}{dx}\left(\frac{df}{dx}g\right)$$

$$= f\frac{d^2 g}{dx^2} + 2\frac{df}{dx}\frac{dg}{dx} + \frac{d^2 f}{dx^2}g.$$

Leibnitz's Rule.   The reader should prove by induction that the nth derivative of a product may be found according to the following rule (Leibnitz's rule):

$$\frac{d^n}{dx^n}(fg) = f\frac{d^n g}{dx^n} + \binom{n}{1}\frac{df}{dx}\frac{d^{n-1}g}{dx^{n-1}}$$

$$+ \binom{n}{2}\frac{d^2 f}{dx^2}\frac{d^{n-2}g}{dx^{n-2}} + \cdots + \binom{n}{n-1}\frac{d^{n-1}f}{dx^{n-1}}\frac{dg}{dx} + \frac{d^n f}{dx^n}g.$$

Here $\binom{n}{1} = n$, $\binom{n}{2} = [n(n-1)]/2!$, etc.,   denote the binomial co-efficients.

Rule 4.   Derivative of a Quotient.   For a quotient

$$\phi(x) = \frac{f(x)}{g(x)}$$

the following rule holds: The function $\phi(x)$ is differentiable at every point at which $g(x)$ does not vanish, and

(4) $$\phi'(x) = \frac{g(x)f'(x) - g'(x)f(x)}{[g(x)]^2} .$$

If $\phi(x) \neq 0$, this can be written as

$$\frac{\phi'(x)}{\phi(x)} = \frac{f'(x)}{f(x)} - \frac{g'(x)}{g(x)} .$$

PROOF.   If we assume the differentiability of $\phi(x)$, we can apply the product rule to $f(x) = \phi(x)g(x)$ and conclude

$$f'(x) = \phi(x)g'(x) + g(x)\phi'(x).$$

By substituting $f(x)/g(x)$ for $\phi(x)$ on the right and solving for $\phi'(x)$, we obtain Rule 4.

We can prove the differentiability of $\phi(x)$ as well as the rule if we write

$$\frac{\phi(x+h) - \phi(x)}{h} = \frac{\dfrac{f(x+h)}{g(x+h)} - \dfrac{f(x)}{g(x)}}{h}$$

$$= \frac{g(x)\dfrac{f(x+h) - f(x)}{h} - \dfrac{g(x+h) - g(x)}{h}f(x)}{g(x)g(x+h)} .$$

If we now let h tend to zero we arrive at the result stated; for by hypothesis the denominator does not tend to zero but to the limit $[g(x)]^2$, and the two terms of the numerator have limits $g(x)f'(x)$ and $g'(x)f(x)$, respectively. This proves both the existence of the limit on the left side and the differentiation formula.

b. Differentiation of the Rational Functions

First, we derive once more the formula

$$\frac{d}{dx} x^n = nx^{n-1}$$

for every positive integer n, invoking the rule for differentiating a product. We think of x^n as a product of n factors, $x^n = x \cdots x$, and thus obtain

$$\frac{d}{dx} x^n = 1 \cdot x^{n-1} + 1 \cdot x^{n-1} + \cdots + 1 \cdot x^{n-1} = nx^{n-1}.$$

The second derivative of the function x^n follows from this formula and eq. (1)

$$\frac{d^2}{dx^2} x^n = n(n-1)x^{n-2}.$$

Continuing, we obtain the higher derivatives

$$\frac{d^3}{dx^3} x^n = n(n-1)(n-2)x^{n-3}$$

$$\cdots\cdots\cdots\cdots\cdots\cdots\cdots\cdots$$

$$\frac{d^n}{dx^n} x^n = 1 \cdot 2 \cdots n = n!.$$

From the last of these formulas it is clear that the nth derivative of x^n is a constant, whereas the $(n+1)$th derivative vanishes everywhere.

By using our first two rules and the rule for differentiating powers we can differentiate any polynomial $y = a_0 + a_1 x + a_2 x^2 + \cdots + a_n x^n$, obtaining

$$y' = a_1 + 2a_2 x + 3a_3 x^2 + \cdots + na_n x^{n-1};$$

furthermore,

$$y'' = 2a_2 + 3 \cdot 2a_3 x + 4 \cdot 3a_4 x^2 + \cdots + n(n-1)a_n x^{n-2},$$

and so on.

The derivative of any rational function can now be found with the help of the quotient rule. In particular, we again deduce the differentiation formula for the function x^n, where $n = -m$ is a negative integer. Application of the quotient rule, together with the fact that the derivative of a constant is equal to zero, gives the result

$$(d/dx)(1/x^m) = -mx^{m-1}/x^{2m} = -m/x^{m+1},$$

or, if we take $m = -n$,

$$\frac{d}{dx} x^n = nx^{n-1},$$

which agrees formally with the result for positive values of n and with the results given earlier (p. 164).

c. Differentiation of the Trigonometric Functions

For the trigonometric functions $\sin x$ and $\cos x$ we have already obtained (p. 165) the differentiation formulas

$$\frac{d}{dx} \sin x = \cos x \qquad \text{and} \qquad \frac{d}{dx} \cos x = -\sin x.$$

The quotient rule now enables us to differentiate the functions

$$y = \tan x = \frac{\sin x}{\cos x} \quad \text{and} \quad y = \cot x = \frac{\cos x}{\sin x}.$$

According to the rule, the derivative of the first of these functions is

$$y' = \frac{\cos^2 x + \sin^2 x}{\cos^2 x} = \frac{1}{\cos^2 x},$$

so that

$$\frac{d}{dx} \tan x = \frac{1}{\cos^2 x} = \sec^2 x = 1 + \tan^2 x.$$

Similarly, we obtain

$$\frac{d}{dx} \cot x = - \frac{1}{\sin^2 x} = - \operatorname{cosec}^2 x = -(1 + \cot^2 x).$$

To the differentiation formulas for $\sin x$, $\cos x$, $\tan x$, and $\cot x$ correspond the following integration formulas:

$$\int \cos x \, dx = \sin x, \qquad \int \sin x \, dx = - \cos x,$$

$$\int \frac{1}{\cos^2 x} \, dx = \tan x, \qquad \int \frac{1}{\sin^2 x} \, dx = - \cot x.$$

From these formulas we obtain by way of the fundamental rule of Section 2.9, p. 190 the value of the definite integral between any limits, the only restriction being that when the last two formulas are used, the interval of integration must not contain any point of discontinuity of the integrand such as an odd multiple of $\pi/2$ in the first case, and an even multiple of $\pi/2$ in the second. For example,

$$\int_a^b \cos x \, dx = \sin x \Big|_a^b = \sin b - \sin a.$$

3.2   The Derivative of the Inverse Function

a. General Formula

We have seen on p. 45 that a continuous function $y = f(x)$ has a continuous inverse in every interval in which it is monotonic. Precisely:

If $a \leq x \leq b$ is an interval in which the continuous function $y = f(x)$ is monotonic, and if $f(a) = \alpha$ and $f(b) = \beta$, then f has an inverse function which in the interval between α and β is continuous and monotonic.

As pointed out on p. 177, the sign of the derivative provides a simple test for seeing when a function is monotonic and therefore has an inverse. A differentiable function is continuous, and is monotonic increasing in an interval where $f'(x)$ is greater than zero, and monotonic decreasing in an interval for which $f'(x)$ is everywhere less than zero.

We shall now characterize the derivative of the inverse function by proving the following theorem.

THEOREM. *If the function* $y = f(x)$ *is differentiable in the interval* $a < x < b$, *and either* $f'(x) > 0$ *or* $f'(x) < 0$ *throughout the interval, then the inverse function* $x = \phi(y)$ *also possesses a derivative at every interior point of its interval of definition: the derivatives of* $y = f(x)$ *and of its inverse* $x = \phi(y)$ *satisfy the relation* $f'(x) \cdot \phi'(y) = 1$ *at corresponding values* x, y.

This relation can also be put in the form

$$(5) \qquad \frac{dy}{dx} = \frac{1}{\dfrac{dx}{dy}}.$$

This last formula again illustrates the suitability of Leibnitz's notation: the symbolic quotient dy/dx can be treated in formulas as if it were an actual fraction.

PROOF. The proof of the theorem is simple. Writing the derivative as the limit of a difference quotient, we have

$$y' = f'(x) = \lim_{\Delta x \to 0} \frac{\Delta y}{\Delta x} = \lim_{x_1 \to x} \frac{y_1 - y}{x_1 - x},$$

where x and $y = f(x)$, and x_1 and $y_1 = f(x_1)$, respectively denote pairs of corresponding values. By hypothesis the first of these limiting values is not equal to zero. Because of the continuity of $y = f(x)$ and $x = \phi(y)$, the relations $y_1 \to y$ and $x_1 \to x$ are equivalent. Therefore the limiting value

$$\lim_{x_1 \to x} \frac{x_1 - x}{y_1 - y} = \lim_{y_1 \to y} \frac{x_1 - x}{y_1 - y}$$

exists and is equal to $1/f'(x)$. On the other hand, the limiting value on the right-hand side is by definition the derivative $\phi'(y)$ of the inverse function $\phi(y)$, and our formula is proved.

The simple geometrical meaning of the formula is clearly shown in Fig. 3.1. The tangent to the curve $y = f(x)$ or $x = \phi(y)$ makes an angle α with the positive x-axis, and an angle β with the positive y-axis; from the geometrical interpretation of the derivative of a function as the slope of the tangent

$$f'(x) = \tan \alpha, \qquad \phi'(y) = \tan \beta.$$

Since the sum of the angles α and β is $\pi/2$, $\tan \alpha \tan \beta = 1$, and this relationship is exactly equivalent to our differentiation formula.

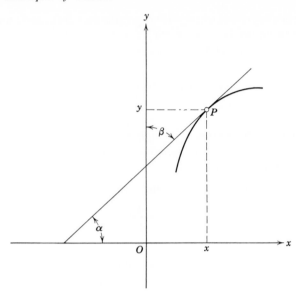

Figure 3.1   Differentiation of the inverse function.

Critical Points

We have hitherto expressly assumed that either $f'(x) > 0$ or $f'(x) < 0$, that is, that $f'(x)$ is never zero. What, then, happens if $f'(x) = 0$? If $f'(x) = 0$ everywhere in an interval, then f is constant there, and consequently has no inverse because the same value of y corresponds to all values of x in the interval. If $f'(x) = 0$ only at isolated "critical" points (and if $f''(x)$ is assumed continuous), then we have two cases, according to whether on passing through these points $f'(x)$ changes sign, or not. In the first case this point separates a point where the function is monotonic increasing from another where it is monotonic decreasing. In the neighborhood of such a point there can be no single-valued inverse function. In the second case the vanishing of the derivative does not contradict the monotonic character of the function $y = f(x)$, so that a single-valued inverse exists. However, the inverse function is no longer differentiable at the corresponding point; in fact, its derivative is infinite there. The functions $y = x^2$ and $y = x^3$ at the point $x = 0$ offer examples of the two types. Figure 3.2 and Fig. 3.3 illustrate the behavior of the two functions upon passing through the origin and at the same time show that the function $y = x^3$ has a single-valued inverse, whereas the other function $y = x^2$ does not.

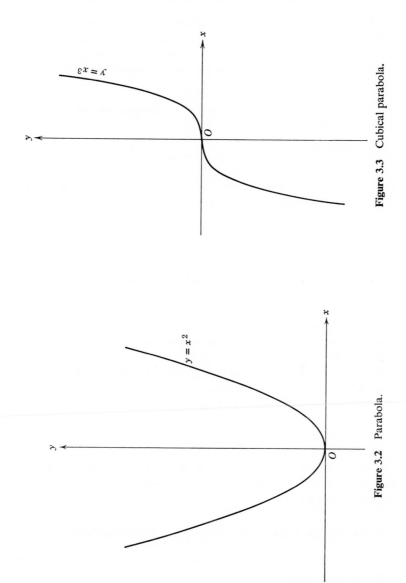

Figure 3.2 Parabola.

Figure 3.3 Cubical parabola.

b. The Inverse of the nth Power: the nth Root

The simplest example is the inverse of the function $y = x^n$ for positive integers n; at first we assume positive values of x, hence also $y > 0$. Under these conditions y' is always positive, so that for all positive values of y we can form the unique inverse function

$$x = \sqrt[n]{y} = y^{1/n}.$$

The derivative of this inverse function is immediately obtained by the above general rule as follows:

$$\frac{d(y^{1/n})}{dy} = \frac{dx}{dy} = \frac{1}{dy/dx} = \frac{1}{nx^{n-1}} = \frac{1}{n} \frac{1}{y^{(n-1)/n}} = \frac{1}{n} y^{(1/n)-1}.$$

If we now change the notation and denote again the independent variable by x, we may finally write

$$\frac{d\sqrt[n]{x}}{dx} = \frac{d}{dx}(x^{1/n}) = \frac{1}{n} x^{(1/n)-1},$$

which agrees with the result obtained on p. 164.

For $n > 1$, the point $x = 0$ requires special consideration. If x approaches zero through positive values, $d(x^{1/n})/dx$ will obviously increase beyond all bounds; this corresponds to the fact that for $n > 1$ the derivative of the nth power $f(x) = x^n$ vanishes at the origin. Geometrically, this means that the curves $y = x^{1/n}$ for $n > 1$ touch the y-axis at the origin (cf. Fig. 1.35, p. 48).

It should be noted that for odd values of n the assumption $x > 0$ can be omitted and the function $y = x^n$ is monotonic and has an inverse over the entire domain of real numbers. The formula

$$\frac{d(\sqrt[n]{y})}{dy} = (1/n)y^{(1/n)-1}$$

still holds for negative values of y, but for $x = 0$, $n > 1$, we have $d(x^n)/dx = 0$, which corresponds to an infinite derivative dx/dy of the inverse function at the point $y = 0$.

c. The Inverse Trigonometric Functions—Multivaluedness

To form the inverses of the trigonometric functions we once again consider the graphs[1] of $\sin x$, $\cos x$, $\tan x$, and $\cot x$. We see at once from Figs. 1.37, p. 50 and 1.38, p. 51, that for each of these functions it

---

[1] The graphical representation will help the reader to overcome the slight difficulties inherent in the discussion of the "multivaluedness" of the inverse functions.

is necessary to select a definite interval if we are to speak of a unique inverse; for the lines $y = c$ parallel to the x-axis cut the curves in an infinite number of points, if at all.

The Inverse Sine and Cosine

For the function $y = \sin x$, for example (Fig. 3.4), the derivative $y' = \cos x$ is positive in the interval $-\pi/2 < x < \pi/2$. In this interval $y = \sin x$ has an inverse function which we denote by[1]

$$x = \text{arc sin } y$$

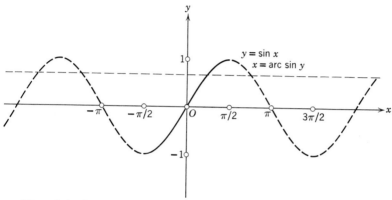

Figure 3.4   Graph of $y = \sin x$ (principal value indicated by solid curve).

(read arc sine y; this means the angle whose sine has the value y). This function increases monotonically from $-\pi/2$ to $+\pi/2$ as y traverses the interval -1 to $+1$. If we wish to emphasize that we are considering the inverse function of the sine in this particular interval, we speak of the *principal value* of the arc sine. For some other interval in which $\sin x$ is monotonic, for example, the interval $+\pi/2 < x < 3\pi/2$, we obtain another inverse or "*branch*" of the arc sine; without the exact statement of the interval in which the values of the inverse function should lie, the symbol arc sine means not one well-defined function but, in fact, denotes an *infinite* number of values.[2]

The multivaluedness of arc sin y is described by the statement: To any one value y of the sine there corresponds not only a specific angle x but also any angle of the form $2k\pi + x$ or $(2k + 1)\pi - x$, where k is any integer (cf. Fig. 3.4).

---

[1] The symbolic notation $x = \sin^{-1} y$ is also used where there is no danger of confusion with the *reciprocal* function $1/\sin x$.

[2] Sometimes loosely called a multiple-valued function.

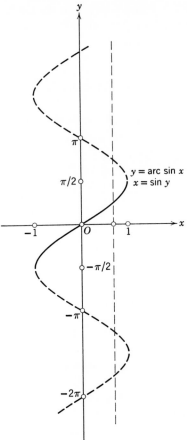

Figure 3.5 Graph of $y = $ arc sin x (principal value indicated by solid curve).

The derivative of the function $x = $ arc sin y is obtained from Eq. (5) as follows:

$$\frac{dx}{dy} = \frac{1}{y'} = \frac{1}{\cos x} = \frac{1}{\pm\sqrt{1 - \sin^2 x}} = \frac{1}{\pm\sqrt{1 - y^2}},$$

where the square root is to be taken as positive if we confine ourselves to the first interval mentioned, that is, $-\pi/2 < x < \pi/2$.[1]

Finally, we change the name of the independent variable from y to the commonly used x (Fig. 3.5); then the derivative of arc sin x is

---

[1] If instead of this we had chosen the interval $\pi/2 < x < 3\pi/2$, corresponding to the substitution of $x + \pi$ for x, we should have had to use the negative square root since cos x is negative in this interval.

expressed by

$$\frac{d}{dx}\arcsin x = \frac{1}{\sqrt{1 - x^2}}.$$

Here it is assumed that arc sine is the principal value which lies between $-\pi/2$ and $+\pi/2$, and the square root sign is chosen positive.

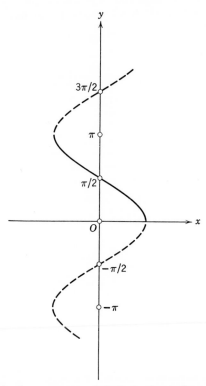

Figure 3.6   Graph of $y = \arccos x$ (principal value indicated by solid curve).

For the inverse function of $y = \cos x$, denoted (after again interchanging the names x and y) by arc cos x, we obtain the formula

$$\frac{d}{dx}\arccos x = \mp \frac{1}{\sqrt{1 - x^2}}$$

in exactly the same way. Here we take the negative sign of the root if the value of arc cos x is taken in the interval between 0 and π (not, as in the case of arc sin x, between $-\pi/2$ and $+\pi/2$) (cf. Fig. 3.6).

The derivatives become infinite on approaching the end points $x = -1$ and $x = +1$, corresponding to the fact that the graphs of the inverse sine and inverse cosine have vertical tangents at these points.

Inverse Tangent and Cotangent

We treat the inverse functions of the tangent and cotangent in an analogous way. The function $y = \tan x$, having an everywhere positive derivative $1/\cos^2 x$ for $x \neq \pi/2 + k\pi$, has a unique inverse

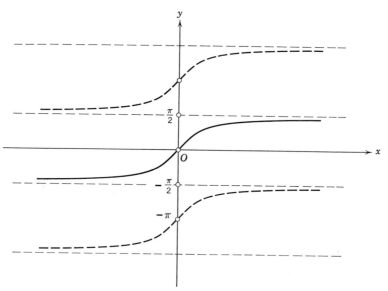

Figure 3.7   Graph of $y = \arc\tan x$ (solid curve for principal value).

in the interval $-\pi/2 < x < \pi/2$. We call this inverse function (the Principal Branch of) $x = \arc\tan y$. We see at once from Fig. 3.7 that for each x we could have chosen instead of y any of the values $y + k\pi$ (where k is an integer). Similarly, the function $y = \cot x$ has an inverse $x = \arc\cot y$ which is uniquely determined if we require that its value shall lie in the interval from 0 to π; otherwise the many-valuedness of $\arc\cot x$ is the same as for $\arc\tan x$.

The differentiation formulas are as follows:

$$x = \arc\tan y, \qquad \frac{dx}{dy} = \frac{1}{dy/dx} = \cos^2 x = \frac{1}{1 + \tan^2 x} = \frac{1}{1 + y^2}$$

$$x = \arc\cot y, \qquad \frac{dx}{dy} = -\sin^2 x = -\frac{1}{1 + \cot^2 x} = -\frac{1}{1 + y^2}$$

or finally, if we denote again the independent variable by x,

$$\frac{d}{dx} \arctan x = \frac{1}{1 + x^2},$$

$$\frac{d}{dx} \operatorname{arc\ cot} x = - \frac{1}{1 + x^2}.$$

d. The Corresponding Integral Formulas

Expressed in terms of indefinite integrals, the formulas which we have just derived are written as follows:

$$\int \frac{1}{\sqrt{1 - x^2}}\, dx = \arcsin x, \qquad \int \frac{1}{\sqrt{1 - x^2}}\, dx = - \arccos x,$$

$$\int \frac{1}{1 + x^2}\, dx = \arctan x, \qquad \int \frac{1}{1 + x^2}\, dx = - \operatorname{arc\ cot} x.$$

Although the two formulas on each line express different functions by identical indefinite integrals, they do not contradict each other. In fact, they illustrate what we learned earlier (see Section 2.9), that all indefinite integrals of the same function differ only by constants; here the constants are $\pi/2$ since $\arccos x + \arcsin x = \pi/2$, $\arctan x + \operatorname{arc\ cot} x = \pi/2$.

The formulas for indefinite integrals may immediately be put to use for finding definite integrals, as on p. 143. In particular,

$$\int_a^b \frac{dx}{1 + x^2} = \arctan x \Big|_a^b = \arctan b - \arctan a.$$

If we put $a = 0$, $b = 1$ and recall that $\tan 0 = 0$ and $\tan \pi/4 = 1$, we obtain the remarkable formula

(6)
$$\frac{\pi}{4} = \int_0^1 \frac{1}{1 + x^2}\, dx.$$

The number π, which originally arose from the consideration of the circle, is brought by this formula into a very simple relationship with the rational function $1/(1 + x^2)$, and represents the area indicated in Fig. 3.8. This formula for π, to which we shall return later (p. 445), constitutes one of the early triumphs of the power of calculus.

More generally, the integral formulas of this section permit us to define the trigonometric functions purely analytically, without any reference to geometric objects such as triangles or circles. For example, the relation between an angle y and its tangent $x = \tan y$ is completely described by the equation

$$y = \int_0^x \frac{du}{1 + u^2}$$

(at least for $-\pi/2 < y < \pi/2$). With this relation we may now *define* without appeal to intuition a numerical value for the angle y in a right triangle with sides a (adjacent) and b (opposite) for which $b/a = x$.

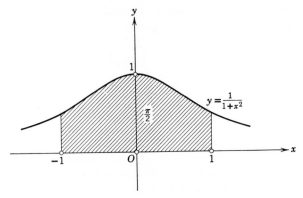

Figure 3.8 $\pi/2$ illustrated by an area.

Such an analytic definition in terms of numerical quantities makes the use of angles and trigonometric functions legitimate in higher analysis irrespective of a definition by geometrical construction.

e. Derivative and Integral of the Exponential Function

In Chapter 2, p. 150, we introduced the exponential function as the inverse of the logarithm. Precisely speaking the relations $y = e^x$ and $x = \log y$ were thus defined to be equivalent. Consequently their derivatives satisfy the relation [see (5) p. 207]

$$\frac{de^x}{dx} = \frac{dy}{dx} = \frac{1}{\dfrac{dx}{dy}} = \frac{1}{\dfrac{d \log y}{dy}} = \frac{1}{\dfrac{1}{y}} = y = e^x.$$

Hence *the exponential function is equal to its derivative*:

$$\frac{de^x}{dx} = e^x.$$

More generally, for any positive a the function $y = a^x$ has as its inverse

$$x = \log_a y = \frac{\log y}{\log a},$$

and the derivative of a^x is

$$\frac{da^x}{dx} = \frac{1}{\dfrac{d \log_a y}{dy}} = (\log a)y = (\log a)a^x.$$

Thus for any positive constant a the derivative of the function $y = a^x$ is proportional to the function itself. The factor of proportionality $\log a$ is 1 when a is the number e. On p. 223 we shall show conversely that any function which is proportional to its derivative must be of the form $y = ce^x$, where c denotes a constant factor.

By the fundamental theorem of calculus we can again translate the formulas for the derivatives of e^x and of a^x into formulas for indefinite integrals:

$$\int e^x \, dx = e^x,$$

$$\int a^x \, dx = \frac{1}{\log a} a^x.$$

3.3 Differentiation of Composite Functions

a. Definitions

The preceding rules allow us to find the derivatives of functions that are obtained as rational expressions in terms of functions with already known derivatives. To find explicit expressions for the derivatives of other functions occurring in analysis we must go further by deriving a general rule for the differentiation of composite or compound functions. We are confronted quite often with functions $f(x)$ built by the process of composition of simpler ones (see Chapter 1, p. 52): $f(x) = g(\phi(x))$, where $\phi(x)$ is defined in a closed interval $a \leq x \leq b$ and has there the range $\alpha \leq \phi \leq \beta$, and where $g(\phi)$ is defined in this latter interval.

In this connection it is useful to remember the interpretation of functions as "operators" or mappings. As in Chapter 1, we write the composite function simply as

$$f = g\phi$$

and call $g\phi$ the (symbolic) "product" of the operators or mappings g and ϕ.

b. The Chain Rule

For functions g and ϕ which are continuous in their respective intervals of definition the compound function $f(x) = g[\phi(x)]$ is continuous as well (see Chapter 1, p. 55).

The functions $\phi(x)$ and $g(\phi)$ are now assumed to be not only continuous but differentiable. We then have the following fundamental theorem, the *chain rule of differentiation*:

The function $f(x) = g[\phi(x)]$ is differentiable, and its derivative is given by the equation

$$(7) \qquad\qquad f'(x) = g'(\phi) \cdot \phi'(x),$$

or, in Leibnitz's notation,

$$\frac{df}{dx} = \frac{df}{d\phi}\frac{d\phi}{dx}.$$

Therefore the derivative of a compound function is the product of the derivatives of its constituent functions. Or: The derivative of the symbolic product of functions is the actual product of their derivatives with respect to their corresponding independent variables.

Intuitively, this chain rule is very plausible. The quantity $\phi'(x) = \lim \Delta\phi/\Delta x$ is the local ratio in which small intervals are magnified by the mapping ϕ. Similarly, $g'(\phi)$ is the magnification given by the mapping g. Applying first ϕ and then g results in magnifying an x-interval first ϕ'fold, and then enlarging the resulting ϕ-interval g'fold, resulting in a total magnification ratio of $g'\phi'$ which must be the magnification ratio for the composite mapping $f = g\phi$.

The theorem follows very easily from the definition of the derivative. In fact, it becomes intuitively almost obvious if we assume $\phi'(x) \neq 0$ in the closed x-interval under consideration. Then for $\Delta x = x_2 - x_1 \neq 0$ we have by the mean value theorem

$$\Delta\phi = \phi_2 - \phi_1 = \phi(x_2) - \phi(x_1) = \phi'(\xi)\Delta x \neq 0 \quad \text{with} \quad x_1 \leq \xi \leq x_2,$$

and, with $\Delta g = g(\phi_2) - g(\phi_1)$ and $\Delta f = f(x_2) - f(x_1)$, we may write

$$\frac{\Delta f}{\Delta x} = \frac{\Delta g}{\Delta\phi}\frac{\Delta\phi}{\Delta x}$$

which is a meaningful identity because $\Delta\phi \neq 0$. Now $\Delta\phi \to 0$ for $\Delta x \to 0$, that is, for $x_2 \to x_1$; therefore for $\Delta x \to 0$ the difference quotients tend to the respective derivatives and the theorem is proved.

To avoid the explicit assumption $\phi'(x) \neq 0$ we can dispense with the division by $\Delta\phi$ in the following slightly more subtle manner:

From the assumption of differentiability of $g(\phi)$ at the point ϕ we know that the quantity $\epsilon = \Delta g/\Delta\phi - g'(\phi)$ as a function of $\Delta\phi$ for fixed ϕ and $\Delta\phi \neq 0$ has the limit zero for $\Delta\phi \to 0$. If we *define* $\epsilon = 0$ for $\Delta\phi = 0$, we have without restriction on $\Delta\phi$

$$\Delta g = [g'(\phi) + \epsilon]\,\Delta\phi.$$

Similarly for fixed x,

$$\Delta\phi = \phi(x + \Delta x) - \phi(x) = [\phi'(x) + \eta]\,\Delta x,$$

where $\lim_{\Delta x \to 0} \eta = 0$. Then for $\Delta x \neq 0$ and $\phi = \phi(x)$,

$$\frac{\Delta g}{\Delta x} = [g'(\phi) + \epsilon]\frac{\Delta\phi}{\Delta x} = [g'(\phi) + \epsilon][\phi'(x) + \eta].$$

For Δx tending to zero through nonzero values we have $\lim_{\Delta x \to 0} \Delta\phi = 0$ and hence $\lim_{\Delta x \to 0} \epsilon = 0$, so that

$$\lim_{\Delta x \to 0}\frac{\Delta g}{\Delta x} = \lim_{\Delta x \to 0}[g'(\phi) + \epsilon]\lim_{\Delta x \to 0}[\phi'(x) + \eta] = g'(\phi)\phi'(x),$$

which proves the chain rule.

By successive application of our rule we immediately extend it to functions arising from the *composition of more than two functions.* If, for example,

$$y = g(u), \qquad u = \phi(v), \qquad v = \psi(x),$$

then $y = f(x) = g[\phi(\psi(x))]$ is a compound function of x; its derivative is given by the rule

$$\frac{dy}{dx} = y' = g'(u)\phi'(v)\psi'(x) = \frac{dy}{du}\cdot\frac{du}{dv}\cdot\frac{dv}{dx}\,;$$

similar relations are true for functions that are compounded of an arbitrary number of functions.

Higher Derivatives of a Composite Function. $y = g[\phi(x)]$ can be found easily by repeated application of the chain rule and the preceding rules:

$$y' = \frac{dy}{d\phi}\frac{d\phi}{dx} = g'\phi',$$

$$y'' = g''\phi'^2 + g'\phi'',$$

$$y''' = g'''\phi'^3 + 3g''\phi'\phi'' + g'\phi'''.$$

Analogous formulas for y'''' etc., can be derived successively.

Finally, let us examine the *composition of two functions inverse* to each other. The function $g(y)$ is the inverse of $y = \phi(x)$ if $f(x) = g[\phi(x)] = x$. It follows that

$$f'(x) = g'(y)\phi'(x) = 1$$

which is exactly the result of Section 3.2, p. 207.

Examples. As a simple but important example of an application of the chain rule we differentiate x^α ($x > 0$) for an arbitrary real power α. In Chapter 2, p. 152, we defined

$$x^\alpha = e^{\alpha \log x};$$

we also proved for $\phi(x) = \log x$, $\psi(u) = \alpha u$, $g(y) = e^y$ that

$$\phi'(x) = \frac{1}{x}, \qquad \psi'(u) = \alpha, \qquad g'(y) = e^y.$$

Now x^α is the compound function $g\{\psi[\phi(x)]\}$. Applying the chain rule we obtain the general formula

$$\frac{d}{dx}(x^\alpha) = g'(y)\psi'(u)\phi'(x)$$

$$= e^y \cdot \alpha \cdot \frac{1}{x}$$

$$= \frac{\alpha e^{\alpha \log x}}{x}$$

$$= \alpha \frac{x^\alpha}{x};$$

hence

$$\frac{d}{dx}(x^\alpha) = \alpha x^{\alpha-1},$$

a result we could prove only with some difficulty had we attempted to proceed directly from the definition of x^α for irrational α as the limit of powers with rational exponents.

An immediate consequence of this differentiation is, again, the integral formula

$$\int x^\alpha \, dx = \frac{x^{\alpha+1}}{\alpha + 1} \qquad (\alpha \neq -1).$$

As a second example, we consider

$$y = \sqrt{1 - x^2} \qquad \text{or} \quad y = \sqrt{\phi},$$

where $\phi = 1 - x^2$ and $-1 < x < 1$. The chain rule yields

$$y' = \frac{1}{2\sqrt{\phi}} \cdot (-2x) = -\frac{x}{\sqrt{1 - x^2}}.$$

Further examples are given by the following brief calculations.

1. $y = \arcsin \sqrt{1 - x^2}$, $(-1 \le x \le 1, x \ne 0)$.

$$\frac{dy}{dx} = \frac{1}{\sqrt{1 - (1 - x^2)}} \frac{d\sqrt{1 - x^2}}{dx}$$

$$= \frac{1}{|x|} \frac{-x}{\sqrt{1 - x^2}} = \frac{-1}{\sqrt{1 - x^2}} \operatorname{sgn}(x).$$

2. $y = \sqrt{\dfrac{1 + x}{1 - x}}$, $(-1 < x < 1)$.

$$\frac{dy}{dx} = \frac{1}{2\sqrt{\dfrac{1 + x}{1 - x}}} \cdot \frac{d\left(\dfrac{1 + x}{1 - x}\right)}{dx}$$

$$= \frac{\sqrt{1 - x}}{2\sqrt{1 + x}} \cdot \frac{2}{(1 - x)^2} = \frac{1}{(1 + x)^{\frac{1}{2}} (1 - x)^{\frac{3}{2}}}.$$

3. $y = \log |x|$. This function[1] can be expressed as $\log x$ for $x > 0$ and as $\log (-x)$ for $x < 0$. For $x > 0$

$$\frac{d \log |x|}{dx} = \frac{d \log x}{dx} = \frac{1}{x}.$$

For $x < 0$ we obtain from the chain rule that

$$\frac{d \log |x|}{dx} = \frac{d \log (-x)}{dx} = \frac{1}{-x} \frac{d(-x)}{dx} = \frac{1}{x}.$$

Hence generally for $x \ne 0$

$$\frac{d \log |x|}{dx} = \frac{1}{x}.$$

4. $y = a^x$. By definition of a^x (see p. 152) we have

$$a^x = e^{\phi(x)},$$

---

[1] The function $\log x$ is defined only for $x > 0$, whereas $\log |x|$ is defined everywhere except for $x = 0$.

where $\phi(x) = (\log a)x$. Then

$$\frac{da^x}{dx} = \frac{de^\phi}{d\phi}\frac{d\phi}{dx} = e^\phi(\log a) = (\log a)a^x.$$

The same result was obtained already on p. 217 from the rule for the derivative of the inverse function.

5. $y = [f(x)]^{g(x)}$. Since

$$[f(x)]^{g(x)} = e^{\phi(x)}$$

with $\phi(x) = g(x) \log [f(x)]$, we find

$$\frac{d}{dx}[f(x)]^{g(x)} = e^\phi\left(g' \log f + g\frac{1}{f}f'\right)$$

$$= [f(x)]^{g(x)}\left(g'(x) \log [f(x)] + \frac{g(x)f'(x)}{f(x)}\right).$$

For example, when $g(x) = f(x) = x$ we have

$$\frac{dx^x}{dx} = x^x(\log x + 1).$$

c. The Generalized Mean Value Theorem of the Differential Calculus

As an application of the chain rule we derive the *generalized mean value theorem* of differential calculus. Consider two functions $F(x)$ and $G(x)$, continuous on a closed interval $[a, b]$ of the x-axis, and differentiable on the interior of that interval. We assume that $G'(x)$ is positive. The ordinary mean value theorem of differential calculus applied separately to F and G furnishes an expression for the difference quotient $\dfrac{F(b) - F(a)}{G(b) - G(a)}$:

$$\frac{F(b) - F(a)}{G(b) - G(a)} = \frac{F'(\xi)(b - a)}{G'(\eta)(b - a)} = \frac{F'(\xi)}{G'(\eta)},$$

where ξ and η are suitable intermediate values in the open interval (a, b). The generalized mean value theorem states that we can write the difference quotient in the simpler form

$$\frac{F(b) - F(a)}{G(b) - G(a)} = \frac{F'(\zeta)}{G'(\zeta)},$$

where F' and G' are evaluated at *the same intermediate value* ζ.

For the proof we introduce $u = G(x)$ as an independent variable in F. From the assumption $G' > 0$ we conclude that the function $u = G(x)$ is monotonic in the interval $[a, b]$, and hence that it has an inverse $x = g(u)$ defined in the interval $[\alpha, \beta]$, where $\alpha = G(a)$, $\beta = G(b)$. The compound function $F[g(u)] = f(u)$ is therefore defined for u in the interval $[\alpha, \beta]$. From the ordinary mean value theorem we find that

$$F(b) - F(a) = f(\beta) - f(\alpha) = f'(\gamma)(\beta - \alpha) = f'(\gamma)[G(b) - G(a)],$$

where γ is a suitable value between α and β. By the chain rule, we infer

$$f'(u) = \frac{d}{du} F[g(u)] = F'[g(u)]g'(u) = \frac{F'(x)}{G'(x)}.$$

To the value $u = \gamma$ there corresponds a value $x = g(\gamma) = \zeta$ in the interval (a, b). Then $f'(\gamma) = F'(\zeta)/G'(\zeta)$, and the generalized mean value theorem follows.

3.4 Some Applications of the Exponential Function

Some miscellaneous problems involving the exponential function will illustrate the fundamental importance of this function in all sorts of applications.

a. Definition of the Exponential Function by Means of a Differential Equation

We can define the exponential function by a simple property, whose use obviates many detailed arguments in particular cases.

If a function $y = f(x)$ satisfies an equation of the form

$$y' = \alpha y,$$

where α is a constant, then y has the form

$$(8) \qquad\qquad y = f(x) = ce^{\alpha x},$$

where c is also a constant; conversely, every function of the form $ce^{\alpha x}$ satisfies the equation $y' = \alpha y$.

Since Eq. (8) expresses a relation between the function and its derivative, it is called the *differential equation* of the exponential function.

It is clear that $y = ce^{\alpha x}$ satisfies this equation for any arbitrary constant c. Conversely, no other function satisfies the differential equation $y' - \alpha y = 0$. For if y is such a function, we consider the function $u = ye^{-\alpha x}$. We then have

$$u' = y'e^{-\alpha x} - \alpha y e^{-\alpha x} = e^{-\alpha x}(y' - \alpha y).$$

However, the right-hand side vanishes, since we have assumed that $y' = \alpha y$; hence $u' = 0$, so that by p. 178 u is a constant c and $y = ce^{\alpha x}$ as we wished to prove.

We shall now apply this theorem to a number of examples.

b. Interest Compounded Continuously.
Radioactive Disintegration

A capital sum, or principal, augmented by its interest at regular periods of time, increases by jumps at these interest periods in the following manner. If 100α is the percent of interest, and furthermore, if the interest accrued is added to the principal at the end of each year, after x years the accumulated amount of an original principal of 1 will be

$$(1 + \alpha)^x.$$

If, however, the principal had the interest added to it not at the end of each year, but at the end of each nth part of a year, after x years the principal would amount to

$$\left(1 + \frac{\alpha}{n}\right)^{nx}.$$

Taking $x = 1$ for the sake of simplicity, we find that the principal 1 has increased after one year to

$$\left(1 + \frac{\alpha}{n}\right)^{n}.$$

If we now let n increase beyond all bounds, that is, if we let the interest be credited at shorter and shorter intervals, the limiting case will mean in a sense that the compound interest is credited at each instant; then the total amount after one year will be e^α times the original principal (see p. 153). Similarly, if the interest is calculated in this manner, an original principal of 1 will have grown after x years to an amount $e^{\alpha x}$; here x may be any number, integral or otherwise,

The discussion in Section 3.4a forms a framework into which examples of this type are readily fitted. We consider a quantity, given by the number y, which increases (or decreases) with time so that the rate at which this quantity increases or decreases is proportional to the total quantity. Then with time as the independent variable x, we obtain a law of the form $y' = \alpha y$ for the rate of increase, where α, the factor of proportionality, is positive or negative depending on whether the quantity is increasing or decreasing. Then in accordance with Section 3.4a the quantity y itself is represented by a formula

$$y = ce^{\alpha x},$$

where the meaning of the constant c is immediately obvious if we consider the instant $x = 0$. At that instant $e^{\alpha x} = 1$, and we find that

$c = y_0$ is the quantity at the beginning of the time considered, so that we may write

$$y = y_0 e^{\alpha x}.$$

A characteristic example is that of *radioactive disintegration*. The rate at which the total quantity y of the radioactive substance is diminishing is proportional at any instant to the total quantity present at that instant; this is a priori plausible, for each portion of the substance decreases as rapidly as every other portion. Therefore the quantity y of the substance expressed as a function of time satisfies a relation of the form $y' = -ky$, where k is to be taken as positive since we are dealing with a diminishing quantity. The quantity of substance is thus expressed as a function of the time by $y = y_0 e^{-kx}$, where y_0 is the amount of the substance at the beginning of the time considered (time $x = 0$).

After a certain time τ the radioactive substance will have diminished to half its original quantity. This so-called *half-life* is given by the equation

$$\tfrac{1}{2} y_0 = y_0 e^{-k\tau},$$

from which we immediately obtain $\tau = (\log 2)/k$.

c. Cooling or Heating of a Body by a Surrounding Medium

Another typical example of the occurrence of the exponential function is the cooling of a body, for example, a metal plate of uniform temperature which is immersed in a very large bath of lower temperature. We assume that the surrounding bath is so large that its temperature is unaffected by the cooling process. We further assume that at each instant all parts of the immersed body are at the same temperature, and that the rate at which the temperature changes is proportional to the difference of the temperature of the body and that of the surrounding medium (Newton's law of cooling).

If we denote the time by x and the temperature difference between the body and the bath by $y = y(x)$, this law of cooling is expressed by the equation

$$y' = -ky,$$

where k is a positive constant (whose value is a physical characteristic of the substance of the body). From this differential equation, which expresses the effect of the cooling process at a given instant, we obtain by means of Eq. (8), p. 223, an "integral law" giving us the temperature at any arbitrary time x in the form

$$y = ce^{-kx}.$$

This shows that the temperature decreases "exponentially" and tends to become equal to the external temperature. The rapidity with which this happens is expressed by the number k. As before, the meaning of the constant c is that of the initial temperature at the instant $x = 0$, $y_0 = c$, so that our law of cooling can be written in the form

$$y = y_0 e^{-kx}.$$

Obviously, the same discussion applies also to the heating of a body. The only difference is that the initial difference of temperature y_0 is in this case negative instead of positive.

d. Variation of the Atmospheric Pressure with the Height above the Surface of the Earth

A further example of the occurrence of the exponential formula is in the variation of atmospheric pressure with height: We make use of (1) the physical fact that the atmospheric pressure is equal to the weight of the column of air vertically above a surface of area one, and (2) of Boyle's law, according to which the pressure p of the air at a given constant temperature is proportional to the density σ of the air. Boyle's law, expressed in symbols, is $p = a\sigma$, where a is a constant depending on a specific physical property of the air. Our problem is to determine $p = f(h)$ as a function of the height h above the surface of the earth.

If by p_0 we denote the atmospheric pressure at the surface of the earth, that is, the total weight of the air column supported by a unit area, by g the gravitational constant, and by $\sigma(\lambda)$ the density of the air at the height λ above the earth, the weight[1] of the column up to the height h is given by the integral $g \int_0^h \sigma(\lambda) \, d\lambda$. The pressure at height h is therefore

$$p = f(h) = p_0 - g \int_0^h \sigma(\lambda) \, d\lambda.$$

By differentiation this yields the following relation between the pressure $p = f(h)$ and the density $\sigma(h)$:

$$g\sigma(h) = -f'(h) = -p'.$$

We now use Boyle's law to eliminate the quantity σ from this equation, thus obtaining an equation $p' = -(g/a)p$ which involves the unknown pressure function only. From Eq. (8) p. 223, it follows that

$$p = f(h) = ce^{-gh/a}.$$

---

[1] $g\sigma(\lambda)$ is the weight of the air per unit volume at the height λ.

If as above we denote the pressure $f(0)$ at the earth's surface by p_0, it follows immediately that $c = p_0$, and consequently

$$p = f(h) = p_0 e^{-gh/a}.$$

Taking the logarithms yields

$$h = \frac{a}{g} \log \frac{p_0}{p}.$$

These two formulas are applied frequently. For example, if the constant a is known, they enable us to find the height of a place from the barometric pressure or to find the difference in height of two places by measuring the atmospheric pressure at each place. Again, if the atmospheric pressure and the height h are known, we can determine the constant a, which is of great importance in gas theory.

e. Progress of a Chemical Reaction

We now consider an example from chemistry, the so-called *unimolecular reaction*. We suppose that a substance is dissolved in a large amount of solvent, say a quantity of cane sugar in water. If a chemical reaction occurs, the chemical law of mass action in this case states that the rate of reaction is proportional to the quantity of reacting substance present. We suppose that the cane sugar is being transformed by catalytic action into invert sugar, and we denote by $u(x)$ the quantity of cane sugar which at time x is still unchanged. The velocity of reaction is then $-du/dx$, and in accordance with the law of mass action an equation of the form

$$\frac{du}{dx} = -ku$$

holds, where k is a constant depending on the substance reacting. From this instantaneous or differential law we immediately obtain, as on p. 223, an integral law, which gives us the amount of cane sugar as a function of the time:

$$u(x) = ae^{-kx}.$$

This formula shows us clearly how the chemical reaction tends asymptotically to its final state $u = 0$, that is, complete transformation of the reacting substance. The constant a is obviously the quantity of cane sugar present at time $x = 0$.

f. Switching an Electric Circuit On or Off

As a final example we consider the growth of a direct electric current when a circuit is completed, or its decay when the circuit is broken. If R is the resistance of the circuit and E the electromotive force (voltage), the current I gradually increases from its original value zero to the steady final value E/R. We have therefore to consider I as a function of the time x. The growth of the current depends on the self-induction of the circuit; the circuit has a characteristic constant L, the *coefficient of self-induction*, of such a nature that, as the current increases, an electromotive force of magnitude $L \, dI/dx$, opposed to the external electromotive force E, is developed. From Ohm's law, asserting that the product of the resistance and the current is at each instant equal to the actual effective voltage, we obtain the relation

$$IR = E - L\frac{dI}{dx}.$$

For

$$f(x) = I(x) - \frac{E}{R}$$

we immediately find $f'(x) = -(R/L)f(x)$, so that by Eq. (8), p. 223, $f(x) = f(0)e^{-Rx/L}$. Recalling $I(0) = 0$, we find $f(0) = -E/R$; thus we obtain the expression

$$I = f(x) + \frac{E}{R} = \frac{E}{R}(1 - e^{-Rx/L})$$

for the current as a function of the time.

This expression shows how the current tends asymptotically to its steady value E/R when the circuit is closed.

3.5   The Hyperbolic Functions

a. Analytical Definition

In many applications the exponential function enters in combinations of the form
$$\tfrac{1}{2}(e^x + e^{-x}) \qquad \text{or} \qquad \tfrac{1}{2}(e^x - e^{-x}).$$

It is convenient to introduce these and similar combinations as special functions; we denote them as follows:

$$(9a) \qquad \sinh x = \frac{e^x - e^{-x}}{2}, \qquad \cosh x = \frac{e^x + e^{-x}}{2},$$

$$(9b) \qquad \tanh x = \frac{e^x - e^{-x}}{e^x + e^{-x}}, \qquad \coth x = \frac{e^x + e^{-x}}{e^x - e^{-x}},$$

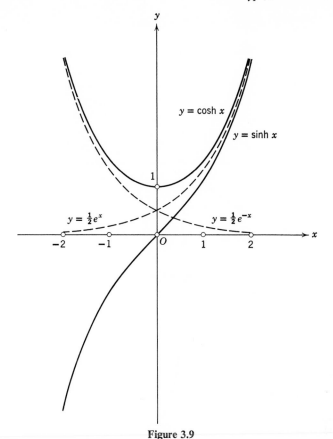

Figure 3.9

and we call them the *hyperbolic sine, hyperbolic cosine, hyperbolic tangent,* and *hyperbolic cotangent* respectively. The functions $\sinh x$, $\cosh x$, and $\tanh x$ are defined for all values of x, whereas for $\coth x$ the point $x = 0$ must be excluded. The names are chosen to express a certain analogy with the trigonometric functions; it is this analogy, which we are about to study in detail, that justifies special consideration of our new functions. In Figs. 3.9, 3.10, and 3.11 the graphs of the hyperbolic functions are shown; the dotted lines in Fig. 3.9 are the graphs of $y = (\frac{1}{2})e^x$ and $y = (\frac{1}{2})e^{-x}$, from which the graphs of $\sinh x$ and $\cosh x$ may easily be constructed.

Cosh x obviously is an *even function*, that is, a function which remains unchanged when x is replaced by $-x$, whereas $\sinh x$ is an *odd function*, that is, a function that changes sign when x is replaced by $-x$ (cf. p. 29).

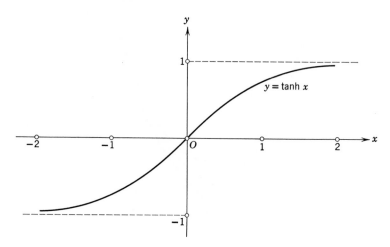

Figure 3.10

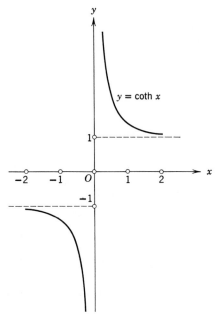

Figure 3.11

By its definition, the function

$$\cosh x = \frac{e^x + e^{-x}}{2}$$

is positive and not less than one for all values of x. It has its least value when $x = 0$: $\cosh 0 = 1$.

The fundamental relation between $\cosh x$ and $\sinh x$

$$\cosh^2 x - \sinh^2 x = 1,$$

follows immediately from the definitions. If we now denote the independent variable by t instead of x and write

$$x = \cosh t, \qquad y = \sinh t,$$

we have

$$x^2 - y^2 = 1;$$

that is, the point with the coordinates $x = \cosh t$, $y = \sinh t$ moves along the rectangular hyperbola $x^2 - y^2 = 1$ as t ranges over the whole scale of values from $-\infty$ to $+\infty$. According to the defining equation, $x \geq 1$, and our formulas make it evident that y runs through the whole scale of values $-\infty$ to $+\infty$ as t does; for if t tends to infinity so does e^t, whereas e^{-t} tends to zero. We may therefore state more exactly: As t runs from $-\infty$ to $+\infty$, the equations $x = \cosh t$, $y = \sinh t$ give us one branch, namely, the right-hand one, of the rectangular hyperbola.

b. Addition Theorems and Formulas for Differentiation

From their definition we obtain the *addition theorems* for the hyperbolic functions:

(10)
$$\cosh (a + b) = \cosh a \cosh b + \sinh a \sinh b,$$
$$\sinh (a + b) = \sinh a \cosh b + \cosh a \sinh b.$$

The proofs are obtained at once if we write

$$\cosh (a + b) = \frac{e^a e^b + e^{-a} e^{-b}}{2}, \qquad \sinh (a + b) = \frac{e^a e^b - e^{-a} e^{-b}}{2},$$

and insert in these equations

$$e^a = \cosh a + \sinh a, \qquad e^{-a} = \cosh a - \sinh a,$$
$$e^b = \cosh b + \sinh b, \qquad e^{-b} = \cosh b - \sinh b.$$

Between these formulas and the corresponding trigonometrical formulas there is a striking analogy. The only difference in the addition theorems is one sign in the first formula.

A corresponding analogy holds for the differentiation formulas. Remembering that $d(e^x)/dx = e^x$, we readily find that

(11)
$$\frac{d}{dx}\cosh x = \sinh x, \qquad \frac{d}{dx}\sinh x = \cosh x,$$

$$\frac{d}{dx}\tanh x = \frac{1}{\cosh^2 x}, \qquad \frac{d}{dx}\coth x = -\frac{1}{\sinh^2 x}.$$

From the first two equations it follows immediately that $y = \cosh x$ and $y = \sinh x$ are solutions of the *differential equation*

(12)
$$\frac{d^2 y}{dx^2} = y,$$

which again only differs in sign from the analogous equation satisfied by the trigonometric functions $\cos x$ and $\sin x$ (see p. 171).

c. The Inverse Hyperbolic Functions

To the hyperbolic functions $x = \cosh t$, $y = \sinh t$, there correspond inverse functions, which we denote[1] by

$$t = \text{ar cosh } x, \qquad t = \text{ar sinh } y.$$

Since the function $\sinh t$ is monotonic increasing[2] throughout the interval $-\infty < t < +\infty$, its inverse function is uniquely determined for all values of y; on the other hand, a glance at the graph (see Fig. 3.9, p. 229) shows that $t = \text{ar cosh } x$ is not uniquely determined, but has an ambiguity of sign, because to a given value of x there corresponds not only the number t but also the number $-t$. Since $\cosh t \geq 1$ for all values of t, its inverse $\text{ar cosh } x$ is defined only for $x \geq 1$.

We can easily express these inverse functions in terms of the logarithm by regarding the quantity $e^t = u$ in the definitions

$$x = \frac{e^t + e^{-t}}{2}, \qquad y = \frac{e^t - e^{-t}}{2}$$

as unknown and solving these (quadratic) equations for u:

$$u = x \pm \sqrt{x^2 - 1}, \qquad u = y + \sqrt{y^2 + 1};$$

since $u = e^t$ can have only positive values, the square root in the second equation must be taken with the positive sign, whereas in the

---

[1] The symbolic notation $\cosh^{-1} x$, etc., is also used; cf. footnote, p. 54.

[2] $(d/dt)\sinh t = \cosh t > 0$.

first either sign is possible (which corresponds to the ambiguity mentioned above). In the logarithmic form, $t = \log u$, and hence

(13)
$$t = \log (x \pm \sqrt{x^2 - 1}) = \text{ar cosh } x,$$
$$t = \log (y + \sqrt{y^2 + 1}) = \text{ar sinh } y.$$

In the case of ar cosh x the variable x is restricted to the interval $x \geq 1$, whereas ar sinh y is defined for all values of y.

Equation (13) gives us two values,

$$\log (x + \sqrt{x^2 - 1}) \qquad \text{and} \qquad \log (x - \sqrt{x^2 - 1})$$

for ar cosh x, corresponding to the two branches of ar cosh x. Since

$$(x + \sqrt{x^2 - 1})(x - \sqrt{x^2 - 1}) = 1,$$

the sum of these two values of ar cosh x is zero, which agrees with the ambiguity in the sign of t mentioned before.

The inverses of the hyperbolic tangent and hyperbolic cotangent can be defined analogously, and can also be expressed in terms of logarithms. These functions we denote by ar tanh x and ar coth x; expressing the independent variable everywhere by x, we readily obtain

(14)
$$\text{ar tanh } x = \frac{1}{2} \log \frac{1 + x}{1 - x} \qquad \text{in the interval } -1 < x < 1,$$
$$\text{ar coth } x = \frac{1}{2} \log \frac{x + 1}{x - 1} \qquad \text{in the intervals } x < -1, x > 1.$$

The differentiation of these inverse functions may be carried out by the reader himself; he may make use of either the rule for differentiating an inverse function or the chain rule in conjunction with these expressions for the inverse functions in terms of logarithms. If x is the independent variable, the results are

(15)
$$\frac{d}{dx} \text{ar cosh } x = \pm \frac{1}{\sqrt{x^2 - 1}}, \qquad \frac{d}{dx} \text{ar sinh } x = \frac{1}{\sqrt{x^2 + 1}},$$
$$\frac{d}{dx} \text{ar tanh } x = \frac{1}{1 - x^2}, \qquad \frac{d}{dx} \text{ar coth } x = \frac{1}{1 - x^2}.$$

The last two formulas do not contradict each other, since the first holds only for $-1 < x < 1$ and the second only for $x < -1$ and $1 < x$. The two values of the derivative $d(\text{ar cosh } x)/dx$, expressed by the sign $\pm$ in the first formula, correspond to the two different branches of the curve $y = \text{ar cosh } x = \log (x \pm \sqrt{x^2 - 1})$.

d. Further Analogies

The similarities between the hyperbolic and the trigonometric functions are no accident. The deeper source of these analogies becomes apparent when we consider these functions for imaginary arguments, as we shall do later in Section 7.7a. We shall then be able to identify $\cosh x$ with $\cos(ix)$ and $\sinh x$ with $(1/i)\sin(ix)$, where $i = \sqrt{-1}$. This fact makes it evident that every relation involving trigonometric functions has its counterpart for hyperbolic functions. Many of those analogies have interesting geometrical or physical interpretations. (See also Chapter 4, p. 363.)

In the above representation of the rectangular hyperbola by the quantity t, we did not ascribe any geometrical meaning to the "parameter" t itself. We shall now return to this subject, and encounter a further analogy between the trigonometric and the hyperbolic functions. If we represent the circle with equation $x^2 + y^2 = 1$ by means of a parameter t in the form $x = \cos t$, $y = \sin t$, we can interpret the quantity t as an angle or as a length of arc measured along the circumference; we may, however, also regard t as twice the area of the circular sector corresponding to that angle, the area being reckoned positive or negative depending on whether the angle is positive or negative.

We now state analogously that for the hyperbolic functions the quantity t is twice the area of the hyperbolic sector for $x^2 - y^2 = 1$ shown shaded in Fig. 3.12.[1] It is this interpretation of t in terms of areas that accounts for the names $t = \operatorname{ar\,cosh} x$ and $t = \operatorname{ar\,sinh} y$ given to the inverse hyperbolic functions.[2] The proof is obtained without difficulty if we refer the hyperbola to its asymptotes as axes by means of the transformation of coordinates

$$x - y = \sqrt{2}\,\xi, \qquad x + y = \sqrt{2}\,\eta,$$

or

$$x = \frac{1}{\sqrt{2}}(\xi + \eta), \qquad y = \frac{1}{\sqrt{2}}(\eta - \xi);$$

with these new coordinates the equation of the hyperbola is $\xi\eta = \tfrac{1}{2}$. Hence the two right triangles OPQ and OAB both have area $\tfrac{1}{4}$, for the lengths of OQ and QP are, respectively, η and $1/2\eta$, and the area in

---

[1] For a different proof, see p. 372.

[2] Just as the notation $t = \operatorname{arc\,cos} x$ refers to an *arc* of the unit circle, so $t = \operatorname{ar\,cosh} x$ refers to an *area* connected with a rectangular hyperbola $x^2 - y^2 = 1$. Incidentally, t is *not* the length of the hyperbolic arc.

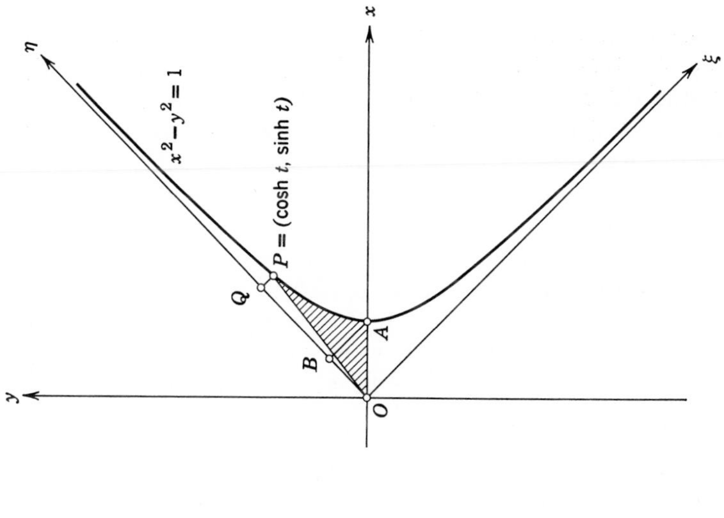

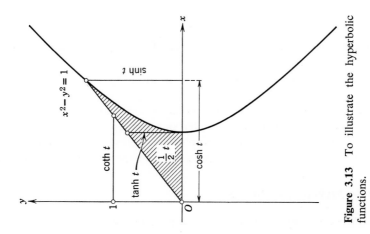

Figure 3.13 To illustrate the hyperbolic functions.

Figure 3.12

question is equal to that of the figure $ABQP$. Obviously the coordinates of the points A and B are

$$\xi = \frac{1}{\sqrt{2}}, \eta = \frac{1}{\sqrt{2}} \quad \text{and} \quad \xi = \frac{x-y}{\sqrt{2}}, \eta = \frac{x+y}{\sqrt{2}}$$

respectively, and for double the area of our figure we thus obtain

$$2 \int_{1/\sqrt{2}}^{(x+y)/\sqrt{2}} (1/2\eta)\, d\eta = \log(x+y) = \log(x \pm \sqrt{x^2-1}),$$

but by Eq. (13), p. 233, this is equal to t, proving our assertion.

In conclusion, it may be pointed out that, as shown in Fig. 3.13, the hyperbolic functions can be graphically represented on the hyperbola, just as the trigonometric functions can be represented on the circle.

3.6   Maxima and Minima

As the first of a great variety of applications we consider the theory of maxima and minima of a function, in conjunction with a geometrical discussion of the second derivative.

a. Convexity and Concavity of Curves

By definition the derivative $f'(x) = df(x)/dx$ represents the slope of the curve $y = f(x)$. The derivative of the function $f'(x)$ or of the slope of the curve $y = f(x)$ is given by the derivative $df'(x)/dx = d^2f(x)/dx^2 = f''(x)$, the *second derivative* of $f(x)$, and so on. If the second derivative $f''(x)$ is positive at a point x—so that owing to continuity (which we assume) it is positive in some neighborhood[1] of this point x—then throughout this neighborhood the derivative $f'(x)$ increases with increasing values of x. Hence the curve $y = f(x)$ turns its convex side downwards or is "open" upwards. We call the function $f(x)$ or the curve $y = f(x)$ *convex*. If $f''(x)$ is negative, the curve and the function are *concave*. Therefore when $f''(x) > 0$, the curve in the neighborhood of the point lies above the tangent while when $f''(x) < 0$,

---

[1] We make use here of the intuitively obvious observation: a *continuous* function $g(x)$ which is positive at a point x_0 also is positive for all points of a sufficiently small neighborhood of x_0 (as far as they belong to the domain of g). The formal proof is simple. From the continuity of g at x_0 we know that for *every* positive ϵ the inequality $|g(x) - g(x_0)| < \epsilon$ holds for all x in a sufficiently small neighborhood $|x - x_0| < \delta$ of the point x_0. Since $g(x_0) > 0$, we are free to choose for ϵ the value $\frac{1}{2}g(x_0)$, so that $|g(x) - g(x_0)| < \frac{1}{2}g(x_0)$ in some neighborhood. Since then $g(x_0) - g(x) \leq |g(x) - g(x_0)| < \frac{1}{2}g(x_0)$, it follows that $g(x) > \frac{1}{2}g(x_0) > 0$.

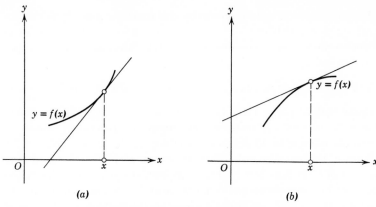

Figure 3.14   (a) $f''(x) > 0$.   (b) $f''(x) < 0$.

it lies below the tangent (see Figs. 3.14a and 3.14b) (cf. Problem 4, p. 200 and Section 5.6).

Point of Inflection

Special consideration is required only in points where $f''(x) = 0$. On passing through such a point the second derivative $f''(x)$ will generally change its sign. Such a point will then be a point of transition between the two cases just indicated; that is, on one side the tangent is above the curve and on the other side below it, whereas at this point it crosses the curve (see Fig. 3.15). Such a point is called a *point of inflection* of the curve, and the corresponding tangent is called an *inflectional tangent*.

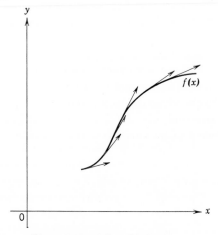

Figure 3.15   Point of inflection.

The simplest example is given by the function $y = x^3$, the cubical parabola, for which the x-axis itself is an inflectional tangent at the inflection point $x = 0$ (see Fig. 3.3, p. 209). Another example is given by the function $f(x) = \sin x$, for which

$$f'(x) = d(\sin x)/dx = \cos x \quad \text{and} \quad f''(x) = d^2(\sin x)/dx^2 = -\sin x.$$

Consequently, $f'(0) = 1$ and $f''(0) = 0$; since the sign of $f''(x)$ changes at $x = 0$, the sine curve has at the origin an inflectional tangent inclined at an angle of 45 degrees to the x-axis.

It must, however, be noted that points can exist where $f''(x) = 0$ and the sign of $f''(x)$ does not change with increasing x, while the tangent does not cut the curve but remains entirely on one side of it. For example, the curve $y = x^4$ lies entirely above the x-axis, although the second derivative $f''(x) = 12x^2$ vanishes for $x = 0$.

b. Maxima and Minima—Relative Extrema.   Stationary Points

A function $f(x)$ has a *maximum* at a point ξ if the value of f at the point ξ is not exceeded by the value of f at any other point x of the domain of f; that is, $f(\xi) \geq f(x)$ for all x where f is defined.[1]   Similarly, f has a *minimum* at ξ if $f(\xi) \leq f(x)$ for all x in the domain. The word *extrema* is used to cover both maxima and minima.

The function $f(x) = \sqrt{1 - x^2}$, for example, which is defined for $-1 \leq x \leq 1$, has minima at $x = \pm 1$ and a maximum at $x = 0$. It is easy to give examples of continuous functions which have no maxima or no minima. Thus the function $f(x) = 1/(1 + x^2)$ (Fig. 3.8, p. 216) in the domain $-\infty < x < +\infty$ has no minimum; the function $f(x) = 1/x$ defined for $0 < x < \infty$ has no extremum points at all. We recall, however, from Chapter 1, p. 101 the theorem of Weierstrass, according to which a continuous function defined in a closed finite interval always has a maximum (and similarly a minimum) there.

Our object is to find a means of locating the extrema of a function or curve. This problem which is encountered very frequently in geometry, mechanics, physics, and other fields was one of the principal incentives for the development of the calculus in the seventeenth century.

Calculus does not furnish a direct method for picking out the extrema of a function $f(x)$, but it permits us to locate the so-called *relative extremum* points, among which the actual maxima and minima have to occur. The point ξ is a *relative* maximum (minimum) of f if f

---

[1] We talk of a *strict maximum point* ξ if $f(\xi) > f(x)$ for all x in the domain of f that are different from ξ.

has its greatest (least) value at ξ when compared not with all possible values of $f(x)$ but just with the values of $f(x)$ for x in some *neighborhood* of ξ. By a neighborhood of the point ξ we mean here any open interval $\alpha < x < \beta$ which contains the point ξ but may be arbitrarily small. A relative extremum point ξ of f is then a point which is an extremum point when f is restricted to all those points of its domain lying sufficiently close to ξ.[1]   Obviously, the extrema of the function are included among the relative extrema. To avoid confusion we shall use

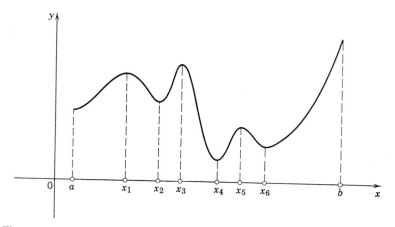

Figure 3.16   Graph of function defined on the interval $[a, b]$ with relative minima at $x = a, x_2, x_4, x_6$, relative maxima at x_1, x_3, x_5, b, absolute maximum at b, and absolute minimum at x_4.

the terms *absolute maxima (minima)* for the maxima and minima of f in its entire domain (see Fig. 3.16).

Geometrically speaking, relative maxima and minima, if not located in the end points of the interval of definition, are respectively the wave crests and troughs of the curve. A glance at Fig. 3.16 shows that the value of a relative maximum at one point x_5 may very well be less than the value of a relative minimum at another point x_2. The diagram also suggests the fact that relative maxima and minima of a continuous function alternate: Between two successive relative maxima there is always located a relative minimum.

Let $f(x)$ be a differentiable function defined in the closed interval $a \leq x \leq b$. We see at once that at a relative extremum point which is

---

[1] The formal definition of a relative maximum point ξ would state that there exists an open interval containing ξ such that $f(\xi) \geq f(x)$ for all x of that interval for which f is defined.

located in the interior of the interval the tangent to the curve must be horizontal. (The formal proof is given below.) Hence the condition

$$f'(\xi) = 0$$

is necessary for a relative extremum at the point ξ with $a < \xi < b$. If, however, $f(\xi)$ is a relative extremum and ξ coincides with one of the end points of the interval of definition, the equation $f'(\xi) = 0$ need not hold. We can only say that if the left-hand end point is a relative maximum (minimum) point, the slope $f'(a)$ of the curve cannot be positive (negative), while if the right-hand end point b is a relative maximum (minimum) then $f'(b)$ cannot be negative (positive).

The points at which the tangent to the curve $y = f(x)$ is horizontal, corresponding to the roots ξ of the equation $f'(\xi) = 0$, are called the *critical* points or *stationary* points of f. All relative extrema of a differentiable function f which are interior points of the domain of f are stationary points. Hence: *an absolute maximum or minimum of the function coincides either with a critical point of the function or with an end point of its domain.* In order to locate the absolute maxima (minima) of the function we have only to compare the values of f in the critical points and in the two end points and to see which of these values are greatest (least). If f fails at a finite number of points to have a derivative, we have only to add those points to the list of possible locations of an extremum and to check also the values of f at those points. Thus the main labor in determining the extrema of a function is reduced to that of finding the zeros of the derivative of the function, which usually are finite in number.

To take a simple example, let us determine the largest and smallest values of the function $f(x) = \frac{1}{10}x^6 - \frac{3}{10}x^2$ in the interval $-2 \le x \le 2$. Here the critical points, the roots of the equation $f'(x) = 6(x^5 - x)/10 = 0$ are located at $x = 0, +1, -1$. Computing the values of f at those points and also at the end points of the interval, we find

x	-2	-1	0	1	2
$f(x)$	5.2	-0.2	0	-0.2	5.2

It is clear that the points $x = \pm 1$ represent relative minima, whereas relative maxima occur at $x = 0$ and $x = \pm 2$. The maximum value of the function, assumed in the end points of the interval, is 5.2; the minimum value, assumed in the points $x = \pm 1$, is -0.2 (see Fig. 3.17).

Without appealing to intuition we can easily prove by purely analytic methods that $f'(\xi) = 0$ whenever ξ is a relative extremum point in the interior of the domain of f provided f is differentiable at ξ. (Compare

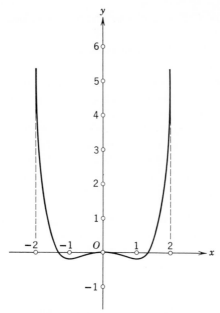

Figure 3.17 $y = (x^6 - 3x^2)/10$.

the exactly analogous considerations for Rolle's theorem, p. 175.) If the function $f(x)$ has a relative maximum at the point ξ, then for all sufficiently small values of h different from zero the expression $f(\xi + h) - f(\xi)$ must be negative or zero. Therefore

$$\frac{[f(\xi + h) - f(\xi)]}{h} \leq 0$$

for $h > 0$, whereas

$$\frac{[f(\xi + h) - f(x)]}{h} \geq 0$$

for $h < 0$. Thus if h tends to zero through positive values, the limit cannot be positive, whereas if h tends to zero through negative values, the limit cannot be negative. However, since we have assumed that the derivative at ξ exists, these two limits must be equal to one another, and, in fact, to the value $f'(\xi)$, which therefore can only be zero; we must have $f'(\xi) = 0$. A similar proof holds for a relative minimum. The proof also shows that if the left-hand end point $\xi = a$ is a relative maximum (minimum) point, then at least $f'(a) \leq 0$ [$f'(a) \geq 0$]; if the right-hand end point b is a relative maximum (minimum) point, then $f'(b) \geq 0$ [$f'(b) \leq 0$].

The condition $f'(\xi) = 0$ characterizing the critical points is by no means sufficient for the occurrence of a relative extremum. There may be points at which the derivative vanishes, that is, at which the tangent is horizontal, although the curve has neither a relative maximum nor minimum there. This occurs if at the given point the curve has a *horizontal inflectional tangent* cutting it, as in the example of the function $y = x^3$ at the point $x = 0$.

The following test gives the conditions under which a critical point is a point of relative maximum or minimum. It applies to a continuous function f, having a continuous derivative f' which vanishes at most at a *finite* number of points or, more generally to differentiable functions f for which f' changes sign at most at a finite number of points:

The function $f(x)$ has a relative extremum at an interior point ξ of its domain if, and only if, the derivative $f'(x)$ changes sign as x passes through this point; in particular, the function has a relative minimum if near ξ the derivative is negative to the left of ξ and positive to the right, whereas in the contrary case it has a maximum.

We prove this rigorously by using the mean value theorem. First, we observe that to the left and right of ξ there exist intervals $\xi_1 < x < \xi$ and $\xi < x < \xi_2$, in each of which $f'(x)$ has only one sign, since f' vanishes only at finite number of points. (Here ξ_1 and ξ_2 can be taken as the points nearest to ξ at which f' vanishes, if such points exist.) If the signs of $f'(x)$ in these two intervals are different, then $f(\xi + h) - f(\xi) = hf'(\xi + \theta h)$ has the same sign for all numerically small values of h, whether h is positive or negative, so that ξ is a relative extremum. If $f'(x)$ has the same sign in both intervals, then $hf'(\xi + \theta h)$ changes sign when h does, so that $f(\xi + h)$ is greater than $f(\xi)$ on one side and less than $f(\xi)$ on the other side, and there is no extreme value. Our theorem is thus proved.

At the same time we see that *the value $f(\xi)$ is the greatest or least value of the function, in every interval containing the point ξ, in which f is differentiable and in which the only change of sign of $f'(x)$ occurs at ξ itself.*

The mean value theorem on which this proof is based can still be used if $f(x)$ is not differentiable at an end point of the interval in which it is applied, provided that $f(x)$ is differentiable at all the other points of the interval; hence this proof still holds if $f'(x)$ does not exist at $x = \xi$. For example, the function $y = |x|$ has a minimum at $x = 0$, since $y' > 0$ for $x > 0$ and $y' < 0$ for $x < 0$ (cf. Fig. 2.24, p. 167). The function $y = \sqrt[3]{x^2}$ likewise has a minimum at the point $x = 0$, even though its derivative $\frac{2}{3}x^{-\frac{1}{3}}$ is infinite there (cf. Fig. 2.27, p. 169).

The simplest method for deciding whether a critical point ξ is a relative maximum or minimum involves the *second derivative* at that point. It is intuitively clear that if $f'(\xi) = 0$, then f has a relative maximum at ξ if $f''(\xi) < 0$, and a relative minimum if $f''(\xi) > 0$. For in the first case the curve in the neighborhood of this point lies completely below the tangent, and in the second case completely above the tangent. This result follows analytically from the preceding test, provided that $f(x)$ and $f'(x)$ are continuous and that $f''(\xi)$ exists. For if $f'(\xi) = 0$ and, say, $f''(\xi) > 0$, we have

$$f''(\xi) = \lim_{h \to 0} \frac{f'(\xi + h) - f'(\xi)}{h} = \lim_{h \to 0} \frac{f'(\xi + h)}{h} > 0.$$

It follows that $f'(\xi + h)/h > 0$ for all $h \neq 0$ which are sufficiently small in absolute value; hence $f'(\xi + h)$ and h have the same sign in a neighborhood of ξ. For x near ξ the derivative $f'(x)$ must be negative for x to the left of ξ, and positive for x to the right of ξ; this implies that there is a relative minimum at ξ.

The situation is particularly simple in case $f''(x)$ is of one and the same sign throughout the interval $[a, b]$ in which f is defined:

A point ξ at which f' vanishes is a maximum point of f if $f''(x) < 0$ throughout the interval (or if its curve is concave), and a minimum point of f if throughout the interval $f''(x) > 0$ (that is, if the curve is convex).

Indeed, if $f''(x) < 0$ the function $f'(x)$ is monotonic decreasing, hence has ξ as its only zero. Moreover, $f' > 0$ for $a \leq x < \xi$, whereas $f' < 0$ for $\xi < x \leq b$. By the mean value theorem this implies again that $f(x) < f(\xi)$ for $x \neq \xi$, so that ξ turns out to be a strict maximum point. The minimum of f must coincide with one of the end points since there is no other critical point besides ξ. The same argument applies when $f'' > 0$ in the interval.

Examples

Example 1. Of all triangles with given base and given area, to find that with the least perimeter.

To solve this problem, we take the x-axis along the given base AB and the middle point of AB as the origin (Fig. 3.18). If C is the vertex of the triangle, h its altitude (which is fixed by the area and the base), and (x, h) are the coordinates of the vertex, then the sum of the two sides AC and BC of the triangle is given by

$$f(x) = \sqrt{(x + a)^2 + h^2} + \sqrt{(x - a)^2 + h^2},$$

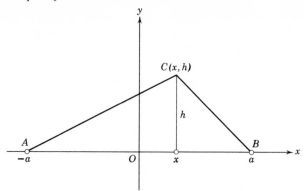

Figure 3.18

where $2a$ is the length of the base. From this we obtain

$$f'(x) = \frac{x + a}{\sqrt{(x + a)^2 + h^2}} + \frac{x - a}{\sqrt{(x - a)^2 + h^2}},$$

$$f''(x) = \frac{-(x + a)^2}{\sqrt{[(x + a)^2 + h^2]^3}} + \frac{1}{\sqrt{(x + a)^2 + h^2}}$$

$$+ \frac{-(x - a)^2}{\sqrt{[(x - a)^2 + h^2]^3}} + \frac{1}{\sqrt{(x - a)^2 + h^2}}$$

$$= \frac{h^2}{\sqrt{[(x + a)^2 + h^2]^3}} + \frac{h^2}{\sqrt{[(x - a)^2 + h^2]^3}}.$$

We see at once (1) that $f'(0)$ vanishes, and (2) that $f''(x)$ is always positive; hence at $x = 0$ there is a least value (see p. 243). This least value is accordingly given by the isosceles triangle.

Similarly, we find that of all the triangles with a given perimeter and a given base the isosceles triangle has the greatest area.

Example 2. To find a point on a given straight line such that the sum of its distances from two given fixed points is a minimum.

Let there be given a straight line and two fixed points A and B on the same side of the line. We wish to find a point P on the straight line such that the distance $PA + PB$ has the least possible value.[1]

We take the given line as the x-axis and use the notation of Fig. 3.19. Then the distance in question is given by

$$f(x) = \sqrt{x^2 + h^2} + \sqrt{(x - a)^2 + h_1^2},$$

---

[1] If A and B lie on opposite sides of the line, P obviously is just the intersection of the line with the segment AB.

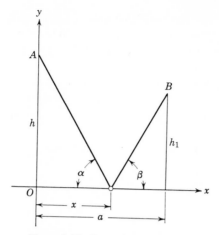

Figure 3.19   Law of reflection.

and we obtain

$$f'(x) = \frac{x}{\sqrt{x^2 + h^2}} + \frac{x - a}{\sqrt{(x - a)^2 + h_1^2}},$$

$$f''(x) = \frac{h^2}{\sqrt{(x^2 + h^2)^3}} + \frac{h_1^2}{\sqrt{[(x - a)^2 + h_1^2]^3}}.$$

The equation $f'(\xi) = 0$ means

$$\frac{\xi}{\sqrt{\xi^2 + h^2}} = \frac{a - \xi}{\sqrt{(\xi - a)^2 + h_1^2}},$$

or

$$\cos \alpha = \cos \beta;$$

hence the two lines PA and PB must form equal angles with the given line. The positive sign of $f''(x)$ shows us that we really have a least value.

The solution of this problem is closely connected with the optical law of reflection. By an important principle of optics, known as Fermat's *principle of least time*, the path of a light ray is determined by the property that the time the light takes to go from a point A to a point B under the given conditions must be the least possible. If the condition is imposed that a ray of light shall on its way from A to B pass through some point on a given straight line (say on a mirror), we see that the shortest time will be taken along the ray for which the "angle of incidence" is equal to the "angle of reflection."

Example 3. The Law of Refraction.[1]　Let there be given two points
A and B on opposite sides of the x-axis. What is the path from A to B
requiring the shortest possible time if the velocity on one side of the
x-axis is c_1 and on the other side c_2?

Clearly, this shortest path must consist of two portions of straight
lines meeting one another at a point P on the x-axis. Using the notation
of Fig. 3.20, we obtain the two expressions $\sqrt{h^2 + x^2}$ and $\sqrt{h_1^2 + (a - x)^2}$

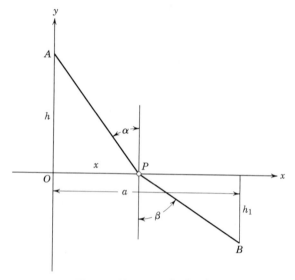

Figure 3.20　Law of refraction.

for the lengths PA, PB, respectively, and we find the time of passage
along this path by dividing the lengths of the two segments by the
corresponding velocities and then adding;

$$f(x) = \frac{1}{c_1} \sqrt{h^2 + x^2} + \frac{1}{c_2} \sqrt{h_1^2 + (a - x)^2}.$$

By differentiation, we obtain

$$f'(x) = \frac{1}{c_1} \frac{x}{\sqrt{h^2 + x^2}} - \frac{1}{c_2} \frac{a - x}{\sqrt{h_1^2 + (a - x)^2}},$$

$$f''(x) = \frac{1}{c_1} \frac{h^2}{\sqrt{(h^2 + x^2)^3}} + \frac{1}{c_2} \frac{h_1^2}{\sqrt{[h_1^2 + (a - x)^2]^3}}.$$

---

[1] While the preceding examples can be treated also by elementary geometry,
this one is not easily disposed of without calculus.

As we readily see from Fig. 3.20, the equation $f'(x) = 0$, that is, the equation

$$\frac{1}{c_1} \frac{x}{\sqrt{h^2 + x^2}} = \frac{1}{c_2} \frac{a - x}{\sqrt{h_1^2 + (a - x)^2}},$$

is equivalent to the condition $(1/c_1) \sin \alpha = (1/c_2) \sin \beta$, or

$$\frac{\sin \alpha}{\sin \beta} = \frac{c_1}{c_2}.$$

The reader should verify the fact that there is only *one* point which satisfies this condition and that this point actually yields the required least value.

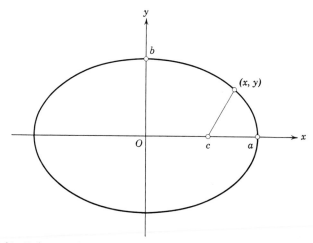

Figure 3.21   Point on ellipse having the least distance from a point on the major axis.

The physical meaning of our example is again given by the optical principle of least time. A ray of light traveling between two points describes the path of shortest time. If c_1 and c_2 are the velocities of light on either side of the boundary of two optical media, the path of the light will be that given by our result, which is a form of *Snell's law of refraction*.

Example 4. Find the point of an ellipse which is closest to a given point on its major axis (Fig. 3.21).

Taking the ellipse in the form

$$\frac{x^2}{a^2} + \frac{y^2}{b^2} = 1 \qquad (b < a)$$

and the given point on the major axis as $(c, 0)$, we find for the distance of any point (x, y) on the ellipse from the point $(c, 0)$, the expression

$$d = \sqrt{(x - c)^2 + b^2(1 - x^2/a^2)},$$

where $-a \leq x \leq a$. The function $f(x) = d^2$ is convex $(f'' > 0)$. It has a minimum for the same x as d itself. The only critical point of f is at $x = c/(1 - b^2/a^2)$. If this point lies in the domain of d, it represents the minimum point; if not, the minimum of d corresponds to the end point of the major axis closest to c. We find accordingly for the minimum distance the values

$$d = b\sqrt{1 - \frac{c^2}{a^2 - b^2}} \qquad \text{if} \quad |c| \leq a\left(1 - \frac{b^2}{a^2}\right),$$

$$d = a - |c| \qquad \text{if} \quad |c| \geq a\left(1 - \frac{b^2}{a^2}\right).$$

*3.7   The Order of Magnitude of Functions

Differences in the behavior of functions for large values of the argument, lead to the notion of the *order of magnitude*. Because of its great importance, this matter deserves a brief discussion here even though it is not directly connected with the idea of the integral or of the derivative.

a. The Concept of Order of Magnitude. The Simplest Cases

If the variable x increases beyond all bounds, then, for $\alpha > 0$, the functions x^α, $\log x$, e^x, $e^{\alpha x}$ also increase beyond all bounds. They increase, however, in essentially different ways. For example, the function x^3 becomes "infinite to a higher order" than x^2; by this we mean: as x increases, the quotient x^3/x^2 itself increases beyond all bounds. Similarly, the function x^α becomes infinite to a higher order than x^β if $\alpha > \beta > 0$, etc.

Quite generally, we shall say of two functions $f(x)$ and $g(x)$, whose absolute values increase with x beyond all bounds, that $f(x)$ *becomes infinite of a higher order than* $g(x)$ if for $x \to \infty$ the quotient $|f(x)/g(x)|$ increases beyond all bounds; we shall say that $f(x)$ becomes infinite of a lower order than $g(x)$ if the quotient $|f(x)/g(x)|$ tends to zero as x increases; and we shall say that the two functions become infinite of the same order of magnitude if as x increases, the quotient $|f(x)/g(x)|$

possesses a limit different from zero or at least remains between two fixed positive bounds. For example, the function $ax^3 + bx^2 + c = f(x)$, where $a \neq 0$, will be of the same order of magnitude as the function $x^3 = g(x)$; for the quotient $|f(x)/g(x)| = |(ax^3 + bx^2 + c)/x^3|$ has the limit $|a|$ as $x \to \infty$; on the other hand the function $x^3 + x + 1$ becomes infinite of a higher order of magnitude than the function $x^2 + x + 1$.

A sum of two functions $f(x)$ and $\phi(x)$, where $f(x)$ is of higher order of magnitude than $\phi(x)$, has the same order of magnitude as $f(x)$. For $|(f(x) + \phi(x))/f(x)| = |1 + \phi(x)/f(x)|$, and by hypothesis this expression tends to one as x increases.

b. The Order of Magnitude of the Exponential Function and of the Logarithm

We might be tempted to measure the order of magnitude of functions by a scale, assigning to the quantity x the order of magnitude one and to the power x^α ($\alpha > 0$) the order of magnitude α. A polynomial of the nth degree then obviously would have the order of magnitude n; a rational function, the degree of whose numerator is higher by h than that of the denominator, would have the order of magnitude h.

It turns out, however, that any attempt to describe the order of magnitude of arbitrary functions by the foregoing scale must fail. For there are functions that become infinite of higher order than the power x^α of x, no matter how large α is chosen; again, there are functions which become infinite of lower order than the power x^α, no matter how small the positive number α is chosen. These functions therefore will not fit in our scale.

Without entering into a detailed theory we state the following theorem.

THEOREM. *If a is an arbitrary number greater than one, then the quotient a^x/x tends to infinity as x increases.*

PROOF. To prove this we construct the function

$$\phi(x) = \log \frac{a^x}{x} = x \log a - \log x;$$

it is obviously sufficient to show that $\phi(x)$ increases beyond all bounds if x tends to $+\infty$. For this purpose we consider the derivative

$$\phi'(x) = \log a - \frac{1}{x}$$

and notice that for $x \geq c = 2/\log a$ this is not less than the positive number $\frac{1}{2} \log a$. Hence it follows that for $x \geq c$

$$\phi(x) - \phi(c) = \int_c^x \phi'(t) \, dt \geq \int_c^x \tfrac{1}{2} \log a \, dt \geq \tfrac{1}{2}(x - c) \log a,$$

$$\phi(x) \geq \phi(c) + \tfrac{1}{2}(x - c) \log a,$$

and the right-hand side becomes infinite for $x \to \infty$.

We give a second proof of this important theorem: with $\sqrt{a} = b = 1 + h$, we have $b > 1$ and $h > 0$. Let n be the integer such that $n \leq x < n + 1$; we may take $x > 1$, so that $n \geq 1$. Applying the lemma of p. 64, we have

$$\sqrt{\frac{a^x}{x}} = \frac{b^x}{\sqrt{x}} = \frac{(1 + h)^x}{\sqrt{x}} > \frac{(1 + h)^n}{\sqrt{n + 1}} > \frac{1 + nh}{\sqrt{n + 1}} > \frac{nh}{\sqrt{2n}} = \frac{h}{\sqrt{2}} \sqrt{n},$$

so that

$$\frac{a^x}{x} > \frac{h^2}{2} \cdot n,$$

and therefore tends to infinity with x.

From the fact just proved many others follow. For example: for every positive index α and every number $a > 1$ the quotient a^x/x^α tends to infinity as x increases; that is,

THEOREM. *The exponential function becomes infinite of a higher order of magnitude than any power of x.*

For the proof we need show only that the αth root of the expression, that is,

$$\frac{a^{x/\alpha}}{x} = \frac{1}{\alpha} \frac{a^{x/\alpha}}{x/\alpha} = \frac{1}{\alpha} \cdot \frac{a^y}{y} \qquad \left(y = \frac{x}{\alpha} \right),$$

tends to infinity. This, however, follows immediately from the preceding theorem when x is replaced by $y = x/\alpha$.

In a similar fashion we prove the following theorem. For every positive value of α the quotient $(\log x)/x^\alpha$ tends to zero for $x \to \infty$; that is

THEOREM. *The logarithm becomes infinite of a lower order of magnitude than any arbitrarily small positive power of x.*

PROOF. The proof follows immediately if we put $\log x = y$ so that our quotient is transformed into $y/e^{\alpha y}$. We then put $e^\alpha = a$; then

$a > 1$, and our quotient y/a^y approaches zero as y tends to infinity. Since y approaches infinity as x does, our theorem is proved.[1]

On the basis of these results we can construct functions of an order of magnitude far higher than that of the exponential function and other functions of an order of magnitude far lower than that of the logarithm. For example, the function $e^{(e^x)}$ is of a higher order than the exponential function, and the function $\log \log x$ is of a lower order than the logarithm; moreover we can iterate these processes as often as we like, piling up the symbols e or log to any extent we please.

All the functions x, $\log x$, $\log (\log x)$, $\log [\log (\log x)]$, etc., eventually become arbitrarily large for sufficiently large x, but with increasing slowness. Taking, for example, for x the tremendous number $x = 10^{100}$ we find that $\log x$ is about 230, whereas $\log (\log x)$ is only about 5.4.

c. General Remarks

These considerations show that it is not possible to assign to all functions definite numbers as orders of magnitude so that of two functions the one with the higher order of magnitude has a higher number. If, for example, the function x is of the order of magnitude one and the function $x^{1+\epsilon}$ of the order of magnitude $1 + \epsilon$, then the function $x \log x$ must be of an order of magnitude that is greater than one and less than $1 + \epsilon$ no matter how small ϵ is chosen. But there is no such number.

In addition, it is easy to see that functions need not possess a clearly defined relative order of magnitude at all. For example, the function $[x^2(\sin x)^2 + x + 1]/[x^2(\cos x)^2 + x]$ approaches no definite limits as x increases; on the contrary, for $x = n\pi$ (where n is an integer) the value is $1/n\pi$, whereas for $x = (n + \frac{1}{2})\pi$ it is $(n + \frac{1}{2})\pi + 1 + 1/(n + \frac{1}{2})\pi$. Although the numerator and denominator both become infinite, the quotient neither remains between positive bounds nor tends to zero nor tends to infinity. The numerator, therefore, is neither of the same order as the denominator, nor of lower order nor of higher order. This apparently startling situation merely means that our definitions are not designed in such a way that we can compare every pair of functions. This is not a defect; we have no desire to compare the orders of such

---

[1] Another simple proof may be suggested: For $x > 1$ and $\epsilon > 0$

$$\log x = \int_1^x \frac{d\xi}{\xi} < \int_1^x \xi^{\epsilon-1}\, d\xi = \frac{1}{\epsilon}(x^\epsilon - 1);$$

if we choose ϵ equal to α and divide both members of this inequality by x^α, then it follows that $(\log x)/x^\alpha \to 0$ as $x \to \infty$.

functions as the numerator and denominator above; knowledge of the value of one of them gives us no useful information about the other.

d. The Order of Magnitude of a Function in the Neighborhood of an Arbitrary Point

Just as we may compare the behavior of functions for $x \to \infty$ we may also compare functions that become infinite at the finite point $x = \xi$.

We say that the function $f(x) = 1/|x - \xi|$ becomes infinite of the first order at the point $x = \xi$, and correspondingly that the function $1/|x - \xi|^\alpha$ becomes infinite of the order α, provided that α is positive.

We recognize then that the function $e^{1/|x-\xi|}$ becomes for $x \to \xi$ infinite of higher order and the function $\log|x - \xi|$ infinite of lower order than all these powers; that is, that the limiting relations

$$\lim_{x \to \xi} (|x - \xi|^\alpha \cdot e^{1/|x-\xi|}) = \infty \quad \text{and} \quad \lim_{x \to \xi} (|x - \xi|^\alpha \cdot \log|x - \xi|) = 0$$

hold.

To confirm this we merely put $1/|x - \xi| = y$; our statements then reduce to the known theorem on p. 249, since

$$|x - \xi|^\alpha \cdot e^{1/|x-\xi|} = \frac{e^y}{y^\alpha} \quad \text{and} \quad |x - \xi|^\alpha \cdot \log|x - \xi| = -\frac{\log y}{y^\alpha}$$

and y increases beyond all bounds as x tends to ξ. (The method of reducing the behavior at a point ξ to the behavior at infinity by the substitution $1/|x - \xi| = y$ frequently proves useful.)

e. The Order of Magnitude (or Smallness) of a Function Tending to Zero

Just as we seek to describe the approach of a function to infinity by means of the concept of order of magnitude, we may also specify the way in which a function approaches zero. We say that as $x \to \infty$ the quantity $1/x$ vanishes to the first order, the quantity $x^{-\alpha}$, where α is positive, to the order α. We find once again that the *function* $1/\log x$ *vanishes to a lower order than an arbitrary power* $x^{-\alpha}$, that is, for every positive α the relation

$$\lim_{x \to \infty} (x^{-\alpha} \cdot \log x) = 0$$

holds.

In the same way we say that for $x = \xi$ the quantity $x - \xi$ vanishes to the first order, the quantity $|x - \xi|^\alpha$ to the order α. With our results it is easy to prove the relations

$$\lim_{x \to 0} (|x|^\alpha \cdot \log |x|) = 0, \qquad \lim_{x \to 0} (|x|^{-\alpha} \cdot e^{-1/|x|}) = 0,$$

which are usually expressed as follows:

The function $1/\log |x|$ *vanishes as* $x \to 0$ *to a lower order than any power of* x; *the exponential function* $e^{-1/|x|}$ *vanishes to a higher order than any power of* x.

f. The "O" and "o" Notation for Orders of Magnitude

A convenient way to indicate that a function $f(x)$ is of lower order of magnitude than a function $g(x)$ is to write $f = o(g)$. This symbolic equation signifies only that the quotient f/g has the limit zero, and can be used to equal advantage for functions vanishing or becoming infinite and for arguments x tending to infinity or approaching a value ξ.[1]

We can rewrite many of the results of the previous section in this notation; for example,

$$x^\alpha = o(x^\beta) \qquad \text{for } \alpha < \beta \qquad \text{as } x \to \infty$$

$$\log x = o(x^\alpha) \qquad \text{for } \alpha > 0 \qquad \text{as } x \to \infty$$

$$e^{-x} = o(x^{-\alpha}) \qquad \text{as } x \to \infty$$

$$e^{-1/x} = o(x^\alpha) \qquad \text{as } x \to 0 \qquad \text{through positive values}$$

$$\log |x| = o(1/x) \qquad \text{as } x \to 0$$

$$1 - \cos x = o(x) \qquad \text{as } x \to 0.$$

This notation, introduced by E. Landau, is useful for indicating the order of magnitude of the error in an approximation formula. For example,

$$\frac{1}{\sqrt{1 + 4x^2}} = \frac{1}{2x} + o\left(\frac{1}{x}\right) \qquad \text{for } x \to \infty$$

stands for the relation

$$\lim_{x \to \infty} \frac{\dfrac{1}{\sqrt{1 + 4x^2}} - \dfrac{1}{2x}}{1/x} = 0.$$

---

[1] The letter o is chosen to suggest the word "order." Observe that the relation $f = o(g)$ for vanishing g means that f vanishes of higher order.

Similarly, the relation between increment and differential of a function f which has a derivative at the point x can be written in the form

$$f(x + h) - f(x) = hf'(x) + o(h) \qquad \text{for} \quad h \to 0.$$

Equally useful is the symbolic notation $f = O(g)$ to indicate that $f(x)$ is *at most* of the order of magnitude of $g(x)$, that is, that the quotient $f(x)/g(x)$ is bounded for the values of x in question.[1]   Use of the symbol O is again very flexible. Thus the phrase "$f = O(g)$ for $x \to \infty$" means that the quotient f/g is bounded for all sufficiently large x as in

$$\sqrt{10x - 1} = O(\sqrt{x}) \qquad \text{for} \quad x \to \infty.$$

Similarly, "$f = O(g)$ for $x \to \xi$" means that f/g is bounded in a sufficiently small neighborhood of the point $x = \xi$ as in

$$e^x - 1 = O(x) \qquad \text{for } x \to 0.$$

More generally we can use the equation $f = O(g)$ to indicate the boundedness of f/g in any domain of the x-axis without requiring x to approach a limit. Thus

$$\log x = O(x) \qquad \text{for } x > 1,$$

$$x = O(\sin x) \qquad \text{for } |x| < \frac{\pi}{2}.$$

Some of the earlier examples involving the symbol o can now be refined to indicate a better estimate of the error with the help of the symbol O. Thus we have for a function f for which f'' is defined and continuous

$$f(x + h) - f(x) = hf'(x) + O(h^2) \qquad \text{for } h \to 0.$$

Other examples are

$$\frac{1}{\sqrt{1 + 4x^2}} = \frac{1}{2x} + O\left(\frac{1}{x^2}\right),$$

$$\cos x = 1 + O(x^2) \qquad \text{for all } x.$$

The same notations can be used for *sequences* a_n, letting the index n tend to infinity. We shall meet some interesting examples of such "asymptotic" formulas with an error term of higher order in the sequel (cf. Stirling's formula for $n!$ on p. 504).   A famous asymptotic law,[2]

---

[1] Notice that $f = O(g)$ does *not* mean that f/g has the limit one or that the quotient necessarily has any limit at all.

[2] The proof cannot be given in this book. See A. E. Ingham, *The Distribution of Primes*, Cambridge University Press, 1932.

already mentioned in Chapter 1, p. 56 states that the number $\pi(n)$ of primes less than n is given approximately by $n/(\log n)$. Here the order of magnitude of the error also has been found and we have more precisely the result

$$\pi(n) = \frac{n}{\log n} + O\left(\frac{n}{\log^2 n}\right).$$

Appendix

The difficulty in appreciating a rigorous development of calculus stems from a basic dilemma: Although the fundamental concepts and procedures, such as continuity, smoothness, etc., are motivated by compelling intuitive needs, they must be made precise in order to have any logical meaning, and the resulting rigorous definitions may cover phenomena beyond those of intuitive character. Thus the rigorous concept of continuity inevitably requires a degree of abstraction not completely reflected in the naive notion of a connected curve, and the concept of differentiability is more restrictive and more abstract than the vague idea of *smoothness* of a curve suggests. Discrepancies of this sort are not avoidable and may tax the patience and understanding of a beginner or of someone for whom logical finesse is not of primary interest. Nevertheless, we want to make the need for precision clearer to the reader by showing that, perhaps unexpectedly, precision and refinement are called for even by simple and intuitively comprehensible examples.

A.1 Some Special Functions

As a rule such examples need not be given in terms of single analytical expressions (see Figs. 2.28, p. 38 and 1.30, p. 39). Here, however, we wish to represent various typical discontinuities and "abnormal" or unexpected phenomena by very simple expressions constructed from the elementary functions. We begin with an example in which no discontinuity is present.

a. The Function $y = e^{-1/x^2}$

This function (cf. Fig. 3.22) is defined in the first instance only for values of x other than zero, and obviously has the limit zero as $x \to 0$. For by the transformation $1/x^2 = \xi$ our function becomes $y = e^{-\xi}$ and

$\lim\limits_{\xi \to \infty} e^{-\xi} = 0$. Hence it is natural to extend our function so that it is continuous for $x = 0$ by defining the value of the function at the point $x = 0$ as $y(0) = 0$.

By the chain rule the derivative of our function for $x \neq 0$ is $y' = -(2/x^3)e^{-1/x^2} = 2\xi^{3/2}e^{-\xi}$. If x tends to zero, this derivative also has the limit zero, as we find immediately from p. 250. At the point $x = 0$ itself the derivative

$$y'(0) = \lim_{h \to 0} \frac{y(h) - y(0)}{h} = \lim_{h \to 0} \frac{e^{-1/h^2}}{h}$$

can also be continuously defined as zero.

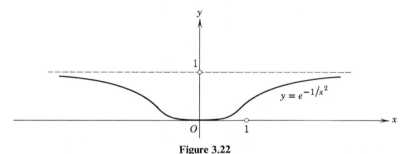

Figure 3.22

For the higher derivatives when $x \neq 0$, we obviously always obtain the product of the function e^{-1/x^2} and a polynomial in $1/x$, and the passage to the limit $x \to 0$ always yields the limit zero. Hence all the higher derivatives vanish, like y', at the point $x = 0$.

Thus our function is continuous everywhere and differentiable as many times as we please, and yet at the point $x = 0$ it vanishes with all its derivatives and yet does not vanish identically. We shall later realize (Appendix I.1 in Chapter 5) how remarkable or "abnormal" this behavior is.

b. The Function $y = e^{-1/x}$

As easily seen, for positive values of x this function behaves in the same way as the function just dealt with; if x tends to zero from one side, through positive values, the function tends to zero, and the same is true of all its derivatives. If we define the value of the function at $x = 0$ as $y(0) = 0$, all the right-hand derivatives at the point $x = 0$ have the value zero. It is quite another matter when x tends to zero through negative values; for then the function and all its derivatives become infinite, and left-hand derivatives at the point $x = 0$ do not exist. At the point $x = 0$, therefore, the function has a remarkable sort of discontinuity, quite unlike the infinite discontinuities of a rational functions considered on pp. 36, 37 (cf. Fig. 3.23).

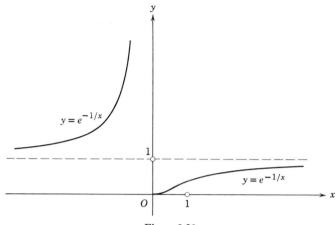

Figure 3.23

c. The Function $y = \tanh \dfrac{1}{x}$

As already seen on p. 65, functions with "jump" discontinuities can be obtained from simple functions by a passage to the limit. The exponential function defined on p. 151 together with the principle of compounding of functions give us another method for constructing functions with such discontinuities from elementary functions, without any further limiting process. An example of this is the function

$$y = \tanh \frac{1}{x} = \frac{e^{1/x} - e^{-1/x}}{e^{1/x} + e^{-1/x}}$$

and its behavior at the point $x = 0$. The function is in the first instance not defined at this point. If we approach the point $x = 0$ through positive values of x, we obviously obtain the limit 1; if, on the other hand, we approach the point $x = 0$ through negative values, we obtain the limit -1. This point $x = 0$ is thus a point of jump discontinuity; as x increases through 0 the value of the function jumps by 2 (cf. Fig. 3.24). On the other hand, the derivative

$$y' = -\frac{1}{\cosh^2(1/x)}\frac{1}{x^2}$$

$$= -\frac{1}{x^2}\frac{4}{(e^{1/x} + e^{-1/x})^2}$$

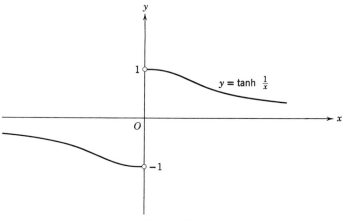

Figure 3.24

approaches the limit zero from both sides, as follows readily from[1] Section 3.76, p. 249.

d. The Function $y = x \tanh 1/x$

In the case of the function

$$y = x \tanh \frac{1}{x} = x \frac{e^{1/x} - e^{-1/x}}{e^{1/x} + e^{-1/x}}$$

the preceding discontinuity is removed by the factor x. This function has the limit zero as $x \to 0$ from either side, so that we can again appropriately define $y(0)$ as equal to zero. Our function is then continuous at $x = 0$, but its first derivative

$$y' = \tanh \frac{1}{x} - \frac{1}{x} \frac{1}{x \cosh^2 (1/x)}$$

has just the same kind of discontinuity as the preceding example. The graph of the function is a curve with a corner (cf. Fig. 3.25); at the point $x = 0$ the function has no actual derivative but a right-hand derivative with the value $+1$ and a left-hand derivative with the value -1.

---

[1] Another example of the occurrence of a "jump" discontinuity is given by the function $y = \text{arc tan } 1/x$ as $x \to 0$.

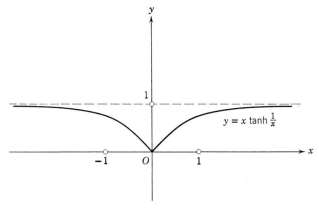

Figure 3.25

e. The Function $y = x \sin 1/x$, $y(0) = 0$

We have already seen that this function is not composed of a finite number of monotonic pieces—as we may say, it is not "sectionally" or "piecewise" monotonic—but that it is nevertheless continuous (p. 40 and Fig. 1.31). Its first derivative

$$y' = \sin \frac{1}{x} - \frac{1}{x} \cos \frac{1}{x} \qquad (x \neq 0),$$

on the contrary, has a discontinuity at $x = 0$; for as x tends to zero this derivative oscillates continually between bounding curves, one positive and one negative, which themselves tend to $+\infty$ and $-\infty$ respectively. At the actual point $x = 0$ the difference quotient is $[y(h) - y(0)]/h = \sin(1/h)$; since this expression swings backward and forward between 1 and -1 an infinite number of times as $h \to 0$, the function possesses neither a right-hand derivative nor a left-hand derivative at $x = 0$.

A.2 Remarks on the Differentiability of Functions

The derivative of a function which is continuous and has a derivative at every point of an interval need not be continuous.

As a simple example we consider the function given by

$$y = f(x) = x^2 \sin \frac{1}{x} \qquad \text{for } x \neq 0$$

and

$$f(0) = 0.$$

This function is defined and continuous everywhere. For all values of x different from zero the derivative is given by the expression

$$f'(x) = -x^2\left(\cos\frac{1}{x}\right)\frac{1}{x^2} + 2x\sin\frac{1}{x} = -\cos\frac{1}{x} + 2x\sin\frac{1}{x}.$$

When x tends to zero, $f'(x)$ has no limit. If, on the other hand, we form the difference quotient $[f(h) - f(0)]/h = (h^2\sin 1/h)/h = h\sin 1/h$,

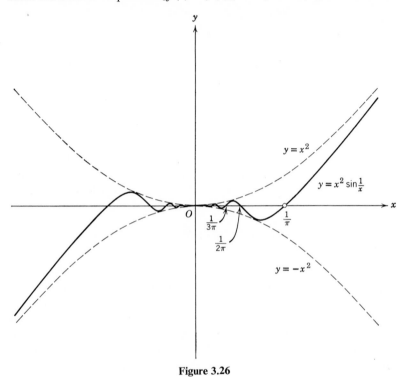

Figure 3.26

we see at once that this tends to zero as h does. The derivative therefore exists for $x = 0$ and has the value 0.

To grasp intuitively the reason for this paradoxical behavior we represent the function graphically (cf. Fig. 3.26). It oscillates between the curves $y = x^2$ and $y = -x^2$, which it touches alternately. Thus the ratio of the heights of the wavecrests of our curve and their distances from the origin steadily becomes smaller. Yet these waves do not become flatter, for their slope, given by the derivative

$$f'(x) = 2x\sin 1/x - \cos 1/x,$$

is equal to -1 at the points $x = 1/2n\pi$ where $\cos 1/x = 1$, and to $+1$ at the points $x = 1/(2n + 1)\pi$ where $\cos 1/x = -1$.

In contrast to the possibility illustrated here (that a derivative exist everywhere and yet not be continuous) we state the following simple theorem, which throws light on a whole series of earlier examples and discussions.

THEOREM. *If we know that in a neighborhood of a point $x = a$, the function $f(x)$ is continuous, and that for $x \neq a$ it also has a derivative $f'(x)$ and if in addition the equation $\lim_{x \to a} f'(x) = b$ holds, then the derivative $f'(x)$ exists at the point a also, and $f'(a) = b$.*

PROOF. The proof follows immediately from the mean value theorem. For we have $[f(a + h) - f(a)]/h = f'(\xi)$, where ξ is a value intermediate between a and $a + h$. If h now tends to zero by hypothesis $f'(\xi)$ tends to b, and our statement follows at once.

A companion theorem may be proved in a similar way: If the function $f(x)$ is continuous in $a \leq x \leq b$ and for $a < x < b$ possesses a derivative which increases beyond all bounds as x tends to a, the right-hand difference quotient $[f(a + h) - f(a)]/h$ also increases beyond all bounds as h tends to zero, so that no finite right-hand derivative exists at $x = a$. The geometrical meaning of this statement is that at the point with the (finite) coordinates $[a, f(a)]$ the curve has a vertical tangent.

Part B   Techniques of Integration

Explicit Functions

A wide class of functions can be constructed from the elementary functions[1] by repeated rational operations, that is, addition, multiplication, division, and furthermore by the operations of forming inverse functions and of compounding functions. The functions thus described form the class of *"explicit" functions* or "closed expressions."[2]

As a result of Part A of this chapter we state the rather general fact:

Every explicit function can be differentiated and its derivative is again an explicit function.

Thus we have attained a fairly complete mastery of the operation or the "algorithm" of differentiation. Yet, the inverse process, that of

---

[1] It should be emphasized that the distinction between "elementary" and "explicit" functions and others is in itself somewhat arbitrary. For us the term "elementary" function includes just the rational functions, the trigonometric and exponential functions, and their inverses.

[2] This name indicates that we shall encounter many other functions which cannot be represented in this fashion but which can be constructed by means of limiting processes such as infinite series.

integration, is generally speaking more important and presents the major challenge. To a certain extent the challenge is met by the fundamental theorem of calculus: To every formula of differentiation $F'(x) = f(x)$ there corresponds an equivalent formula for the primitive functions $F(x)$ to $f(x)$ or the integral:

$$\int f(x)\, dx = F(x).$$

$\left(\text{More precisely we have } F(x) = \int_a^x f(u)\, du + \text{constant}\right)$. Thus as more explicit formulas of differentiation are derived, additional explicit functions can be integrated in terms of explicit functions. A first table of integrals is listed on p. 264; in principle, it would not be difficult, although impractical and confusing, to extend such a table very much.

In the early phases of the development of calculus many mathematicians tried to find, in explicit or closed form, the integral or primitive function for every explicitly given function.

It took some time before it was realized that in principle this problem cannot be solved; on the contrary, for some quite elementary integrands the integral just cannot be expressed in terms of elementary functions (see p. 298). Thus the need for studying new types of functions generated by integration processes from elementary functions became an important stimulus for the development of analysis. Nevertheless the desire to integrate—when feasible—given explicit functions explicitly without getting hopelessly entangled in tedious consultation of tables or numerical computations has led to some simple devices which provide a certain flexibility for transforming given integrals; in fact, these devices permit us to carry out the integration by reduction to one of the elementary integrals in the Table of Integrals.

Section 3.9 will be devoted to the development of such useful devices. In this connection the beginner should be cautioned against merely memorizing the many formulas obtained by using these technical devices. The student should instead direct his efforts toward gaining a clear understanding of the *methods* of integration and learning how to apply them. Moreover, he should remember that even when integration by these devices is impossible, the *integral does exist* (at least for all continuous functions), and can actually be calculated to as high a degree of accuracy as is desired by means of numerical methods which will be further developed later (Section 6.1).

In Part C of this chapter we shall endeavor to extend our conceptions of integration and integral, quite apart from the problem of the technique of integration.

Table of Elementary Integrals

$$F'(x) = f(x) \qquad\qquad\qquad F(x) = \int f(x)\, dx$$

---

1. $x^a \quad (a \neq -1)$ $\qquad\qquad\qquad \dfrac{x^{a+1}}{a+1}$

2. $\dfrac{1}{x}$ $\qquad\qquad\qquad\qquad\quad \log |x|$

3. e^x $\qquad\qquad\qquad\qquad\quad e^x$

4. $a^x \quad (a \neq 1)$ $\qquad\qquad\qquad \dfrac{a^x}{\log a}$

5. $\sin x$ $\qquad\qquad\qquad\qquad -\cos x$

6. $\cos x$ $\qquad\qquad\qquad\qquad \sin x$

7. $\dfrac{1}{\sin^2 x} \quad (= \operatorname{cosec}^2 x)$ $\qquad -\cot x$

8. $\dfrac{1}{\cos^2 x} \quad (= \sec^2 x)$ $\qquad \tan x$

9. $\sinh x$ $\qquad\qquad\qquad\qquad \cosh x$

10. $\cosh x$ $\qquad\qquad\qquad\qquad \sinh x$

11. $\dfrac{1}{\sinh^2 x} \quad (= \operatorname{cosech}^2 x)$ $\qquad -\coth x$

12. $\dfrac{1}{\cosh^2 x} \quad (= \operatorname{sech}^2 x)$ $\qquad \tanh x$

13. $\dfrac{1}{\sqrt{1 - x^2}} \quad (|x| < 1)$ $\qquad \begin{cases} \arcsin x \\ -\arccos x \end{cases}$

14. $\dfrac{1}{1 + x^2}$ $\qquad\qquad\qquad \begin{cases} \arctan x \\ -\operatorname{arccot} x \end{cases}$

15. $\dfrac{1}{\sqrt{1 + x^2}}$ $\qquad\qquad \operatorname{ar\,sinh} x = \log(x + \sqrt{1 + x^2})$

16. $\dfrac{1}{\pm\sqrt{x^2 - 1}} \quad (|x| > 1)$ $\qquad \operatorname{ar\,cosh} x = \log(x \pm \sqrt{x^2 - 1})$

17. $\dfrac{1}{1 - x^2} \begin{cases} |x| < 1 \\[1.5em] |x| > 1 \end{cases}$ $\qquad \begin{aligned} & \operatorname{ar\,tanh} x = \dfrac{1}{2}\log\dfrac{1+x}{1-x} \\[1em] & \operatorname{ar\,coth} x = \dfrac{1}{2}\log\dfrac{x+1}{x-1} \end{aligned}$

3.8 Table of Elementary Integrals

To each of the differentiation formulas proved earlier there corresponds an equivalent integration formula. Since these elementary integrals are used time and again as materials for the art of integration, we collect them in a Table. The right-hand column contains a number of elementary functions and the left-hand column the corresponding derivatives. If we read the table from left to right, we obtain in the right-hand column an indefinite integral of the function in the left-hand column.

We also remind the reader of the fundamental theorems of the differential and integral calculus, proved in Section 2.9, in particular, of the fact that any definite integral is obtained from the indefinite integral $F(x)$ by the formula[1]

$$\int_a^b f(x)\, dx = F(x)\Big|_a^b = F(b) - F(a).$$

In the following sections we shall attempt to reduce the calculation of integrals of given functions in some way or other to the elementary integrals collected in our Table. Apart from special artifices which are learned only from experience, this reduction is based essentially on two useful methods: "substitution" and "integration by parts." Each of these methods enables us to transform a given integral in many ways; the object of such transformations mostly is to reduce the given integral, in one step or in a sequence of steps, to one or more of the elementary integration formulas given above.

3.9 The Method of Substitution

Integrating Compound Functions

The first of these methods is the *introduction of a new variable* (that is, the method of *substitution* or *transformation*). It aims at reducing the integration of composite functions—such as functions of $x - c$ or of $ax + b$—to that of simpler functions.

a. The Substitution Formula. Integral of a Composite Function

The rule for integrating composite functions follows from the corresponding chain rule for differentiation. For a composite function

---

[1] We shall not discuss in this chapter the somewhat different problem of calculating special definite integrals without first finding a general primitive function.

$G(u) = F[\phi(u)]$ we have (see p. 218)

(see p. 218)

(16) $$\frac{dG(u)}{du} = \frac{dF[\phi(u)]}{du} = F'[\phi(u)]\phi'(u).$$

It is sufficient for the validity of this formula that the functions $x = \phi(u)$ and $F(x)$ are continuously differentiable in their arguments u, x respectively, and that $F(x)$ is defined for the values x assumed by the function $x = \phi(u)$ (that is, the range of the function ϕ must belong to the domain of F). Integrating the formula between the limits $u = \alpha$ and $u = \beta$, we find

(17) $G(\beta) - G(\alpha) = F[\phi(\beta)] - F[\phi(\alpha)] = \int_\alpha^\beta F'[\phi(u)]\phi'(u)\, du.$

If here

$$\phi(\beta) = b, \qquad \phi(\alpha) = a,$$

we have

$$F[\phi(\beta)] - F[\phi(\alpha)] = F(b) - F(a) = \int_a^b F'(x)\, dx.$$

Setting $F'(x) = f(x)$ we obtain the *basic substitution formula*

(18) $$\int_a^b f(x)\, dx = \int_\alpha^\beta f[\phi(u)]\phi'(u)\, du, \qquad x = \phi(u)$$

or, written suggestively in Leibnitz's notation with the differential $d\phi = \phi'(x)\, dx$,

(18a) $$\int f(x)\, dx = \int f(\phi)\, d\phi.$$

Here $x = \phi(u)$ may be any function which is defined and has a continuous derivative in the interval J with end points α and β; it maps those end points into $x = a$ and $x = b$ respectively; the function $f(x)$ is assumed to be continuous in an interval I containing the images of all points of J under the mapping ϕ. For $F(x)$ we can take any primitive function of $f(x)$.

As should be noticed the substitution rule (18) does not require that the mapping $x = \phi(u)$ map points between α and β only on points between a and b or that different values u are mapped into different x; all that matters is that α and β are mapped into a and b and that $f(x)$ is defined for the values x taken by $\phi(u)$ for u between α and β.

In terms of indefinite integrals the substitution rule takes the form

(19) $G(u) = \int f[\phi(u)]\phi'(u)\, du = \int f(x)\, dx = F(x) = F[\phi(u)].$

The differential symbols

$$\phi'(u)\,du = \frac{dx}{du}\,du \qquad \text{and} \qquad dx$$

become identical if we formally cancel the symbols du in the numerator and denominator.

Examples. We apply formula (18) to the integrand $f(x) = 1/x$ and make the substitution $x = \phi(u)$, assuming $\phi(u) \neq 0$ in the interval considered; then

$$\int \frac{\phi'(u)}{\phi(u)}\,du = \int \frac{dx}{x} = \log |x| = \log |\phi(u)|$$

or changing the name of the variable u again into x,

(20) $$\int \frac{\phi'(x)}{\phi(x)}\,dx = \log |\phi(x)|.$$

If in this important formula we substitute particular functions, such as $\phi(x) = \log x$, $\phi(x) = \sin x$, or $\phi(x) = \cos x$, we obtain[1]

$$\int \frac{dx}{x \log x} = \log |\log x|,$$

(21)

$$\int \cot x \, dx = \log |\sin x|, \qquad \int \tan x \, dx = -\log |\cos x|.$$

Further Examples.

$$\int \phi(u)\phi'(u)\,du = \int x\,dx = \tfrac{1}{2}x^2 = \tfrac{1}{2}[\phi(u)]^2,$$

where $f(x) = x$. This yields for $\phi(u) = \log u$

(22) $$\int \frac{\log u}{u}\,du = \tfrac{1}{2}(\log u)^2.$$

We finally consider

$$\int \sin^n u \cos u \, du.$$

Here $x = \sin u = \phi(u)$, and hence

$$\int \sin^n u \cos u \, du = \int x^n\,dx = \frac{x^{n+1}}{n+1} = \frac{\sin^{n+1} u}{n+1}.$$

---

[1] These and the following formulas are easily verified by showing that differentiation of the result gives us back the integrand.

The same substitution $x = \sin u$ gives for any function $f(x)$ continuous in the interval $-1 \leq x \leq 1$

$$\int_\alpha^\beta f(\sin u) \cos u \, du = \int_{\sin \alpha}^{\sin \beta} f(x) \, dx.$$

Taking here $\alpha = 0$ and $\beta = 2\pi$ gives us an example for applying the substitution formula to a case where the mapping function $x = \phi(u) = \sin u = x$ is not monotonic throughout the interval $\alpha \leq u \leq \beta$. We find

$$\int_0^{2\pi} f(\sin u) \cos u \, du = \int_0^0 f(x) \, dx = 0.$$

Other Forms of the Rule

In many applications the integral to be evaluated is given in the form

$$F(u) = \int h[\phi(u)] \, du$$

in which the integrand appears as the composite function $h[\phi(u)]$ without the additional factor $\phi'(u)$. We can apply the substitution rule (18) if we succeed in writing the integrand $h[\phi(u)]$ in the form $f[\phi(u)]\phi'(u)$. This can always be achieved under the assumption that the function $x = \phi(u)$ has a continuous derivative $\phi'(u)$ *which does not vanish.* For then there exists an inverse function $u = \psi(x)$ with a continuous derivative $du/dx = \psi'(x) = 1/\phi'(u)$. Taking for $f(x)$ the function $h(x)\psi'(x)$ we have indeed $h[\phi(u)] = f[\phi(u)]/\psi'(x) = f[\phi(u)]\phi'(u)$ and we obtain from the substitution rule

(23)
$$\int h[\phi(u)] \, du = \int f[\phi(u)]\phi'(u) \, du = \int f(x) \, dx$$

$$= \int h(x)\psi'(x) \, dx = \int h(x) \frac{du}{dx} \, dx.$$

The assumption $\phi'(u) \neq 0$ has been introduced in order to prevent the expression dx/du in formula (23) from becoming infinite.

The beginner must never forget that in substituting u for $\psi(x)$ in an integral one must not merely express the old variable x in terms of the new one, u, and then integrate with respect to this new variable; instead, before integrating one must multiply by the derivative of the original variable x with respect to the new variable u. This, of course, is suggested by Leibnitz' notation $h \, dx = h \dfrac{dx}{du} \, du$. In the definite integral

$$\int_a^b h[\psi(x)] \, dx = \int_\alpha^\beta h(u)\phi'(u) \, du$$

we must not forget to change the limits a, b for x into the corresponding limits $\alpha = \psi(a)$ and $\beta = \psi(b)$ for the variable u.

Examples.   In order to calculate $\int \sin 2x \, dx$ we take $u = \psi(x) = 2x$ and $h(u) = \sin u$. We have

$$\frac{du}{dx} = \psi'(x) = 2, \quad \frac{dx}{du} = \frac{1}{2}.$$

If we now introduce $u = 2x$ into the integral as the new variable, then it is transformed, *not* into $\int \sin u \, du$ but into

$$\frac{1}{2} \int \sin u \, du = -\frac{1}{2} \cos u = -\frac{1}{2} \cos 2x;$$

this may, of course, be verified at once by differentiating the right-hand side.

If we integrate with respect to x between the limits zero and $\pi/4$, the corresponding limits for $u = 2x$ are zero and $\pi/2$ and we obtain

$$\int_0^{\pi/4} \sin 2x \, dx = \frac{1}{2} \int_0^{\pi/2} \sin u \, du = -\frac{1}{2} \cos u \Big|_0^{\pi/2} = \frac{1}{2}.$$

Another simple example is the integral $\int_1^4 \frac{dx}{\sqrt{x}}$. Here we take $u = \psi(x) = \sqrt{x}$, from which $x = \phi(u) = u^2$. Since $\phi'(u) = 2u$, we have

$$\int_1^4 \frac{dx}{\sqrt{x}} = \int_1^2 2 \frac{u \, du}{u} = 2 \int_1^2 du = 2.$$

As another example we consider the integral of $\sin 1/x$ for the interval $\frac{1}{2} \leq x \leq 1$. We have for $u = 1/x$ or $x = 1/u$, $dx = -du/u^2$, and hence

$$\int_{1/2}^1 \sin \frac{1}{x} \, dx = -\int_2^1 \frac{\sin u}{u^2} \, du = \int_1^2 \frac{\sin u}{u^2} \, du.$$

*b. An Alternative Derivation of the Substitution Formula

Our integration formula (17) with a slight change of notation can also be interpreted in a direct manner, based on the meaning of the definite integral as a limit of a sum instead of being deduced from the chain rule of differentiation.[1]   To calculate the integral

$$\int_a^b h[\psi(x)] \, dx$$

(for the case $a < b$), we begin with an arbitrary subdivision of the interval $a \leq x \leq b$, and then make the subdivision finer and finer. We choose these subdivisions in the following way. If the function $u = \psi(x)$ is assumed to be monotonic increasing, there is a one-to-one correspondence between the interval $a \leq x \leq b$ on the x-axis and an

---

[1] The result obtained in this way is again restricted to monotonic substitutions and thus is less general than formula (18) furnished by the chain rule (on p. 265).

interval $\alpha \leq u \leq \beta$ of the values of $u = \psi(x)$, where $\alpha = \psi(a)$ and $\beta = \psi(b)$. We divide this x-interval into n parts of length[1] Δx; there is a *corresponding* subdivision of the u-interval into subintervals which, in general, are not all of the same length. We denote the points of division of the x-interval by

$$x_0 = a, x_1, x_2, \ldots, x_n = b$$

and the lengths of the corresponding u-cells by

$$\Delta u_1, \Delta u_2, \ldots, \Delta u_n.$$

The integral we are considering is then the limit of the sum

$$\sum_{v=1}^{n} h\{\psi(\xi_v)\} \, \Delta x,$$

where the value ξ_v is arbitrarily selected from the vth subinterval of the x-subdivision. This sum we now write in the form $\sum_{v=1}^{n} h(v_v) \dfrac{\Delta x}{\Delta u_v} \Delta u_v$, where $v_v = \psi(\xi_v)$. By the mean value theorem of the differential calculus $\Delta x/\Delta u_v = \phi'(\eta_v)$, where η_v is a suitably chosen intermediate value of the variable u in the vth subinterval of the u-subdivision and $x = \phi(u)$ denotes the inverse function of $u = \psi(x)$. If we now select the value ξ_v in such a way that v_v and η_v coincide, that is, $\eta_v = \psi(\xi_v)$ $\xi_v = \phi(\eta_v)$, then our sum takes the form

$$\sum_{v=1}^{n} h(\eta_v)\phi'(\eta_v) \, \Delta u_v.$$

If we here make the passage to the limit letting $n \to \infty$,[2] we obtain the expression

$$\int_{\alpha}^{\beta} h(u) \frac{dx}{du} \, du$$

as the limiting value, that is, as the value of the integral we are considering, in agreement with formula (23) given before.

Thus we arrive at the following result.

THEOREM. *Let $h(u)$ be a continuous function of u in the interval $\alpha \leq u \leq \beta$. Then if the function $u = \psi(x)$ is continuous and monotonic and has a continuous nonvanishing derivative du/dx in $a \leq x \leq b$, and*

---

[1] The assumption that the lengths of these subintervals are all equal is by no means essential for the proof.

[2] This limit exists (for $\Delta x \to 0$) and is the integral, since on account of the uniform continuity of $u = \psi(x)$ the greatest of the lengths Δu_v tends to zero with Δx.

$\psi(a) = \alpha$ *and* $\psi(b) = \beta$, then

$$\int_a^b h\{\psi(x)\}\, dx = \int_a^b h(u)\, dx = \int_\alpha^\beta h(u)\, \frac{dx}{du}\, du.$$

This derivation exhibits the suggestive merit of Leibnitz's notation. In order to carry out the substitution $u = \psi(x)$, we need only write $(dx/du)\, du$ in place of dx, changing the limits from the original values of x to the corresponding values of u.

c. Examples. Integration Formulas

With the help of the substitution rule we can in many cases evaluate a given integral $\int f(x)\, dx$ if we reduce it by means of a suitable substitution $x = \phi(u)$ to one of the elementary integrals in our Table. Whether such substitutions exist and how to find them are questions to which no general answer can be given; this is rather a matter in which practice and ingenuity, in contrast to systematic methods, come into their own.

As an example, we evaluate the integral $\displaystyle\int \frac{dx}{\sqrt{a^2 - x^2}}$ by means of the substitution[1] $x = \phi(u) = au$, $u = \psi(x) = x/a$, $dx = a\, du$, by which, using No. 13 of our Table we obtain

(24)
$$\int \frac{dx}{\sqrt{a^2 - x^2}} = \int \frac{a\, du}{a\sqrt{1 - u^2}} = \text{arc sin } u = \text{arc sin } \frac{x}{a}, \qquad \text{for } |x| < |a|.$$

By the same substitution we similarly obtain

(25) $\displaystyle\int \frac{dx}{a^2 + x^2} = \int \frac{a\, du}{a^2(1 + u^2)} = \frac{1}{a}\,\text{arc tan } u = \frac{1}{a}\,\text{arc tan } \frac{x}{a},$

(26) $\displaystyle\int \frac{dx}{\sqrt{a^2 + x^2}} = \text{ar sinh } \frac{x}{a},$

(27) $\displaystyle\int \frac{dx}{\sqrt{x^2 - a^2}} = \text{ar cosh } \frac{x}{a}, \qquad \text{for } |x| > |a|,$

(28) $\displaystyle\int \frac{dx}{a^2 - x^2} = \begin{cases} \dfrac{1}{a}\,\text{ar tanh } \dfrac{x}{a} & \text{for } |x| < |a|, \\[2mm] \dfrac{1}{a}\,\text{ar coth } \dfrac{x}{a} & \text{for } |x| > |a|, \end{cases}$

---

[1] For the sake of brevity we again take the liberty of writing the symbols dx and du separately, that is, $dx = \phi'(u)\, du$ instead of $dx/du = \phi'(u)$ (cf. p. 180).

formulas which occur very frequently and which can easily be verified by differentiating the right-hand side.

3.10 Further Examples of the Substitution Method

In this section we collect a number of examples which the reader may consider carefully for practice.

By the substitution $u = 1 \pm x^2$, $du = \pm 2x \, dx$, we deduce that

$$(29) \qquad \int \frac{x \, dx}{\sqrt{1 \pm x^2}} = \pm \sqrt{1 \pm x^2},$$

$$(30) \qquad \int \frac{x \, dx}{1 \pm x^2} = \pm \tfrac{1}{2} \log |1 \pm x^2| \, .$$

In these formulas we must take either the plus sign in all three places or the minus sign in all three places.

By the substitution $u = ax + b$, $du = a \, dx$ $(a \neq 0)$, we obtain

$$(31) \qquad \int \frac{dx}{ax + b} = \frac{1}{a} \log |ax + b| \, ,$$

$$(32) \qquad \int (ax + b)^\alpha \, dx = \frac{1}{a(\alpha + 1)} (ax + b)^{\alpha + 1} \qquad (\alpha \neq -1),$$

$$(33) \qquad \int \sin (ax + b) \, dx = -\frac{1}{a} \cos (ax + b);$$

similarly, by means of the substitution $u = \cos x$, $du = -\sin x \, dx$, we obtain

$$(34) \qquad \int \tan x \, dx = -\log |\cos x|,$$

and by means of the substitution $u = \sin x$, $du = \cos x \, dx$,

$$(35) \qquad \int \cot x \, dx = \log |\sin x|$$

[cf. (21) p. 266]. Using the analogous substitutions $u = \cosh x$, $du = \sinh x \, dx$ and $u = \sinh x$, $du = \cosh x \, dx$, we obtain the formulas

$$(36) \qquad \int \tanh x \, dx = \log \cosh x,$$

$$(37) \qquad \int \coth x \, dx = \log |\sinh x| \, .$$

By virtue of the substitution $u = (a/b) \tan x$, $du = (a/b) \sec^2 x \, dx$, we arrive at the two formulas

(38)
$$\int \frac{dx}{a^2 \sin^2 x + b^2 \cos^2 x} = \frac{1}{b^2} \int \frac{1}{(a^2/b^2) \tan^2 x + 1} \frac{dx}{\cos^2 x}$$

$$= \begin{cases} \dfrac{1}{ab} \text{ arc tan } \left(\dfrac{a}{b} \tan x \right) \\[2mm] \dfrac{1}{ab} \text{ arc cot } \left(\dfrac{a}{b} \tan x \right) \end{cases},$$

and

(39)
$$\int \frac{dx}{a^2 \sin^2 x - b^2 \cos^2 x} = \begin{cases} -\dfrac{1}{ab} \text{ ar tanh } \left(\dfrac{a}{b} \tan x \right) \\[2mm] -\dfrac{1}{ab} \text{ ar coth } \left(\dfrac{a}{b} \tan x \right) \end{cases}.$$

We evaluate the integral

$$\int \frac{dx}{\sin x}$$

by writing $\sin x = 2 \sin (x/2) \cos (x/2) = 2 \tan (x/2) \cos^2 (x/2)$, and putting $u = \tan (x/2)$, so that $du = \frac{1}{2} \sec^2 (x/2) \, dx$; the integral then becomes

(40)
$$\int \frac{dx}{\sin x} = \int \frac{du}{u} = \log \left| \tan \frac{x}{2} \right|.$$

If we replace x by $x + \pi/2$, this formula becomes

(41)
$$\int \frac{dx}{\cos x} = \log \left| \tan \left(\frac{x}{2} + \frac{\pi}{4} \right) \right|.$$

The substitution $u = 2x$ yields, if we also apply the known trigonometrical formulas $2 \cos^2 x = 1 + \cos 2x$ and $2 \sin^2 x = 1 - \cos 2x$, the frequently used formulas

(42)
$$\int \cos^2 x \, dx = \tfrac{1}{2}(x + \sin x \cos x)$$

and

(43)
$$\int \sin^2 x \, dx = \tfrac{1}{2}(x - \sin x \cos x).$$

By the substitution $x = \cos u$, equivalent to $u = \text{arc cos } x$, or, more generally, $x = a \cos u \ (a \neq 0)$, we can reduce

$$\int \sqrt{1 - x^2} \, dx \qquad \text{and} \qquad \int \sqrt{a^2 - x^2} \, dx$$

respectively to these formulas. We thus obtain

(44) $$\int \sqrt{a^2 - x^2}\, dx = -\frac{a^2}{2} \arccos \frac{x}{a} + \frac{x}{2}\sqrt{a^2 - x^2}.$$

Similarly, by the substitution $x = a \cosh u$ we obtain the formula

(45) $$\int \sqrt{x^2 - a^2}\, dx = -\frac{a^2}{2} \operatorname{ar\,cosh} \frac{x}{a} + \frac{x}{2}\sqrt{x^2 - a^2}$$

and by the substitution $x = a \sinh u$

(46) $$\int \sqrt{a^2 + x^2}\, dx = \frac{a^2}{2} \operatorname{ar\,sinh} \frac{x}{a} + \frac{x}{2}\sqrt{a^2 + x^2}.$$

The substitution $u = a/x$, $dx = -(a/u^2)\, du$ leads to the formulas

(47) $$\int \frac{dx}{x\sqrt{x^2 - a^2}} = -\frac{1}{a} \arcsin \frac{a}{x},$$

(48) $$\int \frac{dx}{x\sqrt{x^2 + a^2}} = -\frac{1}{a} \operatorname{ar\,sinh} \frac{a}{x},$$

(49) $$\int \frac{dx}{x\sqrt{a^2 - x^2}} = -\frac{1}{a} \operatorname{ar\,cosh} \frac{a}{x}.$$

Finally, we consider the three integrals

$$\int \sin mx \sin nx\, dx, \qquad \int \sin mx \cos nx\, dx, \qquad \int \cos mx \cos nx\, dx,$$

where m and n are positive integers. By well-known trigonometrical formulas we can divide each of these integrals into two parts, writing

$$\sin mx \sin nx = \tfrac{1}{2}[\cos (m - n)x - \cos (m + n)x],$$

$$\sin mx \cos nx = \tfrac{1}{2}[\sin (m + n)x + \sin (m - n)x],$$

$$\cos mx \cos nx = \tfrac{1}{2}[\cos (m + n)x + \cos (m - n)x].$$

If we now make use of the substitutions $u = (m + n)x$ and $u = (m - n)x$ respectively, we obtain directly the following system of formulas:

(50)

$$\int \sin mx \sin nx \, dx = \begin{cases} \dfrac{1}{2} \left\{ \dfrac{\sin(m-n)x}{m-n} - \dfrac{\sin(m+n)x}{m+n} \right\} & \text{if } m \neq n, \\[3mm] \dfrac{1}{2}\left(x - \dfrac{\sin 2mx}{2m} \right) & \text{if } m = n; \end{cases}$$

(51)

$$\int \sin mx \cos nx \, dx = \begin{cases} -\dfrac{1}{2}\left\{ \dfrac{\cos(m+n)x}{m+n} + \dfrac{\cos(m-n)x}{m-n} \right\} & \text{if } m \neq n \\[3mm] -\dfrac{1}{2}\left(\dfrac{\cos 2mx}{2m} \right) & \text{if } m = n; \end{cases}$$

(52)

$$\int \cos mx \cos nx \, dx = \begin{cases} \dfrac{1}{2}\left\{ \dfrac{\sin(m+n)x}{m+n} + \dfrac{\sin(m-n)x}{m-n} \right\} & \text{if } m \neq n, \\[3mm] \dfrac{1}{2}\left(\dfrac{\sin 2mx}{2m} + x \right) & \text{if } m = n. \end{cases}$$

If, in particular, we integrate from $-\pi$ to $+\pi$, we obtain from these formulas the extremely important relations

(53)
$$\int_{-\pi}^{+\pi} \sin mx \sin nx \, dx = \begin{cases} 0 & \text{if } m \neq n, \\ \pi & \text{if } m = n, \end{cases}$$

$$\int_{-\pi}^{+\pi} \sin mx \cos nx \, dx = 0,$$

$$\int_{-\pi}^{+\pi} \cos mx \cos nx \, dx = \begin{cases} 0 & \text{if } m \neq n, \\ \pi & \text{if } m = n. \end{cases}$$

These are the *orthogonality relations* of the trigonometric functions, which we shall encounter again in Section 8.4e.

3.11   Integration by Parts

a. General Formula

The second widely used method for dealing with integration problems expresses in integral form the rule for differentiating a product:

$$(fg)' = f'g + fg'.$$

The corresponding integral formula is (cf. p. 189)

$$f(x)g(x) = \int g(x)f'(x)\,dx + \int f(x)g'(x)\,dx$$

or

(54) $$\int f(x)g'(x)\,dx = f(x)g(x) - \int g(x)f'(x)\,dx.$$

Using Leibnitz's differential notation, this becomes

(54a) $$\int f\,dg = fg - \int g\,df.$$

This formula will be referred to as the formula for *integration by parts*. It reduces the calculation of one integral to the calculation of another integral. Since a given integrand can be regarded as a product $f(x)g'(x)$ in a great many different ways, this formula provides us with an effective tool for the transformation of integrals.

Written as a formula for *definite integration*, the formula for integration by parts is

(54b) $$\int_a^b f(x)g'(x)\,dx = f(x)g(x)\Big|_a^b - \int_a^b g(x)f'(x)\,dx$$

$$= f(b)g(b) - f(a)g(a) - \int_a^b g(x)f'(x)\,dx.$$

This follows either directly by integrating the formula for the derivative of a product between the limits a and b or by forming the difference at the points b and a in formula (54).

We can give a simple geometrical interpretation of formula (54b): Let us suppose that $y = f(x)$ and $z = g(x)$ are monotonic, and that $f(a) = A, f(b) = B, g(a) = \alpha, g(b) = \beta$; we can then form the inverse of the first function and substitute in the second equation, thus obtaining z as a function of y. We assume that this function is monotonic increasing. Since $dy = f'(x)\,dx$ and $dz = g'(x)\,dx$ the formula for integration by parts can be written [cf. the substitution rule (18), p. 265].

$$\int_\alpha^\beta y\,dz + \int_A^B z\,dy = B\beta - A\alpha,$$

in agreement with the relation made clear by Fig. 3.27,

area $NQLK$ + area $PMLQ$ = area $OMLK$ − area $OPQN$.

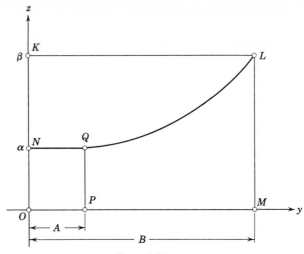

Figure 3.27

The following example may serve as a first illustration:

$$\int \log x \; dx = \int \log x \cdot 1 \; dx.$$

We write the integrand in this way in order to indicate that we put $f(x) = \log x$ and $g'(x) = 1$, so that we have $f'(x) = 1/x$ and $g(x) = x$. Our formula then becomes

(55) $$\int \log x \; dx = x \log x - \int \frac{x}{x} \; dx = x \log x - x.$$

This last expression is therefore the indefinite integral of the logarithm, as may be verified at once by differentiation.

b. Further Examples of Integration by Parts

With $f(x) = x, g'(x) = e^x$, we have $f'(x) = 1, g(x) = e^x$, and

(56) $$\int x \, e^x \; dx = e^x(x - 1).$$

In a similar way we obtain

(57) $$\int x \sin x \; dx = -x \cos x + \sin x$$

and

(58) $$\int x \cos x \; dx = x \sin x + \cos x.$$

For $f(x) = \log x$, $g'(x) = x^a$, we have the relation

(59)
$$\int x^a \log x \, dx = \frac{x^{a+1}}{a+1} \left(\log x - \frac{1}{a+1} \right).$$

Here we must assume $a \neq -1$. For $a = -1$ we obtain

$$\int \frac{1}{x} \log x \, dx = (\log x)^2 - \int \log x \cdot \frac{dx}{x} \; ;$$

transferring the integral on the right-hand side over to the left, we have [cf. (22), p. 266]

(60)
$$\int \frac{1}{x} \log x \, dx = \tfrac{1}{2}(\log x)^2.$$

We calculate the integral $\int$ arc sin $x \, dx$ by taking $f(x) = $ arc sin x, $g'(x) = 1$. Hence

$$\int \text{arc sin } x \, dx = x \text{ arc sin } x - \int \frac{x \, dx}{\sqrt{1 - x^2}} \; .$$

The integration on the right-hand side can be performed as in (29), p. 271; we thus find

(61)
$$\int \text{arc sin } x \, dx = x \text{ arc sin } x + \sqrt{1 - x^2}.$$

In the same way we find

(62)
$$\int \text{arc tan } x \, dx = x \text{ arc tan } x - \tfrac{1}{2} \log (1 + x^2)$$

and many others of a similar type.

The following examples are of a somewhat different nature; here repeated integration by parts brings us back to the original integral, for which we thus obtain an equation.

In this way we obtain

$$\int e^{ax} \sin bx \, dx = -\frac{1}{b} e^{ax} \cos bx + \frac{a}{b} \int e^{ax} \cos bx \, dx$$

$$= -\frac{1}{b} e^{ax} \cos bx + \frac{a}{b^2} e^{ax} \sin bx - \frac{a^2}{b^2} \int e^{ax} \sin bx \, dx;$$

Solving this equation for the integral $\int e^{ax} \sin bx \, dx$,

(63)
$$\int e^{ax} \sin bx \, dx = \frac{1}{a^2 + b^2} e^{ax}(a \sin bx - b \cos bx).$$

In a similar way it follows that

(64)
$$\int e^{ax} \cos bx \, dx = \frac{1}{a^2 + b^2} e^{ax}(a \cos bx + b \sin bx).$$

c. Integral Formula for $f(b) + f(a)$

As a last example we derive a remarkable formula expressing the sum $f(b) + f(a)$ as a definite integral (instead of the difference $f(b) - f(a)$ given by the fundamental formula). Integration by parts will be applied by introducing $1 = g'(x)$, where $g(x) = x - m$ with a constant m at our disposal. Then we have for the indefinite integrals

$$\int f(x)\, dx + \int f'(x)(x - m)\, dx = f(x)(x - m)$$

and for the integral between a and b

$$\int_a^b f(x)\, dx + \int_a^b f'(x)(x - m)\, dx = f(b)(b - m) - f(a)(a - m).$$

If for arbitrary a and b we choose for m the mean value $m = (a + b)/2$, between a and b, we obtain, as the reader will easily verify

$$\frac{b - a}{2}[f(a) + f(b)] = \int_a^b f(x)\, dx + \int_a^b (x - m)f'(x)\, dx.$$

d. Recursive Formulas

In many cases the integrand is not only a function of the independent variable but also depends on an integer index n; on integrating by parts we sometimes obtain, instead of the value of the integral, another similar expression in which the index n has a smaller value. We thus might arrive after a number of steps at an integral which we can deal with by means of the Table of Integrals, p. 263. Such a process is called *recursive*.

The following examples are illustrations: By repeated integration by parts we can calculate the trigonometrical integrals

$$\int \cos^n x\, dx, \qquad \int \sin^n x\, dx, \qquad \int \sin^m x \cos^n x\, dx,$$

provided that m and n are positive integers. For using $f(x) = \cos^{n-1} x$, $g(x) = \sin x$ we find for the first integral that

$$\int \cos^n x\, dx = \cos^{n-1} x \sin x + (n - 1)\int \cos^{n-2} x \sin^2 x\, dx;$$

the right-hand side can be written in the form

$$\cos^{n-1} x \sin x + (n-1) \int \cos^{n-2} x \, dx - (n-1) \int \cos^n x \, dx;$$

thus a recursive relation is obtained:

(65) $$\int \cos^n x \, dx = \frac{1}{n} \cos^{n-1} x \sin x + \frac{n-1}{n} \int \cos^{n-2} x \, dx.$$

This formula enables us to diminish the index in the integrand step by step until we finally arrive at the integral

$$\int \cos x \, dx = \sin x \qquad \text{or} \qquad \int dx = x,$$

depending on whether n is odd or even. In a similar way we obtain the analogous recursive formulas

(66) $$\int \sin^n x \, dx = -\frac{1}{n} \sin^{n-1} x \cos x + \frac{n-1}{n} \int \sin^{n-2} x \, dx$$

and

(67)

$$\int \sin^m x \cos^n x \, dx = \frac{\sin^{m+1} x \cos^{n-1} x}{m+n} + \frac{n-1}{m+n} \int \sin^m x \cos^{n-2} x \, dx.$$

In particular, we calculate the integrals

$$\int \sin^2 x \, dx = \frac{1}{2} (x - \sin x \cos x)$$

and

$$\int \cos^2 x \, dx = \frac{1}{2} (x + \sin x \cos x),$$

as we have already done by the method of substitution [Eqs. (42), (43), p. 272].

It need hardly be mentioned that the corresponding integrals for the hyperbolic functions can be calculated in exactly the same way:

(68) $$\int \sinh^2 x \, dx = \frac{1}{2} (-x + \sinh x \cosh x),$$

(69) $$\int \cosh^2 x \, dx = \frac{1}{2} (x + \sinh x \cosh x).$$

Further recursive formulas are given by the following transformations:

(70) $$\int (\log x)^m \, dx = x(\log x)^m - m \int (\log x)^{m-1} \, dx,$$

(71) $$\int x^m e^x \, dx = x^m e^x - m \int x^{m-1} e^x \, dx,$$

(72) $$\int x^m \sin x \, dx = -x^m \cos x + m \int x^{m-1} \cos x \, dx,$$

(73) $$\int x^m \cos x \, dx = x^m \sin x - m \int x^{m-1} \sin x \, dx,$$

(74) $$\int x^a (\log x)^m \, dx = \frac{x^{a+1}(\log x)^m}{a+1}$$
$$- \frac{m}{a+1} \int x^a (\log x)^{m-1} \, dx \quad (a \neq -1).$$

e. Wallis's Infinite Product for π

The recursive formula for the integral $\int \sin^n x \, dx$ with $n > 1$ leads to a fascinating expression for the number π as an "infinite product." In the formula

$$\int \sin^n x \, dx = -\frac{1}{n} \sin^{n-1} x \cos x + \frac{n-1}{n} \int \sin^{n-2} x \, dx$$

we insert the limits 0 and $\pi/2$, thus obtaining

(75) $$\int_0^{\pi/2} \sin^n x \, dx = \frac{n-1}{n} \int_0^{\pi/2} \sin^{n-2} x \, dx \qquad \text{for } n > 1.$$

If we repeatedly apply the recursive formula, we obtain, distinguishing between the cases $n = 2m$ and $n = 2m + 1$,

(76) $$\int_0^{\pi/2} \sin^{2m} x \, dx = \frac{2m-1}{2m} \frac{2m-3}{2m-2} \cdots \frac{1}{2} \cdot \int_0^{\pi/2} dx,$$

(76a) $$\int_0^{\pi/2} \sin^{2m+1} x \, dx = \frac{2m}{2m+1} \cdot \frac{2m-2}{2m-1} \cdots \frac{2}{3} \cdot \int_0^{\pi/2} \sin x \, dx,$$

whence

(77) $$\int_0^{\pi/2} \sin^{2m} x \, dx = \frac{2m-1}{2m} \cdot \frac{2m-3}{2m-2} \cdots \frac{1}{2} \cdot \frac{\pi}{2},$$

(77a) $$\int_0^{\pi/2} \sin^{2m+1} x \, dx = \frac{2m}{2m+1} \cdot \frac{2m-2}{2m-1} \cdots \frac{2}{3}.$$

By division this yields

$$(78) \quad \frac{\pi}{2} = \frac{2 \cdot 2}{1 \cdot 3} \cdot \frac{4 \cdot 4}{3 \cdot 5} \cdot \frac{6 \cdot 6}{5 \cdot 7} \cdots \frac{2m \cdot 2m}{(2m-1) \cdot (2m+1)} \cdot \frac{\int_0^{\pi/2} \sin^{2m} x \, dx}{\int_0^{\pi/2} \sin^{2m+1} x \, dx}.$$

The quotient of the two integrals on the right-hand side converges to 1 as m increases, as we recognize from the following considerations. In the interval $0 < x < \pi/2$, where $0 < \sin x < 1$, we have

$$0 < \sin^{2m+1} x \le \sin^{2m} x \le \sin^{2m-1} x;$$

consequently,

$$0 < \int_0^{\pi/2} \sin^{2m+1} x \, dx \le \int_0^{\pi/2} \sin^{2m} x \, dx \le \int_0^{\pi/2} \sin^{2m-1} x \, dx.$$

If we here divide each term by $\int_0^{\pi/2} \sin^{2m+1} x \, dx$ and notice that by formula (75)

$$\frac{\int_0^{\pi/2} \sin^{2m-1} x \, dx}{\int_0^{\pi/2} \sin^{2m+1} x \, dx} = \frac{2m+1}{2m} = 1 + \frac{1}{2m},$$

we have

$$1 \le \frac{\int_0^{\pi/2} \sin^{2m} x \, dx}{\int_0^{\pi/2} \sin^{2m+1} x \, dx} \le 1 + \frac{1}{2m},$$

from which the above statement follows.

Consequently, the relation

$$(79) \quad \frac{\pi}{2} = \lim_{m \to \infty} \frac{2}{1} \frac{2}{3} \frac{4}{3} \frac{4}{5} \frac{6}{5} \frac{6}{7} \cdots \frac{2m}{2m-1} \frac{2m}{2m+1}$$

holds.

This product formula (due to Wallis), with its simple law of formation, gives a most remarkable relation between the number π and the integers.

Product for $\sqrt{\pi}$

As an easy consequence we can derive an equally remarkable expression for $\sqrt{\pi}$. If we observe

$$\lim_{m \to \infty} \frac{2m}{2m+1} = 1,$$

we can write

$$\lim_{m \to \infty} \frac{2^2 \cdot 4^2 \cdots (2m - 2)^2}{3^2 \cdot 5^2 \cdots (2m - 1)^2} 2m = \frac{\pi}{2} \, ;$$

taking the square root and then multiplying the numerator and denominator by $2 \cdot 4 \cdots (2m - 2)$, we find

$$\sqrt{\frac{\pi}{2}} = \lim_{m \to \infty} \frac{2 \cdot 4 \cdots (2m - 2)}{3 \cdot 5 \cdots (2m - 1)} \sqrt{2m} = \lim_{m \to \infty} \frac{2^2 \cdot 4^2 \cdots (2m - 2)^2}{(2m - 1)!} \sqrt{2m}$$

$$= \lim_{m \to \infty} \frac{2^2 \cdot 4^2 \cdots (2m)^2}{(2m)!} \frac{\sqrt{2m}}{2m} \, .$$

$$= \lim \frac{(2^2 \cdot 1^2)(2^2 \cdot 2^2)(2^2 \cdot 3^2) \cdots (2^2 \cdot m^2)}{(2m)! \sqrt{2m}}$$

From this we finally obtain

(80) $$\lim_{m \to \infty} \frac{(m!)^2 2^{2m}}{(2m)! \sqrt{m}} = \sqrt{\pi},$$

a form of Wallis's product which will be of use to us later (cf. Chapter 6, Appendix).

*3.12   Integration of Rational Functions

During the seventeenth and eighteenth centuries mathematicians were preoccupied with discovering classes of elementary explicit functions which could be integrated explicitly. A wealth of ingenious devices was invented and at the same time the basis for deeper understanding created. When one later realized that achieving integration of *all* explicit functions in closed form was neither an attainable nor really an important goal, the tedious technicalities which had been developed in connection with such problems were gradually deemphasized. Yet, a significant general result remained:

All rational functions R(x) of a variable x can be integrated explicitly in terms of the elementary integrals listed in Table 3.1.

This general result can be obtained much more easily in the context of the more advanced theory of functions of a complex variable. Yet, it is still worthwhile to sketch an elementary derivation employing only real variables.

The rational functions are those of the form

(81) $$R(x) = \frac{f(x)}{g(x)},$$

where $f(x)$ and $g(x)$ are polynomials:

$$f(x) = a_m x^m + a_{m-1} x^{m-1} + \cdots + a_0,$$

$$g(x) = b_n x^n + b_{n-1} x^{n-1} + \cdots + b_0 \quad (b_n \neq 0).$$

As we recall, every polynomial can be integrated at once and its integral is itself a polynomial. We therefore need consider only those rational functions for which the denominator $g(x)$ is not a constant. Moreover, we can always assume that the degree of the numerator is less than the degree n of the denominator. For otherwise, dividing the polynomial $f(x)$ by the polynomial $g(x)$, we obtain a remainder of degree less than n; in other words, we can write $f(x) = q(x)g(x) + r(x)$, where $q(x)$ and $r(x)$ are also polynomials and $r(x)$ is of lower degree than n. The integration of $f(x)/g(x)$ is then reduced to the integration of the polynomial $q(x)$ and of the "proper" fraction $r(x)/g(x)$. We notice further that the function $f(x)/g(x)$ can be represented as the sum of the functions $a_\nu x^\nu / g(x)$, so that we need only consider integrands of the form $x^\nu / g(x)$.

a. The Fundamental Types

We proceed in steps to the integration of the most general rational function of the type (81), studying first only those functions with denominator $g(x)$ of the particularly simple type

$$g(x) = x^n,$$

or

$$g(x) = (1 + x^2)^n,$$

where n is any positive integer.

To this case we can then reduce the somewhat more general case in which $g(x) = (\alpha x + \beta)^n$, a power of a linear expression $\alpha x + \beta$ ($\alpha \neq 0$), or $g(x) = (ax^2 + 2bx + c)^n$, a power of a definite[1] quadratic

---

[1] A quadratic expression $Q(x) = ax^2 + 2bx + c$ is said to be *definite* if for all real values of x it takes values having one and the same sign, that is, if the equation $Q(x) = 0$ has no real roots. For this it is necessary and sufficient that the "discriminant" $ac - b^2$ is positive. This follows, of course, from the explicit formula $(-b \pm \sqrt{b^2 - ac})/a$ for the roots. Equivalently, a definite quadratic expression is one that cannot be factored into two *real* linear factors.

expression. If $g(x) = (\alpha x + \beta)^n$ we introduce $\xi = \alpha x + \beta$ as a new variable. Then $d\xi/dx = \alpha$, and $x = (\xi - \beta)/\alpha$ is also a linear function of ξ. Each numerator $f(x)$ becomes a polynomial $\phi(\xi)$ of the same degree, and consequently,

$$\int \frac{f(x)}{(\alpha x + \beta)^n}\, dx = \frac{1}{\alpha} \int \frac{\phi(\xi)}{\xi^n}\, d\xi.$$

In the second case, we write

$$ax^2 + 2bx + c = \frac{1}{a}(ax + b)^2 + \frac{d^2}{a} \qquad (d^2 = ac - b^2, d > 0);$$

since we have assumed our expression to be quadratic and definite, $ac - b^2$ must be positive and $a \neq 0$. By introducing the new variable

$$\xi = \frac{ax + b}{d}$$

we arrive at an integral with the denominator $[(d^2/a)(1 + \xi^2)]^n$.

Hence in order to integrate rational functions whose denominators are powers of a linear expression or of a definite quadratic expression it is sufficient to be able to integrate the following types of functions:

$$\frac{1}{x^n}, \qquad \frac{x^{2\nu}}{(x^2 + 1)^n}, \qquad \frac{x^{2\nu+1}}{(x^2 + 1)^n}.$$

We shall, in fact, see that even these types need not be treated in general, for we can reduce the integration of every rational function to the integration of the very special forms of these three functions obtained by taking $\nu = 0$. Accordingly, we now consider the integration of the three expressions

$$\frac{1}{x^n}, \qquad \frac{1}{(x^2 + 1)^n}, \qquad \frac{x}{(x^2 + 1)^n}.$$

b. Integration of the Fundamental Types

Integration of the first type of function, $1/x^n$, immediately yields the expression $\log |x|$ if $n = 1$, and the expression $-1/(n - 1)x^{n-1}$ if $n > 1$, so that in both cases the integral is again an elementary function. Functions of the third type can be integrated immediately by introducing the new variable $\xi = x^2 + 1$, from which we obtain $2x\, dx = d\xi$ and

$$\int \frac{x}{(x^2 + 1)^n}\, dx = \frac{1}{2} \int \frac{d\xi}{\xi^n} = \begin{cases} \frac{1}{2} \log (x^2 + 1) & \text{if } n = 1, \\[2mm] -\dfrac{1}{2(n - 1)(x^2 + 1)^{n-1}} & \text{if } n > 1. \end{cases}$$

Finally, in order to calculate the integral

$$I_n = \int \frac{dx}{(x^2 + 1)^n},$$

where n has any value exceeding one, we make use of a recursive method: If we put

$$\frac{1}{(x^2 + 1)^n} = \frac{1}{(x^2 + 1)^{n-1}} - \frac{x^2}{(x^2 + 1)^n},$$

so that

$$\int \frac{dx}{(x^2 + 1)^n} = \int \frac{dx}{(x^2 + 1)^{n-1}} - \int \frac{x^2\, dx}{(x^2 + 1)^n},$$

we can transform the right-hand side by integrating by parts, using formula (54) on p. 275 with

$$f(x) = x, \qquad g'(x) = \frac{x}{(x^2 + 1)^n}.$$

Then, as we have just found,

$$g(x) = -\frac{1}{2} \frac{1}{(n - 1)(x^2 + 1)^{n-1}},$$

and consequently, we obtain

$$I_n = \int \frac{dx}{(x^2 + 1)^n} = \frac{x}{2(n - 1)(x^2 + 1)^{n-1}} + \frac{2n - 3}{2(n - 1)} \int \frac{dx}{(x^2 + 1)^{n-1}}.$$

The calculation of the integral I_n is thus reduced to that of the integral I_{n-1}. If $n - 1 > 1$ we apply the same process to the latter integral, and continue until we finally arrive at the expression

$$\int \frac{dx}{x^2 + 1} = \text{arc tan } x.$$

We thus see that the integral[1] I_n can be explicitly expressed in terms of rational functions and the function arc tan x.

Incidentally, we could also have integrated the function $1/(x^2 + 1)^n$ directly, using the substitution $x = \tan t$; we should then have obtained $dx = \sec^2 t\, dt$ and $1/(1 + x^2) = \cos^2 t$, so that

$$\int \frac{dx}{(x^2 + 1)^n} = \int \cos^{2n-2} t\, dt,$$

and we have already learned [Eq. (65) p. 279] how to evaluate this integral.

---

[1] The integral of the function $1/(x^2 - 1)^n$ can be calculated in the same way; by the corresponding recurrence method we reduce it to the integral

$$\int \frac{dx}{1 - x^2} = \text{ar tanh } x \qquad \text{(or ar coth } x\text{).}$$

c. Partial Fractions

We are now in a position to integrate the most general rational functions. We make use of the fact that every such function can be represented as the sum of so-called *partial fractions*, that is, as the sum of a polynomial and a finite number of rational functions, each one of which has either a power of a linear expression for its denominator and a constant for its numerator, or else a power of a definite quadratic expression for its denominator and a linear function for its numerator. If the degree of the numerator $f(x)$ is less than that of the denominator $g(x)$, the polynomial does not occur. We know already how to integrate each partial fraction. For according to p. 284 the denominator can be reduced to one of the special forms x^n and $(x^2 + 1)^n$, and the fraction is then a combination of the fundamental types integrated on p. 284.

We shall not give the general proof of the possibility of this resolution into partial fractions. We shall merely confine ourselves to making the statement of the theorem intelligible to the reader and to showing by examples how the resolution into partial fractions can be carried out in typical cases. In practice only comparatively simple functions are dealt with, for otherwise the computations become too cumbersome.

As we know from elementary algebra, every real polynomial $g(x)$ can be written in the form[1]

$$g(x) = a(x - \alpha_1)^{l_1}(x - \alpha_2)^{l_2} \cdots$$

$$\cdots (x^2 + 2b_1x + c_1)^{r_1}(x^2 + 2b_2x + c_2)^{r_2} \cdots .$$

Here the distinct numbers $\alpha_1, \alpha_2, \ldots$ are the real roots of the equation $g(x) = 0$, and the positive integers $l_1, l_2, \ldots$ indicate the multiplicity of these roots; the factors $x^2 + 2b_\nu x + c_\nu$ indicate definite quadratic expressions, of which no two are the same, with conjugate complex roots, and the positive integers $r_1, r_2, \ldots$ give the multiplicity of these roots.

We assume that the denominator is either given to us in this form or that we have brought it to this form by calculating the real and imaginary roots. Let us further suppose that the numerator $f(x)$ is of lower degree than the denominator (cf. p. 283). Then the theorem on resolution into partial fractions can be stated as follows: For each

---

[1] The actual *proof* of this so-called fundamental theorem of algebra does not belong to algebra. It is achieved most easily by methods belonging to the theory of functions of a *complex variable*.

factor $(x - \alpha)^l$, where α is any one of the real roots of multiplicity l, one can determine an expression of the form

$$\frac{A_1}{x - \alpha} + \frac{A_2}{(x - \alpha)^2} + \cdots + \frac{A_l}{(x - \alpha)^l},$$

and for each quadratic factor $Q(x) = x^2 + 2bx + c$ in our product which is raised to the power r we can determine an expression of the form

$$\frac{B_1 + C_1 x}{Q} + \frac{B_2 + C_2 x}{Q^2} + \cdots + \frac{B_r + C_r x}{Q^r},$$

in such a way that the function $f(x)/g(x)$ is the sum of all these expressions (A_v, B_v, C_v, are constants). In other words, the quotient $f(x)/g(x)$ can be represented as a sum of fractions, each of which belongs to one of the types integrated above.[1]

In particular cases the decomposition into partial fractions can be done easily by inspection. If, for example, $g(x) = x^2 - 1$, we see at once that

$$\frac{1}{x^2 - 1} = \frac{1}{2} \frac{1}{(x - 1)} - \frac{1}{2} \frac{1}{(x + 1)},$$

so that

$$\int \frac{dx}{x^2 - 1} = \tfrac{1}{2} \log \left| \frac{x - 1}{x + 1} \right|.$$

---

[1] We give a brief sketch of a method by which the possibility of this decomposition into partial fractions can be proved without using the theory of functions of complex variables, once $g(x)$ can be factored completely into linear factors. If $g(x) = (x - \alpha)^k h(x)$ and $h(\alpha) \neq 0$, then on the right-hand side of the equation

$$\frac{f(x)}{g(x)} - \frac{f(\alpha)}{h(\alpha)(x - \alpha)^k} = \frac{1}{h(\alpha)} \frac{f(x)h(\alpha) - f(\alpha)h(x)}{(x - \alpha)^k h(x)}$$

the numerator obviously vanishes for $x = \alpha$; it is therefore of the form $h(\alpha)(x - \alpha)^m f_1(x)$, where $f_1(x)$ is also a polynomial, the integer $m \geq 1$, and $f_1(\alpha) \neq 0$. Writing $f(\alpha)/h(\alpha) = \beta$, this gives us

$$\frac{f(x)}{g(x)} - \frac{\beta}{(x - \alpha)^k} = \frac{f_1(x)}{(x - \alpha)^{k-m} h(x)}.$$

Continuing the process, we can keep on diminishing the degree of the power of $(x - \alpha)$ occurring in the denominator until finally no such factor is left. On the remaining fraction we repeat the process for some other root of $g(x)$, and do this as many times as $g(x)$ has distinct factors. By doing this not only for the real but also for the complex roots, and by combining conjugate complex fractions we eventually arrive at the complete decomposition into partial fractions.

More generally, if $g(x) = (x - \alpha)(x - \beta)$, that is, if $g(x)$ is a nondefinite quadratic expression with two real zeros α and β, we have

$$\frac{1}{(x - \alpha)(x - \beta)} = \frac{1}{(\alpha - \beta)} \frac{1}{(x - \alpha)} - \frac{1}{(\alpha - \beta)} \frac{1}{(x - \beta)}$$

so that

$$\int \frac{dx}{(x - \alpha)(x - \beta)} = \frac{1}{\alpha - \beta} \log \left| \frac{x - \alpha}{x - \beta} \right|.$$

d. Examples of Resolution into Partial Fractions. Method of Undetermined Coefficients

If $g(x) = (x - \alpha_1)(x - \alpha_2) \cdots (x - \alpha_n)$, where $\alpha_i \neq \alpha_k$ if $i \neq k$, that is, if the equation $g(x) = 0$ has only simple real roots, and if $f(x)$ is any polynomial of degree $< n$, the expression in terms of partial fractions has the simple form

$$\frac{f(x)}{g(x)} = \frac{a_1}{x - \alpha_1} + \frac{a_2}{x - \alpha_2} + \cdots + \frac{a_n}{x - \alpha_n}.$$

We obtain explicit expressions for the coefficients $a_1, a_2, \ldots$ if we multiply both sides of this equation by $(x - \alpha_1)$, cancel the common factor $(x - \alpha_1)$ in the numerator and denominator on the left and in the first term on the right, and then put $x = \alpha_1$. This gives

$$a_1 = \frac{f(\alpha_1)}{(\alpha_1 - \alpha_2)(\alpha_1 - \alpha_3) \cdots (\alpha_1 - \alpha_n)}$$

The reader will observe from the rule for the derivative of a product that the denominator on the right is $g'(\alpha_1)$, that is, the derivative of the function $g(x)$ at the point $x = \alpha_1$. Similar formulas for $a_2, a_3, \ldots$, obtained in this way, lead to the explicit partial fraction expansion

$$\frac{f(x)}{g(x)} = \frac{f(\alpha_1)}{g'(\alpha_1)(x - \alpha_1)} + \frac{f(\alpha_2)}{g'(\alpha_2)(x - \alpha_2)} + \cdots + \frac{f(\alpha_n)}{g'(\alpha_n)(x - \alpha_n)}.$$

As a typical example of a denominator $g(x)$ with multiple roots, we consider the function $1/[x^2(x - 1)]$. It has a representation

$$\frac{1}{x^2(x - 1)} = \frac{a}{x - 1} + \frac{b}{x} + \frac{c}{x^2}$$

in accordance with p. 287. If we multiply both sides of this equation by $x^2(x - 1)$, we obtain the equation

$$1 = (a + b)x^2 - (b - c)x - c,$$

true for all values of x, from which we have to determine the coefficients a, b, c. This condition cannot hold unless all the coefficients of the polynomial $(a + b)x^2 - (b - c)x - c - 1$ are zero; that is, we must have $a + b = b - c = c + 1 = 0$ or $c = -1$, $b = -1$, $a = 1$. We thus obtain the resolution

$$\frac{1}{x^2(x - 1)} = \frac{1}{x - 1} - \frac{1}{x} - \frac{1}{x^2},$$

and consequently,

$$\int \frac{dx}{x^2(x - 1)} = \log |x - 1| - \log |x| + \frac{1}{x}.$$

Next we decompose the function $1/[x(x^2 + 1)]$ whose denominator has complex zeros in accordance with the equation

$$\frac{1}{x(x^2 + 1)} = \frac{a}{x} + \frac{bx + c}{x^2 + 1}.$$

For the coefficients we obtain $a + b = c = a - 1 = 0$ so that

$$\frac{1}{x(x^2 + 1)} = \frac{1}{x} - \frac{x}{x^2 + 1},$$

and therefore

$$\int \frac{dx}{x(x^2 + 1)} = \log |x| - \tfrac{1}{2} \log (x^2 + 1).$$

As a third example we consider the function $1/(x^4 + 1)$, whose integration was a challenge even in Leibnitz' time. We can represent the denominator as the product of two quadratic factors:[1]

$$x^4 + 1 = (x^2 + 1)^2 - 2x^2 = (x^2 + 1 + \sqrt{2}x)(x^2 + 1 - \sqrt{2}x).$$

We know therefore that the resolution into partial fractions will have the form

$$\frac{1}{x^4 + 1} = \frac{ax + b}{x^2 + \sqrt{2}x + 1} + \frac{cx + d}{x^2 - \sqrt{2}x + 1}.$$

To determine the coefficients a, b, c, d, we use the equation

$$(a + c)x^3 + (b + d - a\sqrt{2} + c\sqrt{2})x^2$$
$$+ (a + c - b\sqrt{2} + d\sqrt{2})x + (b + d - 1) = 0,$$

---

[1] The factorization of $x^4 + 1$ into real quadratic factors corresponds to the factorization into conjugate complex linear factors

$$x^4 + 1 = [(x - \epsilon)(x - \epsilon^{-1})][(x - \epsilon^3)(x - \epsilon^{-3})],$$

where

$$\epsilon = \cos \frac{\pi}{4} + i \sin \frac{\pi}{4} = \frac{1}{2} \sqrt{2}(1 + i)$$

is one of the eighth roots of $+1$, and a fourth root of -1 (see p. 105).

which is satisfied by the values

$$a = \frac{1}{2\sqrt{2}}, \quad b = \tfrac{1}{2}, \quad c = -\frac{1}{2\sqrt{2}}, \quad d = \tfrac{1}{2}.$$

We therefore have

$$\frac{1}{x^4 + 1} = \frac{1}{2\sqrt{2}} \cdot \frac{x + \sqrt{2}}{x^2 + \sqrt{2}x + 1} - \frac{1}{2\sqrt{2}} \cdot \frac{x - \sqrt{2}}{x^2 - \sqrt{2}x + 1},$$

and, applying the method given on p. 284, we obtain

$$\int \frac{dx}{x^4 + 1} = \frac{1}{4\sqrt{2}} \log |x^2 + \sqrt{2}x + 1| - \frac{1}{4\sqrt{2}} \log |x^2 - \sqrt{2}x + 1|$$

$$+ \frac{1}{2\sqrt{2}} \arctan(\sqrt{2}x + 1) + \frac{1}{2\sqrt{2}} \arctan(\sqrt{2}x - 1),$$

which may easily be verified by differentiation.

The preceding examples illustrate a general method of integrating a rational function $f(x)/g(x)$. We first divide and are reduced to the case where the degree of f is less than that of g. We factor $g(x)$ into linear and definite quadratic factors, grouping the product into powers of such factors. We write down the *appropriate* partial fraction representation for f/g with *indeterminate* coefficients $a, b, c, \ldots$. Multiplying through with $g(x)$ and *comparing coefficients* of equal powers in the resulting polynomial identity, we obtain a system of linear equations for the unknown coefficients that should just be adequate for determining those coefficients, if we really have the correct form for the partial fraction expansion. We are then ready to integrate any of the resulting partial fractions by the rules discussed before.

3.13   Integration of Some Other Classes of Functions

a. Preliminary Remarks on the Rational Representation[1] of the Circle and the Hyperbola

The integration of some other general classes of functions can be reduced to the integration of rational functions. We shall be able to better understand this reduction by first stating certain elementary facts about the trigonometric and hyperbolic functions. If we put $t = \tan(x/2)$, elementary trigonometry yields the simple formulas

$$\sin x = \frac{2t}{1 + t^2}, \quad \cos x = \frac{1 - t^2}{1 + t^2};$$

---

[1] Sometimes called "uniformization."

indeed, from

$$\frac{1}{1+t^2} = \cos^2 \frac{x}{2} \quad \text{and} \quad \frac{t^2}{1+t^2} = \sin^2 \frac{x}{2},$$

and from the elementary formulas

$$\sin x = 2 \cos^2 \frac{x}{2} \tan \frac{x}{2} \quad \text{and} \quad \cos x = \cos^2 \frac{x}{2} - \sin^2 \frac{x}{2},$$

we obtain these equations. They show that sin x *and* cos x *can both be expressed rationally in terms of the quantity* $t = \tan x/2$. By differentiation we have

$$\frac{dt}{dx} = \frac{1}{2\cos^2 x/2} = \frac{1+t^2}{2},$$

so that

(82)
$$\frac{dx}{dt} = \frac{2}{1+t^2};$$

hence the derivative dx/dt is also a rational expression in t.

*The geometrical representation of our formulas and their geometrical meaning are given in Fig. 3-28. Here the circle $u^2 + v^2 = 1$ in a u,v-plane is shown. If x denotes the angle TOP in the figure, then $u = \cos x$ and $v = \sin x$. The angle OSP with its vertex at the point $u = -1$, $v = 0$ is equal to $x/2$, by

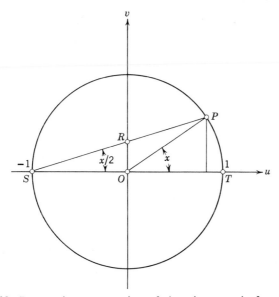

Figure 3.28 Parametric representation of the trigonometric functions.

a theorem in elementary geometry, and we can read off the geometrical meaning of the parameter t from the figure; $t = \tan \frac{1}{2}x = OR$ where R is the "projection" from S of the point P of the circle onto the v-axis. If the point P starts from S and describes once the circle in the positive direction, that is, if x runs through the interval from $-\pi$ to $+\pi$, the quantity t will run through the whole range of values from $+\infty$ to $+\infty$ exactly once. (Notice that the point S itself corresponds to $t = \pm\infty$). We have here a representation of the general point (u, v) of the circle $u^2 + v^2 = 1$ in terms of the rational functions $u = (1 - t^2)/(1 + t^2)$, and $v = 2t/(1 + t^2)$ of the parameter t. These formulas then define a rational mapping of the t-line onto the circle in the u,v-plane (which incidentally is the two-dimensional analogue of the stereographic projection of a sphere mentioned on p. 21). At the basis of this rational representation of the circle lies obviously the identity

$$(t^2 - 1)^2 + (2t)^2 = (t^2 + 1)^2.$$

Curiously enough, this formula is of interest also in number theory since it generates for each integer t *Pythagorean integers* $a = t^2 - 1$, $b = 2t$, and $c = t^2 + 1$ which satisfy the identity $a^2 + b^2 = c^2$, that is, determine a right triangle with commensurable sides. Thus $t = 2$ gives rise to the well-known triple $a = 3$, $b = 4$, $c = 5$; for $t = 4$ we obtain $a = 15$, $b = 8$, $c = 17$, etc. It is remarkable, and, of course, no accident, that the same algebraic identity is of significance in such diverse contexts as integration in closed form, geometry, and number theory. Linking different fields in such a manner is the typical trend in modern mathematics, although our particular example goes back to antiquity.

Similarly we may express the hyperbolic functions

$$\cosh x = \tfrac{1}{2}(e^x + e^{-x})$$

and $\sinh x = \frac{1}{2}(e^x - e^{-x})$ as rational functions of a third quantity. The most obvious way is to put $e^x = \tau$, so that we have

$$\cosh x = \frac{1}{2}\left(\tau + \frac{1}{\tau}\right), \qquad \sinh x = \frac{1}{2}\left(\tau - \frac{1}{\tau}\right),$$

which are rational expressions for $\sinh x$ and $\cosh x$. Here again $dx/d\tau = 1/\tau$ is rational in τ. However, we obtain a closer analogy with the trigonometric functions by introducing the quantity $t = \tanh(x/2) = (\tau - 1)/(\tau + 1)$; we then arrive at the formulas

$$\cosh x = \frac{1 + t^2}{1 - t^2}, \qquad \sinh x = \frac{2t}{1 - t^2}.$$

By differentiating $t = \tanh(x/2)$ we obtain, as in Eq. (82) on p. 291, the rational expression

$$(83) \qquad\qquad \frac{dx}{dt} = \frac{2}{1 - t^2}$$

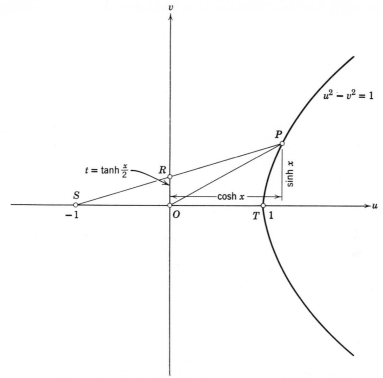

Figure 3.29   Parametric representation of the hyperbolic functions.

for the derivative dx/dt. Here again the quantity t has a geometrical meaning similar to that which it has for the trigonometric functions, as we see at once from Fig. 3.29.

We have here a rational representation of the hyperbola $u^2 - v^2 = 1$ in the u,v-plane by means of the equations $u = (1 + t^2)/(1 - t^2)$ $v = 2t/(1 - t^2)$. The points on the right-hand branch of the curve are of the form $u = \cosh x$, $v = \sinh x$ and correspond to values of t with $|t| < 1$. The other branch is obtained for $|t| > 1$.

We now proceed to our integration problems.

*b. Integration of $R(\cos x, \sin x)$

Let $R(\cos x, \sin x)$ denote an expression which is rational in the two functions $\sin x$ and $\cos x$, that is, an expression which is formed rationally from these two functions and constants, such as

$$\frac{3 \sin^2 x + \cos x}{3 \cos^2 x + \sin x}.$$

If we apply the substitution $t = \tan x/2$, the integral

$$\int R(\cos x, \sin x)\, dx$$

is transformed into the integral

$$\int R\left(\frac{1-t^2}{1+t^2}, \frac{2t}{1+t^2}\right)\frac{2}{1+t^2}\, dt,$$

and under the integral sign we now have a rational function of t. Thus we have in principle obtained the integral of our expression, since we can now perform the integration by the methods of the preceding section.

c. Integration of $R(\cosh x, \sinh x)$

In the same way, if $R(\cosh x, \sinh x)$ is an expression which is rational in terms of the hyperbolic functions $\cosh x$ and $\sinh x$, we can effect its integration by means of the substitution $t = \tanh x/2$. Recalling Eq. (83), we have

$$\int R(\cosh x, \sinh x)\, dx = \int R\left(\frac{1+t^2}{1-t^2}, \frac{2t}{1-t^2}\right)\frac{2}{1-t^2}\, dt.$$

(According to a previous remark we could also have introduced $\tau = e^x$ as a new variable and expressed $\cosh x$ and $\sinh x$ in terms of τ.) The integration is once again reduced to that of a rational function.

*d. Integration of $R(x, \sqrt{1-x^2})$

The integral $\int R(x, \sqrt{1-x^2})\, dx$ can be reduced to the type treated in Section 3.13b by using the substitution

$$x = \cos u, \qquad \sqrt{1-x^2} = \sin u, \qquad dx = -\sin u\, du;$$

from this stage the transformation $t = \tan u/2$ brings us to the integration of a rational function. Incidentally, we could have carried out the reduction in one step instead of two by using the substitution

$$t = \sqrt{\frac{1-x}{1+x}}\,; \qquad x = \frac{1-t^2}{1+t^2}\,;$$

$$\sqrt{1-x^2} = \frac{2t}{1+t^2}\,; \qquad \frac{dx}{dt} = \frac{-4t}{(1+t^2)^2}\,;$$

that is, we could have introduced $t = \tan u/2$ directly as the new variable and thereby obtained a rational integrand.

*e. Integration of $R(x, \sqrt{x^2 - 1})$

The integral $\int R(x, \sqrt{x^2 - 1})\, dx$ is transformed by the substitution $x = \cosh u$ into the type treated in Section 3.13c. Here again we can arrive at our goal directly by introducing

$$t = \sqrt{\frac{x-1}{x+1}} = \tanh \frac{u}{2}.$$

*f. Integration of $R(x, \sqrt{x^2 + 1})$

The integral $\int R(x, \sqrt{(x^2 + 1)})\, dx$ is reduced by the transformation $x = \sinh u$ to the type considered in Section 3.13c (p. 294) and can therefore be integrated in terms of elementary functions. Instead of the further reduction to the integral of a rational function by the substitution $e^u = \tau$ or $\tanh u/2 = t$, we could have reached the integral of a rational function in a single step by either of the substitutions

$$\tau = x + \sqrt{x^2 + 1}, \qquad t = \frac{-1 + \sqrt{x^2 + 1}}{x}.$$

*g. Integration of $R(x, \sqrt{ax^2 + 2bx + c})$

The integral $\int R(x, \sqrt{ax^2 + 2bx + c})\, dx$ of an expression which is rational in terms of x and the square root of an arbitrary polynomial of the second degree in x can immediately be reduced to one of the types just treated. We write (cf. p. 284)

$$ax^2 + 2bx + c = \frac{1}{a}(ax + b)^2 + \frac{ac - b^2}{a}.$$

If $ac - b^2 > 0$ we introduce a new variable ξ by means of the transformation $\xi = (ax + b)/\sqrt{ac - b^2}$, whereupon the surd takes the form $\sqrt{(ac - b^2)(\xi^2 + 1)}/a$. Hence our integral when expressed in terms of ξ is of the type of Section 3.13f. The constant a must here be positive in order that the square root may have real values.

If $ac - b^2 = 0$, and $a > 0$, then by way of the formula

$$\sqrt{ax^2 + 2bx + c} = \sqrt{a}\left(x + \frac{b}{a}\right)$$

we see that the integrand was rational in x to begin with.

If, finally, $ac - b^2 < 0$, we put $\xi = (ax + b)/\sqrt{b^2 - ac}$ and obtain for the surd the expression $\sqrt{(ac - b^2)(\xi^2 - 1)}/a$. If a is positive, our integral is thus reduced to the type of Section 3.13e; if, on the other hand, a is negative, we write the surd in the form

$$\sqrt{(b^2 - ac)(1 - \xi^2)/(-a)}$$

and see that the integral is thus reduced to the type of Section 3.13d.

*h. Further Examples of Reduction to Integrals of Rational Functions

Of other types of functions which can be integrated by reduction to rational functions we shall briefly mention two: (1) rational expressions involving two different square roots of linear expressions, $R(x, \sqrt{ax + b}, \sqrt{\alpha x + \beta})$; (2) expressions of the form $R(x, \sqrt[n]{(ax + b)/(\alpha x + \beta)})$, where a, b, α, β are constants. In the first type we introduce the new variable $\xi = \sqrt{\alpha x + \beta}$, so that $\alpha x + \beta = \xi^2$, and consequently

$$x = \frac{\xi^2 - \beta}{\alpha} \quad \text{and} \quad \frac{dx}{d\xi} = \frac{2\xi}{\alpha} \; ;$$

then

$$\int R(x, \sqrt{ax + b}, \sqrt{\alpha x + \beta}) \, dx$$

$$= \int R\left\{\frac{\xi^2 - \beta}{\alpha}, \sqrt{\frac{1}{\alpha}[a\xi^2 - (a\beta - b\alpha)]}, \xi\right\} \frac{2\xi}{\alpha} \, d\xi,$$

which is of the type discussed in Section 3.13g.

If in the second type we introduce the new variable

$$\xi = \sqrt[n]{\frac{ax + b}{\alpha x + \beta}},$$

we have

$$\xi^n = \frac{ax + b}{\alpha x + \beta}, \qquad x = \frac{-\beta\xi^n + b}{\alpha\xi^n - a}, \qquad \frac{dx}{d\xi} = \frac{a\beta - b\alpha}{(\alpha\xi^n - a)^2} \, n\xi^{n-1},$$

and we immediately arrive at the formula

$$\int R\left(x, \sqrt[n]{\frac{ax + b}{\alpha x + \beta}}\right) dx = \int R\left(\frac{-\beta\xi^n + b}{\alpha\xi^n - a}, \xi\right) \frac{a\beta - b\alpha}{(\alpha\xi^n - a)^2} \, n\xi^{n-1} \, d\xi,$$

which is the integral of a rational function.

i. Remarks on the Examples

The preceding discussions are chiefly of theoretical interest. In complicated expressions the actual calculations would be far too involved. It is therefore expedient to take advantage, when possible, of the special form of the integrand to simplify the work. For example, to integrate $1/(a^2 \sin^2 x + b^2 \cos^2 x)$ it is better to use the substitution $t = \tan x$ instead of that given on p. 294; for $\sin^2 x$ and $\cos^2 x$ can be expressed rationally in terms of $\tan x$, and it is therefore unnecessary to go back to $t = \tan x/2$. The same is true for every expression formed rationally from[1] $\sin^2 x$, $\cos^2 x$, and $\sin x \cos x$. Moreover, for the calculation of many integrals a trigonometrical form is to be preferred to a rational one, provided that the trigonometrical form can be evaluated by some simple recurrence method. For example, although the integrand in $\int x^n \left(\sqrt{1 - x^2}\right)^m dx$ can be reduced to a rational form, it is better to write $x = \sin u$ and bring it to the form $\int \sin^n u \cos^{m+1} u \, du$, since this can easily be treated by the recurrence method on p. 279 (or by using the addition theorems to reduce the powers of the sine and cosine to sines and cosines of multiple angles).

For the evaluation of the integral

$$\int \frac{dx}{a \cos x + b \sin x} \qquad (a^2 + b^2 > 0),$$

instead of referring to the general theory we write

$$A = \sqrt{a^2 + b^2}, \qquad \sin \theta = \frac{a}{A}, \qquad \cos \theta = \frac{b}{A}.$$

The integral then takes the form

$$\frac{1}{A} \int \frac{dx}{\sin (x + \theta)},$$

and on introducing the new variable $x + \theta$ we find [(cf. Eq. (40), p. 272)] that the value of the integral is

$$\frac{1}{A} \log \left| \tan \frac{x + \theta}{2} \right|.$$

---

[1] For $\sin x \cos x = \tan x \cos^2 x$ can, of course, be expressed rationally in terms of $\tan x$.

Part C   Further Steps in the Theory of Integral Calculus

3.14   Integrals of Elementary Functions

a. Definition of Functions by Integrals.
Elliptic Integrals and Functions

With the examples already given of types of functions which can be integrated by reduction to rational functions, we have practically exhausted the list of functions which are integrable in terms of elementary functions. Attempts to express indefinite integrals such as (for $n > 2$)

$$\int \frac{dx}{\sqrt{a_0 + a_1 x + \cdots + a_n x^n}},$$

$$\int \sqrt{a_0 + a_1 x + \cdots + a_n x^n}\, dx,$$

or

$$\int \frac{e^x}{x}\, dx$$

in terms of elementary functions have failed; in the nineteenth century it was finally proved that it is actually impossible to carry out these integrations in terms of elementary functions.

If therefore the object of the integral calculus were to integrate functions explicitly, we should have come to a definite halt. However, such a restricted objective has no intrinsic justification; it is of an artificial nature. We know that the integral of every continuous function *exists* as a limit and is itself a continuous function of the upper limit whether or not the integral can be expressed in terms of elementary functions. The distinguishing features of the elementary functions are based on the fact that their properties are easily recognized, that their application to numerical problems is facilitated by convenient tables, or that they can easily be calculated with as great a degree of accuracy as we please.

Whenever the integral of a function cannot be expressed by means of functions with which we are already acquainted, there is no objection to introducing this integral as a new "higher" function, which really means no more than giving the integral a name. Whether the introduction of such a new function is convenient depends on the properties which it possesses, the frequency with which it occurs, and the ease with which it can be manipulated in theory and in practice. In this sense the process of integration is a general principle for the generation of new functions.

We are already acquainted with this principle from our dealings with the elementary functions. Thus we were forced (p. 145) to introduce the integral of $1/x$ as a new function, which we called the logarithm and whose properties we could easily derive. We could have introduced the trigonometric functions in a similar way, making use only of the rational functions, the process of integration, and the process of inversion. For this purpose we need only take one or other of the equations

$$\text{arc tan } x = \int_0^x \frac{dt}{1 + t^2}$$

or

$$\text{arc sin } x = \int_0^x \frac{dt}{\sqrt{1 - t^2}}$$

as the *definition* of the function arc tan x or arc sin x respectively, and then obtain the trigonometric functions by inversion. By this process the definition of these functions is divorced from intuitive geometry, (in particular, from the intuitive notion of "angle"), but we are left with the task of developing their properties, independently of geometry.[1] (Later, in Section 3.16 we shall give another purely analytic discussion of the trigonometric functions.)

*Elliptic Integrals

The first important example which leads beyond the set of elementary functions is given by the *elliptic integrals*. These are integrals in which the integrand depends rationally on the square root of a polynomial of third or fourth degree. Among these integrals the function

$$u(s) = \int_0^s \frac{dx}{\sqrt{(1 - x^2)(1 - k^2 x^2)}}$$

has become particularly important. Its inverse function $s(u)$ similarly plays an important role.[2] This function $s(u)$ has been as thoroughly examined and tabulated as the elementary functions.[3]

---

[1] We shall not go into the development of these ideas here. The essential step is to prove the addition theorems for the inverse functions, that is, for the sine and the tangent.

[2] For the special value $k = 0$ we obtain $u(s) = $ arc sin x and $s(u) = \sin u$ respectively.

[3] The function $s(u)$, one of the so-called *Jacobian elliptic functions*, is usually denoted by the symbol sn u to indicate that it is a generalization of the ordinary sine-function.

It is the prototype of the so-called *elliptic functions* which occupy a central position in the theory of functions of a complex variable and occur in many physical applications (for example, in connection with the motion of a simple pendulum; see p. 410).

The name "elliptic integral" arises from the fact that such integrals enter into the problem of determining the length of an arc of an ellipse (cf. Chapter 4, p. 378).

We point out further that integrals which at first glance have quite a different appearance turn out to be elliptic integrals after a simple substitution. As an example, the integral

$$\int \frac{dx}{\sqrt{\cos \alpha - \cos x}}$$

is transformed by means of the substitution $u = \cos x/2$ into the integral

$$-k\sqrt{2} \int \frac{du}{\sqrt{(1 - u^2)(1 - k^2 u^2)}}, \qquad k = \frac{1}{\cos (\alpha/2)} ;$$

the integral

$$\int \frac{dx}{\sqrt{\cos 2x}}$$

by means of the substitution $u = \sin x$ becomes

$$\int \frac{du}{\sqrt{(1 - u^2)(1 - 2u^2)}} ;$$

and finally the integral

$$\int \frac{dx}{\sqrt{1 - k^2 \sin^2 x}}$$

is transformed by the substitution $u = \sin x$ into

$$\int \frac{du}{\sqrt{(1 - u^2)(1 - k^2 u^2)}} .$$

b. On Differentiation and Integration

Another remark on the relation between differentiation and integration should be inserted. Differentiation may be considered a more elementary process than integration, because it does not lead us out of the domain of "known" functions. On the other hand, we must remember that the differentiability of an arbitrary continuous function is by no means a foregone conclusion but a stringent assumption. In fact, as we have seen, there are continuous functions which are not differentiable at certain isolated points, whereas since Weierstrass' time many examples of continuous functions have been constructed which do not possess a derivative anywhere.[1] In contrast, even though integration in terms of elementary functions is generally not possible, we are certain at least that the integral of a continuous function exists.

Taken all in all, integration and differentiation cannot be contrasted simply as more elementary and less elementary operations; from some points of view the former and from other points of view the latter could be thought of as more elementary.

Insofar as the concept of integral is concerned, we shall free ourselves in the next section from the assumption that the integrand is everywhere continuous; we shall see that it may be extended to wide classes of functions which have discontinuities.

3.15   Extension of the Concept of Integral

a. Introduction. Definition of "Improper" Integrals

In Chapter 2, p. 128, we defined $\int_a^b f(x)\,dx$ by forming the "Riemann sums"

$$F_n = \sum_{i=1}^n f(\xi_i)\,\Delta x_i$$

based on a subdivision of the interval $[a, b]$ into n subintervals of lengths Δx_i and a choice of intermediate points ξ_i in those subintervals. If the sequences F_n tend to the same limit F_a^b, for any sequence of subdivisions and intermediate points, as long as the largest value Δx_i tends to zero, we define $\int_a^b f(x)\,dx$ to be that limit F_a^b. This limit was shown to exist when $f(x)$ is continuous in $[a, b]$. However, we are often confronted with the need for defining an integral when $f(x)$ is not defined, or not continuous, in all points of the closed interval I or when the interval of integration extends to infinity. We would wish, for example, to attach an appropriate meaning to expressions such as

$$\int_0^1 \frac{1}{\sqrt{x}}\,dx \quad \text{or} \quad \int_0^1 \sin\frac{1}{x}\,dx,$$

$$\int_0^\infty e^{-x}\,dx \quad \text{or} \quad \int_1^\infty \frac{\sin x}{x^2}\,dx, \qquad \text{etc.}$$

---

[1] Compare Titchmarsh, *The Theory of Functions*, Oxford, 1932, Sections 11.21 to 11.23, pp. 350–354.

We first of all extend the concept of the integral to functions that are continuous in the open interval (a, b) but are not necessarily defined or continuous at the endpoints a, b. For any numbers α, b with $a < \alpha < \beta < b$ the ordinary ("proper") integral $f(x)\, dx$ is then defined. If

$$F = \lim_{\epsilon \to 0} \int_{\alpha_\epsilon}^{\beta_\epsilon} f(x)\, dx$$

exists when $a < \alpha_\epsilon < \beta_\epsilon < b$ and $\lim_{\epsilon \to 0} \alpha_\epsilon = a$, $\lim_{\epsilon \to 0} \beta_\epsilon = b$, and if F is independent of the particular choice of α_ϵ and β_ϵ we say that the improper integral $\int_a^b f(x)\, dx$ converges and has the value F.

Sectionally Continuous Integrand.   If, more generally, $f(x)$ is defined and continuous in (a, b) with the possible exception of a finite number of intermediate points $c_1, c_2 \cdots c_n$ and f is continuous in each of the open intervals (a, c_1), (c_1, c_2), ..., (c_n, b) we define $\int_a^b f(x)\, dx$ as the sum of the improper integrals over the subintervals, provided each of those converges.

The improper integral $\int_a^b f(x)\, dx$ always converges when f is continuous and *bounded* in the open interval (a, b). For example, the integral

$$\int_0^1 \sin \frac{1}{x}\, dx = \lim_{\epsilon \to 0} \int_\epsilon^1 \sin \frac{1}{x}\, dx$$

converges. To prove this general statement we may assume, for brevity, that f is continuous at b, but not necessarily at a. Then by definition

$$\int_a^b f(x)\, dx = \lim_{\alpha \to a} F(\alpha),$$

where $F(\alpha)$ for $a < \alpha < b$ is defined as $\int_\alpha^b f(x)\, dx$. If M is an upper bound for $|f|$ and α_n a sequence tending to a, we have by the mean value theorem of integral calculus $|F(\alpha_n) - F(\alpha_m)| \le M\, |\alpha_n - \alpha_m|$; hence, by Cauchy's convergence test $\lim_{n \to \infty} F(\alpha_n)$ exists. Since this is the case for any sequence α_n converging to a, it follows that $\lim_{\alpha \to a} F(\alpha)$ exists.

As a matter of fact, when f is continuous and bounded in (a, b) we can assign to f any values at the endpoints a, b and also obtain $\int_a^b f(x)\, dx$ directly as a "proper" integral defined as the limit of Riemann sums. It is easily seen that for continuous bounded f both definitions apply and lead to the same value, independently of the choice of $f(a)$ and $f(b)$.

The same is true more generally for bounded functions that are defined and continuous in (a, b) with the possible exception of a finite number of points. In particular, $\int_a^b f(x)\,dx$ always exists when f is continuous except for a finite number of jump discontinuities. Altogether con-

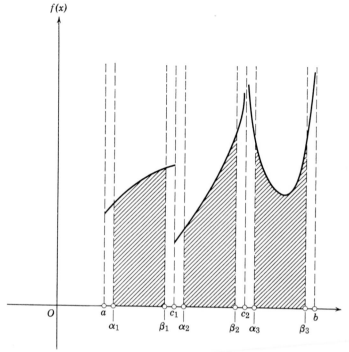

$f(x)$

Figure 3.30   Integral of a function with discontinuities.

vergence of the improper integral of a function over a finite interval demands attention only when f becomes infinite.

We note that the geometrical interpretation of the integral as the area under the curve is unchanged from the interpretation for a continuous f (Fig. 3.30).

b. Functions with Infinite Discontinuities

We begin with the integral

$$J = \int_0^1 \frac{dx}{x^\alpha},$$

where α is a positive number. The integrand $1/x^\alpha$ becomes infinite for $x \to 0$. We therefore must define J by taking the integral J_ϵ from the positive limit ϵ

to the limit 1, and finally letting ϵ tend to zero. According to the elementary rules of integration, we obtain, provided $\alpha \neq 1$,

$$J_\epsilon = \int_\epsilon^1 \frac{dx}{x^\alpha} = \frac{1}{1-\alpha}(1 - \epsilon^{1-\alpha}).$$

We immediately recognize the following possibilities: (1) α is greater than 1; then for $\epsilon \to 0$ the right-hand side tends to infinity. (2) α is less than 1; then the right-hand side tends to the limit $1/(1 - \alpha)$. In the second case, therefore, we shall simply have to take this limiting value as the integral $J = \int_0^1 \frac{dx}{x^\alpha}$. In the first case we shall say that the integral from 0 to 1 does not exist or *diverges*. (3) In the third case, where $\alpha = 1$, the integral is equal to $-\log \epsilon$ and therefore for $\epsilon \to 0$ does not approach a limit, but tends to infinity; that is, the integral $\int_0^1 \frac{dx}{x} = J$ does not exist or is divergent.

Another example for an integrand with an infinite discontinuity is given by $f(x) = 1/\sqrt{1 - x^2}$. We find

$$\int_0^{1-\epsilon} \frac{dx}{\sqrt{1 - x^2}} = \text{arc sin } (1 - \epsilon).$$

For $\epsilon \to 0$, the right-hand side converges to the limit, $\pi/2$; this therefore is the value of the integral

$$\frac{\pi}{2} = \int_0^1 \frac{dx}{\sqrt{1 - x^2}},$$

although the integrand becomes infinite at the point $x = 1$.

c. Interpretation as Areas

Improper integrals can be interpreted as areas of regions extending to infinity defined by means of a passage to the limit from bounded regions. For example, the preceding results for the function $1/x^\alpha$ assert that the area bounded by the x-axis, the line $x = 1$, the line $x = \epsilon$, and the curve $y = 1/x^\alpha$ tends to a finite limit as $\epsilon \to 0$, provided that $\alpha < 1$, and that it tends to infinity if $\alpha \geq 1$. This fact may be simply expressed as follows: The area between the x-axis, the y-axis, the curve $y = 1/x^\alpha$, and the line $x = 1$ is finite or infinite according as $\alpha < 1$ or $\alpha \geq 1$.

Intuition can, of course, give us no reliable information about the finiteness or infiniteness of the area of a region stretching to infinity. Figure 3.31 illustrates the fact that for $\alpha < 1$ the area under our curve remains finite, whereas for $\alpha \geq 1$ it is infinite, a fact which is certainly not suggested by geometrical intuition.

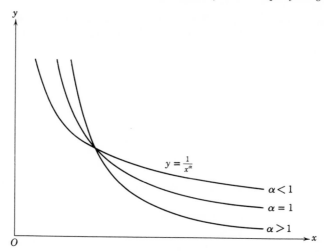

Figure 3.31   To illustrate the convergence or divergence of improper integrals.

d. Tests for Convergence

To check the convergence of an integral of a function $f(x)$ with an infinite discontinuity at the point $x = b$ we can often use the following criterion.

Let the function $f(x)$ be continuous in the interval $a \leq x < b$, and let $\lim_{x \to b} f(x) = \infty$. Then the integral $\int_a^b f(x)\, dx$ *converges* if there exist both a positive number μ less than 1 and a fixed number M independent of x, such that everywhere in the interval $a \leq x < b$ the inequality $|f(x)| \leq M/(b - x)^\mu$ is true; in other words, *if at the point $x = b$ the function $f(x)$ becomes infinite of a lower order than the first*: $f(x) = 0[1/(b - x)^\mu]$ for some $\mu < 1$. On the other hand, the integral *diverges* if there exist both a number $\nu \geq 1$ and a fixed number N, such that everywhere in the interval $a \leq x < b$ the inequality $f(x) \geq N/(b - x)^\nu$ is true; in other words, *if at the point $x = b$ the positive function $f(x)$ becomes infinite of the first order at least.*

The proof follows almost immediately by comparison with the very simple special case just discussed. In order to prove the first part of the theorem we observe that for $0 < \epsilon < b - a$ we have

$$0 \leq \frac{M}{(b - x)^\mu} + f(x) \leq \frac{2M}{(b - x)^\mu}$$

and hence also

$$0 \leq \int_a^{b-\epsilon} \left[\frac{M}{(b - x)^\mu} + f(x) \right] dx \leq \int_a^{b-\epsilon} \frac{2M}{(b - x)^\mu}\, dx \, .$$

As $\epsilon \to 0$ the integral on the right, which is obtained from the integral $\int dx/x^\mu$ by a simple substitution of $b - x$ for x, has a limit and therefore stays bounded. Moreover, the values of the integral in the middle increase monotonically as $\epsilon \to 0$; since they are also bounded, they must possess a limit and the integral

$$\int_a^b \left(\frac{M}{(b-x)^\mu} \, dx + f(x) \right) dx$$

$$= \lim_{\epsilon \to 0} \left(\int_a^{b-\epsilon} \frac{M}{(b-x)^\mu} \, dx + \int_a^{b-\epsilon} f(x) \, dx \right)$$

converges. The convergence of the integral of $M/(b-x)^\mu$ then also implies that of $\int_a^b f(x) \, dx$.

The proof of the second part of the theorem is left as an exercise for the reader.

We likewise see at once that exactly analogous theorems hold where the *lower* boundary of the integral is a point of infinite discontinuity. If a point of infinite discontinuity lies in the interior of the interval of integration, we merely separate the interval into two subintervals by this point and then apply these considerations to each of these.

As an example we consider the elliptic integral

$$\int_0^1 \frac{dx}{\sqrt{(1-x^2)(1-k^2x^2)}} \qquad (k^2 < 1).$$

From the identity $1 - x^2 = (1-x)(1+x)$ we see at once that as $x \to 1$ the integrand becomes infinite only of order $\frac{1}{2}$, from which it follows that the improper integral converges. (For $k = 1$ the integral diverges.)

e. Infinite Interval of Integration

Another important extension of the concept of integral concerns an infinite interval of integration. For a precise formulation, we introduce the following notation: If the integral

$$\int_a^A f(x) \, dx,$$

with a fixed, tends to a definite limit for $A \to \infty$, we define the integral of $f(x)$ over the infinite interval $x \geq a$, as

$$\lim_{A \to \infty} \int_a^A f(x) \, dx = \int_a^\infty f(x) \, dx.$$

Again, such an integral is called *convergent*.

Examples. Simple examples of the various possibilities are again given by the functions $f(x) = 1/x^\alpha$,

$$\int_1^A \frac{dx}{x^\alpha} = \frac{1}{1-\alpha} (A^{1-\alpha} - 1).$$

Here we see that, if we again exclude the case $\alpha = 1$, the integral to infinity exists for the case $\alpha > 1$, and, in fact,

$$\int_1^\infty \frac{dx}{x^\alpha} = \frac{1}{\alpha - 1} \, ;$$

when $\alpha < 1$, the integral no longer exists. For the case $\alpha = 1$ the integral again clearly fails to exist since $\log x$ tends to infinity as x does. We see therefore that with regard to integration over an infinite interval the functions $1/x^\alpha$ do not behave in the same way as for integration up to the origin. This statement also is made plausible by a glance at Fig. 3.31. For obviously, the larger α is, the more closely do the curves draw towards the x-axis for $x \to \infty$; thus it is plausible that the area under consideration tends to a definite limit for sufficiently large values of α.

The following criterion for the existence of an integral with an infinite limit is often useful. (We again assume that for sufficiently large values of x, say for $x \geq a$, the integrand is continuous.)

Criterion of Convergence

The integral $\int_a^\infty f(x)\, dx$ *converges if the function* $f(x)$ *vanishes at infinity to a higher order than the first,* that is, if there is a number $\nu > 1$ such that for all values of x, that are sufficiently large, the relation $|f(x)| \leq M/x^\nu$ is true, where M is a fixed number independent of x. In symbols: $f(x) = 0\left(\dfrac{1}{x^\nu}\right)$.

Again, *the integral diverges if the function remains positive and vanishes at infinity to an order not higher than the first,* that is, if there is a fixed number $N > 0$ such that $xf(x) \geq N$.

The proof of these criteria is exactly parallel to the previous argument and can be left to the reader.

A very simple example is the integral $\int_a^\infty \frac{1}{x^2}\, dx$ $(a > 0)$. The integrand vanishes at infinity to the second order. We see at once that the integral converges, for $\int_a^A \frac{1}{x^2}\, dx = \frac{1}{a} - \frac{1}{A}$, and therefore

$$\int_a^\infty \frac{1}{x^2}\, dx = \frac{1}{a} \, .$$

Another equally simple example is

$$\int_0^\infty \frac{1}{1 + x^2}\, dx = \lim_{A \to \infty} (\text{arc tan } A - \text{arc tan } 0) = \frac{\pi}{2} \, .$$

Then obviously also

$$\int_{-\infty}^{+\infty} \frac{1}{1 + x^2}\, dx = \pi,$$

since the integrand is an even function. It is curious that the area between the curve and $y = 1/(1 + x^2)$ and the x-axis (see Fig. 3.8, p. 216) that extends to infinity turns out to be the same as that of a circle of radius one.

f. The Gamma Function

A further example of particular importance in analysis is that of the so-called gamma function

$$\Gamma(n) = \int_0^\infty e^{-x} x^{n-1}\, dx \qquad (n > 0).$$

Splitting up the interval of integration into one part from $x = 0$ to $x = 1$ and another one from $x = 1$ to $x = \infty$, we see that the integral over the first part clearly converges, since $0 < e^{-x} x^{n-1} < 1/x^\mu$ with $\mu = 1 - n < 1$. For the integral over the second, infinite part, the criterion of convergence is also satisfied; for example, for $\nu = 2$, we have $\lim\limits_{x \to \infty} x^2 e^{-x} x^{n-1} = 0$, since the exponential function e^{-x} tends to zero to a higher order than any power $1/x^m$ $(m > 0)$ (see p. 253). This gamma function which we consider as a function of the number n (not necessarily an integer) satisfies a remarkable relation obtained by integration by parts as follows. First, we have (with $f(x) = x^{n-1}$, $g'(x) = e^{-x}$)

$$\int e^{-x} x^{n-1}\, dx = -e^{-x} x^{n-1} + (n - 1)\int e^{-x} x^{n-2}\, dx.$$

If we take this integral relation between 0 and A and then let A increase beyond all bounds, we immediately obtain

$$\Gamma(n) = (n - 1)\int_0^\infty e^{-x} x^{n-2}\, dx = (n - 1)\Gamma(n - 1) \qquad \text{for} \quad n > 1,$$

and by this recurrence formula, provided μ is an integer and $0 < \mu < n$, it follows that

$$\Gamma(n) = (n - 1)(n - 2)\cdots(n - \mu)\int_0^\infty e^{-x} x^{n-\mu-1}\, dx.$$

In particular, if n is a positive integer, we have for $\mu = n - 1$

$$\Gamma(n) = (n - 1)(n - 2)\cdots 3 \cdot 2 \cdot 1 \int_0^\infty e^{-x}\, dx,$$

and since

$$\int_0^\infty e^{-x}\, dx = 1,$$

we have

$$\Gamma(n) = (n - 1)(n - 2)\cdots 2 \cdot 1 = (n - 1)!,$$

a most useful expression of a factorial by an integral.

Other Examples. The integrals

$$\int_0^\infty e^{-x^2}\,dx, \qquad \int_0^\infty x^n e^{-x^2}\,dx$$

also converge, as we may easily deduce from our criterion. The first one is identical with $\frac{1}{2}\Gamma(\frac{1}{2})$, the second one with $\frac{1}{2}\Gamma[(n+1)/2]$ for $n > -\frac{1}{2}$, as is seen by the substitution $x^2 = u$, $dx = (1/2\sqrt{u})\,du$.

g. The Dirichlet Integral

In many applications we encounter integrals whose convergence does not follow directly from our criterion. An important example is furnished by the integral

$$I = \int_0^\infty \frac{\sin x}{x}\,dx$$

investigated by Dirichlet. If the upper limit is not infinite but finite, the integral is convergent since the function $(\sin x)/x$ is continuous for all finite x; $\left(\text{for } x = 0 \text{ it is given by } \lim \dfrac{\sin x}{x} = 1 \text{ for } x \to 0\right)$. The convergence of the integral I is due to the periodic change in sign of the integrand, which causes contributions to the integral from neighboring intervals of length π almost to cancel one another (Fig. 3.32). Thus the sum of the infinitely many areas

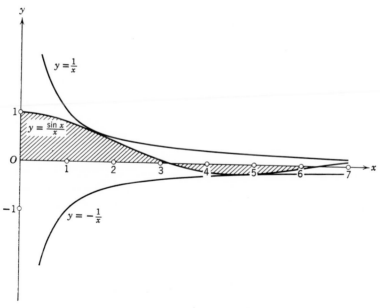

Figure 3.32 Graph of $y = \dfrac{\sin x}{x}$.

between the x-axis and the curve $y = \dfrac{\sin x}{x}$ converges, if we count areas above the x-axis as positive and below the x-axis as negative. (On the other hand, the sum of the numerical values of all areas, that is, the integral,

$$\int_0^\infty \frac{|\sin x|}{x}\, dx,$$

can easily be shown to diverge.)

The alternating character of the function $\sin x$ accounts for the fact that its indefinite integral

$$\int \sin x\, dx = 1 - \cos x$$

is bounded for all x. We make use of this fact in estimating the expression

$$I_{AB} = \int_A^B \frac{\sin x}{x}\, dx = \int_A^B \frac{1}{x} \frac{d(1 - \cos x)}{dx}\, dx$$

Integration by parts shows that

$$I_{AB} = \frac{1 - \cos B}{B} - \frac{1 - \cos A}{A} + \int_A^B \frac{1 - \cos x}{x^2}\, dx.$$

Hence

$$\int_0^\infty \frac{\sin x}{x}\, dx = \lim_{\substack{A \to 0 \\ B \to \infty}} I_{AB} = \int_0^\infty \frac{1 - \cos x}{x^2}\, dx,$$

where the integral on the right-hand side clearly is convergent. In other words, the integral I exists. In Section 8.4c we shall establish further the remarkable fact that I has the value $\pi/2$.

h. Substitution. Fresnel Integrals

Obviously, all rules for the substitution of new variables, etc., remain valid for convergent improper integrals. Often such transformations can lead to different, more tractable expressions for the integral.

As an example, to calculate

$$\int_0^\infty x e^{-x^2}\, dx$$

we introduce the new variable $u = x^2$ and obtain

$$\int_0^\infty x e^{-x^2}\, dx = \frac{1}{2} \int_0^\infty e^{-u}\, du = \lim_{A \to \infty} \frac{1}{2}(1 - e^{-A}) = \frac{1}{2}.$$

Another example in the investigation of improper integrals is given by the *Fresnel integrals*, which occur in the theory of diffraction of light:

$$F_1 = \int_0^\infty \sin (x^2) \, dx, \qquad F_2 = \int_0^\infty \cos (x^2) \, dx.$$

The substitution $x^2 = u$ yields

$$F_1 = \frac{1}{2} \int_0^\infty \frac{\sin u}{\sqrt{u}} \, du, \qquad F_2 = \frac{1}{2} \int_0^\infty \frac{\cos u}{\sqrt{u}} \, du.$$

Integrating by parts, we find

$$\int_A^B \frac{\sin u}{\sqrt{u}} \, du = \frac{1 - \cos B}{\sqrt{B}} - \frac{1 - \cos A}{\sqrt{A}} + \frac{1}{2} \int_A^B \frac{1 - \cos u}{u^{3/2}} \, du.$$

As A and B tend to zero and infinity respectively, we see by the same argument as for the Dirichlet integral that the integral F_1 converges. The convergence of the integral F_2 is proved in exactly the same way.

These Fresnel integrals show that an improper integral may exist even if the integrand does not tend to zero as $x \to \infty$. In fact, an improper integral can exist even when the integrand is unbounded, as is shown by the example

$$\int_0^\infty 2u \cos (u^4) \, du.$$

When $u^4 = n\pi$, that is, when $u = \sqrt[4]{n\pi}$, $n = 0, 1, 2, \ldots$ the integrand becomes $2\sqrt[4]{n\pi} \cos n\pi = \pm 2\sqrt[4]{n\pi}$, so that the integrand is unbounded. By the substitution $u^2 = x$, however, the integral is reduced to

$$\int_0^\infty \cos (x^2) \, dx,$$

which we have just shown to be convergent.

By means of a substitution an improper integral may often be transformed into a proper one. For example, the transformation $x = \sin u$ gives

$$\int_0^1 \frac{dx}{\sqrt{1 - x^2}} = \int_0^{\pi/2} du = \frac{\pi}{2}.$$

On the other hand, integrals of continuous functions may be transformed into improper integrals; this occurs if the transformation $u = \phi(x)$ is such that at the end of the interval of integration the derivative $\phi'(x)$ vanishes, so that dx/du is infinite.

3.16 The Differential Equations of the Trigonometric Functions

a. Introductory Remarks on Differential Equations

Integration is merely the first step into a much more extensive field: Instead of inverting differentiation by integration, that is of solving the equation $y' = f(x)$ with given $f(x)$ for $y = F(x)$, we might aim at finding functions $y = F(x)$ which satisfy more general relationships between y and derivatives of y. Such "differential equations" occur everywhere in applications as well as in strictly theoretical contexts. Penetrating studies far beyond the framework of this book are made of these equations: we shall return to some elementary aspects of the theory of differential equations later in this and the following volume. At this stage we confine ourselves to a quite simple, yet significant, example. We shall discuss the differential equations of the functions $\sin x$ and $\cos x$, which we have already mentioned on p. 171.

Although in elementary trigonometry these functions and their properties were taken from a geometric standpoint, we now discard the reliance on geometric intuition and put the trigonometric functions in a simple way on a precise, analytical basis, in accordance with the general trend of development mentioned before.

b. Sin x and cos x Defined by a Differential Equation and Initial Conditions

We consider the differential equation

$$u'' + u = 0$$

with the aim of characterizing solutions $u(x)$ which we shall identify with the sine and cosine functions. Any function $u = F(x)$ satisfying the equation, that is for which $F''(x) + F(x) = 0$, is called a *solution*.[1]

At once we realize that together with a solution $u = F(x)$ the function $u = F(x + h)$ for arbitrary constant h is also a solution, as immediately verified by differentiating $F(x + h)$ twice with respect to x. Similarly, it is immediately seen that with $F(x)$ the derivative $F'(x) = u$ is also a solution, as is of course, $cF(x)$ with a constant factor c. In addition, together with $F_1(x)$ and $F_2(x)$ any linear combination $c_1 F_1(x) + c_2 F_2(x) = F(x)$ with constants c_1 and c_2 is a solution.

---

[1] Of course, it is always understood that the functions under consideration are sufficiently differentiable.

To single out from the multitude of solutions of the differential equation a specific one, we impose "initial conditions" stipulating that for $x = 0$ the values of $u = F(0)$ and $u' = F'(0)$ be prescribed as a and b respectively. We state first:

The solution is uniquely determined by these initial values.

For the proof we start with a general remark valid for any solution u. By multiplying the differential equation with $2u'$ we find because of $2u''u' = (u'^2)'$ and $2u'u = (u^2)'$ the equation

$$0 = 2u''u' + 2u'u = [(u')^2 + u^2]',$$

which can be integrated at once and implies

$$u'^2 + u^2 = c,$$

where c is a constant, that is, does not depend on x; therefore c must have the same value as the left-hand side for $x = 0$. Thus we have for any solution u

$$u'^2(0) + u^2(0) = c.$$

Now, suppose we have two solutions u_1 and u_2 with the same initial conditions: Then the difference $z = u_1 - u_2$ is a solution with $z'(0) = z(0) = 0$. Hence we have $c = 0$ and for all x $z'^2 + z^2 = 0$; this means that $z = 0$ and $z' = 0$ which obviously proves our statement.

We now define the functions $\sin x$ and $\cos x$ as those solutions of the differential equation $u''(x) + u(x) = 0$ for which the initial conditions are, respectively, for $u = \sin x$,

$$u(0) = a = 0, \qquad u'(0) = b = 1,$$

and for $u = \cos x$,

$$u(0) = a = 1, \qquad u'(0) = b = 0.$$

We take for granted here the fact that such solutions exist and are arbitrarily often differentiable, since its proof will be given later anyway in a more general context (see Section 9.2).[1]

The only solution u of $u'' + u = 0$ for which $u = a$, $u' = b$ for $x = 0$ is then the function $u = a \cos x + b \sin x$. This proves that every solution of the differential equation is a linear combination of $\cos x$ and $\sin x$.

Now we obtain the basic properties of the trigonometric functions from our differential equation $u'' + u = 0$ applied, for example, to the

---

[1] Incidentally, we can infer these facts immediately from the equation $u'^2 + u^2 = 1$, which is valid for $\sin x$ as well as for $\cos x$ and from whose equivalent form $dx/du = 1/\sqrt{1 - u^2}$ the inverse functions of $\sin x$ and $\cos x$ are immediately obtained by integrations.

function $u = \sin x$. Obviously, with u also $v = u'$ is a solution: $v'' + v = 0$. Because of $u'' + u = v' + u = 0$ we have $v'(0) = -u(0) = 0$ whereas $v(0) = u'(0) = 1$. Hence

$$v(x) = \cos x = \frac{d}{dx} \sin x.$$

Similarly, we derive $(d/dx) \cos x = -\sin x$.

The central theorem of trigonometry is the addition theorem

$$\cos(x + y) = \cos x \cos y - \sin x \sin y.$$

It now follows immediately from our approach: First, the function $\cos(x + y)$ as a function of x, with y remaining constant for the moment, is a solution $u(x)$ of the differential equation $u'' + u = 0$ satisfying for $x = 0$ the initial conditions $u(0) = \cos y = a$ and $u'(0) = -\sin y = b$. Now, as verified immediately the solution—according to the preceding statement, the only one—for which $u(0) = a$ and $u'(0) = b$ is $a \cos x + b \sin x$. Hence we have at once for our solution $\cos(x + y)$ the expression

$$\cos(x + y) = \cos x \cos y - \sin x \sin y,$$

as we wanted to prove.

The remarks in this section should suffice to indicate how trigonometric functions can be introduced in an entirely analytical manner without any reference to geometry.

Without going into further details we mention the following.

The number $\frac{1}{2}\pi$ could now be defined as the smallest positive value of x for which $\cos x = 0$.

The periodicity of the trigonometric functions likewise follows easily from the analytic approach.

We shall return to the analytical construction of the trigonometric functions by infinite power series (see Section 5.56).

PROBLEMS

SECTION 3.1, page 201

1. Let $P(x) = a_0 + a_1 x + a_2 x^2 + \cdots + a_n x^n$.

(a) Calculate the polynomial $F(x)$ from the equation

$$F(x) - F'(x) = P(x).$$

*(b) Calculate $F(x)$ from the equation

$$c_0 F(x) + c_1 F'(x) + c_2 F''(x) = P(x).$$

2. Find the limit as $n \to \infty$ of the absolute value of the nth derivative of $1/x$ at the point $x = 2$.

3. Prove if $f^{(n)}(x) = 0$ for all x, then f is a polynomial of degree at most $n - 1$, and conversely.

4. Determine the form of a rational function r for which

$$\lim_{x \to \infty} \frac{xr'(x)}{r(x)} = 0.$$

5. Prove by induction that the nth derivative of a product may be found according to the following rule (Leibnitz's rule):

$$\frac{d^n}{dx^n}(fg) = f\frac{d^ng}{dx^n} + \binom{n}{1}\frac{df}{dx}\frac{d^{n-1}g}{dx^{n-1}} + \binom{n}{2}\frac{d^2f}{dx^2}\frac{d^{n-2}g}{dx^{n-2}} + \cdots$$

$$+ \binom{n}{n-1}\frac{d^{n-1}f}{dx^{n-1}}\frac{dg}{dx} + \frac{d^nf}{dx^n}g.$$

Here $\binom{n}{1} = n$, $\binom{n}{2} = \frac{n(n-1)}{2!}$, etc.; denote binomial coefficients.

6. Prove that $\sum\limits_{i=1}^{n-1} ix^{i-1} = \dfrac{(n-1)x^n - nx^{n-1} + 1}{(x-1)^2}$.

SECTION 3.2, page 206

1. Let $y = e^x(a \sin x + b \cos x)$. Show that y'' can be expressed as a linear combination of y and y', that is,

$$y'' = py' + qy,$$

where p and q are constants. Express all higher derivatives as linear combinations of y' and y.

*2. Find the nth derivative of arc sin x at $x = 0$, and then of (arc sin x)^2 at $x = 0$.

SECTION 3.3, page 217

1. Find the second derivative of $f[g\{h(x)\}]$.

2. Differentiate the following function: $\log_{v(x)} u(x)$, [that is, the logarithm of $u(x)$ to the base $v(x)$; $v(x) > 0$].

3. What conditions must the coefficients α, β, a, b, c satisfy in order that

$$\frac{\alpha x + \beta}{\sqrt{(ax^2 + 2bx + c)}}$$

shall everywhere have a finite derivative that is never zero?

4. Show that $d^n(e^{x^2/2})/dx^n = u_n(x)e^{x^2/2}$, where $u_n(x)$ is a polynomial of degree n. Establish the recurrence relation

$$u_{n+1} = xu_n + u_n'.$$

*5. By applying Leibnitz's rule to

$$\frac{d}{dx}(e^{x^2/2}) = xe^{x^2/2},$$

obtain the recurrence relation

$$u_{n+1} = xu_n + nu_{n-1}.$$

*6. By combining the recurrence relations of Problems 4 and 5, obtain the differential equation

$$u_n'' + xu_n' - nu_n = 0$$

satisfied by $u_n(x)$.

7. Find the polynomial solution

$$u_n(x) = x^n + a_1 x^{n-1} + \cdots + a_n$$

of the differential equation $u_n'' + xu_n' - nu_n = 0$.

*8. If $P_n(x) = \dfrac{1}{2^n n!} \dfrac{d^n}{dx^n} (x^2 - 1)^n$, prove the relations

(a) $P'_{n+1} = \dfrac{x^2 - 1}{2(n + 1)} P_n'' + \dfrac{(n + 2)x}{n + 1} P_n' + \dfrac{n + 2}{2} P_n$.

(b) $P'_{n+1} = xP_n' + (n + 1)P_n$.

(c) $\dfrac{d}{dx} [(x^2 - 1)P_n'] - n(n + 1)P_n = 0$.

9. Find the polynomial solution

$$P_n = \frac{(2n)!}{2^n (n!)^2} x^n + a_1 x^{n-1} + \cdots + a_n$$

of the differential equation

$$\frac{d}{dx} [(x^2 - 1)P_n'] - n(n + 1)P_n = 0.$$

10. Determine the polynomial $P_n(x) = \dfrac{1}{2^n n!} \dfrac{d^n}{dx^n} (x^2 - 1)^n$ by using the binomial theorem.

*11. Let $\lambda_{n,p}(x) = \dbinom{p}{n} x^n (1 - x)^{p-n}$, $n = 0, 1, 2, \ldots, p$. Show that

$$1 = \sum_{n=0}^{p} \lambda_{n,p}(x).$$

$$x^k = \sum_{n=k}^{p} \frac{\dbinom{n}{k}}{\dbinom{p}{k}} \lambda_{n,p}(x).$$

$$\cdots \cdots \cdots$$

$$x^p = \lambda_{p,p}(x).$$

$$\cdots \cdots \cdots$$

SECTION 3.4, page 223

1. The function $f(x)$ satisfies the equation

$$f(x + y) = f(x)f(y).$$

(a) If $f(x)$ is differentiable, either $f(x) \equiv 0$ or $f(x) = e^{ax}$.

*(b) If $f(x)$ is continuous, either $f(x) \equiv 0$ or $f(x) = e^{ax}$.

2. If a differentiable function $f(x)$ satisfies the equation

$$f(xy) = f(x) + f(y),$$

then $f(x) = \alpha \log x$.

3. Prove that if $f(x)$ is continuous and

$$f(x) = \int_0^x f(t) \, dt,$$

then $f(x)$ is identically zero.

SECTION 3.5, page 228

1. Prove the formula

$$\sinh a + \sinh b = 2 \sinh \left(\frac{a+b}{2}\right) \cosh \left(\frac{a-b}{2}\right).$$

Obtain similar formulas for $\sinh a - \sinh b$, $\cosh a + \cosh b$, $\cosh a - \cosh b$.

2. Express $\tanh (a + b)$ in terms of $\tanh a$ and $\tanh b$.
Express $\coth (a \pm b)$ in terms of $\coth a$ and $\coth b$.
Express $\sinh \frac{1}{2}a$ and $\cosh \frac{1}{2}a$ in terms of $\cosh a$.

3. Differentiate
(a) $\cosh x + \sinh x$; (b) $e^{\tanh x + \coth x}$,
(c) $\log \sinh (x + \cosh^2 x)$; (d) $\operatorname{ar} \cosh x + \operatorname{ar} \sinh x$ (e) $\operatorname{ar} \sinh (\alpha \cosh x)$;
(f) $\operatorname{ar} \tanh (2x/(1 + x^2))$.

4. Calculate the area bounded by the catenary $y = \cosh x$, the ordinates $x = a$ and $x = b$, and the x-axis.

SECTION 3.6, page 236

1. Determine the maxima, minima, and points of inflection of $x^3 + 3px + q$. Discuss the nature of the roots of $x^3 + 3px + q = 0$.

2. Given the parabola $y^2 = 2px$, $p > 0$, and a point $P(x = \xi, y = \eta)$ within it ($\eta^2 < 2p\xi$), find the shortest path (consisting of two line segments) leading from P to a point Q on the parabola and then to the focus $F(x = \frac{1}{2}p, y = 0)$ of the parabola. Show that the angle FQP is bisected by the normal to the parabola, and that QP is parallel to the axis of the parabola (principle of the parabolic mirror).

3. Among all triangles with given base and given vertical angle, the isosceles triangle has the maximum area.

4. Among all triangles with given base and given area, the isosceles triangle has the maximum vertical angle.

*5. Among all triangles with given area, the equilateral triangle has the least perimeter.

*6. Among all triangles with given perimeter the equilateral triangle has the maximum area.

*7. Among all triangles inscribed in a circle the equilateral triangle has the maximum area.

8. Prove that if $p > 1$ and $x > 0$, $x^p - 1 \geq p(x - 1)$.

9. Prove the inequality $1 > (\sin x)/x \geq 2/\pi, 0 \leq x \leq \pi/2$.

10. Prove that (a) $\tan x \geq x, 0 \leq \pi/2$.
(b) $\cos x \geq 1 - x^2/2$.

***11.** Given $a_1 > 0, a_2 > 0, \ldots, a_n > 0$, determine the minimum of

$$\frac{\dfrac{a_1 + \cdots + a_{n-1} + x}{n}}{\sqrt[n]{a_1 a_2 \cdots a_{n-1} x}}$$

for $x > 0$. Use the result to prove by mathematical induction that (cf. Problem 13, p. 109)

$$\sqrt[n]{a_1 a_2 \cdots a_n} \leq \frac{a_1 + \cdots + a_n}{n}.$$

12. (a) Given n fixed numbers $a_1, \ldots, a_n$, determine x so that $\sum_{i=1}^{n} (a_i - x)^2$ is a minimum.

***(b)** Minimize $\sum_{i=1}^{n} |a_i - x|$.

***(c)** Minimize $\sum_{i=1}^{n} \lambda_i |a_i - x|$, where $\lambda_i > 0$.

13. Sketch the graph of the function

$$y = (x^2)^x, y(0) = 1.$$

Show that the function is continuous at $x = 0$. Has the function maxima, minima, or points of inflection?

***14.** Find the least value α such that

$$\left(1 + \frac{1}{x}\right)^{x+\alpha} > e$$

for all positive x. (*Hint:* It is known that $[1 + (1/x)]^{x+1}$ decreases monotonically and $[1 + (1/x)]^x$ increases monotonically to the limit e at infinity.)

***15.** (a) Find the point such that the sum of the distances to the three sides of a triangle is a minimum.

(b) Find the point for which the sum of the distances to the vertices is a minimum.

16. Prove the following inequalities:
(a) $e^x > 1/(1 + x), x > 0$.
(b) $e^x > 1 + \log(1 + x), x > 0$.
(c) $e^x > 1 + (1 + x)\log(1 + x), x > 0$.

17. Suppose $f''(x) < 0$ on (a, b). Prove:
(a) Every arc of the graph within the interval lies above the chord joining its endpoints.

(b) The graph lies below the tangent at any point within (a, b).

***18.** Let f be a function possessing a second derivative on (a, b).
(a) Show that either condition a or b of Problem 22 is sufficient for $f''(x) \leq 0$.

(b) Show that the condition

$$f\left(\frac{x+y}{2}\right) \geq \frac{f(x) + f(y)}{2}$$

for all x and y in (a, b) is sufficient for $f''(x) \leq 0$.

*19. Let a, b be two positive numbers, p and q any nonzero numbers $p < q$. Prove that

$$\frac{[\theta a^p + (1 - \theta)b^p]^{1/p}}{[\theta a^q + (1 - \theta)b^q]^{1/q}} \leq 1$$

for all values of θ in the interval $0 < \theta < 1$.

(This is Jensen's inequality, which states that the pth power mean $[\theta a^p + (1 - \theta)b^p]^{1/p}$ of two positive qualities a, b is an increasing function of p.)

20. Show that the equality sign in the above inequality holds if, and only if, $a = b$.

21. Prove that $\lim\limits_{p \to 0} [\theta a^p + (1 - \theta)b^p]^{1/p} = a^\theta b^{1-\theta}$.

22. Defining the zeroth power mean of a, b as $a^\theta b^{1-\theta}$, show that Jensen's inequality applies to this case, and becomes ($a \neq b$),

$$a^\theta b^{1-\theta} \gtrless [\theta a^q + (1 - \theta)b^q]^{1/q} \quad \text{according to whether} \quad q \lessgtr 0$$

For $q = 1$, $\qquad\qquad a^\theta b^{1-\theta} \leq \theta a + (1 - \theta)b$.

23. Prove the inequality

$$a^\theta b^{1-\theta} \leq \theta a + (1 - \theta)b,$$

$a, b > 0, 0 < \theta < 1$, without reference to Jensen's inequality, and show that equality holds only if $a = b$. (This inequality states that the $\theta, 1 - \theta$ geometric mean is less than the corresponding arithmetic mean.)

*24. Let f be continuous and positive on $[a, b]$ and let M denote its maximum value. Prove

$$M = \lim_{n \to \infty} \sqrt[n]{\int_a^b [f(x)]^n \, dx}.$$

SECTION 3.7, page 248

1. Let $f(x)$ be a continuous function vanishing, together with its first derivative, for $x = 0$. Show that $f(x)$ vanishes to a higher order than x as $x \to 0$.

2. Show that $f(x) = \dfrac{a_0 x^n + a_1 x^{n-1} + \cdots + a_n}{b_0 x^m + b_1 x^{m-1} + \cdots + b_m}$,

when $a_0, b_0 \neq 0$, is of the same order of magnitude as x^{n-m}, when $x \to \infty$.

*3. Prove that e^x is not a rational function.

*4. Prove that e^x cannot satisfy an algebraic equation with polynomials in x as coefficients.

5. If the order of magnitude of the positive function $f(x)$ as $x \to \infty$ is higher, the same, or lower than that of x^m, prove that $\int_a^x f(\xi) \, d\xi$ has the corresponding order of magnitude relative to x^{m+1}.

6. Compare the order of magnitude as $x \to \infty$ of $\int_a^x f(\xi)\, d\xi$ relative to $f(x)$ for the following functions $f(x)$:

(a) $\dfrac{e\sqrt{x}}{\sqrt{x}}$.

(c) xe^{x^2} .

(b) e^x .

(d) $\log x$.

SECTION 3.8, page 263

1. Find the limit as $n \to \infty$ of $a_n = \dfrac{1}{n+1} + \dfrac{1}{n+2} + \cdots + \dfrac{1}{2n}$.

*2. Find the limit of

$$b_n = \frac{1}{\sqrt{n^2-0}} + \frac{1}{\sqrt{n^2-1}} + \frac{1}{\sqrt{n^2-4}} + \cdots + \frac{1}{\sqrt{n^2-(n-1)^2}}$$

*3. If α is any real number greater than -1, evaluate

$$\lim_{n \to \infty} \frac{1^\alpha + 2^\alpha + 3^\alpha + \cdots + n^\alpha}{n^{\alpha+1}} .$$

SECTION 3.11, page 274

1. Show that for all odd positive values of n the integral $\int e^{-x^2} x^n\, dx$ can be evaluated in terms of elementary functions.

2. Show that if n is even, the integral $\int e^{-x^2} x^n\, dx$ can be evaluated in terms of elementary functions and the integral $\int e^{-x^2}\, dx$ (for which tables have been constructed).

3. Prove that

$$\int_0^x \left[\int_0^u f(t)\, dt \right] du = \int_0^x f(u)(x-u)\, du.$$

*4. Problem 3 gives a formula for the second iterated integral. Prove that the nth iterated integral of $f(x)$ is given by

$$\frac{1}{(n-1)!} \int_0^x f(u)(x-u)^{n-1}\, du.$$

5. Prove for the binomial coefficient $\dbinom{n}{k}$ that

$$\binom{n}{k} = \left[(n+1) \int_0^1 x^k (1-x)^{n-k}\, dx \right]^{-1}.$$

6. Obtain a recursive formula for

$$\int x^p (ax^n + b)^q\, dx$$

and use this relation to integrate

$$\int x^3 (x^7 + 1)^4\, dx.$$

*7. (a) Let $P_n(x) = \dfrac{1}{2^n n!} \dfrac{d^n}{dx^n} (x^2 - 1)^n$. Show that

$$\int_{-1}^{1} P_n(x) P_m(x)\, dx = 0, \qquad \text{if} \quad m \neq n.$$

(b) Prove that $\displaystyle \int_{-1}^{1} P_n{}^2(x)\, dx = \dfrac{2}{2n + 1}$.

(c) Prove that $\displaystyle \int_{1}^{-1} x^m P_n(x)\, dx = 0$, if $m < n$.

(d) Evaluate $\displaystyle \int_{-1}^{1} x^n P_n(x)\, dx$.

SECTION 3.12, page 282

*1. Integrate

$$\int \frac{dx}{x^6 + 1}.$$

2. Use the partial fraction expansion to prove Newton's formulas

$$\frac{\alpha_1{}^k}{g'(\alpha_1)} + \frac{\alpha_2{}^k}{g'(\alpha_2)} + \cdots + \frac{\alpha_n{}^k}{g'(\alpha_n)} = \begin{cases} 0 & \text{for } k = 0, 1, 2, \ldots, n - 2 \\ 1 & \text{for } k = n - 1, \end{cases}$$

where $g(x)$ is a polynomial of the form $x^n + \alpha_1 x^{n-1} + \cdots$ with distinct roots $\alpha_1, \ldots, \alpha_n$.

SECTION 3.14, page 298

*1. Prove that the substitution $x = (\alpha t + \beta)/(\gamma t + \delta)$ with $\alpha\delta - \gamma\beta \neq 0$, transforms the integral

$$\int \frac{dx}{\sqrt{ax^4 + bx^3 + cx^2 + dx + e}}$$

into an integral of similar type, and that if the biquadratic

$$ax^4 + bx^3 + cx^2 + dx + e$$

has no repeated factors, neither has the new biquadratic in t which takes its place. Prove that the same is true for

$$\int R(x, \sqrt{ax^4 + bx^3 + cx^2 + dx + e})\, dx,$$

where R is a rational function.

2. The function

$$\phi(x) = \int_{0}^{x} \frac{du}{\sqrt{1 - k^2 \sin^2 u}}$$

is known as the *elliptic integral of the first kind*.

(a) Show that ϕ is continuous and increasing and hence has a continuous inverse.

(b) Let $am(x)$ denote the inverse of $\phi(x)$. Prove $sn(x) = \sin[am(x)]$, where $sn(x)$ is defined on p. 299, footnote 3.

SECTION 3.15, page 301

*1. Prove that $\displaystyle\int_0^\infty \sin^2\left[\pi\left(x + \frac{1}{x}\right)\right] dx$ does not exist.

*2. Prove that $\displaystyle\lim_{k\to\infty} \int_0^\infty \frac{dx}{1 + kx^{10}} = 0$.

3. For what values of s is (a) $\displaystyle\int_0^\infty \frac{x^{s-1}}{1 + x} dx$, (b) $\displaystyle\int_0^\infty \frac{\sin x}{x^s} dx$ convergent?

*4. Does $\displaystyle\int_0^\infty \frac{\sin t}{1 + t} dt$ converge?

*5. (a) If a is a fixed positive number, prove that

$$\lim_{h\to 0+} \int_{-a}^a \frac{h}{h^2 + x^2} dx = \pi.$$

(b) If $f(x)$ is continuous in the interval $-1 \le x \le 1$, prove that

$$\lim_{h\to 0} \int_{-1}^1 \frac{h}{h^2 + x^2} f(x)\, dx = \pi f(0).$$

*6. Prove that $\displaystyle\lim_{x\to\infty} e^{-x^2} \int_0^x e^{t^2}\, dt = 0$.

7. Assuming that $|\alpha| \ne |\beta|$, prove that

$$\lim_{T\to\infty} \frac{1}{T} \int_0^T \sin \alpha x \sin \beta x\, dx = 0.$$

*8. If $\displaystyle\int_a^\infty \frac{f(x)}{x} dx$ converges for any positive value of a, and if $f(x)$ tends to a limit L as $x \to 0$, show that $\displaystyle\int_0^\infty \frac{f(\alpha x) - f(\beta x)}{x} dx$ converges for α and β positive and has the value $L \log \dfrac{\beta}{\alpha}$.

9. By reference to the Problem 8, show that

(a) $\displaystyle\int_0^\infty \frac{e^{-\alpha x} - e^{-\beta x}}{x} dx = \log \frac{\beta}{\alpha}$.

(b) $\displaystyle\int_0^\infty \frac{\cos \alpha x - \cos \beta x}{x} dx = \log \frac{\beta}{\alpha}$.

*10. If $\displaystyle\int_a^b \frac{f(x)}{x} dx$ converges for any positive values of a and b, and if $f(x)$ tends to a limit M as $x \to \infty$ and a limit L as $x \to 0$, show that

$$\int_0^\infty \frac{f(\alpha x) - f(\beta x)}{x} dx = (L - M) \log \frac{\beta}{\alpha}.$$

11. Obtain the following expressions for the gamma function:

$$\Gamma(n) = 2 \int_0^\infty x^{2n-1} e^{-x^2}\, dx,$$

$$\Gamma(n) = \int_0^1 \left(\log \frac{1}{x}\right)^{n-1} dx.$$

SECTION 3.16, page 312

1. Obtain the addition formula for $\sin(x + y)$.

2. Without using the addition formulas prove that $\cos x$ is an even function and $\sin x$, odd.

3. $(a)^*$ Prove for some positive h that $\cos x < 1$ for $0 < x < h$.
(b) Prove if $\cos z > 0$ for $0 \le z \le 2^n x$ that

$$\cos(2^{n+1}x) < 2^n(\cos x - 1) + 1.$$

(c) Combining the results (a) and (b) prove that $\cos x$ has a zero.

4. Let a be the smallest positive zero of $\cos x$. Prove that

$$\sin(x + 4a) = \sin x,$$
$$\cos(x + 4a) = \cos x.$$

5. Fill in the steps of the following indirect proof that $\cos x$ has a zero:
(a) If $\cos x$ has no zeros, then $\sin x$ is monotonically increasing for $x \ge 0$.
(b) The functions $\sin x$ and $\cos x$ are bounded from above and below.
(c) The limit of $\sin x$ as x tends to infinity exists and is positive.
(d) The equation

$$\cos x = 1 - \int_0^x \sin t\, dt$$

stands in contradiction to (b).

MISCELLANEOUS PROBLEMS

1. Prove

$$\frac{d^n}{dx^n} f(\log x) = x^{-n} \frac{d}{dt}\left(\frac{d}{dt} - 1\right)\left(\frac{d}{dt} - 2\right) \cdots \left(\frac{d}{dt} - n + 1\right) f(t)$$

when $t = \log x$. Here, we employ

$$\left(\frac{d}{dt} - k\right)\phi = \frac{d\phi}{dt} - k\phi,$$

where ϕ is any function of t and k is a constant.

2. A smooth closed curve C is said to be convex if it lies wholly to one side of each tangent. Show that for the triangle of minimum area circumscribed about $\mathbf{Z}$ that each side is tangent to C at its midpoint.

4

Applications in
Physics and Geometry

4.1 Theory of Plane Curves

a. Parametric Representation

Definition

The representation of a curve by an equation $y = f(x)$ imposes a serious geometrical restriction: A curve so represented must not be intersected at more than one point by any parallel to the y-axis. Usually, this restriction can be overcome by decomposing the curve into portions each representable in the form $y = f(x)$. Thus a circle of radius a about the origin is given by the two functions $y = \sqrt{a^2 - x^2}$ and $y = -\sqrt{a^2 - x^2}$ defined for $-a \leq x \leq a$. However, for as simple a curve as a parallel to the y-axis this device does not work.

More flexibility is obtained by an *implicit* representation through an equation $\phi(x, y) = 0$ which involves a function ϕ of two independent variables. For example, the circle of radius a about the origin is completely described by $\phi(x, y) = x^2 + y^2 - a^2 = 0$. Any straight line in the plane has an implicit equation of the form $ax + by + c = 0$, where a, b, c are constants and a and b do not both vanish; for $b = 0$ we obtain a parallel to the y-axis.

The implicit description of a curve has the disadvantage that to find points (x, y) of the curve at all, say for a given x, we must *solve* the equation $\phi(x, y) = 0$. This problem we shall discuss in detail in Volume II.

The most direct and most flexible description of a curve is a *parametric representation*. Instead of considering one of the rectangular coordinates y or x as a function of the other we think of both coordinates x and y as functions of a *third* independent variable t, a so-called *parameter*;[1] the point with coordinates x and y then describes the curve as t traverses a corresponding interval. Such parametric representations have already been encountered; for example, the circle $x^2 + y^2 = a^2$ has the parametric representation $x = a \cos t$, $y = a \sin t$. Here t denotes the angle at the center of the circle.

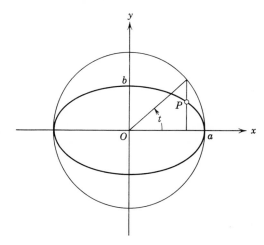

Figure 4.1

For the ellipse $x^2/a^2 + y^2/b^2 = 1$ we have the similar parametric representation $x = a \cos t, y = b \sin t$, where t is the so-called eccentric angle, that is, the angle at the center corresponding to the point of the circumscribed circle lying vertically above or below the point $P = (a \cos t, b \sin t)$ of the ellipse. We assume here that $b < a$ (see Fig. 4.1). In both cases the point with the coordinates x, y describes the complete circle or ellipse as the parameter t traverses the interval $0 \leq t < 2\pi$.

In general, curves C are parametrically represented by two functions of a parameter t,

$$x = \phi(t) = x(t), \qquad y = \psi(t) = y(t);$$

---

[1] This word denotes an auxiliary variable which we do not want to emphasize primarily.

the shorter notation $x(t)$ and $y(t)$ will be used when there is no danger of confusion.[1]

We assume throughout that ϕ and ψ possess continuous derivatives unless the contrary is said.

Mapping of Parameter Interval on Curve—Sense of Direction

For a given curve these two functions $\phi(t)$ and $\psi(t)$ must be determined in such a way that the set of pairs of functional values $x(t)$ and $y(t)$ corresponding to a certain interval of values t defines all the points on the curve and no other points. We have then a correspondence between the points of the curve and the values of t in an interval of the t-axis. The parameter representation defines a *mapping of the t-axis onto the curve*, the original point t on the t-axis being mapped onto the point $x = \phi(t)$, $y = \psi(t)$ of C.

Since $x(t)$ and $y(t)$ are assumed continuous, neighboring points on the t-axis correspond to neighboring points on the curve. Since the points of the t-axis are ordered, we may in an obvious manner assign an order or "sense" to the points of C by saying that the point onto which the number t_1 is mapped *precedes* the point onto which t_2 is mapped if $t_1 < t_2$ (see p. 334). The parametric representation thus gives precise meaning to the vague intuitive notion of a curve as a set of points in which the points are arranged in the same order as on a straight line.

b. Change of Parameters

The values of the parameter t serve to distinguish the different points on the curve C; they play the role of "names" for the individual points of the curve.

The same curve C admits of many different parameter representations. Any quantity that varies continuously along the curve and has different values in different points of the curve can serve as parameter.

If, say, the curve originally is given by an equation $y = f(x)$, we can choose for the parameter t the variable x and describe the curve by the functions $x = t$, $y = f(t)$. Similarly, for a curve described by giving x as a function of y, say $x = g(y)$, we can use y as parameter t and write $x = g(t)$, $y = t$.

---

[1] The notation $x = \phi(t)$, etc., puts emphasis on the specific functional connection between the dependent and independent variable; the notation $x(t)$, etc., just means that t is to be considered as the independent variable which determines the function value of x in some prescribed way.

For a curve given by an equation $r = h(\theta)$ in *polar coordinates r, θ* (see Chapter 1, p. 101) we can choose θ as parameter t and obtain the parametric representation

$$x = r \cos \theta = h(t) \cos t = \phi(t),$$
$$y = r \sin \theta = h(t) \sin t = \psi(t).$$

From a given parametric representation $x = \phi(t)$, $y = \psi(t)$ of a curve C we can always derive many other parameter representations. For that purpose we take an arbitrary function $\tau = \chi(t)$ which is monotonic and continuous in that t-interval corresponding to the points of C; the function χ has then a monotone and continuous inverse $t = \sigma(\tau)$ in a corresponding τ-interval. The coordinates of the points (x, y) of C can then be represented in the form

$$x = \phi[\sigma(\tau)] = \alpha(\tau), \qquad y = \psi[\sigma(\tau)] = \beta(\tau).$$

The functions $\alpha(\tau)$ and $\beta(\tau)$ are again continuous; moreover, different points of C correspond to different values of t and hence, because of the monotone character of the function σ, to different values of τ. The total effect of the change of parameter from t to τ is that of "renaming" the points on C.

Thus the line $y = x$ has the parameter representation $x = t$, $y = t$, where $-\infty < t < \infty$. Substituting $\tau = t^3$ gives rise to the parameter representation $x = \tau^{1/3}$, $y = \tau^{1/3}$ for the same line.

Similarly, the ellipse $x^2/a^2 + y^2/b^2 = 1$ admits of the parameter representation $x = a \cos t$, $y = b \sin t$, where $0 \leq t < 2\pi$. Defining $t = c\zeta + d$, for c, d real numbers $(c \neq 0)$ yields another representation $x(\zeta) = a \cos(c\zeta + d)$, $y(\zeta) = b \sin(c\zeta + d)$ for the same ellipse, with ζ varying in the interval $-d/c \leq \zeta < (2\pi - d)/c$ for $c > 0$, and $(2\pi - d)/c < \zeta \leq -d/c$, for $c < 0$. The substitution $\tau = \tan(t/2)$ leads to the "*rational*" *parameter representation* (see p. 292)

$$x = \frac{a(1 - \tau^2)}{1 + \tau^2}, \qquad y = \frac{2b\tau}{1 + \tau^2}$$

for the ellipse; as τ runs through all real values we obtain all points of the ellipse with the exception of the point $S = (-a, 0)$.

Singularities in ordinary representation may disappear if a suitable parameter is used. For example, we can represent the curve $y = \sqrt[3]{x^2}$ by the smooth functions $x = t^3$, $y = t^2$. The point with coordinates x, y then describes the whole curve (semicubical parabola) as t varies from $-\infty$ to $+\infty$.

This flexibility in the choice of the parameter often permits us to simplify the study of geometrical properties which, of course, do not depend on specific representations.

In particular, we may sometimes find it convenient to use a representation $y = f(x)$ for C or part of C. Such a representation is always possible for a portion of the curve $t_0 \leq t \leq t_1$ in which one of the functions ϕ, ψ, say $x = \phi(t)$, is monotonic. Indeed for this portion we have a unique inverse function $t = \gamma(x)$ and thus $y = \psi[\gamma(x)]$.[1]

c. Motion along a Curve. Time as the Parameter. Example of the Cycloid

Motion along a Curve

Very often the parameter t has the natural physical meaning of time. Any motion of a point in the plane may be expressed by representing its coordinates x and y as functions of the time such that at the time t, the point (x, y) is at $(x(t), y(t))$. These two functions therefore determine the motion along a path or trajectory C in parametric form; they constitute a mapping of the time scale onto the trajectory.[2]

The Cycloids and Trochoids

An example is furnished by the *cycloids*, the paths of points on a circle rolling uniformly without slipping along a straight line or another circle. In the simplest case a circle of radius a rolls along the x-axis; the path of a point P on its circumference is a "common" cycloid. We choose the origin of the coordinate system and the initial time in such a way that for time $t = 0$ the point P is at the origin and that at the time t the circle has turned from its original orientation by the angle t. This means that the circle turns clockwise with "angular velocity" one. The circle is assumed to roll uniformly along the x-axis without sliding so that at the time t the distance of the point of contact from the origin is exactly equal to the length of the arc from the point of contact to P. Thus at the time t the center M of the rolling circle must be at the point (at, a); the center moves with constant velocity a to the right. For the

---

[1] This is, of course, merely a statement about a property "in the small" of a curve, meaning a statement made only for a suitably small portion. Usually (for example, in the case of a circle), the variable x cannot be used as a parameter throughout the whole curve but only on a portion.

[2] To a change of the parameter t there would correspond then a change in the time scale according to which the curve C is described by the moving point.

coordinates of P at the time t we find then (see Fig. 4.2) the parametric representation

(1) $$x = a(t - \sin t), \qquad y = a(1 - \cos t).$$

By eliminating the parameter t we can obtain the equation of the curve in nonparametric form, at the cost, however, of neatness of

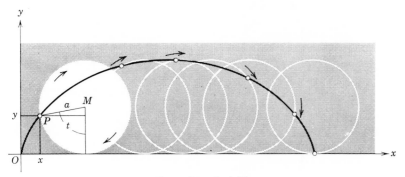

Figure 4.2   Cycloid.

expression. We have

$$\cos t = \frac{a - y}{a}, \qquad t = \text{arc cos} \frac{a - y}{a}, \qquad \sin t = \pm\sqrt{1 - \frac{(a - y)^2}{a^2}},$$

and hence

(1a) $$x = a \ \text{arc cos} \frac{a - y}{a} \mp \sqrt{y(2a - y)},$$

thus obtaining x as a function of y.

Epicycloid

Our next example is that of an *epicycloid*, defined as the path of a point P fixed on the circumference of a circle of radius c, as it rolls at a uniform speed along the circumference and outside of a second circle of radius a. Let the fixed circle be centered at the origin of the x,y-plane. Suppose the moving circle is rolling along the fixed one in such a way that its center has rotated about the origin to an angle t at time t (Fig. 4.3). Then we find for the position at the time t of the point $P = (x(t), y(t))$, which at the time $t = 0$ is the point of contact $(a, 0)$, the parametric equations

(2)
$$x(t) = (a + c) \cos t - c \cos \left(\frac{a + c}{c} t \right),$$
$$y(t) = (a + c) \sin t - c \sin \left(\frac{a + c}{c} t \right).$$

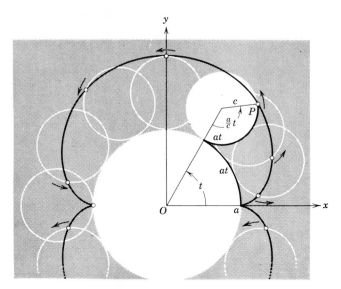

Figure 4.3   Epicycloid.

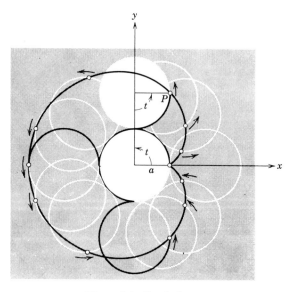

Figure 4.4   Cardiod.

When $a = c$, the curve formed is called a cardiod, (Fig. 4.4) and is given by the parametric equations

(3)
$$x(t) = 2a \cos t - a \cos (2t),$$
$$y(t) = 2a \sin t - a \sin (2t).$$

A third variety of cycloids is obtained as the locus of a point attached to the circumference of one circle rolling along the circumference of another fixed circle, but interior to it. To find the parametric equations for this "*hypocycloid*," let a be the radius of the fixed circle and c that of the rolling circle. Let the point P on the circumference of the moving circle be located

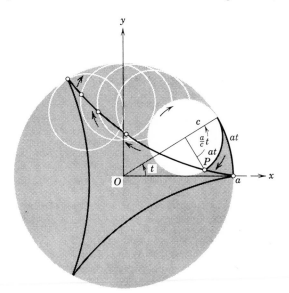

Figure 4.5 Hypocycloid.

at $(a, 0)$ at time $t = 0$. Suppose that the rolling circle is moving along the fixed one in such a way that at time t its center has rotated about the origin through an angle t (Fig. 4.5). Then we find the parametric equations for the hypocycloid to be

(4)
$$x(t) = (a - c) \cos t + c \cos \left(\frac{a - c}{c} t \right),$$
$$y(t) = (a - c) \sin t - c \sin \left(\frac{a - c}{c} t \right).$$

In the special case when the fixed circle has twice the radius of the moving one, $c = \frac{1}{2}a$, we find

$$x(t) = a \cos t,$$
$$y(t) = 0,$$

and the hypocycloid degenerates into the diameter of the fixed circle, described back and forth. The interesting feature of this example is that it provides a mechanical solution of the problem of drawing a straight line by using merely circular motions (Fig. 4.6).

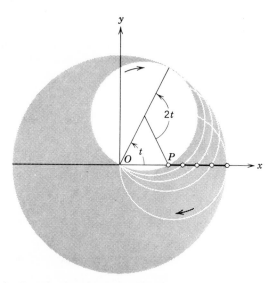

Figure 4.6 A point P on the rim of a circle rolling inside a circle of twice the radius describes a straight line segment.

If the radius of the fixed circle is three times that of the moving one, then $c = a/3$, and

$$x(t) = \tfrac{2}{3}a \cos t + \tfrac{1}{3}a \cos (2t),$$

$$y(t) = \tfrac{2}{3}a \sin t - \tfrac{1}{3}a \sin (2t).$$

By an elementary computation we find

$$x^2 + y^2 = \tfrac{5}{9}a^2 + \tfrac{4}{9}a^2 \cos (3t),$$

so that the hypocycloid meets the fixed circle at exactly three points and the curve appears as shown in Fig. 4.5.

Trochoids

More general curves called *trochoids* (epitrochoids, hypotrochoids) are obtained if we consider the motion of a point P attached to a circle (but not necessarily on its rim) when that circle rolls along a straight line or along the outside or inside of another circle (see Fig. 4.7). The same type of curve arises as the path of a point moving uniformly on a circle while the center of the circle itself moves uniformly along a line or circle. These curves play a

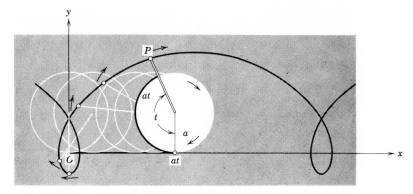

Figure 4.7 Trochoid.

central role in the Ptolomaic description of the apparent motion of the planets.

Some of the remarkable properties of cycloids will be discussed later on in this chapter (p. 428).

d. Classifications of Curves. Orientation

Definitions

Among the most obvious features of a curve are the number of separate pieces or branches and the number of loops which it has. A hyperbola is an example of a curve consisting of two disjoint branches; another such example is the curve $y^2 = (4 - x^2)(x^2 - 1)$ which consists of two separate ovals. We shall be concerned mainly with curves consisting of one piece, the *connected* curves. A connected curve can *intersect* itself, like the trochoid (Fig. 4.7) or the "lemniscate" of Fig. 153, p. 103.

A connected curve without self-intersections is called *simple*. Among the simple curves we can still distinguish the *closed* curves, such as circles or ellipses, from the ones that are not closed, such as parabolas or straight-line segments. We shall not attempt here to give either a rigorous or a complete classification of curves, but only point out certain *"topological"* features of a curve relevant for parameter representation.

Simple Arcs

A parameter representation of a curve C by two continuous functions $x = \phi(t)$, $y = \psi(t)$ defines a mapping of the t-axis or of a portion of it onto C. We call C a *simple arc* if it can be represented in such a way that the parameter t describes a closed interval $[a, b]$ on the t-axis, forming the domain of the functions $\phi(t)$, $\psi(t)$, and if in addition different t in the interval correspond to different points P on C. An example is the parabolic arc $x = t$, $y = t^2$ for $0 \leq t \leq 1$.

The same arc C (that is, the same points in the plane) can be represented parametrically in many ways. Any monotone continuous function $\tau = \chi(t)$ for $a \leq t \leq b$ defines a parameter τ such that x and y are continuous functions of τ in a suitable closed interval $[\alpha, \beta]$, different values of τ corresponding to different P. As a matter of fact, as is easily seen the continuous monotone substitutions $\tau = \chi(t)$ provide the *most general* continuous parameter representations of a simple arc that assign different points of the arc to different parameter values. (See the remarks on p. 55 about one-to-one continuous mappings.)

To a special parameter representation $x = x(t)$, $y = y(t)$ of a simple arc C belongs a definite *sense* on C corresponding to the direction of *increasing* t. Given any two distinct points P_0, P_1 we say that P_1 *follows* P_0 if P_1 belongs to the larger value of the parameter t. If we introduce a new parameter τ by a continuous *increasing* function $\tau = \chi(t)$ the order of the pairs of points with respect to τ is the same; the parameter τ defines the *same sense* on C. If $\chi(t)$ is decreasing, the sense is reversed.

Direction or Orientation of Arcs

A *directed* or *oriented* simple arc is one on which a definite sense has been selected (for example, that sense corresponding to an increase in a particular choice of the parameter t); that sense is then called the *positive sense* on the arc. The positive sense is completely specified, if we know which of the two end points of the arc *follows* the other one. We call the end point that follows, the *final* point of the arc, and the other one the *initial* point. Given any parameter representation $x = x(\tau)$, $y = y(\tau)$ of the oriented arc, where $a \leq \tau \leq b$, the positive sense will be that of increasing τ if the parameter value $\tau = a$ corresponds to the initial point and $\tau = b$ to the final point; otherwise the sense of increasing τ will be the negative sense on the arc (Fig. 4.8).

Any two distinct points P_0, P_1 on a simple arc C define a sub-arc with end points P_0, P_1, which consists of the points with parameter values between those of P_0 and P_1. If C is a directed arc and P_1 follows P_0 in the positive sense on C, we obtain a directed sub-arc with initial point P_0 and final point P_1. A finite number of points of subdivision on a directed simple arc C breaks up that arc into a sequence of directed

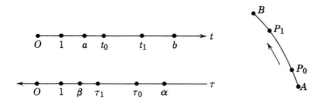

Figure 4.8   Sense and parameter representation.

sub-arcs, the initial point of one sub-arc being the final point of the preceding one.

Often it is impractical to restrict oneself to simple arcs and to insist that different parameter values t shall belong to different points of the curve. If, for example, the equations $x = x(t)$, $y = y(t)$ give the position of a moving particle P at the time t, there is no reason why the particle should not stand still for a while or why its path should not be allowed to cross itself so that the particle returns to the same position at a later time.

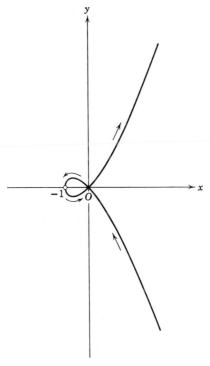

Figure 4.9   A curve with a loop: $x = t^2 - 1$, $y = t^3 - t$ with sense of increasing t.

An example is the curve $x = t^2 - 1$, $y = t^3 - t$ [which also could be described completely by the cubic equation $y^2 - x^2(1 + x) = 0$]. As t varies from $-\infty$ to $+\infty$ the curve crosses the origin twice, for $t = -1$ and $t = +1$ (Fig. 4.9). We verify easily that all other points of the curve belong to a unique value of t. Geometrically, the interval $-1 < t < +1$ corresponds to a *loop* of the curve. Here again the sense of increasing t defines a certain *order among the points of the curve*, at least if we visualize in some way the points corresponding to $t = -1$ and $t = +1$ as distinct, one lying "on top" of the other one. The whole oriented cubic curve can be decomposed into directed simple arcs, for example, into the arcs corresponding to $n \leq t \leq n + 1$, where n runs over all integers.

Closed Curves

The standard example of a parameter representation in which different t correspond to the same point on the curve is given by the formulas

$$x = a \cos t, \qquad y = a \sin t,$$

which describe the uniform motion of a point on a circle with t as the time. As t varies from $-\infty$ to $+\infty$ the point $P = (x, y)$ describes the circle infinitely often in the counterclockwise sense. We can cause the points of the circle to be described exactly once by restricting t to any half-open interval of length 2π:

$$\alpha \leq t < \alpha + 2\pi.$$

The end points α and $\alpha + 2\pi$ of the interval correspond to the same point on the circle. Here the end points of the parameter interval have no special geometrical significance for the curve.

Generally, a pair of continuous functions $x = \phi(t)$, $y = \psi(t)$ defined in a closed interval $a \leq t \leq b$ will represent a *closed* curve if $\phi(a) = \phi(b)$, $\psi(a) = \psi(b)$. The closed curve will be simple if different t-values with $a \leq t < b$ correspond to different points (x, y).

The point corresponding to $t = a$ and $t = b$ could be any point on the curve; it is just the point at which we "break" the curve to make its points correspond to those of an interval on the axis.

Closed Curves Represented by Periodic Functions

Just as in the example of the circle we can avoid distinguishing any particular break by taking for $\phi(t)$ and $\psi(t)$ *periodic* functions with period $p = b - a$. It is of value here to make some general remarks about periodic functions to which we will turn more extensively in Chapter 8.

A function $f(t)$ is called *periodic* with period p if it is defined for all t and satisfies the equation $f(t) = f(t + p)$. Thus, for example, the trigonometric functions $\sin t$ and $\cos t$ are periodic with period 2π. (Any multiple $2n\pi$ where n is an integer is then also a period.) Geometrically interpreted $f(t)$ has the period p if a shift of its graph by p units to the right leads to the same graph again.

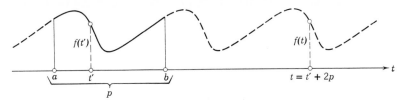

Figure 4.10 Graph of a periodic function $f(t)$.

Since then $f(t)$ "repeats" itself, a function $f(t)$ of period p is determined for all t if it is known merely in a single interval $a \leq t < b$ of length $p = b - a$ (Fig. 4.10). Indeed for every t there exists a value t' in the interval $a \leq t' < b$ such that $t - t' = np$, where n is an integer [one only has to take for n the largest integer that does not exceed $(t - a)/p$]. Then $f(t) = f(t')$ is known.

As a matter of fact we can start with *any continuous* function $f(t)$ in a half-open interval $a \leq t < b$; the extended function will clearly be

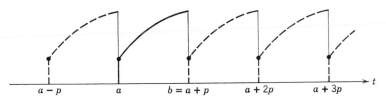

Figure 4.11 Periodic continuation of a function $f(t)$ from the interval $a \leq t < b$.

continuous for all t which are not of the form $t = a + np$ with an integer n (Fig. 4.11).

For example, extending the function $f(t)$ defined by $f(t) = t$ for $0 \leq t < 1$ periodically, leads to a function of period $p = 1$ which we can call the "fractional part of t," and which is discontinuous at points t which are integers (Fig. 4.12a). Generally, at $t = a + np$ the periodically extended function f will have the value $f(a)$; this will also be the limit of f on approaching the point from the right, whereas the limit of f from the left will be the same as at the point b. In the case of greatest

interest to us at present we start out with a function defined and
continuous in the *closed* interval $a \leq t \leq b$ which moreover has the
same value in the end points $f(a) = f(b)$. Continuing such a function
periodically always leads to a function $f(t)$ of period $p = b - a$ which
is continuous for all t. (Fig. 4.12b.)

Continuous periodic functions are ideal for representing closed
curves C. Let C be given parametrically by $x = \phi(t), y = \psi(t)$, for ϕ, ψ
continuous in the interval $a \leq t \leq b$ and having the same values in both
end points. We can extend the definition of these functions to all values

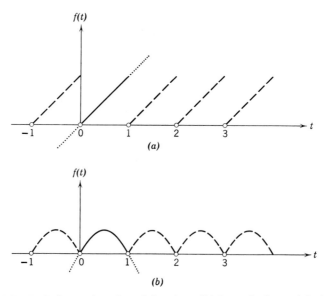

Figure 4.12   Periodic continuation of functions $f(t)$ from the interval $0 \leq t < 1$.
Here (a) $f(t) = t$, (b) $f(t) = 2t - 2t^2$.

of t in such a way that ϕ and ψ have period $b - a = p$ and are continuous for all t. For any t the extended parameter representation only
yields points of C, since we have $t = t' + np$ with n an integer and
$a \leq t' \leq b$. The point corresponding to t is then the same as the one
corresponding to t', which lies on C. As t varies from $-\infty$ to $+\infty$ the
point (x, y) traverses the curve C infinitely often, just as in the circle
$x = a \cos t, y = a \sin t$. Here the distinguished role of the parameter
value $t = a$ is removed. For any α the whole curve is already represented by $x = \phi(t), y = \psi(t)$ when t runs from α to $\alpha + p$.

A portion of the closed curve C corresponding to the parameter
values t in an interval $\alpha \leq t \leq \beta$ forms a simple arc if different t-values

in that interval lead to different points (x, y). The whole closed curve C is a simple curve if different t in the same interval $\alpha \leq t < \alpha + p$ always lead to different points on C. Thus any closed parameter interval of length less than p gives a simple arc.

Closed Curves Composed of Simple Arcs. Order of Points

The closed curves which we shall consider can all be decomposed into simple arcs. If the whole closed curve C is simple, it can be decomposed into two simple arcs $t_0 \leq t \leq t_1$ and $t_1 \leq t \leq t_0 + p$ which have only their end points P_0, P_1 in common. The sense of increasing t determines a positive sense or *orientation* on C by fixing a positive direction on each simple arc of C. Any two distinct points P_0, P_1 on the simple closed curve C divide C into two simple arcs. In the sense of increasing t exactly one of the two arcs will have P_0 as the initial point and P_1 as the end point; we will call it P_0P_1: the reverse holds for the other arc.

Orientation and Order

The positive orientation of C can also be characterized by an ordered triple of points $P_0P_1P_2$ of C if we specify that P_2 does not lie on the

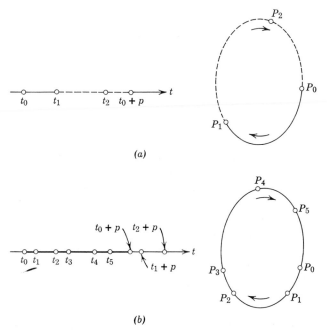

(a)

(b)

Figure 4.13 Orientation of closed curves in the sense of increasing t.

simple directed arc with initial point P_0 and final point P_1. The triples $P_1P_2P_0$ and $P_2P_0P_1$ obtained by a cyclic permutation from $P_0P_1P_2$ describe the same orientation (Fig. 4.13*a*).

*Quite generally, any n distinct points on the oriented closed simple curve C always follow each other in a certain order $P_1P_2 \cdots P_n$ determined up to cyclic permutations[1], and divide C into directed simple arcs, $P_1P_2, \ldots, P_{n-1}P_n, P_nP_1$. We can always choose parameter values $t_1, t_2, \ldots, t_n$ for the points $P_1, P_2, \ldots, P_n$ such that the t_i form a monotone increasing sequence and are all contained in one and the same parameter interval of length equal to the period p (Fig. 4.13*b*).

Orientation of Curves and Angles

As already emphasized in Chapter 1 we are forced to make use of the sign plus or minus to establish satisfactory relations between

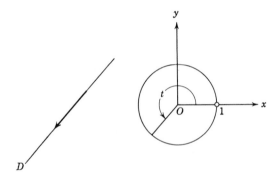

Figure 4.14   Angle of inclination t of a direction D.

geometric entities and analytic concepts expressed by numbers. *Directed lines*, such as the number axis, are the simplest instances. Which direction on a line we define as positive is arbitrary at the beginning. A positive sense corresponding to increasing t can be associated with any particular parameter representation $x = at + b$, $y = ct + d$ of the line. A line oriented in this way *points* in a certain *direction*. Two parallel directed lines have either the same or the opposite direction. A direction can also be determined by a *ray* issuing from a point P_0, that is, by a half-line which consists of the points on a line which "follow" a given point P_0 in the positive sense.

---

[1] That is, $P_2P_3 \cdots P_nP_1$, $P_3P_4 \cdots P_nP_1P_2$, $\ldots$, $P_nP_1 \cdots P_{n-1}$ give the same orientation.

Any direction in the plane can be represented by a ray from the origin or also by the point P on the circle of radius 1 about the origin that lies on that ray. If we represent this unit circle parametrically by $x = \cos t$, $y = \sin t$, we have associated with every direction certain values t, differing from each other by multiples of 2π. We call them the *angles* of *inclination* of the direction or the angles the direction makes with the positive x-axis. There is always exactly one angle of inclination t for which $0 \le t < 2\pi$ (Fig. 4.14).

The *angles between two directions* are simply the differences of their angles of inclination. More precisely, since the order in which we take the two directions matters, we say that *a direction with inclination t' forms with a direction with inclination t'' an angle $\alpha = t' - t''$* (Fig. 4.15).

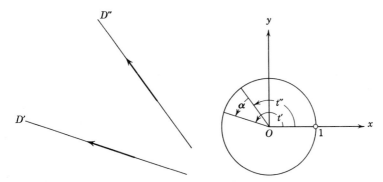

Figure 4.15 Angle α the direction D' forms with the direction D''.

Since t and t' can be changed by integral multiples of 2π, the same change is permissible for the angle one direction makes with another one.

Sense of Rotation

We also say that the direction with angle of inclination t'' passes into that with direction t' by a *rotation* through the angle α. The intuitive idea of rotation here is that of a *continuous motion*, by which the direction with inclination t'' goes into that with inclination t' by passing through directions with all possible inclinations t intermediate between t'' and t'. We call the rotation *positive* or *counterclockwise* if $\alpha = t' - t''$ is positive, and *negative* or *clockwise* in the opposite case. Of course, there are many different rotations both clockwise and counterclockwise that will take a given direction into another given one unless we insist that the angle of rotation α satisfies $-\pi < \alpha \le \pi$.

Ultimately then, the positive sense of rotation is associated with a particular parameter representation $x = \cos t$, $y = \sin t$ of the circle which we have chosen. If as usual, the x-axis points to the right and the y-axis upwards, then the positive sense of rotation coincides with the sense opposite to that of the hands on a conventional clock.[1]

Positive and Negative Sides of a Curve

A curve separates the points of the plane near one of its points P into two classes. Locally at least we can distinguish two "sides" of

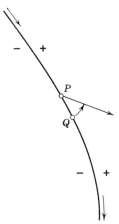

Figure 4.16 Positive and negative side of oriented arc.

the curve. If the curve C is oriented, we can define a *positive* (or "left") and a *negative* (or "right") side[2] as follows: Consider a ray issuing from P. We say that this ray points to the positive side of the curve if there are points Q on the curve arbitrarily close to P and following P in the sense given to the curve, such that the angle through which a line from P to Q must be rotated in the counterclockwise sense to reach the given ray, lies between 0 and π (Fig. 4.16). The points on the ray close to P are then said to lie on the positive side of the curve. In the opposite case the ray is said to point to the negative side of C, and the points on it are said to lie on the negative side of the curve. If the curve C is a simple closed curve, it divides all points of the plane into two classes, those *interior* to C and those *exterior* to C.[3] We say that C has the *counterclockwise orientation* if its interior lies on the positive (that is, left) side (Fig. 4.17).

If the closed curve C, however, consists of several loops, then it is not always possible to describe C so that all enclosed regions are on the positive side of C (see Fig. 4.18).

---

[1] This sense, in turn, is suggested by the motion of the shadow on the ground in a sun dial in the northern hemisphere.

[2] The terms "left" and "right" side correspond to the ordinary usage of the words "left bank" and "right bank" for a river oriented by its direction of flow.

[3] These concepts as well as the division of the plane by a simple closed continuous curve into two parts are analyzed precisely in topology and must be accepted here on an intuitive basis.

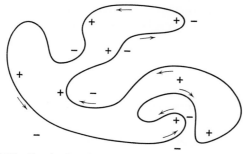

Figure 4.17   Simple closed curve with counterclockwise orientation.

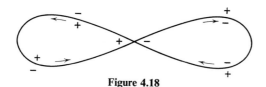

Figure 4.18

e. Derivatives, Tangent and Normal, in Parametric Representation

Direction and Speed

For a curve C given in parameter representation with the time parameter t

$$x = x(t) = \phi(t), \qquad y = y(t) = \psi(t)$$

we denote the derivatives, as Newton did, by a dot:

$$\dot{x} = \frac{d\phi}{dt} = \dot{\phi}, \qquad \dot{y} = \frac{d\psi}{dt} = \dot{\psi}.$$

The derivatives $\dot{x}, \dot{y}$ are often conveniently visualized as the "velocity components" or the "speeds" of the coordinates of a point P moving along C.

Whenever $\dot{x} \neq 0$, it is possible to represent the corresponding portion of C by an equation $y = f(x)$ by first calculating t as a function of x from the first equation and then substituting the resulting expression for t into the second equation. By the chain rule of differentiation and the rule for the derivative of the inverse of a function (see p. 207) we find then for the slope of the tangent to the curve

$$\frac{dy}{dx} = \frac{dy}{dt}\frac{dt}{dx} = \frac{\dfrac{dy}{dt}}{\dfrac{dx}{dt}} = \frac{\dot{y}}{\dot{x}}.$$

The equivalent formula $dx/dy = \dot{x}/\dot{y}$ holds if $\dot{y} \neq 0$.

Unless the contrary is stated we always assume that $\dot{x}$ and $\dot{y}$ do not vanish simultaneously or, concisely written, we assume

$$\dot{x}^2 + \dot{y}^2 \neq 0.$$

Then the tangent always exists;[1] it is horizontal if $\dot{y} = 0$ and vertical if $\dot{x} = 0$.

For the *cycloid*, for example, [see Eq. (1), p. 329] we have

$$\dot{x} = a(1 - \cos t) = 2a \sin^2 \frac{t}{2},$$

$$\dot{y} = a \sin t = 2a \sin \frac{t}{2} \cos \frac{t}{2},$$

$$\frac{dy}{dx} = \cot \frac{t}{2}.$$

These formulas show that $\dot{x}^2 + \dot{y}^2 \neq 0$ except for $t = 0, \pm 2\pi, \pm 4\pi, \ldots$. Moreover, the cycloid has a cusp (that is, a point where it reverses direction), with a vertical tangent at those exceptional points at which it also meets the x-axis, that is, when $y = 0$; for on approaching these points, the derivative $y' = \dot{y}/\dot{x} = \cot (t/2)$ becomes infinite.

Tangent, Normal, and Direction Cosines

The equation of the tangent to the curve at the point x, y is

$$\eta - y = \frac{dy}{dx} (\xi - x),$$

where ξ and η are the "running" coordinates corresponding to an arbitrary point on the tangent, whereas x, y, and dy/dx have the fixed values belonging to the point of contact. Substituting $\dot{y}/\dot{x}$ for dy/dx we can write the equation of the tangent in the form

(5) $$(\xi - x)\dot{y} - (\eta - y)\dot{x} = 0.$$

Exactly the same equation is obtained under the assumption, $\dot{y} \neq 0$; we only have to express x as a function of y. In the exceptional points where both $\dot{x}$ and $\dot{y}$ vanish for the same t the equation becomes meaningless, since it is satisfied for all ξ, η.

---

[1] We observe that the condition $\dot{x}^2 + \dot{y}^2 \neq 0$, although sufficient, is not necessary to guarantee a nonparametric representation. Thus we may define the curve $y = x^2$ by means of the parametric equations $x = t^3, y = t^6$. At the origin of the t-axis, the condition of positivity for $\dot{x}^2 + \dot{y}^2$ fails, but still the curve has a definite and well-defined nonparametric representation.

The *normal to the curve*, that is, the straight line through a point of the curve perpendicular to the tangent at that point, has the slope $-dx/dy$. This leads to the equation

$$(6) \qquad (\xi - x)\dot{x} + (\eta - y)\dot{y} = 0$$

for the normal.

If a point of C corresponds to several values of t, then in general a different tangent exists for each of the branches of the curve passing through the point, or for each value of t. For example, the curve $x = t^2 - 1$, $y = t^3 - t$ (Fig. 4.9, p. 335) passes through the origin for $t = -1$ and $t = +1$. For $t = -1$ we find for the equation of the tangent $\xi + \eta = 0$, whereas the tangent for $t = +1$ is given by $\xi - \eta = 0$.

From the definition of derivative we have

$$\frac{dy}{dx} = \frac{\dot{y}}{\dot{x}} = \tan \alpha,$$

where α is the angle the tangent makes with the x-axis. This means a rotation by the angle α applied to the x-axis (counterclockwise if $\alpha > 0$, clockwise if $\alpha < 0$) will cause it to be parallel to the tangent. Rotations by the angles $\alpha \pm \pi$, $\alpha \pm 2\pi$, ... will then also make the x-axis parallel to the tangent. Hence the angle α is determined only to within a multiple of π, whereas $\tan \alpha$ is determined uniquely. From the relations $\dot{y}/\dot{x} = (\sin \alpha)/(\cos \alpha)$ and $\dot{x}^2 + \dot{y}^2 \neq 0$ we find

$$\cos \alpha = \pm \frac{\dot{x}}{\sqrt{\dot{x}^2 + \dot{y}^2}}, \qquad \sin \alpha = \pm \frac{\dot{y}}{\sqrt{\dot{x}^2 + \dot{y}^2}}$$

(where the same sign must be taken in both formulas). We call $\cos \alpha$ and $\sin \alpha$ the *direction cosines* of the tangent.[1]

Assigning Directions to Tangent and Normal

The two possible choices for the direction cosines correspond to the two directions in which we can traverse the tangent; the corresponding angles α differ by an odd multiple of π. One of the two directions on the tangent corresponds to increasing t, the other one to decreasing t. Assume that the sense on the curve is that of increasing t: Then, by definition, the positive direction on the tangent, or the one that corresponds to increasing values of t, is the one that forms with the positive

---

[1] One thinks here of $\sin \alpha$ as $\cos \beta$, where $\beta = \pi/2 - \alpha$ is the angle the y-axis forms with the tangent.

x-axis an angle α for which $\cos \alpha$ has the same sign as $\dot{x}$ and $\sin \alpha$ the same sign as $\dot{y}$. The direction cosines of that direction on the tangent are then, without ambiguity,

$$(7) \qquad \cos \alpha = \frac{\dot{x}}{\sqrt{\dot{x}^2 + \dot{y}^2}}, \qquad \sin \alpha = \frac{\dot{y}}{\sqrt{\dot{x}^2 + \dot{y}^2}}.$$

If, say, $\dot{x} = dx/dt > 0$, then the direction of increasing t on the tangent is that of increasing x; the angle that direction forms with the positive

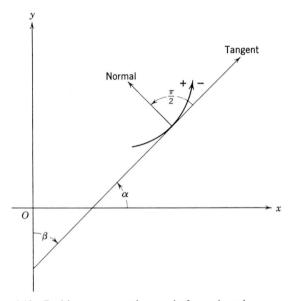

Figure 4.19   Positive tangent and normal of an oriented curve.

x-axis has then a positive cosine. Similarly, that normal direction obtained by rotating the direction of the positive tangent corresponding to increasing t in the positive (counterclockwise) sense by $\pi/2$ has the unambiguous direction cosines

$$\cos \left(\alpha + \frac{\pi}{2} \right) = \frac{-\dot{y}}{\sqrt{\dot{x}^2 + \dot{y}^2}}, \qquad \sin \left(\alpha + \frac{\pi}{2} \right) = \frac{\dot{x}}{\sqrt{\dot{x}^2 + \dot{y}^2}}.$$

It is called the positive normal direction and points to the "positive side" of the curve (Fig. 4.19).

If we introduce a new parameter $\tau = \chi(t)$ on the curve, then the values of $\cos \alpha$ and $\sin \alpha$ stay unchanged if $d\tau/dt > 0$ and they change sign if $d\tau/dt < 0$; that is, if we change the sense of the curve, then the positive sense of tangent and normal likewise is changed.

Critical Points

If $\dot{x}$ and $\dot{y}$ are continuous and $\dot{x}^2 + \dot{y}^2 > 0$, the quantities $\cos \alpha$ and $\sin \alpha$ which determine the direction of the tangent will vary continuously with t. The tangent, whose equation is

$$(\xi - x) \sin \alpha - (\eta - y) \cos \alpha = 0,$$

then changes continuously along the curve, as does the normal.

If both $\dot{x}$ and $\dot{y}$ vanish for a certain value of t, the direction cosines of the tangent are not defined by our formulas; a tangent may fail to exist altogether or it may not be determined uniquely. Such a point is called a "critical" point or a "stationary" point. We illustrate by examples various possibilities that arise at critical points.

One example is furnished by the curve $y = |x|$ with the parameter representation $x = t^3$, $y = |t|^3$; this curve has a corner for $t = 0$ although both $\dot{x}$ and $\dot{y}$ stay continuous. In the example of the cycloid, discussed on p. 344, the "stationary" points at which $\dot{x} = \dot{y} = 0$ correspond to cusps. On the other hand, the vanishing of $\dot{x}$ and $\dot{y}$ in some cases is merely inherent in a specific parameter representation and not connected with the behavior of the curve, as for the straight line represented by $x = t^3$, $y = t^3$ for the parameter value $t = 0$.

Corners

Curves consisting of several smooth arcs meeting at corners are represented conveniently in parameter representation by functions $x(t)$, $y(t)$ which are continuous but have derivatives $\dot{x}$, $\dot{y}$ with jump discontinuities. This is illustrated by the trivial example of the broken

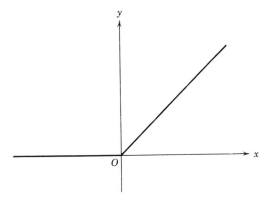

Figure 4.20   Graph of $x = t$, $y = \frac{1}{2}(t + |t|)$.

line represented by

$$x = t, \qquad y = 0 \qquad \text{for} \quad t \leq 0$$

and

$$x = t, \qquad y = t \qquad \text{for} \quad t \geq 0.$$

Here $\dot{x} = 1$, $\dot{y} = 0$ for $t < 0$ and $\dot{x} = 1$, $\dot{y} = 1$ for $t > 0$. At $t = 0$ the tangent is indeterminate (see Fig. 4.20).

f. The Length of a Curve

The Length as an Integral

Two different types of geometrical properties or quantities are associated with curves. The first type depends only on the behavior of the curve *in the small*, that is, in the immediate neighborhood of a point; such properties are those which can be expressed by means of derivatives at the point. Properties of the second type or *properties in the large* depend on the whole configuration of the curve or of a portion of the curve, and are usually expressed analytically by means of the concept of integral. We shall begin by considering a quantity of the second type, the length of a curve.

Of course, we have an intuitive notion of what we mean by the length of a curve. However, just as in the classical case of circular arcs, a precise mathematical meaning must be given to the intuitive concept. Guided by intuition we define the length of an arbitrary curve as the limit of the lengths of approximating polygons, in particular, inscribed polygons. The lengths of polygons, in turn, are immediately defined as soon as a unit of length is chosen. The final result will be the expression of length by an integral.

We assume our curve given in the form $x = x(t)$, $y = y(t)$, $\alpha \leq t \leq \beta$. In the interval between α and β we choose intermediate points $t_1, t_2, \ldots, t_{n-1}$ such that

$$\alpha = t_0 < t_1 < t_2 < \cdots < t_{n-1} < t_n = \beta.$$

We join the points $P_0, P_1, \ldots, P_n$ on the curve corresponding to these values t_i in order, by line segments, thus obtaining an inscribed polygon. The length of the perimeter of this inscribed polygon depends on the way in which the points t_i, or the vertices P_i of the polygon, are chosen. We now let the number of the points t_i increase beyond all bounds in such a way that at the same time the length of the longest subinterval (t_i, t_{i+1}) tends to zero. The length of the curve is then defined to be the limit of the perimeters of these inscribed polygons, *provided* that such a limit exists and is independent of the particular way in which the polygons are chosen. When this assumption (assumption of *rectifiability*) is fulfilled, we can speak of the length of the curve.

We assume that the functions $x(t)$ and $y(t)$ have continuous derivatives $\dot{x}(t)$ and $\dot{y}(t)$ for $\alpha \le t \le \beta$. The inscribed polygon corresponding to the subdivision of the t-interval by points t_i with $\Delta t_i = t_{i+1} - t_i$ has vertices $P_i = (x(t_i), y(t_i))$; its total length is given by the expression

$$S = \sum_{i=0}^{n-1} \overline{P_i P_{i+1}} = \sum_{i=0}^{n-1} \sqrt{[x(t_{i+1}) - x(t_i)]^2 + [y(t_{i+1}) - y(t_i)]^2}$$

according to the theorem of Pythagoras (cf. Fig. 4.21, p. 356). By the mean value theorem of differential calculus

$$x(t_{i+1}) - x(t_i) = \dot{x}(\xi_i)\,\Delta t_i, \qquad y(t_{i+1}) - y(t_i) = \dot{y}(\eta_i)\,\Delta t_i$$

where ξ_i and η_i are intermediate values in the interval $t_i < t < t_{i+1}$. This leads to the expression

$$S_n = \sum_{i=0}^{n-1} \sqrt{[\dot{x}^2(\xi_i)]^2 + [\dot{y}^2(\eta_i)]^2}\,\Delta t_i$$

for the length of the polygon, where we have made use of the fact that the differences Δt_i are positive. If the number n of points of subdivision t_i increases beyond all bounds while at the same time the largest value Δt_i tends to zero, the sum S_n tends to the integral

$$L = \int_{\alpha}^{\beta} \sqrt{\dot{x}^2 + \dot{y}^2}\,dt.$$

This fact is a direct consequence of the existence theorems for integrals in Chapter 2.[1]

This proves that for continuous $\dot{x}$, $\dot{y}$ the curve actually has a length and that this length is given analytically by the expression

$$(8) \qquad\qquad L = \int_{\alpha}^{\beta} \sqrt{\dot{x}^2 + \dot{y}^2}\,dt.$$

The same is true if $\dot{x}$ and $\dot{y}$ are allowed to be discontinuous at isolated points, where then the curve may not have a unique tangent; the

---

[1] Since the intermediate points ξ_i and η_i need not coincide, we make use of the more general approximating sums that were shown to converge to the integral on p. 195.

integral of course must then be considered as an "improper" one (see Chapter 3, p. 301). More general "rectifiable" curves, for which our integral is meaningful, will not be discussed in this volume.

Alternative Definition of Length

We add an interesting observation: The perimeter S of any inscribed polygon π can never exceed the length L of the curve. (In particular, the distance of the end points of the curve cannot exceed L; for the straight line joining the end points is the *shortest* curve joining those points.) Indeed we may obtain L as limit of the perimeters of a special sequence of inscribed polygons, in which we start with the polygon π of perimeter S and obtain the following ones by adding successively more and more vertices. Inserting an additional vertex between two successive vertices of an inscribed polygon can never lead to a decrease in perimeters, because one side of a triangle can never exceed the sum of the other two. Thus L is the limit of a non-decreasing sequence of perimeters that starts with S. Hence $S \leq L$. Instead of defining therefore L as limit of the perimeters of a sequence of inscribed polygons corresponding to finer and finer subdivisions of the t-interval, we could also have defined L as the least upper bound of the perimeters of *all* inscribed polygons. It is interesting that the length can be defined without formally invoking any passage to the limit.

Invariance of Length under Parameter Changes

From its definition it is clear that the length L of a curve c cannot depend on the particular parametric representation we use for C. Hence, if we introduce a new parameter $\tau = \chi(t)$, where $d\tau/dt > 0$, our integral formula for L must give the same value whether t or τ is used as parameter. This can be verified immediately from the chain rule of differentiation and the substitution law for integrals. We have indeed

$$\sqrt{\dot{x}^2 + \dot{y}^2} = \sqrt{\left(\frac{dx}{dt}\right)^2 + \left(\frac{dy}{dt}\right)^2} = \sqrt{\left(\frac{dx}{d\tau}\frac{d\tau}{dt}\right)^2 + \left(\frac{dy}{d\tau}\frac{d\tau}{dt}\right)^2}$$

$$= \sqrt{\left(\frac{dx}{d\tau}\right)^2 + \left(\frac{dy}{d\tau}\right)^2}\frac{d\tau}{dt} \; ;$$

hence, if $\chi(\alpha) = a$, $\chi(\beta) = b$,

$$L = \int_\alpha^\beta \sqrt{\dot{x}^2 + \dot{y}^2}\, dt = \int_\alpha^\beta \sqrt{\left(\frac{dx}{d\tau}\right)^2 + \left(\frac{dy}{d\tau}\right)^2}\frac{d\tau}{dt}\, dt$$

$$= \int_a^b \sqrt{\left(\frac{dx}{d\tau}\right)^2 + \left(\frac{dy}{d\tau}\right)^2}\, d\tau,$$

so that the expression for length based on the parameter τ leads to the same value L. If, instead, $d\tau/dt < 0$ we find similarly

$$\int_\alpha^\beta \sqrt{\dot{x}^2 + \dot{y}^2}\, dt = -\int_a^b \sqrt{\left(\frac{dx}{d\tau}\right)^2 + \left(\frac{dy}{d\tau}\right)^2}\, d\tau$$

$$= \int_b^a \sqrt{\left(\frac{dx}{d\tau}\right)^2 + \left(\frac{dy}{d\tau}\right)^2}\, d\tau;$$

the right-hand side is again the correct integral for the length of C referred to the parameter τ since now $b < a$ because $\chi(t)$ is a decreasing function.

For a curve given nonparametrically by a function $y = f(x)$, $a \le x \le b$, we can introduce x as parameter t. Then $\dot{x} = 1, \dot{y} = dy/dx$. The length of the curve is then given by

$$(9) \qquad L = \int_a^b \sqrt{1 + \left(\frac{dy}{dx}\right)^2}\, dx.$$

Examples. As an example we find for the length of a segment of the parabola $y = \frac{1}{2}x^2$ corresponding to the interval $a \le x \le b$:

$$L = \int_a^b \sqrt{1 + x^2}\, dx.$$

Here the substitution $x = \sinh t$ (see Chapter 3, p. 273) leads to

$$\int_{ar\,\sinh a}^{ar\,\sinh b} \cosh^2 t\, dt = \frac{1}{2}\int_{ar\,\sinh a}^{ar\,\sinh b} (1 + \cosh 2t)\, dt$$

$$= \frac{1}{2}(t + \sinh t \cosh t)\,\Big|_{ar\,\sinh a}^{ar\,\sinh b}$$

$$= \frac{1}{2}(ar\,\sinh b + b\sqrt{1 + b^2} - ar\,\sinh a - a\sqrt{1 + a^2}).$$

For a curve given by an equation $r = r(\theta)$, $\alpha \le \theta \le \beta$ in polar coordinates, we have the representation $x = r(\theta)\cos\theta$, $y = r(\theta)\sin\theta$. Choosing θ as parameter, we have

$$\dot{x} = \dot{r}\cos\theta - r\sin\theta, \qquad \dot{y} = \dot{r}\sin\theta + r\cos\theta, \qquad \dot{x}^2 + \dot{y}^2 = r^2 + \dot{r}^2.$$

This leads to the expression

$$(10) \qquad L = \int_\alpha^\beta \sqrt{r^2 + \left(\frac{dr}{d\theta}\right)^2}\, d\theta$$

for the length of a curve in polar coordinates. We have, for example, for the circle of radius a about the origin, the equation $r = \text{constant} = a$ $0 \le \theta \le 2\pi$. This gives for the total length of the circle

$$L = \int_0^{2\pi} a\, d\theta = 2\pi a.$$

Additivity of Length

Let C be a curve given by $x = x(t)$, $y = y(t)$, $\alpha \leq t \leq \beta$, where $\dot{x}$ and $\dot{y}$ are continuous. Let γ be any intermediate value between α and β. From the general rules for integrals we have

$$\int_\alpha^\beta \sqrt{\dot{x}^2 + \dot{y}^2}\, dt = \int_\alpha^\gamma \sqrt{\dot{x}^2 + \dot{y}^2}\, dt + \int_\gamma^\beta \sqrt{\dot{x}^2 + \dot{y}^2}\, dt.$$

The integrals on the right, respectively represent the lengths of the portions into which C is divided by the point corresponding to $t = \gamma$. Hence the length of the whole curve equals the sum of the lengths of its parts.

It is not necessary that $\dot{x}$ and $\dot{y}$ are continuous. The integrals exist just as well when $\dot{x}$ and $\dot{y}$ have a finite number of jump discontinuities, as would occur in a curve with corners. The total length of the curve is then the sum of the lengths of the smooth portions between the corners. Even more singular behavior of $\dot{x}$ and $\dot{y}$ is permitted as long as the expression for the length is meaningful as an improper integral.

g. The Arc Length as a Parameter

We have seen that one and the same curve permits many different parameter representations $x = x(t)$, $y = y(t)$. Any monotone function of t can be used as parameter instead of t. For many purposes, however, it is of advantage to refer curves C to some "standard parameter" which in some way is distinguished geometrically. The abscissa x or the polar angle θ are not suitable for that purpose if curves are to be described in the large; moreover, they depend on the choice of coordinate system. The possibility of measuring lengths along a curve provides us with a natural geometrically defined parameter to which points P of a rectifiable curve can be referred, namely, the length of the portion of the curve between P and some fixed point P_0.

We start out with an arbitrary parameter representation $x = x(t)$, $y = y(t)$, $\alpha \leq t \leq \beta$ of C. Differentiation with respect to t is indicated by a dot. We introduce the "arc length" s by the indefinite integral

$$(11) \qquad s = \int \sqrt{\dot{x}^2 + \dot{y}^2}\, dt$$

or more precisely s as a function of t by

$$(11a) \qquad s = s(t) = c + \int_{t_0}^t \sqrt{\dot{x}^2(\tau) + \dot{y}^2(\tau)}\, d\tau,$$

where c is a constant, t_0 a value between α and β, and where we have written τ for the variable of integration to distinguish it from the upper limit t. Clearly, for any values t_1 and t_2 in the parameter interval the difference

$$(12) \qquad s(t_2) - s(t_1) = \int_{t_1}^{t_2} \sqrt{\dot{x}^2 + \dot{y}^2}\, d\tau$$

is equal to the length of the portion of the curve bounded by the points corresponding to $t = t_1$ and $t = t_2$, provided $t_1 < t_2$. For $t_1 > t_2$ the difference $s(t_2) - s(t_1)$ is the negative of the length of that portion. Thus the knowledge of any indefinite integral s permits us to calculate the length of any part of the curve.

The Sign of Arc Length

If the constant c has the value 0 we can interpret $s(t)$ itself as the length of the arc of the curve (or the "distance along the curve") between the point P_0 with parameter t_0 and the point P with parameter t; here the length is counted positive in the case where the arc with initial point P_0 and end point P has the orientation corresponding to increasing t.[1]

The integral form of the definition of s is equivalent to the relation

$$(12a) \qquad \frac{ds}{dt} = \sqrt{\left(\frac{dx}{dt}\right)^2 + \left(\frac{dy}{dt}\right)^2}.$$

Using the symbolic notation for differentials (p. 180) $ds = (ds/dt)\, dt$, etc., we can write this relation in the suggestive form

$$ds = \sqrt{dx^2 + dy^2}$$

for the "element of length" ds.

Speed of Motion along a Curve

If t is interpreted as the time and $x(t)$, $y(t)$ as coordinates of the position of a moving point at the time t, we have in

$$\dot{s} = \frac{ds}{dt} = \lim_{h \to 0} \frac{s(t + h) - s(t)}{h}$$

the rate of change of the distance moved by the point *along its path* with respect to the time, that is, the *speed* of the particle. For a particle

---

[1] Notice that the variable s is not completely unique; it depends on the choice of P_0 and c and also on the orientation of the curve induced by the parameter t. However, any other arc length is expressible in terms of s in the form (s + constant) or ($-s$ + constant).

moving with uniform speed along the curve $\dot{s}$ is a constant and s is a linear function of the time t.

If our usual assumption

$$\dot{x}^2 + \dot{y}^2 \neq 0$$

is satisfied, we have $ds/dt \neq 0$ and can introduce s itself as parameter. Many formulas and calculations then simplify. The quantities

$$\frac{dx}{ds} = \frac{dx}{dt}\frac{dt}{ds} = \frac{\dot{x}}{\sqrt{\dot{x}^2 + \dot{y}^2}}, \qquad \frac{dy}{ds} = \frac{dy}{dt}\frac{dt}{ds} = \frac{\dot{y}}{\sqrt{\dot{x}^2 + \dot{y}^2}}$$

are then just the direction cosines of the tangent pointing in the direction of increasing s (see (7), p. 346). The relation

$$(13) \qquad \left(\frac{dx}{ds}\right)^2 + \left(\frac{dy}{ds}\right)^2 = 1$$

characterizes the parameter s as the arc length along the curve.

h. Curvature

Definition by Rate of Change of Direction

We discuss next a basic concept which refers only to the *local behavior* of a curve in the neighborhood of a point, the concept of *curvature*.

As we describe the curve, the angle α of inclination of the curve will vary at a definite rate per unit arc length traversed; this rate of change of α we call *the curvature of the curve*. Accordingly the curvature is defined as

$$(14) \qquad \kappa = \frac{d\alpha}{ds}.$$

Parametric Expressions. Let the curve be given parametrically by functions $x = x(t)$, $y = y(t)$ having continuous first and second derivatives with respect to t, for which $\dot{x}^2 + \dot{y}^2 \neq 0$. In calculating the rate of change of the direction angle α at the point P we have to take into account that α is not defined uniquely. However, the trigonometric function of α, $\tan \alpha = \dot{y}/\dot{x}$ (or $\cot \alpha = \dot{x}/\dot{y}$ for $\dot{x} = 0$) has a definite value. In forming $d\alpha/ds$ we can always assume that the parameter values belonging to points in a neighborhood of P all lie in an interval throughout which one of the quantities $\dot{x}$, $\dot{y}$ stays different from zero. If, say, $\dot{x} \neq 0$ we can assign to α a value that varies continuously with t

throughout the interval by taking

$$\alpha = \alpha(t) = \arctan \frac{\dot{y}}{\dot{x}} + n\pi,$$

where n is a fixed, possibly negative integer, and "arc tan" stands for the principal value of the function (cf. p. 214), lying between $-\pi/2$ and $\pi/2$. Similarly, if $\dot{y} \neq 0$ in the interval we can take for α the expression

$$\alpha(t) = \text{arc cot} \frac{\dot{x}}{\dot{y}} + n\pi = \frac{\pi}{2} - \arctan \frac{\dot{x}}{\dot{y}} + n\pi.^1$$

In either case we find by direct differentiation for any parameter representation

$$\dot{\alpha} = \frac{d\alpha}{dt} = \frac{\dot{x}\ddot{y} - \ddot{x}\dot{y}}{\dot{x}^2 + \dot{y}^2}.$$

Since (see (12a), p. 353) also

$$\dot{s} = \frac{ds}{dt} = \sqrt{\dot{x}^2 + \dot{y}^2},$$

we obtain for the curvature $d\alpha/ds = \dot{\alpha}/\dot{s}$ of the curve the expression

(15) $$\kappa = \frac{d\alpha}{ds} = \frac{\dot{\alpha}}{\dot{s}} = \frac{\dot{x}\ddot{y} - \dot{y}\ddot{x}}{(\dot{x}^2 + \dot{y}^2)^{3/2}}.$$

Choosing in particular, the arc length s as the parameter t we have

$$\dot{x}^2 + \dot{y}^2 = 1$$

[see Eq. (13), p. 354] and hence we obtain the simplified result

(15a) $$\kappa = \dot{x}\ddot{y} - \dot{y}\ddot{x}.$$

Sign and Absolute Value of Curvature

Intoducing a new parameter $\tau = \tau(t)$ instead of t does not affect the direction of the tangent, and hence, does not affect changes in α. Similarly, the absolute value of the difference of the s-values in two points has a geometric meaning independent of the choice of parameter, namely that of distance measured along the curve. However, the *sign* of the difference must always be taken as the same as the sign of the difference of the corresponding parameter values, since we defined s as

---

[1] We could define $\alpha(t)$ as a continuous function for *all* parameter values t by dissecting the whole parameter interval into subintervals in each of which either $\dot{x} \neq 0$ or $\dot{y} \neq 0$. In each of the subintervals we can define then $\alpha(t)$ by one of the above expressions, choosing for each interval the constant integer n in such a way that the values of α in the common end point of two adjacent intervals, as determined from the expressions for those intervals, coincide.

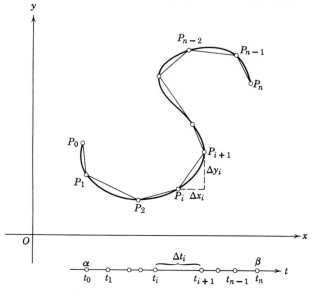

Figure 4.21 Rectification of curves.

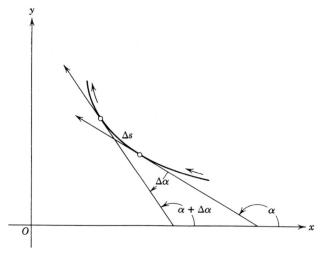

Figure 4.21(a) Curvature $\kappa = \lim \Delta\alpha/\Delta s$ of a curve. (In the case illustrated we have $\kappa < 0$.)

an increasing function of t. Thus the absolute value of the curvature $|\kappa| = |d\alpha/ds|$ does not depend on choice of parameter, whereas the sign of κ depends on the sense on the curve corresponding to increasing t. Obviously, $\kappa > 0$ means that α increases with s, that is, that the tangent turns counterclockwise as we proceed along the curve with increasing s or t (see Fig. 4.21a). In this case the orientation of the curve C is such that the positive side of C also is the "inner" side of C, that is, the side toward which C curves.

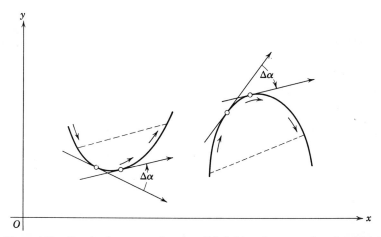

Figure 4.22   Graph of a convex function $f(x)$ (left) and concave function (right).

If the curve is given by an equation $y = f(x)$, we have, using x as parameter,

$$(16) \qquad \kappa = \frac{y''}{(1 + y'^2)^{3/2}} \, ,$$

where y' and y'' are the derivatives of y with respect to the variable x. Here the sign of the curvature is that corresponding to increasing x. Obviously, κ is positive for $y'' > 0$; in this case the tangent turns counterclockwise as x increases; we call the function $f(x)$ *convex*. The portion of the curve joining any two points lies below the straight line joining them. For $y'' < 0$ the tangent turns clockwise for increasing x, and the function f is called *concave*. (Fig. 4.22.) Here the curve lies above the chord joining two of its points. The intermediate case where the curvature has the value zero corresponds (generally speaking) to a *point of inflection* at which $y'' = 0$ (see p. 237).

Examples. For the curvature of the circle of radius a given by $x = a \cos t$, $y = a \sin t$ we find the constant value $1/a$ from the general formula (15). Thus the curvature of a circle described in the counterclockwise sense is the reciprocal of the radius. This result assures us that our definition of curvature is really a suitable one; for in a circle we naturally think of the reciprocal of the radius as a measure of its curvature.

A second example is the curve defined by the function $y = x^3$. The curvature is

$$\kappa = \frac{6x}{(1 + 9x^4)^{3/2}}.$$

For $x < 0$, the function $y = x^3$ is concave, since $\kappa < 0$, and the tangent is turning in a clockwise sense, whereas at $x = 0$, we have a point of inflection, and for $x > 0$ the function becomes convex.

A function whose curvature is identically equal to zero is a straight line as is easily seen by our definition, and the straight line is the only such curve.

Circle of Curvature and Center of Curvature

We introduce $\rho = 1/\kappa$. The quantity $|\rho| = 1/|\kappa|$ is called the *radius of curvature* at the point in question. (It is infinite at a point of inflection where $\kappa = 0$.) For a circle the radius of curvature at any point is just the radius of the circle.

To any point $P = (x, y)$ of the curve C we assign a circle tangent to C and P and having the same curvature as C when we traverse the curve and the circle in the same sense at P. This circle is called the *circle of curvature* of the curve C at the point P. Its center is the *center of curvature* of the curve C corresponding to the point P (Fig. 4.23). Since C and the circle have the same radius of curvature the radius of the circle must be the radius of curvature $|\rho|$ of C, and the center (ξ, η) of the circle must lie on the normal of C at P, and a distance $|\rho|$ away from P. Since C and the circle curve toward the same side, the center lies along the normal direction to the curve at P, on the positive or negative side according as the curvature κ is positive or negative.

The direction from P to the center of curvature forms an angle $\alpha + \pi/2$ with the positive x-axis, if $\kappa > 0$. Thus, if ξ, η are the coordinates of the center of curvature and x, y those of P, we have [see Equation (7), p. 346]

$$\frac{\xi - x}{\rho} = \cos\left(\alpha + \frac{\pi}{2}\right) = -\sin\alpha = \frac{-\dot{y}}{\sqrt{\dot{x}^2 + \dot{y}^2}},$$

$$\frac{\eta - y}{\rho} = \sin\left(\alpha + \frac{\pi}{2}\right) = \cos\alpha = \frac{\dot{x}}{\sqrt{\dot{x}^2 + \dot{y}^2}}.$$

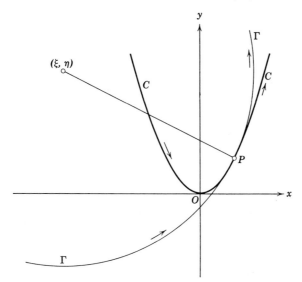

Figure 4.23   Circle of curvature Γ and center of curvature (ξ, η) corresponding to point P of curve C.

Hence for $\kappa > 0$,

$$(17) \qquad \xi = x - \frac{\rho \dot{y}}{\sqrt{\dot{x}^2 + \dot{y}^2}}, \qquad \eta = y + \frac{\rho \dot{x}}{\sqrt{\dot{x}^2 + \dot{y}^2}} .$$

If arc length s is used as parameter t, we obtain the simple expressions

$$(17a) \qquad \xi = x - \rho \dot{y}, \qquad \eta = y + \rho \dot{x}.$$

The same formulas for ξ, η are obtained for $\kappa < 0$, in which case the radius of curvature is $-\rho$ and moreover the direction from P to the center forms an angle $\alpha - \pi/2$ with the positive x-axis.

Circle of Curvature as Osculating Circle

Formulas (17) give an expression for the center of curvature in terms of the parameter t of the point P on the curve. As t ranges over all values in the parameter interval the center of curvature describes a curve, the so-called *evolute* of the given curve; since, with x and y, we have to regard $\dot{x}$, $\dot{y}$, and ρ as known functions of t, the foregoing formulas give parametric equations for this evolute. Examples and a discussion of geometrical properties of the evolute will be found in Appendix I, p. 424.

Any two curves are said to "*osculate*" at a point P or to have "*contact of order two*" at P, if they pass through P, have the same tangent at P, and also the same curvature, when oriented the same way. Obviously, two osculating curves have the same circle of curvature and center of curvature at P. If the curves are given by equations $y = f(x)$ and $y = g(x)$ in nonparametric form, it is easy to express the condition that they have a point of contact P and the same tangent and curvature at P. If x is the abscissa of the point of contact P, we have $f(x) = g(x)$, $f'(x) = g'(x)$; the equality of curvature is expressed by

$$\frac{f''(x)}{[1 + f'^2(x)]^{3/2}} = \frac{g''(x)}{[1 + g'^2(x)]^{3/2}},$$

and hence also $f''(x) = g''(x)$. Thus the condition for a point of contact with equal curvatures is that the values of f and g together with those of their first and second derivatives agree at the point.

Consider a curve C: $y = f(x)$ and its circle of curvature Γ at P represented by $y = g(x)$ in a neighborhood of P. Since the circle Γ coincides with its circle of curvature, we see that C and Γ have the same circle of curvature, hence osculate at P. Consequently, at the point of contact $f(x) = g(x)$, $f'(x) = g'(x)$, $f''(x) = g''(x)$. We say this circle is the "best fitting" circle to the curve at the point P of contact, since no other circle meeting the curve at the point of contact has "contact of order two" with C at the point. The circle of curvature is the osculating circle. (See also Chapter 6, p. 459.)

Incidentally, just as the tangent to a curve is the limit for $P_1 \to P$ of a line through two consecutive points P and P_1 on C, one can show that the circle of curvature at P is the limit of the circles through three points P, P_1, P_2 for $P_1 \to P$ and $P_2 \to P$. The proof is left to the reader. (See Problem 4, p. 437.)

i. Change of Coordinate Axes. Invariance

Properties inherent in a geometrical or physical situation do not depend on the specific coordinate system or "frame of reference" with respect to which they are formulated; the intrinsic character of properties such as distance or length or angle must be reflected in statements showing that the respective formulas remain unchanged or are *invariant* if one passes from one coordinate system to another. A few brief remarks concerning this subject are appropriate in this section.

We use the general equations connecting the coordinates x, y of a point P in one coordinate system with the coordinates ξ, η of the same point P in any other system. The relative position of the second set of coordinate axes to the first set is characterized by the coordinates a, b that the origin of the second system has in the first system, and by the angle γ which the positive ξ-axis makes with the positive x-axis.[1] The coordinates (x, y) and (ξ, η) of the same point in the two systems are (cf. Fig. 4.24) connected by the transformation

(18)
$$x = \xi \cos \gamma - \eta \sin \gamma + a,$$
$$y = \xi \sin \gamma + \eta \cos \gamma + b.$$

For $\gamma = 0$ no rotation of the axes but only a *parallel displacement* or *translation* is involved, and the formulas take the simple form $x = \xi + a$, $y = \eta + b$.

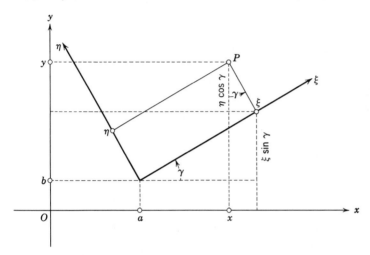

Figure 4.24   Change of coordinate axes.

Solving for ξ, η in terms of x, y we find

(18a)
$$\xi = (x - a) \cos \gamma + (y - b) \sin \gamma,$$
$$\eta = -(x - a) \sin \gamma + (y - b) \cos \gamma.$$

If x and y are functions of a parameter t defining a curve, we obtain immediately from these formulas expressions for ξ and η as functions of t, giving the parameter representation of the same curve in the ξ,η-system. Differentiating with respect to t (the quantities a, b, γ which fix the relative position of the two coordinate systems do not depend on t) yields the transformation of the "velocity components," that is,

---

[1] We restrict ourselves to "right-handed" coordinate systems in which the positive direction of the second axis of a system is obtained by a counterclockwise 90° rotation from that of the first axis.

for the derivatives of the coordinates with respect to t,

$$\dot{x} = \dot{\xi} \cos \gamma - \dot{\eta} \sin \gamma, \qquad \dot{y} = \dot{\xi} \sin \gamma + \dot{\eta} \cos \gamma.[1]$$

We confirm

$$\dot{x}^2 + \dot{y}^2 = \dot{\xi}^2 + \dot{\eta}^2.$$

Thus the expression $\sqrt{\dot{x}^2 + \dot{y}^2}$ has the same value in all coordinate systems; this invariance property is, of course, obvious from the interpretation of this quantity as rate of change ds/dt of the length along the curve with respect to t. The reader may verify by an easy

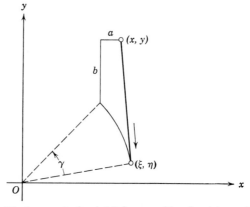

Figure 4.25   Displacement of point P from position (x, y) to position (ξ, η).

calculation that also the expression $\kappa = (\dot{x}\ddot{y} - \ddot{x}\dot{y})(\dot{x}^2 + \dot{y}^2)^{-\frac{3}{2}}$ for the curvature is invariant. (This, of course, follows also directly from the fact that the angles the tangent makes respectively with the ξ- and x-axes differ only by the constant value γ, so that $\kappa = d\alpha/ds$ cannot change.)

Equations (18) relating the coordinates x, y to the coordinates ξ, η are often interpreted in a different way as describing a displacement. In this interpretation the *points P* are shifted instead of the coordinate axes (Fig. 4.25). Only one coordinate system is used. The point with

---

[1] In some physical applications, where t stands for time, the relative position of the two coordinate systems also depends on time; let the quantities x, y stand for the coordinates of a particle in a coordinate system that is at rest, whereas ξ, η are the coordinates of the same particle referred to a moving coordinate system, for example, axes that are attached to the moving earth. The functions $x(t)$, $y(t)$ describe the path of the particle as it looks to an observer at rest, whereas $\xi(t)$, $\eta(t)$ describe the path as it looks to a moving observer. The formulas connecting $\dot{x}$, $\dot{y}$ with $\dot{\xi}$, $\dot{\eta}$ have to include then also the obvious terms arising from differentiation of a, b, and γ.

coordinates (x, y) in that system is mapped onto the point with co-ordinates (ξ, η) in the same system. Invariance of length or curvature of a curve now means that these quantities do not change when the whole curve undergoes a rigid motion.

*j. Uniform Motion in the Special Theory of Relativity

As pointed out on p. 234 there are far reaching analogies between the trigonometric and the hyperbolic functions which have their geometric counterpart in the correspondence between properties of ellipses and hyperbolas. The relationship will become clear when we shall be able to define the trigonometric functions for an imaginary argument and to verify that $\cos(it) = \cosh t$, $\sin(it) = i \sinh t$ in Section 7.7a. As an application of this analogy we consider the "hyperbolic rotations" of the plane which can be identified with the *Lorentz-transformations* of a line in Einstein's special theory of relativity.

We saw in (18a), p. 361, that a rotation of coordinate axes by an angle γ which leaves the origin fixed can be described by the equations

(18b) $\xi = x \cos \gamma + y \sin \gamma$, $\eta = -x \sin \gamma + y \cos \gamma$

connecting the coordinates x, y of a point P in the first system with its co-ordinates ξ, η in the second system. The distance of P from the origin is given by the same expression in both systems:

$$OP = \sqrt{x^2 + y^2} = \sqrt{\xi^2 + \eta^2}.$$

This follows also immediately from the transformation equations if we make use of the identity $\cos^2 \gamma + \sin^2 \gamma = 1$.

We now consider the analogous transformation with coefficients that are hyperbolic instead of trigonometric functions:

(19) $\xi = x \cosh \alpha - t \sinh \alpha$, $\tau = -x \sinh \alpha + t \cosh \alpha$;

these formulas can be obtained from the formulas (18b) for rotations by taking for the rotation angle γ and the y- and η-coordinates, pure imaginary quantities:

$$\gamma = i\alpha, \qquad y = it, \qquad \eta = i\tau.$$

We notice that for a real value of α (which would mean an imaginary angle of rotation γ in the original interpretation) formulas (19) define ξ and τ as real linear functions of x and t. These functions have the special property that

$$\xi^2 - \tau^2 = (x \cosh \alpha - t \sinh \alpha)^2 - (-x \sinh \alpha + t \cosh \alpha)^2$$
$$= x^2 - t^2$$

as a consequence of the identity $\cosh^2 \alpha - \sinh^2 \alpha = 1$. (This follows, of course, also from the observation that $x^2 - t^2 = x^2 + y^2$ is the square of the distance from the origin in the x,y-plane.) We now interpret t as the time and

x as a space-coordinate describing the location of a point in a one-dimensional space, that is, on a straight line. Any event takes place at a certain point at a certain time. These two pieces of information are provided by the two numbers x, t giving respectively the (signed) distance x of the point from the origin O and the time t that has elapsed from the time 0. In the theory of relativity we take the point of view that the measured values of this distance and of elapsed time depend on the frame of reference used by the observer, that is, on the special coordinate system in the space-time continuum. The quantities ξ, τ obtained from the formulas (19) will describe the same event in a different frame of reference in which distances and lengths of time intervals can have different values. The quantity that is unchanged in the transition from one reference frame to another one (known as "Lorentz-transformation") is

$$\sqrt{x^2 - t^2} = \sqrt{\xi^2 - \tau^2},$$

the "space-time distance" of the event from the origin. For an observer using the second system the quantity ξ is the space distance measured from the origin $\xi = 0$. That origin is a point for which

$$x \cosh \alpha - t \sinh \alpha = 0$$

or $x/t = \tanh \alpha$. Thus the origin of the second system is a point which in the first system appears to move with uniform velocity $v = dx/dt = \tanh \alpha$ relative to the origin of the first system. Hence the Lorentz transformation relates the values of distances and times as they appear for observers in two systems moving with constant velocity v relative to each other. Here

$$v = \frac{\sinh \alpha}{\cosh \alpha} = \frac{e^\alpha - e^{-\alpha}}{e^\alpha + e^{-\alpha}}$$

lies necessarily between -1 and $+1$ so that we are restricted to relative velocities of the two systems that lie numerically below the value 1. The value 1 here represents for suitable choice of units the velocity c of light which cannot be exceeded by v.

For a constant u the equation $x = ut$ corresponds to a point which in the first system moves with the velocity u, starting at $x = 0$ at the time $t = 0$. In the second system the same point will have the velocity

$$\omega = \frac{d\xi}{d\tau} = \left(\frac{d\xi}{dt}\right) \bigg/ \left(\frac{d\tau}{dt}\right) = \frac{u - \tanh \alpha}{1 - u \tanh \alpha} = \frac{u - v}{1 - uv}.$$

This result, valid in Einstein's special theory of relativity, differs from what we would obtain in classical kinematics where the velocity ω of a point with respect to a system moving with velocity v would simply be given by $\omega = u - v$. The relativistic formula shows that $\omega = u$ when $u = +1$ or -1; this corresponds to the fact suggested by the famous Michelson-Morley experiment that the velocity of light is the same for observers moving with different velocities.

k. *Integrals Expressing Area within Closed Curves*

In Chapter 2 the concept of integral was motivated by reference to "area under a curve," that is, the area of a strip of special shape. This specialization to areas under a curve is not quite satisfactory since the areas actually encountered most frequently are those of domains inside closed curves *C*, and are of more general shape than the strips whose area can be represented by integrals of the form $\int_a^b f(x)\,dx$.

The Basic Formula

We shall now derive an elegant general integral representation for the area bounded by a closed curve *C* which is given in parametric representation, by breaking up the area into special strip areas. This representation will be independent of the parameter representation and likewise independent of the coordinate system. Furthermore, it will express the *oriented area* within the curve in accordance with the sense of direction assigned to the boundary *C*; that is it will assign to an area within a simple closed curve *C* the negative or positive sign according as the sense of the boundary curve is clockwise or counterclockwise.

Assume that the simple closed oriented curve *C* is given by $x = x(t)$, $y = y(t)$, where *t* varies over the interval $\alpha \leq t \leq \beta$ and the sense of increasing *t* determines the sense on *C*. We assume that *x* and *y* are continuous functions of *t* (with the same value at $t = \alpha$ and $t = \beta$) and that their first derivatives $\dot{x}$ and $\dot{y}$ are continuous, with the possible exception of a finite number of jump-discontinuities if *C* has corners. Under these assumptions we shall prove the basic formula

$$(20) \qquad A = -\int_\alpha^\beta y\dot{x}\,dt = \int_\alpha^\beta x\dot{y}\,dt = \frac{1}{2}\int_\alpha^\beta (x\dot{y} - y\dot{x})\,dt$$

for the oriented area *A* within *C*.

That the three integral representations in the formula are equivalent follows directly if we integrate the first one by parts and use the periodicity conditions $x(\alpha) = x(\beta)$, $y(\alpha) = y(\beta)$; the third, more symmetric representation is just the arithmetic mean of the first two.

The expressions (20) do not depend on the location of the coordinate system in the plane. In fact, the symmetric expression

$$A = \frac{1}{2}\int_\alpha^\beta (x\dot{y} - y\dot{x})\,dt$$

shows clearly that the value of A is independent of the choice of the coordinate system. As we saw on p. 361 a change of coordinates from an xy-system to a $\xi\eta$-system is achieved by a substitution of the form

$$x = \xi \cos \gamma - \eta \sin \gamma + a,$$
$$y = \xi \sin \gamma + \eta \cos \gamma + b,$$

with constant a, b, γ. Differentiation of these formulas with respect to t yields

$$\dot{x} = \dot{\xi} \cos \gamma - \dot{\eta} \sin \gamma, \qquad \dot{y} = \dot{\xi} \sin \gamma + \dot{\eta} \cos \gamma$$

and consequently,

$$x\dot{y} - y\dot{x} = \xi\dot{\eta} - \eta\dot{\xi} + a\dot{y} - b\dot{x}.$$

Thus the expression $x\dot{y} - y\dot{x}$ is invariant under rotations about the origin (that is, when $a = b = 0$). Even when a or b do not vanish, the value of the integral for A is not affected, since

$$\int_{\alpha}^{\beta} (a\dot{y} - b\dot{x}) \, dt = (ay - bx) \Big|_{\alpha}^{\beta} = 0$$

for the closed curve C.

Proof of the Basic Formula (20). *Line Integrals over Simple Arcs.* The basic formula (20) is proved in some easy steps.

First, let C be a simple *oriented* arc with initial point P_0 and final point P_1. Let $x = x(t)$, $y = y(t)$ be any parameter representation of C with P_0, P_1 corresponding respectively to $t = t_0, t_1$. (Here t_0 may be larger or smaller than t_1.) Then the integral

$$A = - \int_{t_0}^{t_1} y \frac{dx}{dt} \, dt$$

depends only on C and not on the particular parameter representation. This is an obvious consequence of the substitution rule; if we introduce a new parameter τ by the monotone function $\tau = \chi(t)$ where $\tau_0 = \chi(t_0)$, $\tau_1 = \chi(t_1)$ the corresponding integral is[1]

$$- \int_{\tau_0}^{\tau_1} y \frac{dx}{d\tau} \, d\tau = - \int_{t_0}^{t_1} y \frac{dx}{d\tau} \frac{d\tau}{dt} \, dt = - \int_{t_0}^{t_1} y \frac{dx}{dt} \, dt = A.$$

---

[1] We assume not only that $x(t)$, $y(t)$ but also $\tau(t)$ are continuous functions and that their derivatives are continuous with the possible exception of a finite number of jump-discontinuities.

It is therefore justified to drop from the expression for the integral A the reference to any special parameter t and simply to write

$$A = A_C = -\int_C y \, dx.$$

Here A_C for a simple oriented arc C is to be computed by referring the arc to a parameter t, using $dx = (dx/dt) \, dt$, and taking as limits for the t-integration the parameter values for the end points of C in the order determined by the orientation of C.[1]

If C' is the arc obtained from C by changing its orientation, that is, the arc with initial point P_1 and final point P_0 we have, using the same parameter representation for C',

$$A_{C'} = -\int_{t_1}^{t_0} y \frac{dx}{dt} \, dt = +\int_{t_0}^{t_1} y \frac{dx}{dt} \, dt = -A_C.$$

Hence changing the orientation of an arc C changes the sign of the integral A_C.

If the oriented simple arc C is broken up into oriented subarcs $C_1, C_2, \ldots, C_n$, each with the same orientation as C, we obviously have

$$A_C = A_{C_1} + A_{C_2} + \cdots + A_{C_n}.$$

For in a parameter representation of C where, say, the sense of C is that of increasing t, this decomposition corresponds to a subdivision of the parameter interval $t_0 \le t \le t_n$ for C into subintervals $t_0 \le t \le t_1$, $t_1 \le t \le t_2, \ldots, t_{n-1} \le t \le t_n$ corresponding to $C_1, \ldots, C_n$. The result then follows from the additivity of integrals.

The additivity of the integrals A_C makes it much easier to compute the value of A_C in cases where C consists of several smooth arcs C_1, $C_2, \ldots$, each with its own parameter representation. We do not need to construct artificially a common parameter representation for the whole curve C, but instead compute each A_{C_i} separately from its parameter representation and then take the sum. Moreover, the A_{C_i} can be added in any order; we only have to make sure that all C_i have the same orientation as C.

The Basic Line Integral for Closed Curves

We can now define A_C for any oriented, simple closed curve C by breaking up C into simple arcs $C_1, \ldots, C_n$ with orientations agreeing

---

[1] The integral $\int_C y \, dx$ is an example of the general *line integrals* $\int_C p \, dx + q \, dy$ which will be discussed in Volume II.

with that on C and forming the sum of the A_{C_i}.[1] If the whole closed curve C has the parameter representation $x = x(t)$, $y = y(t)$ for $\alpha \leq t \leq \beta$, where the sense of increasing t gives the orientation of C and where $t = \alpha$ and $t = \beta$ correspond to the same point then A_C is again given by

$$A_C = -\int_\alpha^\beta y \, \frac{dx}{dt} \, dt.$$

In the same way we can define A_C for nonsimple oriented curves C by decomposition into simple oriented arcs, even when C consists of several disjoint pieces, as long as each portion of C has a definite sense.

The Basic Integral as Area

We now turn to the main point; that is, we identify the expressions A_C for a closed curve with the intuitive geometric quantity of oriented area within C.

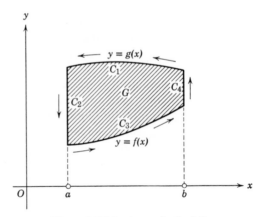

Figure 4.25(a)   Area of a "cell."

We consider first a domain G bounded from above by an arc C_1: $y = g(x)$ for $a \leq x \leq b$; below by an arc C_3: $y = f(x)$ for $a \leq x \leq b$: and laterally by line segments C_2, C_4 given by $x = a$ and $x = b$ (Fig. 4.25a). Here C_2 and C_4 are permitted to shrink into points. If we give to

---

[1] That the value of A_C obtained in this fashion does not depend on the particular way in which we divide up C into simple arcs follows easily: first the additivity property of A for simple arcs shows that refining a given subdivision by introducing additional dividing points does not change the resulting value of A_C; moreover, any two subdivisions can be replaced by one that is a refinement of both without changing the value of A_C.

C the counterclockwise orientation, the arc C_1 will be described in the sense of decreasing x, and the arc C_3 in that of increasing x. In forming A_C as the sum of the four A_{C_i} the portions C_2 and C_4 along which x is constant make no contribution since there $dx/dt = 0$. Using x as parameter on the arcs C_1 and C_3, we find

$$A_C = A_{C_1} + A_{C_3} = -\int_b^a g(x)\, dx - \int_a^b f(x)\, dx$$

$$= \int_a^b g(x)\, dx - \int_a^b f(x)\, dx.$$

This clearly is the positive area of the domain G, if G lies completely above the x-axis, being the difference of the areas lying respectively below the curves C_1 and C_3. We can always guarantee that G lies above the axis by replacing y by $y + c$ with a suitable constant c, that is, by a translation in the y-direction. This does not change areas and also does not affect the value of $A_C = -\int_C y\, dx$ for a closed curve C as we saw before. Hence for domains G of the type described which have a boundary C intersected in no more than two points by parallels to the y-axis, the integral A_C represents the area, taken positive if C is oriented counterclockwise, negative if clockwise. We obtain the same result for areas bounded by a curve C intersected by parallels to the x-axis in at most two points; we have only to write A_C in the form $\int_C x\, dy$ and to interchange x and y in the preceding argument. We call domains G of one of these two types "cells." We shall talk of "oriented cells" when their boundary curves are given one or the other orientation.

We now consider a domain G with the oriented boundary C, which is composed of a number of simple cells $G_1, G_2, \ldots, G_n$ with the boundaries $C_1, \ldots, C_n$ respectively; all these cells are assumed to have the same orientation, say counterclockwise. Then, as indicated in Fig. 4.26, the parts of the boundaries of the cells which are common to two adjacent cells are described in a different sense according as they are considered boundary arcs of one or the other of the adjacent cells. Therefore, if we add the integrals $A_{C_i} = -\int_{C_i} y\, dx$ for the different cells, the contributions of all the interior cell boundaries cancel out and we obtain

$$A = \sum_{i=1}^n A_{C_i} = \sum_{i=1}^n \left(-\int_{C_i} y\, dx \right) = -\int_C y\, dx = A_C$$

where A is the oriented area of the total domain G.

Thus the formulas (20) for the area A of an oriented domain G within a closed curve is proved for all domains which can be decomposed into simple cells, for example, by drawing parallels to the coordinate axes.

For all domains that we shall encounter, this assumption will be obviously satisfied, as for example, polygonal domains.

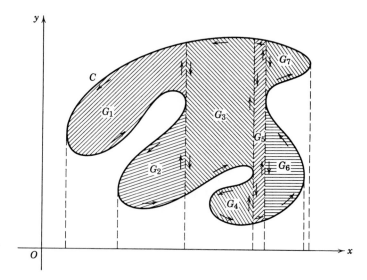

Figure 4.26   Decomposition of oriented domain into oriented cells.

Supplementary Remarks

Finally, it might be added that the validity of the formula for area follows in the same way, even for multiply connected domains, such as ring-shaped domains, which can be decomposed into a finite number of simple cells. Then all the boundary curves have to be described consistently in such a sense that the interior of G is always either on the "left" side or always on the "right" side.

The formulas for A remain meaningful even when C is not a simple curve but is allowed to intersect itself, dividing the plane into more than two regions. In this case we may consider the formula as a guide to interpreting area suitably as an additive combination of the oriented areas of the various connected pieces of the plane bounded by C. We shall discuss this matter in Appendix II to this chapter.

Examples. As an example we can find the *area enclosed by the ellipse* $x^2/a^2 + y^2/b^2 = 1$. Using the counterclockwise orientation for the ellipse, we find from the parameter representation

$$x = a \cos t, \qquad y = b \sin t \qquad \text{for} \quad 0 \leq t \leq 2\pi$$

that

$$A = \frac{1}{2} \int_0^{2\pi} (x\dot{y} - y\dot{x}) \, dt = \frac{1}{2} \int_0^{2\pi} ab \, dt = \pi ab.$$

Area in Polar Coordinates. To express area in polar coordinates r and θ, we consider first the area A of the region bounded by a curve segment $r = f(\theta)$ and the radii $\theta = \alpha$ and $\theta = \beta$. We assume that $\alpha < \beta$ and that θ can be used as parameter along the curve (that is, that different points have different polar angles). We use for A the expression

$$A = \frac{1}{2} \int (x \, dy - y \, dx) = \frac{1}{2} \int (x\dot{y} - y\dot{x}) \, dt,$$

which then has to be extended over the curved part of the boundary and over the two radii. On the radii $\theta = \alpha$ and $\theta = \beta$ we can use r as parameter and find from $x = r \cos \theta$, $y = r \sin \theta$, and $\theta = \text{constant}$ that $\dot{x} = \cos \theta$, $\dot{y} = \sin \theta$, and thus $x\dot{y} - y\dot{x} = 0$. On the curved part we use θ as parameter. Then

$$\dot{x} = \frac{dr}{d\theta} \cos \theta - r \sin \theta, \qquad \dot{y} = \frac{dr}{d\theta} \sin \theta + r \cos \theta,$$

and thus $x\dot{y} - y\dot{x} = r^2$. Consequently,

$$(21) \qquad A = \frac{1}{2} \int_\alpha^\beta r^2 \, d\theta = \frac{1}{2} \int_\alpha^\beta f^2(\theta) \, d\theta.$$

For a simple closed curve C which contains the origin in its interior and is intersected by every ray from the origin in exactly one point we can use θ as parameter for $0 \leq \theta \leq 2\pi$ and find for the enclosed area

$$(22) \qquad A = \frac{1}{2} \int_0^{2\pi} r^2 \, d\theta.$$

Formula (21) for area in polar coordinates can also be derived directly from the definition of integrals. For that purpose we divide our domain into sectors by drawing radii from the origin (Fig. 4.27). Each sector is described by inequalities

$$\theta_{i-1} < \theta < \theta_i, \qquad 0 < r < f(\theta).$$

Obviously, the area of the sector lies between the areas of the inscribed and circumscribed circular sectors; the area of a sector of the domain

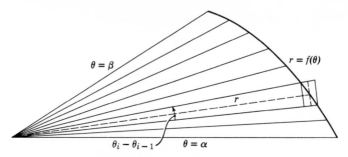

Figure 4.27   Area in polar coordinates.

is then equal to $\frac{1}{2}r^2(\theta_i - \theta_{i-1})$, where r lies between the largest and smallest values of $f(\theta)$ for the interval $\theta_{i-1} < \theta < \theta_i$. As we refine the subdivision, the sum of the areas of the sectors of our domain clearly converges to the integral $\dfrac{1}{2}\displaystyle\int_\alpha^\beta r^2\,d\theta$.

Area in a Lemniscate

As an example of Equation (21) we consider the area bounded by a loop of a lemniscate. The equation of the lemniscate (cf. p. 103) is $r^2 = 2a^2 \cos 2\theta$; one loop is obtained by having θ vary from $-\pi/4$ to $+\pi/4$. This gives us the expression

$$a^2 \int_{-\pi/4}^{\pi/4} \cos 2\theta\,d\theta = a^2$$

for the area. Of course, the other loop has the same absolute but negative value of area.

Area Bounded by a Hyperbola

We now consider the area of a sector bounded by the hyperbola $x^2 - y^2 = 1$, which we computed already on p. 234 in a rather cumbersome fashion (see Fig. 3.12). For the hyperbola (or rather for its right-hand branch) we have the parameter representation $x = \cosh t$, $y = \sinh t$. We find indeed for twice the area bounded by the hyperbola and the radii leading to the points with parameters 0 and t the value

$$2A = \int_0^t (x\dot{y} - y\dot{x})\,d\tau = \int_0^t (\cosh^2 \tau - \sinh^2 \tau)\,d\tau$$

$$= \int_0^t d\tau = t.$$

(There is again no contribution to the integral from the radii.)

1. Center of Mass and Moment of a Curve

We now turn to some ideas arising in mechanics. We consider a system of n-particles in a plane having the masses $m_1, m_2, \ldots, m_n$ and the respective ordinates $y_1, y_2, \ldots, y_n$. We then call

$$T = \sum_{\nu=1}^{n} m_\nu y_\nu = m_1 y_1 + m_2 y_2 + \cdots + m_n y_n$$

the *moment of the system of particles with respect to the x-axis*. The expression $\eta = T/M$, where M denotes the total mass $m_1 + m_2 + \cdots + m_n$ of the system, defines the *height of the center of mass* of the system of particles above the x-axis, or its ordinate. It is just the *weighted average* of $y_1, y_2, \ldots, y_n$ using the "weight factors" m_1, $m_2, \ldots, m_n$ (see p. 142). Hence η is the *average height* of the masses. We define similarly the moment with respect to the y-axis and the abscissa of the center of mass.

We can now easily extend these definitions of the moment to a curve along which a mass is uniformly distributed, and thus define the coordinates ξ and η of the center of mass of such a curve. (The assumption of a constant density, say μ, along the curve is not essential: Any continuous distribution could be discussed equally well.)

In a procedure typical for mechanics we start with a system of a finite number n of particles, and then pass to a limit for $n \to \infty$. For this purpose we introduce the length of arc s as a parameter on the curve, and subdivide the curve by $(n - 1)$ points of division into arcs of lengths $\Delta s_1, \Delta s_2, \ldots, \Delta s_n$. We represent the mass $\mu \, \Delta s_i$ of each arc Δs_i as if it is concentrated at an arbitrary point of the arc, say that with the ordinate y_i.

By definition the moment of this system of particles with respect to the x-axis is

$$T = \mu \sum y_i \, \Delta s_i.$$

If now the largest of the quantities Δs_i tends to zero, this sum tends to a limit given by the integral

(23)
$$T = \mu \int_{s_0}^{s_1} y \, ds = \mu \int_{x_0}^{x_1} y \sqrt{1 + y'^2} \, dx,$$

which is therefore naturally accepted as the definition of the moment of the curve with respect to the x-axis. Since the total mass of the curve is equal to its length multiplied by μ,

$$\mu \int_{s_0}^{s_1} ds = \mu(s_1 - s_0),$$

we are immediately led to the following expressions for the coordinates of the center of mass of the curve:

$$(24) \qquad \xi = \frac{\displaystyle\int_{s_0}^{s_1} x\,ds}{s_1 - s_0}, \qquad \eta = \frac{\displaystyle\int_{s_0}^{s_1} y\,ds}{s_1 - s_0}.$$

These statements are actually *definitions* of the moment and center-of-mass of a curve; but they are such straightforward extensions of the simpler case of a finite number of particles that we naturally expect that—as is actually the case—any statement in mechanics involving the center-of-mass or the moment of a system of particles will be valid also for continuous mass distribution along curves.

m. Area and Volume of a Surface of Revolution

Guldins Rule

If we rotate a curve $y = f(x)$ for which $f(x) \geq 0$, about the x-axis, it describes a so-called *surface of revolution*. The area of this surface, whose abscissas we suppose to lie between the bounds x_0 and $x_1 > x_0$, is obtained by a discussion analogous to that above. For if we replace the curve by an inscribed polygon, instead of the curved surface, we have a figure composed of a number of thin truncated cones. Intuition suggests that we should define the area of the surface of revolution as the limit of the areas of these conical surfaces when the length of the longest side of the inscribed polygon tends to zero. From elementary geometry we know that the area of each truncated cone is equal to the length of the slanted straight generating side multiplied by the circumference of the circular section of mean radius. (Fig. 4.28). If we add these expressions and then carry out the passage to the limit, we obtain the expression

$$(25) \qquad A = 2\pi \int_{s_0}^{s_1} y\,ds = 2\pi \int_{x_0}^{x_1} y\sqrt{1 + y'^2}\,dx = 2\pi\eta(s_1 - s_0)$$

for the area. Expressed in words, this result states that the area of a surface of revolution is equal to the length of the generating curve multiplied by the distance traversed by the center of mass (*Guldin's rule*).

In the same way we find that the volume interior to the surface of revolution and bounded at the ends by the planes $x = x_0$ and $x = x_1 > x_0$ is given by the expression

$$(26) \qquad V = \pi \int_{x_0}^{x_1} y^2\,dx.$$

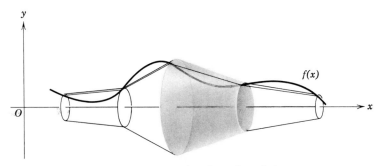

Figure 4.28   Area of surface of revolution.

This formula is obtained by following the intuitive suggestion that the volume in question is the limit of the volumes of the earlier mentioned figures consisting of truncated cones. The rest of the proof is left to the reader.

n. Moment of Inertia

In the study of the rotation of an object an important role is played by certain quantities called *moments of inertia*. These expressions will be briefly mentioned here.

We suppose that a particle m at a distance y from the x-axis rotates uniformly about that axis with angular velocity ω (that is, in unit time it rotates through an angle ω). The *kinetic energy* of the particle, expressed by half the product of the mass and the square of the velocity, is obviously

$$\frac{m}{2}(y\omega)^2.$$

We call the coefficient of $\frac{1}{2}\omega^2$, that is, the quantity my^2, the *moment of inertia of the particle about the x-axis*.

Similarly, if we have n-particles with masses $m_1, m_2, \ldots, m_n$ and ordinates $y_1, y_2, \ldots, y_n$, we call the expression

$$T = \sum_i m_i y_i^2$$

the moment of inertia of the system of masses about the x-axis. The moment of inertia is a quantity that belongs to the system of masses itself, without reference to its state of motion. Its importance lies in the fact that under rigid rotation of the system about an axis, without change of the distance between pairs of particles, the kinetic energy is obtained by multiplying the moment of inertia about that axis by half the square of the angular velocity. Thus the moment of inertia about an axis plays the same part in rotation about an axis as is played by the mass in rectilinear motion.

Suppose now that we have an arbitrary curve $y = f(x)$ lying between the abscissas x_0 and $x_1 > x_0$, along which a mass is uniformly distributed with unit density. In order to define the moment of inertia of this curve we proceed just as in the preceding section, arriving at an expression for the moment of inertia about the x-axis,

$$(27) \qquad T_x = \int_{s_0}^{s_1} y^2 \, ds = \int_{x_0}^{x_1} y^2 \sqrt{1 + y'^2} \, dx.$$

For the moment of inertia about the y-axis we have correspondingly

$$(28) \qquad T_y = \int_{s_0}^{s_1} x^2 \, ds = \int_{x_0}^{x_1} x^2 \sqrt{1 + y'^2} \, dx.$$

4.2 Examples

From the great variety of plane curves we choose a few typical examples to illustrate the concepts discussed.

a. The Common Cycloid

From the equations (cf. (1), p. 329) $x = a(t - \sin t), y = a(1 - \cos t)$, we obtain $\dot{x} = a(1 - \cos t), \dot{y} = a \sin t$, and find for the length of arc

$$s = \int_0^\alpha \sqrt{\dot{x}^2 + \dot{y}^2} \, dt = \int_0^\alpha \sqrt{2a^2(1 - \cos t)} \, dt.$$

Since $1 - \cos t = 2 \sin^2 t/2$ the integrand is equal to $2a \sin t/2$, and hence for $0 \le \alpha \le 2\pi$

$$s = 2a \int_0^\alpha \sin (t/2) \, dt = -4a \cos \frac{t}{2} \Big|_0^\alpha = 4a\left(1 - \cos \frac{\alpha}{2}\right) = 8a \sin^2 \frac{\alpha}{4}.$$

If, in particular, we consider the length of arc between two successive cusps, we must put $\alpha = 2\pi$, since the interval $0 \le t \le 2\pi$ of the values of the parameter corresponds to one revolution of the rolling circle. We thus obtain the value $8a$; that is, the length of arc of the cycloid between successive cusps is equal to *four* times the diameter of the rolling circle.

Similarly, we calculate the area bounded by one arch of the cycloid and the x-axis:

$$I = \int_0^{2\pi} y\dot{x}\, dt = a^2 \int_0^{2\pi} (1 - \cos t)^2\, dt$$

$$= a^2 \int_0^{2\pi} (1 - 2\cos t + \cos^2 t)\, dt$$

$$= a^2 \left(t - 2\sin t + \frac{t}{2} + \frac{\sin 2t}{4} \right) \Big|_0^{2\pi} = 3a^2\pi.$$

This area is therefore three times the area of the rolling circle.

For the radius of curvature $|\rho| = 1/|\kappa|$ we have by Eq. (15), p. 355,

$$\rho = \frac{(\dot{x}^2 + \dot{y}^2)^{3/2}}{\ddot{y}\dot{x} - \dot{y}\ddot{x}} = -2a\sqrt{2(1 - \cos t)} = -4a \left| \sin\frac{t}{2} \right|;$$

at the points $t = 0$, $t = \pm 2\pi, \ldots$ this expression has the value zero. These are actually the cusps, where the cycloid meets the x-axis at right angles.

The area of the surface of revolution formed by rotating an arch of the cycloid about the x-axis is given according to our formula (25), p. 374, by

$$A = 2\pi \int_0^{8a} y\, ds = 2\pi \int_0^{2\pi} a(1 - \cos t) \cdot 2a \sin\frac{t}{2}\, dt$$

$$= 8a^2\pi \int_0^{2\pi} \sin^3\frac{t}{2}\, dt = 16a^2\pi \int_0^{\pi} \sin^3 u\, du$$

$$= 16a^2\pi \int_0^{\pi} (1 - \cos^2 u)\sin u\, du.$$

The last integral can be evaluated by means of the substitution $\cos u = v$; we find

$$A = 16a^2\pi \left(-\cos u + \frac{1}{3}\cos^3 u \right) \Big|_0^{\pi} = \frac{64a^2\pi}{3}.$$

As an exercise the reader may calculate for himself the height η of the center-of-mass of the cycloid above the x-axis and also the moment of inertia T_x. The results are

$$\eta = \frac{4}{3}a = \frac{A}{2\pi s} \quad \text{and} \quad T_x = \frac{256}{15}a^3.$$

b. The catenary

The catenary[1] is the curve defined by the equation $y = \cosh x$. The length of the catenary between the abscissas $x = a$ and $x = b$ is

$$s = \int_a^b \sqrt{1 + \sinh^2 x}\, dx = \int_a^b \cosh x\, dx = \sinh b - \sinh a.$$

For the area of the surface of revolution obtained by rotating the catenary about the x-axis, the so-called *catenoid*, we find

$$A = 2\pi \int_a^b \cosh^2 x\, dx = 2\pi \int_a^b \frac{1 + \cosh 2x}{2}\, dx$$

$$= \pi(b - a + \tfrac{1}{2}\sinh 2b - \tfrac{1}{2}\sinh 2a).$$

From this we further obtain the height of the center-of-mass of the arc from a to b:

$$\eta = \frac{A}{2\pi s} = \frac{b - a + \tfrac{1}{2}\sinh 2b - \tfrac{1}{2}\sinh 2a}{2(\sinh b - \sinh a)}.$$

Finally, for the curvature we have

$$\kappa = \frac{y''}{(1 + y'^2)^{3/2}} = \frac{\cosh x}{\cosh^3 x} = \frac{1}{\cosh^2 x}.$$

c. The Ellipse and the Lemniscate

The lengths of arc of these two curves cannot be reduced to elementary functions, but belong to the class of "elliptic integrals" mentioned on p. 299.

For the ellipse $y = (b/a)\sqrt{a^2 - x^2}$ we obtain

$$s = \frac{1}{a} \int \sqrt{\frac{a^4 - (a^2 - b^2)x^2}{a^2 - x^2}}\, dx = a \int \sqrt{\frac{1 - \eta^2 \xi^2}{1 - \xi^2}}\, d\xi,$$

where we have put $x/a = \xi$, $1 - b^2/a^2 = \eta^2$. By the substitution $\xi = \sin \phi$ this integral can be expressed in the form

$$s = a \int \sqrt{1 - \eta^2 \sin^2 \phi}\, d\phi.$$

---

[1] The name derives from the fact that a chain suspended from its ends will assume the shape of this curve. Curiously enough the same curve arises in a quite different physical application. A soap film, bounded by two circles in space that lie in parallel planes and have centers on the same perpendicular to those planes, has the same shape as the surface of revolution obtained by rotating the catenary about the x-axis.

Here, to obtain the semiperimeter of the ellipse, we must let x traverse the interval from $-a$ to $+a$, which corresponds to the interval

$$-1 \leq \xi \leq +1 \qquad \text{or} \qquad -\frac{\pi}{2} \leq \phi \leq +\frac{\pi}{2}.$$

For the lemniscate, whose equation in polar coordinates r, t is $r^2 = 2a^2 \cos 2t$, we similarly obtain

$$s = \int \sqrt{r^2 + \dot{r}^2}\, dt = \int \sqrt{2a^2 \cos 2t + 2a^2 \frac{\sin^2 2t}{\cos 2t}}\, dt$$

$$= a\sqrt{2} \int \frac{dt}{\sqrt{\cos 2t}} = a\sqrt{2} \int \frac{dt}{\sqrt{1 - 2\sin^2 t}}.$$

If we introduce $u = \tan t$ as an independent variable in the last integral, we have

$$\sin^2 t = \frac{u^2}{1 + u^2}, \qquad dt = \frac{du}{1 + u^2},$$

and consequently,

$$s = a\sqrt{2} \int \frac{du}{\sqrt{1 - u^4}}.$$

In a complete loop of the lemniscate u runs from -1 to $+1$, and the length of arc is therefore equal to

$$a\sqrt{2} \int_{-1}^{+1} \frac{du}{\sqrt{1 - u^4}},$$

a special elliptic integral which played a great part in the researches of Gauss.

4.3 Vectors in Two Dimensions

For the discussions of plane curves and of many other topics in geometry, mechanics, and physics, vector notation constitutes a convenient and almost indispensable tool. We shall develop and apply in this chapter the concept of a vector in two dimensions, leaving extensions to higher dimensions to Volume II.

Intuitive Explanation

Many mathematical and physical objects are characterized completely by a single number, called a "scalar" since it measures the object on a given scale. Examples are angles, lengths, areas, times, masses, and temperatures. There are other objects, however, for which such a characterization is not possible, for example, the shape of a triangle, the location of a point in space, the acceleration or direction of motion of a particle, and the state of tension in a body. Several numbers are required to identify each of these objects. Gradually, mathematical concepts beyond the continuum of real numbers have been developed which permit us to represent such objects by a single symbol.[1] *Vectors* in a plane are objects that can be described by two items of information: A *length* and a *direction*. Of this type are, for example, the relative position of two points, the velocity and acceleration of a particle, and the force acting on a particle.[2]

Geometrically, or intuitively, a vector is essentially a directed straight line segment in the plane (or in space), characterized by its length or magnitude and by its direction. Ordinarily, vectors are indicated by *arrows* of the given length and pointing in the given direction. Unless a restriction is explicitly imposed, the *vector is "free"*, that is, the location of the beginning of the directed line, or arrow, is not an inherent part of the specifications for the vector.

While physical concepts, such as velocity, acceleration, and force, are primary instances of vectors in applications, we shall define vectors geometrically, by means of "translations" or "parallel displacements."

Vector analysis starts simply by giving a name, "vector," to such directed line segments or parallel displacements. However, its decisive significance is not that a unifying name was introduced, but that these entities, the vectors, (similarly as the complex numbers) can be combined with each other or with scalars (that is, ordinary numbers) by a set of rules, called vector algebra or vector analysis, in ways that have natural interpretations in the various applications, as, for example, the super-position of two velocities or the work done by a displacement against a force. In the intuitively appealing language of vectors many mathematical and physical relations can be expressed concisely and clearly.

a. Definition of Vectors by Translation. Notations

The simplest type of transformation of the plane is a *translation* or *parallel displacement*. A translation shifts or maps any point $P = (x, y)$ into the point $P' = (x', y')$ with coordinates

$$x' = x + a, \qquad y' = y + b,$$

---

[1] Of course, complex numbers $a + bi = z$ are such symbols representing the pair of real numbers, a, b; it is indeed sometimes convenient to use complex numbers rather than vectors.

[2] Vectors are insufficient for some purposes; to describe, for example, tensions or curvature of spaces, more general entities called "tensors" are used.

where a and b are constants. The translation is completely determined by the constants a and b which we call the *components* of the translation. We shall use the term "vector" as another name for "translation." Employing boldface type to denote vectors or translations we write $\mathbf{R} = (a, b)$ for the vector with components a, b (Fig. 4.29).

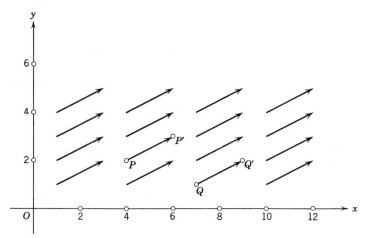

Figure 4.29 The translation $x' = x + 2$, $y' = y + 1$ corresponding to the vector $\mathbf{R} = \overrightarrow{PP'} = \overrightarrow{QQ'} = (2, 1)$.

The components of the vector $\mathbf{R}$ are determined by one pair of corresponding points $P = (x, y)$ and $P' = (x', y')$, since then

$$a = x' - x, \qquad b = y' - y.$$

Clearly, for any points P and P' it is always possible to find a translation $\mathbf{R}$ which takes P into P'. We denote it as the vector $\mathbf{R} = \overrightarrow{PP'}$. Any ordered pair of points $P = (x, y)$, $P' = (x', y')$, that is, any oriented line segment, thus determines the vector $\mathbf{R} = \overrightarrow{PP'} = (x' - x, y' - y)$. We observe that a second pair of points $Q = (\xi, \eta)$, $Q' = (\xi', \eta')$ defines the same vector if $\xi' - \xi = x' - x$ and $\eta' - \eta = y' - y$; the same translation $\mathbf{R}$ takes then P into P' and Q into Q'. Vectors $\mathbf{R}$ are determined by two numbers, the components, just as points are by two coordinates in the plane; the basic distinction is that a vector is represented geometrically by a *pair* of points. In the representation $\mathbf{R} = \overrightarrow{PP'}$ we call P the *initial point* and P' the *end point*. For given $\mathbf{R}$ one of the points, say the initial point $P = (x, y)$, can be chosen arbitrarily; the end point $P' = (x', y')$ is then determined uniquely

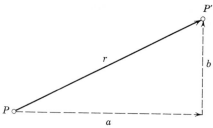

Figure 4.30   Components a, b and length r of a vector $\mathbf{R} = \overrightarrow{PP'}$.

by the relations $x' = x + a$, $y' = y + b$. Interchanging initial and end point leads to the *opposite* vector $\overrightarrow{P'P} = (-a, -b)$.

If we choose for the initial point the origin $O = (0, 0)$, we can associate uniquely a vector $\mathbf{R}$ with every point $Q = (x, y)$ by taking $\mathbf{R} = \overrightarrow{OQ}$. The vector $\mathbf{R}$ with the fixed initial point O is then called the *position vector* of Q. The components of the position of Q are simply the coordinates x, y of Q.

The vector $\mathbf{R}$ with components $a = 0$, $b = 0$ is called the *null vector* and is denoted by $\mathbf{O}$. It corresponds to a translation that leaves every point fixed:

$$\mathbf{O} = (0, 0) = \overrightarrow{PP}.$$

The distance r of two points $P = (x, y)$, $P' = (x', y')$ depends only on the vector $\mathbf{R} = (a, b) = \overrightarrow{PP'}$, since

$$r = \sqrt{(x' - x)^2 + (y' - y)^2} = \sqrt{a^2 + b^2}.$$

We call it the *length of the vector* $\mathbf{R}$ and write $r = |\mathbf{R}|$. The length of $\mathbf{R}$ is always a positive number unless $\mathbf{R} = \mathbf{O}$ (see Fig. 4.30).

We define the product of a vector $\mathbf{R} = (a, b)$ by a number or a "scalar" λ as the vector

$$\mathbf{R}^* = \lambda\mathbf{R} = (\lambda a, \lambda b).$$

For $\lambda = -1$ we have in $\mathbf{R}^* = (-a, -b)$ the vector *opposite* to $\mathbf{R}$ (Fig. 4.31).

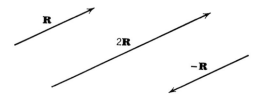

Figure 4.31   Scalar multiples of a vector $\mathbf{R}$.

If $\mathbf{R} = \overrightarrow{PP'} = (a, b)$ with $P = (x, y)$, $P' = (x', y')$, we can represent $\mathbf{R}^* = \lambda\mathbf{R}$ as $\overrightarrow{PP''}$, where $P'' = (x'', y'') = (x + \lambda a, y + \lambda b)$ (see Fig. 4.32). For $a = b = 0$ we have, of course, $P'' = P' = P$. For a and b not both zero the point $P'' = (x'', y'') = (x + \lambda a, y + \lambda b)$ traverses for varying λ the whole line

$$x''b - y''a = xb - ya.$$

The value $\lambda = 0$ gives $P'' = P$, whereas $\lambda = 1$ gives $P'' = P'$. Thus P'' lies on the line through P and P'; for $\lambda > 0$ the points P'' and P' lie on the *same side* of P, for $\lambda < 0$ they lie on *opposite sides*.

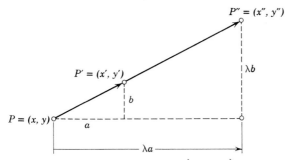

Figure 4.32   The vector relation $\mathbf{R}^* = \overrightarrow{PP''} = \lambda \overrightarrow{PP'}$ for $\lambda = \frac{8}{3}$.

The two vectors $\mathbf{R} = (a, b)$ and $\mathbf{R}^* = (a^*, b^*)$ are said to have the *same direction* if $\mathbf{R}^* = \lambda\mathbf{R}$ with a positive λ and opposite directions if $\lambda < 0$. If $\mathbf{R} = \mathbf{O}$, this means that also $\mathbf{R}^* = \mathbf{O}$. If $\mathbf{R} \neq \mathbf{O}$, the necessary and sufficient condition for $\mathbf{R}^*$ to have the same direction as $\mathbf{R}$ is that

$$\frac{a}{\sqrt{a^2 + b^2}} = \frac{a^*}{\sqrt{a^{*2} + b^{*2}}}, \qquad \frac{b}{\sqrt{a^2 + b^2}} = \frac{b^*}{\sqrt{a^{*2} + b^{*2}}}.$$

We call the quantities

$$\xi = \frac{a}{\sqrt{a^2 + b^2}} = \frac{a}{|\mathbf{R}|} = \frac{a}{r}, \qquad \eta = \frac{b}{\sqrt{a^2 + b^2}} = \frac{b}{|\mathbf{R}|} = \frac{b}{r}$$

which determine the direction of the vector $\mathbf{R}$ the *direction cosines* of $\mathbf{R}$; they are, of course, not defined for $\mathbf{R} = \mathbf{O}$. Since $\xi^2 + \eta^2 = 1$, we can always find an angle α and a corresponding angle $\beta = \pi/2 - \alpha$ such that

$$\xi = \cos \alpha, \qquad \eta = \sin \alpha = \cos \beta.$$

The angle α is called a *direction angle* of **R** (Fig. 4.33). It is determined uniquely only to within an even multiple of π. For $\mathbf{R} = \overrightarrow{PP'}$ we have

$$\cos \alpha = \frac{x' - x}{r}, \qquad \sin \alpha = \frac{y' - y}{r}.$$

Obviously, α is the angle between the positive x-axis and the line from P to P'. More precisely a rotation of the positive x-axis about the origin by the angle α (counted positive if we turn counterclockwise, negative if clockwise) will give the axis the direction from P to P'.

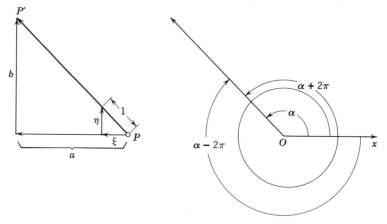

Figure 4.33   Direction cosines ξ, η, and direction angles for a vector $\overrightarrow{PP'}$.

The opposite vector $-\mathbf{R} = (-a, -b)$ has direction cosines $-\xi$ and $-\eta$ and direction angles differing from α by an odd multiple of π. If the initial point P of the vector $\mathbf{R} = \overrightarrow{PP'}$ is the origin, the direction angle α of **R** is simply the polar angle θ of P'.

b. Addition and Multiplication of Vectors

Sums of Vectors

Vectors have been defined by translations, that is as certain mappings of points in the plane. There is a perfectly general way of combining any two mappings by applying them *successively*. If the first mapping carries a point P into the point P' and the second one carries P' into P'', the combined mapping is the one that carries P into P''. In the case of two vectors $\mathbf{R} = (a, b)$ and $\mathbf{R}^* = (a^*, b^*)$ the vector **R** will map the point $P = (x, y)$ onto the point $P' = (x + a, y + b)$ and $\mathbf{R}^*$ will map P' onto $P'' = (x + a + a^*, y + b + b^*)$. The resulting mapping from P onto P'' is again a translation; we call it the *sum* or the *resultant of the vectors* $\mathbf{R} = \overrightarrow{PP'}$ and $\mathbf{R}^* = \overrightarrow{P'P''}$, and denote it by $\mathbf{R} + \mathbf{R}^*$ (Fig. 4.34).[1] The components of the sum are $a + a^*$ and $b + b^*$. Thus our definition of the sum of two vectors is

$$\overrightarrow{PP'} + \overrightarrow{P'P''} = \overrightarrow{PP''},$$

or, if we describe the vectors by their components,

$$(a, b) + (a^*, b^*) = (a + a^*, b + b^*).$$

If $\mathbf{R}^*$ is taken from the same initial point as $\mathbf{R}$, say $\mathbf{R}^* = \overrightarrow{PP'''}$, the points P, P''', P'', and P' form the vertices of a parallelogram. The

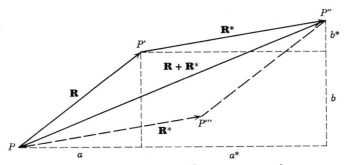

Figure 4.34   Addition of the vectors $\overrightarrow{PP'} = (a, b)$ and $\overrightarrow{P'P''} = (a^*, b^*)$.

two sides from P represent the vectors $\mathbf{R}$ and $\mathbf{R}^*$; the sum $\mathbf{R} + \mathbf{R}^*$ is represented by the *diagonal* from P *("parallelogram construction" for the sum of vectors*).

Sums of vectors obey the commutative and associative laws of arithmetic, since addition of vectors just amounts to addition of corresponding components (Fig. 4.35). They obey moreover the *distributive laws for multiplication of a sum of two vectors by a number* λ *and of a vector by the sum of two numbers* λ, μ:

$$\lambda(\mathbf{R} + \mathbf{R}^*) = \lambda\mathbf{R} + \lambda\mathbf{R}^*, \qquad (\lambda + \mu)\mathbf{R} = \lambda\mathbf{R} + \mu\mathbf{R}.[2]$$

---

[1] This "sum" is really the "symbolic product" of the two mappings as defined on p. 52. The sum notation is here more natural because it corresponds to addition of the components.

[2] To distinguish vectors from numbers in an equation we always let the number precede the vector in writing products; the combination $\mathbf{R}\lambda$ will not be used, although it could be defined by $\lambda\mathbf{R} = \mathbf{R}\lambda$.

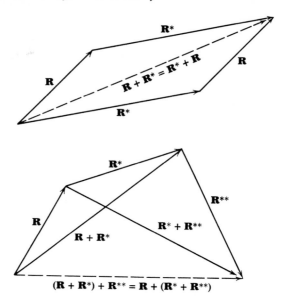

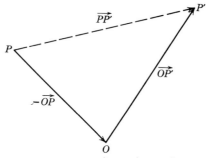

Figure 4.35   Commutative and associate laws of vector addition.

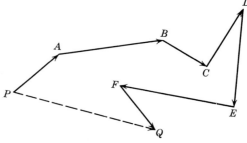

Figure 4.36 $\overrightarrow{PP'} = \overrightarrow{OP'} - \overrightarrow{OP}.$

Figure 4.37 $\overrightarrow{PQ} = \overrightarrow{PA} + \overrightarrow{AB} + \overrightarrow{BC} + \cdots + \overrightarrow{FQ}.$

These rules permit us to express a vector $\overrightarrow{PP'}$ in terms of the position vectors $\overrightarrow{OP}$ and $\overrightarrow{OP'}$ of the points P and P' (Fig. 4.36):

$$\overrightarrow{PP'} = \overrightarrow{PO} + \overrightarrow{OP'} = \overrightarrow{OP'} + \overrightarrow{PO} = \overrightarrow{OP'} - \overrightarrow{OP}.$$

It is important to realize that generally if we go from a point P to a point Q by way of points $A, B, C, \ldots, E, F$, then the vector $\overrightarrow{PQ}$ is the sum of the vectors $\overrightarrow{PA}, \overrightarrow{AB}, \overrightarrow{BC}, \ldots, \overrightarrow{EF}, \overrightarrow{FQ}$ (Fig. 4.37).

Angle between Vectors

The angle θ formed by a vector $\mathbf{R}^* = (a^*, b^*)$ with the vector $\mathbf{R} = (a, b)$ is defined as the difference of their direction angles: $\theta = \alpha^* - \alpha$. (It is assumed here that neither $\mathbf{R}$ nor $\mathbf{R}^*$ is a zero vector.) The angle θ again is determined only to within integer multiples

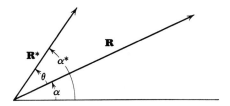

Figure 4.38 Angle θ the vector $\mathbf{R}^*$ forms with $\mathbf{R}$.

of 2π (Fig 4.38) A rotation by the angle θ (with the sign of θ indicating the sense of rotation) will take the direction of $\mathbf{R}$ into that of $\mathbf{R}^*$. The quantities $\cos\theta$ and $\sin\theta$, which are determined uniquely, can be expressed immediately in terms of the direction cosines of $\mathbf{R}$ and $\mathbf{R}^*$:

$$\cos\theta = \cos(\alpha^* - \alpha) = \cos\alpha\cos\alpha^* + \sin\alpha\sin\alpha^*$$

$$= \frac{aa^* + bb^*}{\sqrt{a^2 + b^2}\sqrt{a^{*2} + b^{*2}}},$$

$$\sin\theta = \sin(\alpha^* - \alpha) = \cos\alpha\sin\alpha^* - \sin\alpha\cos\alpha^*$$

$$= \frac{ab^* - a^*b}{\sqrt{a^2 + b^2}\sqrt{a^{*2} + b^{*2}}}.$$

The denominator in each expression is just the product rr^* of the length of the vectors. We introduce the expressions occurring in the numerators as "products" of the two vectors.

Inner Product and Exterior Product of Two Vectors

We define the "scalar" or "inner" or "dot" product of the vectors $R = (a, b)$ and $R^* = (a^*, b^*)$ by

$$R \cdot R^* = aa^* + bb^* = rr^* \cos \theta,$$

and the "outer" or "exterior" or "cross" product by

$$R \times R^* = ab^* - a^*b = rr^* \sin \theta.^{[1]}$$

As immediately confirmed *inner and outer products obey the distributive and associative laws:*

$$R \cdot (R^* + R^{**}) = R \cdot R^* + R \cdot R^{**},$$
$$R \times (R^* + R^{**}) = R \times R^* + R \times R^{**},$$
$$\lambda(R \cdot R^*) = (\lambda R) \cdot R^* = R \cdot (\lambda R^*),$$
$$\lambda(R \times R^*) = (\lambda R) \times R^* = R \times (\lambda R^*).$$

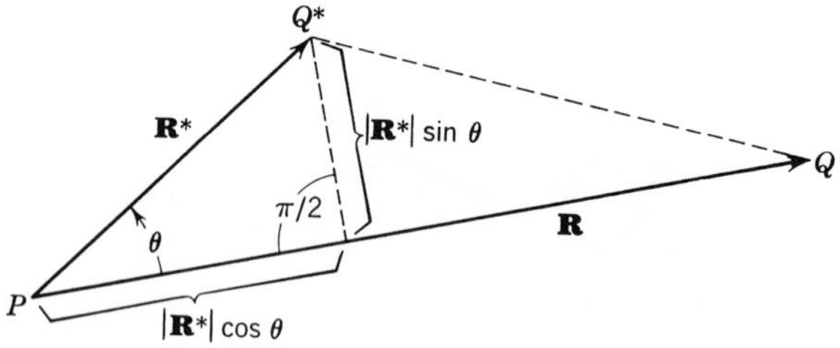

Figure 4.39 The vector product $R \times R^* = |R|\,|R^*| \sin \theta$ as twice the area of the triangle PQQ^*.

The commutative law of multiplication also holds for inner products

$$R \cdot R^* = R^* \cdot R;$$

for exterior products however, the sign is changed if the factors are interchanged:

$$R \times R^* = -R^* \times R.$$

Giving R and R^* the same initial point, $R = \overrightarrow{PQ}$, $R^* = \overrightarrow{PQ^*}$ we can interpret $R \cdot R^*$ as the product of the projection $r^* \cos \theta$ of the segment PQ^* onto the segment PQ, with the length r of that segment. The outer product $R \times R^*$ is simply twice the area of the oriented triangle PQQ^*, taken with the positive sign if the vertices PQQ^* are in counterclockwise order, with the negative sign if in clockwise order (Fig. 4.39).

---

[1] With our definition both inner and exterior products are actually "scalars." The term "scalar product" is reserved for the inner product because in three dimensions the analogue of the exterior product is a vector.

For any vector $\mathbf{R} = (a, b)$

$$\mathbf{R} \cdot \mathbf{R} = a^2 + b^2 = |\mathbf{R}|^2$$

is the square of the length of the vector. Thus $\mathbf{R} \cdot \mathbf{R}$ is positive unless $\mathbf{R} = \mathbf{O}$. On the other hand, $\mathbf{R} \times \mathbf{R}$ is always zero. The condition for two nonzero vectors to be orthogonal to each other is that $\mathbf{R} \cdot \mathbf{R}^* = 0$ while they are parallel (that is, have the same or opposite directions) if $\mathbf{R} \times \mathbf{R}^* = 0$.

Equation of Straight Line

We can easily write the equation of a line through two points and that of a line through a given point with a given direction, in vector

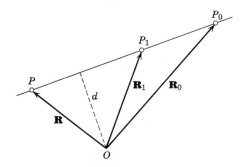

Figure 4.40   Line in vector notation.

notation. If $P = (x, y)$, $P_0 = (x_0, y_0)$, and $P_1 = (x_1, y_1)$ are three points with $P_0 \neq P_1$ then P lies on the line through P_0 and P_1 if the vectors $\overrightarrow{P_0P}$ and $\overrightarrow{P_0P_1}$ are parallel, that is,

$$\overrightarrow{P_0P} \times \overrightarrow{P_0P_1} = \mathbf{0}.$$

If $\mathbf{R} = \overrightarrow{OP}$, $\mathbf{R}_0 = \overrightarrow{OP_0}$, and $\mathbf{R}_1 = \overrightarrow{OP_1}$ are the position vectors of the three points, the condition takes the form

$$(\mathbf{R} - \mathbf{R}_0) \times (\mathbf{R}_1 - \mathbf{R}_0) = \mathbf{0}$$

or

$$(\mathbf{R}_1 - \mathbf{R}_0) \times \mathbf{R} = \mathbf{R}_1 \times \mathbf{R}_0.$$

Substituting the coordinates of the points for the position vectors, we obtain the equation of the line in the usual form (Fig. 4.40):

$$(x_1 - x_0)y - (y_1 - y_0)x = x_1y_0 - y_1x_0.$$

Instead of prescribing two points of the line we can prescribe one point P_0 and require that the line is to be parallel to a vector $\mathbf{S} = (a, b)$. Obviously, the equation of the line is then

$$(\mathbf{R} - \mathbf{R}_0) \times \mathbf{S} = 0$$

or

$$(x - x_0)b - (y - y_0)a = 0.$$

For $\mathbf{S} = \overrightarrow{P_0P_1}$ we obtain the previous equation.

The distance d of the line from the origin can also be expressed in vector notation. Obviously, d multiplied with the length of the vector $\overrightarrow{P_0P_1}$ is twice the area of the triangle OP_0P_1. Hence

$$d = \frac{1}{|\overrightarrow{P_0P_1}|} (\overrightarrow{OP_0} \times \overrightarrow{OP_1}) = \frac{\mathbf{R}_0 \times \mathbf{R}_1}{|\mathbf{R}_1 - \mathbf{R}_0|}$$

$$= \frac{x_0y_1 - x_1y_0}{\sqrt{(x_1 - x_0)^2 + (y_1 - y_0)^2}}.$$

Here d is taken with the positive sign if the points O, P_0, P_1 follow each other in counterclockwise order.

Coordinate Vectors.   A vector $\mathbf{R} = (a, b)$ trivially can be represented in the form

(29) $$\mathbf{R} = a\mathbf{i} + b\mathbf{j},$$

where we denote by $\mathbf{i}$ and $\mathbf{j}$ the "*coordinate vectors*"

(30) $$\mathbf{i} = (1, 0), \qquad \mathbf{j} = (0, 1).$$

In this way $\mathbf{R}$ is split into two vectors $a\mathbf{i}$ and $b\mathbf{j}$ pointing respectively in the direction of the x-axis and y-axis. The components a and b of $\mathbf{R}$ are just the (signed) lengths of these two vectors.

In applications one is often called upon to represent a vector $\mathbf{R}$ as resultant of vectors with two given orthogonal (that is, mutually perpendicular) directions. For that purpose it is best to introduce two *unit vectors* (that is, vectors of length 1) $\mathbf{I}$ and $\mathbf{J}$ with the given directions. The required decomposition of $\mathbf{R}$ is achieved if we can represent $\mathbf{R}$ in the form

(31) $$\mathbf{R} = A\mathbf{I} + B\mathbf{J}$$

with suitable scalars A, B (cf. Fig. 4.40). It is easy to find the values of A and B if such a representation of $\mathbf{R}$ exists. For, by assumption, the vectors $\mathbf{I}$ and $\mathbf{J}$ are orthogonal unit vectors of length 1, so that

(32) $$\mathbf{I} \cdot \mathbf{I} = \mathbf{J} \cdot \mathbf{J} = 1, \qquad \mathbf{I} \cdot \mathbf{J} = 0.$$

Forming the scalar product of Eq. (31) with **I**, **J** respectively we find immediately that A and B must have the values

(33) $$A = \mathbf{R} \cdot \mathbf{I}, \qquad B = \mathbf{R} \cdot \mathbf{J};$$

in words, A and B are the (signed) lengths of the projections of the segment representing **R** in the given directions.

The possibility of writing **R** as a linear combination (31) of **I** and **J** follows from the representation (29) of **R** in terms of **i** and **j**, if we can

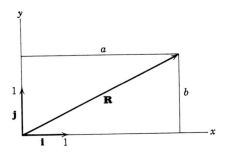

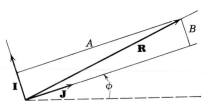

Figure 4.40

show that **i** and **j** themselves can be expressed in terms of **I** and **J**. However, $\mathbf{I} = (\alpha, \beta)$, $\mathbf{J} = (\gamma, \delta)$, can be written as,

(34) $$\mathbf{I} = \alpha \mathbf{i} + \beta \mathbf{j}, \qquad \mathbf{J} = \gamma \mathbf{i} + \delta \mathbf{j}.$$

Because of (32) the quantities α, β, γ, δ must satisfy the so-called orthogonality relations

(35) $$\alpha^2 + \beta^2 = \gamma^2 + \delta^2 = 1, \qquad \alpha\gamma + \beta\delta = 0.$$

If we multiply the first of the equations (34) by δ, the second one by β, and subtract we find

(36) $$(\alpha\delta - \beta\gamma)\mathbf{i} = \delta\mathbf{I} - \beta\mathbf{J}$$

and similarly,

(37) $$(\alpha\delta - \beta\gamma)\mathbf{j} = -\gamma\mathbf{I} + \alpha\mathbf{J}.$$

Here for the mutually perpendicular unit vectors $\mathbf{I}$ and $\mathbf{J}$

(38) $$\alpha\delta - \beta\gamma = \mathbf{I} \times \mathbf{J} = \pm 1,$$

where the upper or lower sign holds depending on the counterclockwise or clockwise sense of the 90° rotation that takes $\mathbf{I}$ into $\mathbf{J}$. In either case formulas (36) and (37) express $\mathbf{i}$ and $\mathbf{j}$ in terms of $\mathbf{I}$ and $\mathbf{J}$; substituting these expressions into (29) justifies the representation formula (31) for an arbitrary vector $\mathbf{R}$.

Formula (31) also can be interpreted as the representation of the vector $\mathbf{R}$ in a new coordinate system with axes pointing respectively in the directions of $\mathbf{I}$ and $\mathbf{J}$. The components of a unit vector are at the same time the direction cosines of the direction angle of the vector. Let $\mathbf{I}$ and $\mathbf{J}$ have direction angles ϕ and ψ respectively. Then

$$\alpha = \cos\phi, \qquad \beta = \sin\phi, \qquad \gamma = \cos\psi, \qquad \delta = \sin\psi.$$

Here either $\psi = \phi + \frac{1}{2}\pi$ or $\psi = \phi - \frac{1}{2}\pi$. In the first case (which corresponds to a right-handed system of coordinate vectors $\mathbf{I}, \mathbf{J}$), we have $\gamma = -\beta$, $\delta = \alpha$, $\alpha\delta - \beta\gamma = +1$ so that

(39) $$\mathbf{I} = (\cos\phi, \sin\phi), \qquad \mathbf{J} = (-\sin\phi, \cos\phi).$$

The formulas (33) giving the components of $\mathbf{R}$ referred to coordinate vectors $\mathbf{I}, \mathbf{J}$ then take the form

(40) $$A = a\cos\phi + b\sin\phi, \qquad B = -a\sin\phi + b\cos\phi.$$

These formulas express the relations between the components of one and the same vector $\mathbf{R}$ in two right-handed coordinate systems obtained one from the other by a rotation of axes by the angle ϕ. If we assume that the coordinate systems have the same origin O and that $\mathbf{R}$ is the position vector $\overrightarrow{OP}$ of an arbitrary point P we have in (40) the formulas for changes of coordinate systems already derived on p. 361, Equation (18). The components a, b and A, B are then respectively the coordinates of P in the two systems.

c. Variable Vectors, Their Derivatives, and Integrals

It is natural to consider vectors $\mathbf{R} = (a, b)$ whose components a, b are functions of a variable t, say $a = a(t), b = b(t)$. For any t we then have a vector

$$\mathbf{R} = \mathbf{R}(t) = (a(t), b(t))$$

and we say that $\mathbf{R}(t)$ is a vector function of t. An example is furnished by the position vector of a point that moves with the time t.

We say that $\mathbf{R}(t)$ has the limit $\mathbf{R}^* = (a^*, b^*)$ for $t \to t_0$ if $a(t)$ has the limit a^* and $b(t)$ the limit b^* for $t \to t_0$. In that case the length of $\mathbf{R}(t)$ tends toward that of $\mathbf{R}^*$, and in case $\mathbf{R}^* \neq \mathbf{O}$ the direction of $\mathbf{R}(t)$ tends toward that of $\mathbf{R}^*$ (this means that the direction cosines of $\mathbf{R}$ tend toward those of $\mathbf{R}^*$). The vector $\mathbf{R}(t)$ is said to depend continuously on t, if

$$\lim_{t \to t_0} \mathbf{R}(t) = \mathbf{R}(t_0),$$

that is, if the components of $\mathbf{R}$ are continuous functions of t. The length and, if $\mathbf{R}(t_0) \neq \mathbf{O}$, also the direction of a continuous vector vary continuously with t.

To introduce the derivative of a vector we form for two values t and $t + h$ of the parameter the difference quotient

$$\frac{1}{h}[\mathbf{R}(t + h) - \mathbf{R}(t)] = \left[\frac{a(t + h) - a(t)}{h}, \frac{b(t + h) - b(t)}{h}\right],$$

and define the derivative of $\mathbf{R}$ as the limit of the difference quotient for $h \to 0$:

$$\dot{\mathbf{R}} = \frac{d\mathbf{R}}{dt} = \lim_{h \to 0}\frac{1}{h}[\mathbf{R}(t + h) - \mathbf{R}(t)] = \left(\frac{da}{dt}, \frac{db}{dt}\right) = (\dot{a}, \dot{b}).$$

The derivative of a vector is formed by differentiating the components.

Derivatives of products of vectors are easily seen to obey the ordinary rules

$$(\mathbf{RS})^{\cdot} = \frac{d\mathbf{R} \cdot \mathbf{S}}{dt} = \frac{d\mathbf{R}}{dt} \cdot \mathbf{S} + \mathbf{R} \cdot \frac{d\mathbf{S}}{dt} = \dot{\mathbf{R}}\mathbf{S} + \mathbf{R}\dot{\mathbf{S}}$$

$$(\mathbf{R} \times \mathbf{S})^{\cdot} = \frac{d\mathbf{R} \times \mathbf{S}}{dt} = \dot{\mathbf{R}} \times \mathbf{S} + \mathbf{R} \times \dot{\mathbf{S}},$$

where for outer products, factors have to be taken in the original order.

We define similarly the integral of the vector $\mathbf{R}(t)$ in terms of the integrals of its components:

$$\int_\alpha^\beta \mathbf{R}(t)\,dt = \left(\int_\alpha^\beta a(t)\,dt, \int_\alpha^\beta b(t)\,dt\right).$$

The fundamental theorem of calculus implies

$$\frac{d}{dt}\int_\alpha^t \mathbf{R}(s)\,ds = \mathbf{R}(t).$$

d. Application to Plane Curves. Direction, Speed, and Acceleration

Velocity Vector

In Section 4.1 we represented a curve C by two functions $x = \phi(t)$ and $y = \psi(t)$. Each t in the domain of these functions determines a point $P = (x, y)$ on C; here t may be considered as time and P as a moving point whose position at the time t is given by $x(t)$ and $y(t)$. If we identify x and y with the components of the position vector $\mathbf{R} = \overrightarrow{OP}$ of P,

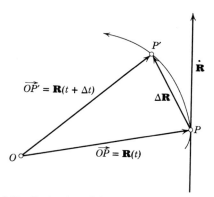

Figure 4.41   Derivative of the position vector for a curve.

then C is described by the end point of the position vector

$$\mathbf{R} = \mathbf{R}(t) = (x(t), y(t))$$

(Fig. 4.41). For two points P and P' of C corresponding to the parameter values t and $t + \Delta t$ we have in

$$\overrightarrow{PP'} = \overrightarrow{OP'} - \overrightarrow{OP} = \mathbf{R}(t + \Delta t) - \mathbf{R}(t) = \Delta\mathbf{R}$$

the vector represented by the directed secant of C with end points P, P'. If here Δt is positive, that is, if the point P' follows P on C in the direction of increasing t, then the vector

$$\frac{1}{\Delta t} (\mathbf{R}(t + \Delta t) - \mathbf{R}(t))$$

has the same direction as the vector $\mathbf{R}(t + \Delta t) - \mathbf{R}(t) = \overrightarrow{PP'}$; its length is the distance of the points P and P' divided by Δt. For Δt tending to zero we obtain in the limit the vector

$$\dot{\mathbf{R}} = \dot{\mathbf{R}}(t) = (\dot{x}(t), \dot{y}(t)),$$

where again the dot is used to denote differentiation with respect to the parameter t. The direction of $\dot{\mathbf{R}}$ is the limit of the direction of the secants PP' and hence is the direction of the tangent at the point P. More precisely $\dot{\mathbf{R}}$ points in that direction on the tangent that corresponds to increasing t on C, provided $\dot{\mathbf{R}} \neq \mathbf{O}$. The direction cosines of $\dot{\mathbf{R}}$ are the quantities

$$\cos \alpha = \frac{\dot{x}}{\sqrt{\dot{x}^2 + \dot{y}^2}}, \qquad \sin \alpha = \frac{\dot{y}}{\sqrt{\dot{x}^2 + \dot{y}^2}},$$

introduced on p. 346 as direction cosines of the tangent. The length of $\dot{\mathbf{R}}$

$$|\dot{\mathbf{R}}| = \sqrt{\dot{x}^2 + \dot{y}^2}$$

can be interpreted as ds/dt, the rate of change of the length s along the curve with respect to the parameter t. If t stands for the time, we have in $|\dot{\mathbf{R}}|$ the *speed* with which the point travels along the curve.

In mechanics one must consider the velocity of a particle not only as having a certain *magnitude* (the "speed") but also a certain *direction*. *Velocity* is then represented by the *vector* $\dot{\mathbf{R}} = (\dot{x}, \dot{y})$, whose length is the speed and whose direction is the instantaneous direction of motion, that is, the direction of the tangent in the sense of increasing t.

Acceleration

Similarly the *acceleration* of the particle is defined as the vector $\ddot{\mathbf{R}} = (\ddot{x}, \ddot{y})$. Vanishing acceleration means that $\ddot{x} = \ddot{y} = 0$; if $\ddot{\mathbf{R}} = \mathbf{O}$ along a whole t-interval, the velocity components have constant values $\dot{x} = a$, $\dot{y} = b$; the components of the position vector itself are then linear functions of t: $x = at + c$, $y = bt + d$. The particle in this case moves with constant speed along a straight line.

All our previous results pertaining to curves are easily expressible in vector notation if the curve is described by the position vector $\mathbf{R} = \mathbf{R}(t) = (x(t), y(t))$, with $\alpha \leq t \leq \beta$. We find for the length [cf. Eq. (8), p. 349]

$$\int_{\alpha}^{\beta} |\dot{\mathbf{R}}| \, dt,$$

while for the signed area enclosed by a curve [cf. Eq. (20), p. 365]

$$A = \tfrac{1}{2} \int_{\alpha}^{\beta} \mathbf{R} \times \dot{\mathbf{R}} \, dt$$

(the sign of this quantity depending again on the orientation of the curve). Finally, we have for the curvature κ the formula [cf. Eq. (15), p. 355]

$$\kappa = \frac{\dot{\mathbf{R}} \times \ddot{\mathbf{R}}}{|\dot{\mathbf{R}}|^3}.$$

Tangential and Normal Components of Acceleration

These formulas have interesting implications if we interpret t again as the time. Let γ be the angle formed by the vector $\ddot{\mathbf{R}}$ with the vector $\dot{\mathbf{R}}$, that is, with the instantaneous direction of motion. The quantity $|\ddot{\mathbf{R}}| \cos \gamma$ represents the projection of $\ddot{\mathbf{R}}$ onto the direction of $\dot{\mathbf{R}}$; we call it the *tangential component* of acceleration. Similarly, $|\ddot{\mathbf{R}}| \sin \gamma$ is the projection of $\ddot{\mathbf{R}}$ onto the normal (more precisely onto that normal

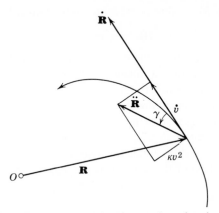

Figure 4.42 Tangential and normal acceleration.

obtained by a 90° counterclockwise rotation from $\dot{\mathbf{R}}$); this is the *normal component* of acceleration (see Fig. 4.42). By definition of inner and outer products

$$|\ddot{\mathbf{R}}| \cos \gamma = \frac{\dot{\mathbf{R}} \cdot \ddot{\mathbf{R}}}{|\dot{\mathbf{R}}|}, \qquad |\ddot{\mathbf{R}}| \sin \gamma = \frac{\dot{\mathbf{R}} \times \ddot{\mathbf{R}}}{|\dot{\mathbf{R}}|}.$$

Now

$$\dot{\mathbf{R}} \cdot \ddot{\mathbf{R}} = \frac{1}{2}(\dot{\mathbf{R}} \cdot \ddot{\mathbf{R}} + \ddot{\mathbf{R}} \cdot \dot{\mathbf{R}}) = \frac{1}{2}\frac{d}{dt}(\dot{\mathbf{R}} \cdot \dot{\mathbf{R}}) = \frac{1}{2}\frac{dv^2}{dt} = v\frac{dv}{dt},$$

where $v = ds/dt = |\dot{\mathbf{R}}| = \sqrt{\dot{\mathbf{R}} \cdot \dot{\mathbf{R}}}$ is the *speed* of the point. Hence

(41)
$$|\ddot{\mathbf{R}}| \cos \gamma = \frac{dv}{dt} = \dot{v};$$

Thus the tangential component of acceleration is identical with the rate of change of speed with respect to time. For the normal acceleration the formula for the curvature yields

(42)
$$|\ddot{\mathbf{R}}| \sin \gamma = \kappa |\dot{\mathbf{R}}|^2 = \kappa v^2,$$

that is, the product of the square of the speed with the curvature.

For a particle moving with *constant* speed v along a curve the tangential acceleration $\dot{v}$ vanishes. The acceleration vector then is perpendicular to the curve. More precisely it points toward the "inner" side of the curve, the side toward which the curve turns (this is seen, for example, from the fact that $\sin \gamma > 0$ when $\kappa > 0$, that is, when the tangent turns counterclockwise). In moving along a curve at constant speed therefore, a point experiences an acceleration toward the inside of the curve which is proportional to the curvature and also to the square of the speed. This fact is of obvious significance because as a result of Newton's law (to be discussed later) a force proportional to the acceleration is needed to hold the point P on the curve.

4.4 Motion of a Particle under Given Forces

The early development of calculus was decisively stimulated not only by geometry but just as much by the concepts of mechanics. Mechanics rests on certain basic principles first laid down by Newton; the statement of these principles involves the concept of the derivative, and their application requires the theory of integration. Without analyzing Newton's principles in detail, we shall illustrate by some simple examples how calculus is applied in mechanics.

a. Newton's Law of Motion

We shall restrict ourselves to the consideration of a single particle, that is, of a point at which a mass m is imagined to be concentrated. We shall further assume that the motion takes place in the x,y-plane, in which the position of the particle at the time t is specified by its coordinates $x = x(t), y = y(t)$, or, equivalently, by its "position vector" $\mathbf{R} = \mathbf{R}(t) = (x(t), y(t))$. A dot above a quantity indicates differentiation with respect to the *time t*. The *velocity* and *acceleration* of the particle are then represented by the vectors

$$\dot{\mathbf{R}} = (\dot{x}, \dot{y}) \quad \text{and} \quad \ddot{\mathbf{R}} = (\ddot{x}, \ddot{y}).$$

In mechanics one relates the motion of a point to the concept of *forces* of definite direction and magnitude acting on the point. A force is then also described by a vector $\mathbf{F} = (\rho, \sigma)$. The effect of several forces $\mathbf{F}_1, \mathbf{F}_2, \ldots$ acting on the same particle is the same as that of a single force $\mathbf{F}$, the *resultant* force, which is simply the vector sum $\mathbf{F} = \mathbf{F}_1 + \mathbf{F}_2 + \cdots$ of the individual forces.

Newton's fundamental law states: *The mass m multiplied by the acceleration is equal to the force acting on the particle*, in symbols

(43) $$m\ddot{\mathbf{R}} = \mathbf{F}.$$

If we write this vector equation which expresses the fundamental law in terms of the components of those vectors, we obtain the equivalent pair of equations

(44) $m\ddot{x} = \rho, \quad m\ddot{y} = \sigma.$

Since acceleration and force differ only by the positive factor m, the *direction of the acceleration is the same as that of the force*. If no force acts, that is, $\mathbf{F} = \mathbf{O}$, the acceleration vanishes, the velocity is constant, and x and y become linear functions of t. This is *Newton's first law*: A particle on which no force acts moves with constant velocity along a straight line.

Newton's law $m\ddot{\mathbf{R}} = \mathbf{F}$ is in the first instance nothing more than a quantitative definition of the concept of force. The left-hand side of this relation can be determined by observation of the motion, by means of which we then obtain the force.

However, Newton's law has a far deeper meaning, due to the fact that in many cases we can determine the acting force from other physical considerations, without any knowledge of the corresponding motion. This fundmental law is then no longer a *definition of force*, but it instead is a relation from which we can hope to determine the motion. This decisive turn in using Newton's law comes into play in all the numerous instances where physical considerations permit us to express the force $\mathbf{F}$ or its components ρ, σ in an explicit way as functions of the position and velocity of the particle and of the time t. The law of motion then is not a tautology, but furnishes two equations expressing $m\ddot{x}$, $m\ddot{y}$ in terms of x, y, $\dot{x}$, $\dot{y}$, and t, the so-called *equations of motion*. These equations are *differential equations*, that is, relations between functions and their derivatives. *Solving* these differential equations, that is, finding all pairs of functions $x(t)$, $y(t)$ for which the equations of motion are valid, yields all possible motions of a particle under the prescribed force.

b. Motion of Falling Bodies

The simplest example of a known force is that of *gravity* acting on a particle near the surface of the earth. It is known from direct observation that (aside from effects of air resistance) every falling body has an acceleration which is directed vertically downward, and which has the same magnitude g for all bodies. Measured in feet per second per second, g has the approximate value 32.16.[1] If we choose an

---

[1] The precise value of g, which also includes in addition to gravitational attraction, effects of the rotation of the earth, depends on the location on the earth.

x,y-coordinate system in which the y-axis points vertically upward while the x-axis is horizontal, the acceleration $\ddot{\mathbf{R}} = (\ddot{x}, \ddot{y})$ has the components

$$\ddot{x} = 0, \qquad \ddot{y} = -g.$$

By Newton's fundamental law the vector $\mathbf{F}$ representing the force of gravity acting on a particle of mass m must then be

$$\mathbf{F} = (0, -mg).$$

This force vector is likewise directed vertically downward; its magnitude, the *weight* of the body near the surface of the earth, is mg.

When we cancel out the factor m, the equations of motion of a particle under gravity take the form

$$\ddot{x} = 0, \qquad \ddot{y} = -g.$$

From these equations we can easily obtain a description of the most general motion possible for a falling body. Integrating with respect to t yields

$$\dot{x} = a, \qquad \dot{y} = -gt + b,$$

where a and b are constants. A further integration then shows that

$$x = at + c, \qquad y = -\tfrac{1}{2}gt^2 + bt + d,$$

where c and d are constants. Thus the general solution of our equations of motion depends on four un-specified constants a, b, c, d. We can immediately relate the values of these constants for an individual motion to the *initial conditions* for that motion. If the particle at the initial time $t = 0$ is at the point (x_0, y_0), then setting $t = 0$ we find

$$c = x_0, \qquad d = y_0.$$

The velocity $\dot{\mathbf{R}} = (\dot{x}, \dot{y}) = (a, -gt + b)$ reduces for $t = 0$ to (a, b). Thus (c, d) and (a, b) represent respectively initial position and initial velocity of the particle. Any choice of these initial conditions leads uniquely to a motion.

In case $a \neq 0$, that is, in case the initial velocity is not vertical, we can eliminate t and obtain a nonparametric representation for the orbit of the particle. Solving the first equation for t and substituting into the second yields

$$y = -\frac{g}{2a^2}(x - c)^2 + \frac{b}{a}(x - c) + d.$$

Hence the path is a parabola. For $a = 0$ we have $x = c = $ constant, and the whole motion takes place along a vertical straight line.

c. Motion of a Particle Constrained to a Given Curve

In most problems of mechanics the forces acting on a particle depend on the position and velocity of the particle. As a rule, the equations of motion are too complicated to permit us to determine all possible motions. Considerable simplification arises if we may consider the curve C described by the particle as known and only have to determine the motion of the particle along the curve. In a large class of mechanical problems the particle is *constrained* (by means of some mechanical device) to move on a given curve C. The simplest example is the plane pendulum where a mass m is joined by an inextensible string of length L to a point P_0 and moves under the influence of gravity on a circle of radius L about P_0.

Along the curve C we use the arc length s as parameter. The curve is then given by $x = x(s)$, $y = y(s)$. Finding the motion of the particle along C then amounts to finding s as a function of t. An equation of motion along the curve is obtained as follows.

We form the inner product of both sides of Newton's formula $m\ddot{\mathbf{R}} = \mathbf{F}$ with a vector $\boldsymbol{\xi}$:

$$m\ddot{\mathbf{R}} \cdot \boldsymbol{\xi} = \mathbf{F} \cdot \boldsymbol{\xi}.$$

If we take for $\boldsymbol{\xi}$ the vector of length 1 whose direction is that of the tangent to C in the sense of increasing s, that is, $\boldsymbol{\xi} = d\mathbf{R}/ds$, we have in $\mathbf{F} \cdot \boldsymbol{\xi} = f$ the *tangential component* of the force, or *the force acting in the direction of the motion*. According to Equation (41), p. 396, the tangential component $\ddot{\mathbf{R}} \cdot \boldsymbol{\xi}$ of the acceleration is just $dv/dt = d^2s/dt^2$, that is, the acceleration of the particle along the curve. Newton's law then yields the formula

$$(45) \qquad m\ddot{s} = f,$$

that is, *the mass of the particle multiplied with the acceleration of the particle along its path equals the force acting on the particle in the direction of motion*.

In applying this equation to a particle *constrained* to move along C we assume that the constraints make no contribution to f.[1] For a force $\mathbf{F} = (\rho, \sigma)$ we have then by Equation (44), p. 398,

$$(46) \qquad f = \rho \frac{dx}{ds} + \sigma \frac{dy}{ds}$$

---

[1] Actually, the mechanism of constraint has to supply a force that holds the particle on C (in the simple pendulum this is provided by the tension of the string). We assume that this "reaction" force is perpendicular to the curve and thus has no tangential component; this would be the case for frictionless sliding of the particle along a curve.

since the vector ξ has the components $\dfrac{dx}{ds}, \dfrac{dy}{ds}$ (see p. 394). For a known

curve C the direction cosines $\dfrac{dx}{ds}$ and $\dfrac{dy}{ds}$ of the tangent can be considered

as known functions of s. If likewise the force $\mathbf{F} = (\rho, \sigma)$ depends only on the position of the particle, we have in f a known function of s. The motion of the particle along C then has to be determined from the relatively simple differential equation $m\ddot{s} = f(s)$.

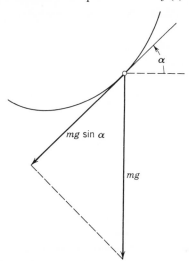

Figure 4.43   Motion on a given curve under gravity.

Specifically, for the gravitational force $\mathbf{F} = (0, -mg)$ we have

$$(46a) \qquad\qquad f = -mg\,\frac{dy}{ds}\,;$$

thus the equation of motion of a particle constrained to move on a curve C under the influence of gravity becomes

$$(47) \qquad\qquad \frac{d^2s}{dt^2} = -g\,\frac{dy}{ds}.$$

If α denotes the inclination angle of the curve, we have $dy/ds = \sin \alpha$ (see Fig. 4.43), and the equation of motion becomes

$$\frac{d^2s}{dt^2} = -g \sin \alpha.$$

For a particle constrained to move on a circle of radius L about the origin ("simple pendulum")

$$x = L \sin \theta, \qquad y = -L \cos \theta,$$

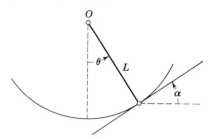

Figure 4.44   The simple pendulum.

where $\theta = s/L$ is the polar angle counted from the downward direction. Here (see Fig. 4.44) $\alpha = \theta$ and thus

$$\frac{d^2s}{dt^2} = -g\,\frac{dy}{d\theta}\frac{d\theta}{ds} = -g \sin \theta$$

or

$$\frac{d^2\theta}{dt^2} = -\frac{g}{L} \sin \theta.$$

4.5   Free Fall of a Body Resisted by Air

We start with two examples of the motion of a particle along a straight line. We consider only cases where the force acts in the direction of the line so that no mechanism of constraint is necessary.

The path of a body falling freely downward can be described parametrically by $x = $ constant, $y = s$. If gravity is the only force acting, we have the equation of motion

$$m\ddot{s} = -mg.$$

For a particle released at the time $t = 0$ from the altitude $y_0 = s_0$ with initial velocity v_0 (counted positive if upward), we find then by integration

$$s = -\tfrac{1}{2}gt^2 + v_0 t + s_0.$$

If we wish to take account of the effect of the *friction* or *air resistance* acting on the particle, we have to consider this as a force whose direction is opposite to the direction of motion and concerning which we must make definite physical assumptions.[1]   We shall work out the results of

---

[1] These assumptions must be chosen to suit the particular physical system under consideration; for example, the law of resistance for low speeds is not the same as that for high ones (such as bullet velocities).

different physical assumptions: (*a*) The resistance is proportional to the velocity, and is given by an expression of the form $-r\dot{s}$, where r is a positive constant; (*b*) the resistance is proportional to the square of the velocity, and is of the form $-r\dot{s}^2$ for positive $\dot{s}$ and $r\dot{s}^2$ for negative $\dot{s}$. In accordance with Newton's law we obtain the equations of motion

(*a*) $$m\ddot{s} = -mg - r\dot{s},$$

(*b*) $$m\ddot{s} = -mg + r\dot{s}^2,$$

where we have assumed in (*b*) that the body is falling ($\dot{s} < 0$). If we first consider $\dot{s} = v(t)$ as the function sought, we have

(*a*) $$m\dot{v} = -mg - rv,$$

(*b*) $$m\dot{v} = -mg + rv^2.$$

Instead of determining v as a function of t by these equations, we determine t as a function of v, writing our differential equations in the form

(*a*) $$\frac{dt}{dv} = -\frac{1}{g(1 + k^2 v)},$$

(*b*) $$\frac{dt}{dv} = -\frac{1}{g(1 - k^2 v^2)},$$

where we have put $\sqrt{r/mg} = k$. With the help of the methods given in Chapter 3 we can immediately carry out the integrations and obtain

(*a*) $$t = -\frac{1}{gk^2}\log(1 + k^2 v) + t_0,$$

(*b*) $$t = \frac{1}{2gk}\log\frac{1 - kv}{1 + kv} + t_0.$$

Solving these equations for v, we have

(*a*) $$v = -\frac{1}{k^2}(1 - e^{-gk^2(t-t_0)}).$$

(*b*) $$v = -\frac{1}{k}\frac{1 - e^{-2gk(t-t_0)}}{1 + e^{-2gk(t-t_0)}} = -\frac{1}{k}\tanh[gk(t - t_0)].$$

These equations at once reveal an important property of the motion. The velocity does not increase with time beyond all bounds, but tends to a definite limit depending on the mass m and the constant r (which, in turn, depends on the shape of the falling body and the air density).

For

(a)
$$\lim_{t \to \infty} v(t) = -\frac{1}{k^2} = -\frac{mg}{r},$$

(b)
$$\lim_{t \to \infty} v(t) = -\frac{1}{k} = -\sqrt{\frac{mg}{r}}.$$

For the limiting velocities frictional resistance just balances gravitational attraction. A second integration performed on our expressions for $v(t) = \dot{s}$, with the help of the methods of Chapter 3, gives the results (which may be verified by differentiation)

(a)
$$s(t) = -\frac{1}{k^2}(t - t_0) - \frac{1}{gk^4}(e^{-gk^2(t-t_0)} - 1) + c$$

(b)
$$s(t) = -\frac{1}{gk^2} \log \left[\cosh gk(t - t_0) \right] + c,$$

where c is a constant of integration. Here t_0 is the time at which the particle would have had velocity 0 and c its altitude at the time t_0. The two constants c and t_0 can also be related easily to the velocity and position at any other time t_1, if we consider those quantities as initial conditions.

4.6   The Simplest Type of Elastic Vibration—Motion of a Spring

As a second example—of major significance—we consider the motion of a particle which moves along the x-axis and is pulled back toward the origin by an elastic force. As regards the elastic force we assume that it is always directed toward the origin and that its magnitude is proportional to the distance from the origin. In other words, we take the force as equal to $-kx$, where the coefficient k is a measure of the stiffness of the elastic connection. Since k is assumed positive, the force is negative when x is positive and positive when x is negative. Newton's law now tells us that

(48)
$$m\ddot{x} = -kx.$$

This differential equation by itself does not determine the motion completely, but for a given instant of time, say $t = 0$, we can arbitrarily assign the initial position $x(0) = x_0$ and the initial velocity $\dot{x}(0) = v_0$; that is, in physical language, that we can start off the particle from an arbitrary position with an arbitrary velocity; thereafter the motion is determined by the differential equation. Mathematically, this is expressed by the fact that the general solution of our differential equation contains two constants of integration, at first undetermined, whose values we find by means of the initial conditions. This fact we shall prove immediately.

We can easily state such a solution directly. If we put $\omega = \sqrt{k/m}$, our differential equation becomes $d^2x/dt^2 = -\omega^2 x$. The substitution $\tau = \omega t$ for the independent variable reduces this equation to the form $d^2x/d\tau^2 = -x$, discussed in Chapter 3, p. 312. Thus our differential equation is satisfied by all the functions

$$x(t) = c_1 \cos \omega t + c_2 \sin \omega t,$$

which may also be verified at once by differentiation (where c_1 and c_2 denote constants chosen arbitrarily). In Chapter 3, p. 313, we saw that there are no other solutions of our differential equation and hence that every such motion under the influence of an elastic force is given by this expression. This can easily be put in the form

$$x(t) = a \sin \omega(t - \delta) = -a \sin \omega \delta \cos \omega t + a \cos \omega \delta \sin \omega t;$$

we need only write $-a \sin \omega \delta = c_1$ and $a \cos \omega \delta = c_2$, thus introducing instead of c_1 and c_2 the new constants a and δ. Motions of this type are said to be *sinusoidal* or *simple harmonic*. They are periodic; any state [that is, position $x(t)$ and velocity $\dot{x}(t)$] is repeated after the time $T = 2\pi/\omega$, which is called the *period*, since the functions $\sin \omega t$ and $\cos \omega t$ have the period T. The number a is called the *maximum displacement* or *amplitude* of the oscillation. The number $1/T = \omega/2\pi$ is called the *frequency* of the oscillation; it measures the number of oscillations per unit time. We shall return to the theory of oscillations in Chapter 8.

*4.7 Motion on a Given Curve

a. The Differential Equation and Its Solution

We now turn to the general form of the problem of motion along a given curve under an arbitrary preassigned force $mf(s)$. We shall determine the function $s(t)$ as a function of t by means of the differential equation [Eq. (45), p. 400]

$$\ddot{s} = f(s),$$

where $f(s)$ is a given function.[1] This differential equation in s can be solved completely by the following device.

---

[1] Our original equation of motion along a curve was $m\ddot{s} = f(s)$; we can, however, always write the function $f(s)$ in the form $mf(s)$, obtaining the simpler form of the equation used here.

We consider any primitive function $F(s)$ of $f(s)$, so that $F'(s) = f(s)$, and multiply both sides of the equation $\ddot{s} = f(s) = F'(s)$ by $\dot{s}$. We can then write the left-hand side in the form $d(\dot{s}^2/2)/dt$, as we see at once by differentiating the expression $\dot{s}^2$; the right-hand side $F'(s)\dot{s}$, however, by the chain rule of differentiation is the derivative of $F(s)$ with respect to the time t, if in $F(s)$ we regard the quantity s as a function of t. Hence we immediately have

$$\frac{d}{dt}\left(\frac{1}{2}\,\dot{s}^2\right) = \frac{d}{dt}\,F(s),$$

or by integration

$$\tfrac{1}{2}\dot{s}^2 = F(s) + c,$$

where c denotes a constant yet to be determined.

We have now arrived at an equation which only involves the function $s(t)$ and its *first* derivative. (Later on we shall interpret this equation as expressing the conservation of energy during the motion.) Let us write this equation in the form $ds/dt = \sqrt{2[F(s) + c]}$. We see that from this we cannot immediately find s as a function of t by integration. However, we arrive at a solution of the problem if we at first content ourselves with finding the inverse function $t(s)$, that is, the time taken by the particle to reach a definite position s. For $t(s)$ we have the equation

$$\frac{dt}{ds} = \frac{1}{\sqrt{2[F(s) + c]}};$$

thus the derivative of the function $t(s)$ is known, and we have

$$t = \int \frac{ds}{\sqrt{2[F(s) + c]}} + c_1,$$

where c_1 is another constant of integration. As soon as we have performed this last integration we have solved the problem, for although we have not determined the position s as a function of t, we have inversely found the time t as a function of the position s. The fact that the two constants of integration c and c_1 are still available enables us to make the general solution fit special initial conditions.

The general discussion can be illustrated by our earlier example of elastic vibrations if we identify x with s; here $f(s) = -\omega^2 s$ and correspondingly, say, $F(s) = -\tfrac{1}{2}\omega^2 s^2$. We therefore obtain

$$\frac{dt}{ds} = \frac{1}{\sqrt{2c - \omega^2 s^2}},$$

and furthermore,

$$t = \int \frac{ds}{\sqrt{2c - \omega^2 s^2}} + c_1.$$

This integral, however, can easily be evaluated by introducing $\omega s / \sqrt{2c}$ as a new variable: we thus obtain

$$t = \frac{1}{\omega} \arcsin \frac{\omega s}{\sqrt{2c}} + c_1,$$

or, forming the inverse function,

$$s = \frac{\sqrt{2c}}{\omega} \sin \omega(t - c_1).$$

We are thus led to exactly the same formula for the solution as before.

From this example we also see what the constants of integration mean and how they are to be determined. If, for example, we require that at the time $t = 0$ the particle shall be at the point $s = 0$ and at that instant shall have the velocity $\dot{s}(0) = 1$, we obtain the two equations

$$0 = \frac{\sqrt{2c}}{\omega} \sin \omega c_1, \qquad 1 = \sqrt{2c} \cos \omega c_1,$$

from which we find that the constants have the values $c_1 = 0$, $c = \frac{1}{2}$. The constants of integration c and c_1 can be determined in exactly the same way when the initial position s_0 and the initial velocity $\dot{s}_0$ (at time $t = 0$) are prescribed arbitrarily.

b. Particle Sliding down a Curve

The case of a particle sliding down a frictionless curve under the influence of gravity can be treated very simply by the method just described. We found already on p. 401 the equation of motion corresponding to this case:

$$\ddot{s} = -g \frac{dy}{ds},$$

where dots indicate differentiation with respect to the time t. The right-hand side of this equation is a known function of s, since we know the curve and we can therefore regard the quantities x and y as known functions of s.

As in the last section, we multiply both sides of this equation by $\dot{s}$. The left-hand side then becomes the derivative of $\frac{1}{2}\dot{s}^2$ with respect to t. If in the function $y(s)$ we regard s as a function of t, the right-hand side of our equation is the derivative of $-gy$ with respect to t. On integrating, we therefore have

$$\tfrac{1}{2}\dot{s}^2 = -gy + c,$$

where c is a constant of integration. To find the interpretation of this constant, we suppose that at the time $t = 0$ our particle is at the point of the curve for which the coordinates are x_0 and y_0 and that at this instant its velocity is zero, that is, $\dot{s}(0) = 0$. Then putting $t = 0$ we immediately have $-gy_0 + c = 0$, so that

$$\tfrac{1}{2}\dot{s}^2 = g(y_0 - y).$$

Since $\dot{s}^2$ could never be negative, we see that the altitude y of the particle never exceeds the value y_0, and only reaches it at those instants when the velocity of the particle is zero. The velocity is larger as the particle is lower. Now instead of regarding s as a function of t we shall consider the inverse function $t(s)$. For this we at once obtain

$$\frac{dt}{ds} = \pm \frac{1}{\sqrt{2g(y_0 - y)}},$$

which is equivalent to

$$t = c_1 \pm \int \frac{ds}{\sqrt{2g(y_0 - y)}},$$

where c_1 is a new constant of integration. As regards the sign of the square root, which is the same as the sign of $\dot{s}$, we notice that if the particle moves along an arc which is lower than y_0 everywhere except at the ends, the sign cannot change. For the sign of $\dot{s}$ can change only where $\dot{s} = 0$, that is, where $y - y_0 = 0$. Thus the particle can only "turn back" at points of maximum elevation y_0 on the curve. Instead of the arc length s the curve can also be referred to any parameter θ, so that $x = \phi(\theta)$, $y = \psi(\theta)$. Introducing θ as independent variable, we obtain

$$t = c_1 \pm \int \frac{ds}{d\theta} \frac{d\theta}{\sqrt{2g(y_0 - y)}} = c_1 \pm \int \sqrt{\frac{x'^2 + y'^2}{2g(y_0 - y)}}\, d\theta,$$

where the functions $x' = \phi'(\theta)$, $y' = \psi'(\theta)$, and $y = \psi(\theta)$ are known. In order to determine the constant of integration c_1 we note that for $t = 0$ the parameter θ will have a value θ_0. This immediately gives us our solution in the form

$$(49) \qquad\qquad t = \pm \int_{\theta_0}^{\theta} \sqrt{\frac{x'^2 + y'^2}{2g(y_0 - y)}}\, d\theta.$$

We see that this equation represents the time taken by the particle to move from the parameter value θ_0 to the parameter value θ. The inverse function $\theta(t)$ of this function $t(\theta)$ enables us to describe the motion completely; for at each instant t we can determine the point $x = \phi[\theta(t)]$, $y = \psi[\theta(t)]$ which the particle is then passing.

c. Discussion of the Motion

From the equations just found, even without an explicit expression for the result of the integration we can deduce the general nature of the motion by simple intuitive reasoning. We suppose that our curve is of

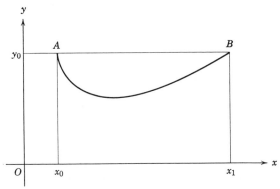

Figure 4.45

the type shown in Fig. 4.45, that is, that it consists of an arc convex downward; we take s as increasing from left to right. If we initially release the particle at the point A with coordinates $x_0 = \phi(\theta_0)$, $y_0 = \psi(\theta_0)$, corresponding to $\theta = \theta_0$, the velocity increases, for the acceleration $\ddot{s}$ is positive. The particle travels from A to the lowest point with ever-increasing velocity. After the lowest point is passed, however, the acceleration is negative, since the right-hand side $-g\,dy/ds$ of the equation of motion is negative. The velocity therefore decreases. From the equation $\dot{s}^2 = 2g(y_0 - y)$ we see at once that the velocity reaches the value zero when the particle reaches the point B whose height is the same as that of the initial position A. Since the acceleration is still negative, the motion of the particle must be reversed at this point, so that the particle will swing back to the point A; this action will repeat itself indefinitely. (The reader will recall that friction has been disregarded.) In this oscillatory motion the time which the point takes to return from B to A must clearly be the same as the time taken to move from A to B, since at equal heights we have equal values of $|\dot{s}|$. If we denote the time required for a complete journey from A to B and back again by T, the motion will obviously be periodic with period T. If θ_0 and θ_1 are the values of the parameter corresponding to the points A and B, respectively, the half-period is given by the expression

(50)
$$
\begin{aligned}
\frac{T}{2} &= \frac{1}{\sqrt{2g}} \left| \int_{\theta_0}^{\theta_1} \sqrt{\frac{x'^2 + y'^2}{y_0 - y}} \, d\theta \right| \\
&= \frac{1}{\sqrt{2g}} \left| \int_{\theta_0}^{\theta_1} \sqrt{\frac{\phi'^2(\theta) + \psi'^2(\theta)}{\psi(\theta_0) - \psi(\theta)}} \, d\theta \right|.
\end{aligned}
$$

If θ_2 is the value of the parameter corresponding to the lowest point of the curve, the time which the particle takes to fall from A to this lowest point is

$$
\frac{1}{\sqrt{2g}} \left| \int_{\theta_0}^{\theta_2} \sqrt{\frac{x'^2 + y'^2}{y_0 - y}} \, d\theta \right|.
$$

d. The Ordinary Pendulum

The simplest example is given by the so-called simple pendulum. Here the curve under consideration is a circle of fixed radius L:

$$
x = L \sin \theta, \qquad y = -L \cos \theta,
$$

where the angle θ is measured in the positive sense from the position of rest. From the general expression (50) we at once obtain using the addition theorem for the cosine,

$$
T = \sqrt{\frac{2L}{g}} \int_{-\theta_0}^{\theta_0} \frac{d\theta}{\sqrt{\cos \theta - \cos \theta_0}} = \sqrt{\frac{L}{g}} \int_{-\theta_0}^{\theta_0} \frac{d\theta}{\sqrt{\sin^2 \dfrac{\theta_0}{2} - \sin^2 \dfrac{\theta}{2}}},
$$

where θ_0 $(0 < \theta_0 < \pi)$ denotes the amplitude of oscillation of the pendulum, that is, the angular position from which the particle is released at time $t = 0$ with velocity zero.[1] By the substitution

$$
u = \frac{\sin (\theta/2)}{\sin (\theta_0/2)}, \qquad \frac{du}{d\theta} = \frac{\cos (\theta/2)}{2 \sin (\theta_0/2)}
$$

our expression for the period of oscillation of the pendulum becomes

$$
T = 2 \sqrt{\frac{L}{g}} \int_{-1}^{1} \frac{du}{\sqrt{(1 - u^2)\left(1 - u^2 \sin^2 \left(\dfrac{\theta_0}{2}\right)\right)}}.
$$

---

[1] We have assumed here that the velocity does become equal to zero at some time during the motion. This excludes the type of tumbling motion of the pendulum in which θ is not periodic and varies monotonically for all t.

We have therefore expressed the period of oscillation of the pendulum by an *elliptic integral* (see p. 299).

If we assume that the amplitude of the oscillation is small, so that we may with sufficient accuracy replace the second factor under the square root sign by 1, we obtain the expression

$$2\sqrt{\frac{L}{g}}\int_{-1}^{1}\frac{du}{\sqrt{1-u^2}}$$

as an approximation for the period of oscillation. We can evaluate this last integral by formula 13 in our table of integrals (p. 263) and obtain the expression $2\pi\sqrt{L/g}$ as an approximate value for T. To this order of approximation the period is independent of θ_0, that is, of the amplitude of the oscillation of the pendulum. Clearly, the exact period is larger and increases with θ_0. Since in the interval of integration

$$1 \geq 1 - u^2\sin^2\frac{\theta_0}{2} \geq 1 - \sin^2\frac{\theta_0}{2} = \cos^2\frac{\theta_0}{2},$$

we find for the period the estimates

$$2\pi\sqrt{\frac{L}{g}} \leq T \leq \frac{1}{\cos(\theta_0/2)}\,2\pi\sqrt{\frac{L}{g}}.$$

For angles $\theta_0 < 10°$ we have $1/(\cos\theta_0/2) \leq \sec 5° < 1.004$, so that the period will be given by the formula $2\pi\sqrt{L/g}$ with a relative error of less than $\frac{1}{2}\%$. For finer approximation of the elliptic integral for T see Section 7.6f.

e. The Cycloidal Pendulum

The fact that the period of oscillation of the ordinary pendulum is not strictly independent of the amplitude of oscillation caused Christian Huygens, in his prolonged efforts to construct accurate clocks, to seek a curve C for which the period of oscillation is independent of the position on C at which the oscillating particle begins its motion.[1] Huygens recognized that the cycloid is such a curve.

In order that a particle may actually be able to oscillate on a cycloid the cusps of the cycloid must point in the direction opposite to that of the force of gravity; that is, we must rotate the cycloid considered previously (p. 328) about the x-axis (cf. Fig. 4.2, p. 329). We therefore

---

[1] The oscillations are then said to be *isochronous*.

write the equations of the cycloid in the form

$$x = a(\theta + \pi + \sin \theta),$$

$$y = -a(1 + \cos \theta),$$

which also involves a change of the parameter t into $\theta + \pi$ (Fig. 4.46).

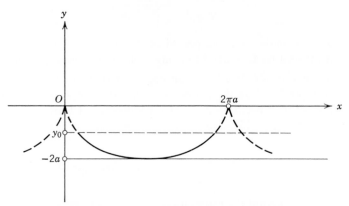

Figure 4.46 Path described by a cycloidal pendulum.

The time which the particle takes to travel from a point at the height

$$y_0 = -a(1 + \cos \theta_0) \qquad (0 < \theta_0 < \pi)$$

down to the lowest point, and up again to the height y_0, by formula (50) of p. 410, is

$$\frac{T}{2} = \sqrt{\frac{1}{2g}} \int_{-\theta_0}^{\theta_0} \sqrt{\frac{x'^2 + y'^2}{y_0 - y}} \, d\theta = \sqrt{\frac{2a}{g}} \int_{-\theta_0}^{\theta_0} \frac{\cos(\theta/2)}{\sqrt{\cos \theta - \cos \theta_0}} \, d\theta.$$

Using exactly the same substitutions as for the period of the simple pendulum, we arrive at the integral

$$\frac{T}{2} = 2\sqrt{\frac{a}{g}} \int_{-1}^{1} \frac{du}{\sqrt{1 - u^2}}$$

and we therefore obtain

$$T = 4\pi \sqrt{\frac{a}{g}}.$$

The period of oscillation T, therefore, is indeed independent of the amplitude θ_0. A simple way of actually constraining a particle by a string to move on a cycloid will be described on p. 428.

*4.8   Motion in a Gravitational Field

As an example of unconstrained motion we consider a particle moving in the gravitational field of an attracting mass.

a. Newton's Universal Law of Gravitation

Kepler's description of the motion of the planets, which was based on the precise observations of Tycho Brahe, led Newton to formulate his general law for the gravitational attraction between any two particles. Let $P_0 = (x_0, y_0)$ and $P = (x, y)$ be two particles of masses m_0 and m, respectively. Let $r = \sqrt{(x - x_0)^2 + (y - y_0)^2}$ be the distance between the particles. Then P_0 exerts on P a force $\mathbf{F}$ which has the direction of $\overrightarrow{PP_0}$ and the magnitude $|\mathbf{F}| = \gamma m_0 m / r^2$, where γ is the "*universal gravitational constant.*" Since $\mathbf{F}$ can then only differ by a positive factor from the vector $\overrightarrow{PP_0}$, which itself has magnitude r, we must have

$$\mathbf{F} = \frac{\gamma m_0 m}{r^3} \overrightarrow{PP_0} = \left(\frac{\gamma m_0 m (x_0 - x)}{r^3}, \frac{\gamma m_0 m (y_0 - y)}{r^3} \right).$$

This law of attraction refers to *particles*, that is, to bodies that can be considered to be concentrated in points, neglecting the actual extent of the bodies (Fig. 4.47). The validity of such an assumption is plausible enough for celestial bodies whose mutual distances are tremendous when compared with their diameters. Newton vastly increased the range of application of this law by showing that the same law of attraction also describes the attraction of a body of mass m_0 of considerable extent on a particle of mass m, provided that the body is a sphere of constant density, or, more generally, provided that the body is made up of concentric spherical shells of constant density; in that case the attraction of the body on a particle P located outside the body is the same as if the total mass m_0 of the body were located at its center P_0 (Fig. 4.47). The earth can with fair accuracy be thought of as made up of concentric shells of constant density, so that the attraction of the earth on a particle of mass m on its surface is directed toward the center P_0 of the earth (that is, vertically downward for an observer) and has magnitude $\gamma m_0 m / R^2$, where R is the radius of the earth and m_0 its mass. We can identify then $\gamma m_0 m / R^2$ with mg, where g is the gravitational acceleration (see p. 398). In other words, we have $g = \gamma m_0 / R^2$.

From Newton's fundamental law we find for a particle P of mass m moving under the influence of the attraction of a mass m_0 located at P_0

the equations of motion

$$\ddot{x} = \frac{\gamma m_0(x_0 - x)}{r^3}, \qquad \ddot{y} = \frac{\gamma m_0(y_0 - y)}{r^3}.$$

We now make the further simplifying assumption that m_0 is so much larger than m that the effects of the attraction of P on P_0 can be neglected

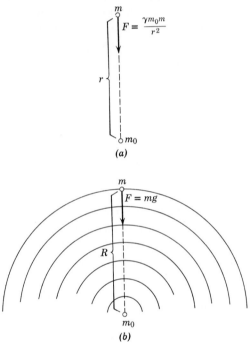

(a)

(b)

Figure 4.47 (*a*) Newtonian attraction of two particles. (*b*) Gravitational attraction of the earth.

and P_0 can be considered at rest. This would, for example, be the situation for a pair of bodies like the sun and a planet or the earth and a body on its surface. Taking the origin of coordinates at P_0 we then have for $P = (x, y)$ the equations of motion

$$(51) \qquad \ddot{x} = -\frac{\gamma m_0 x}{r^3}, \qquad \ddot{y} = -\frac{\gamma m_0 y}{r^3}$$

with $r = \sqrt{x^2 + y^2}$.

b. Circular Motion about the Center of Attraction

We shall not attempt to find the most general solution of these differential equations (which, as is well known, would correspond to motion along a path of the form of a conic section, with one focus at the attracting center). Instead, we shall just consider the simplest types of motion consistent with these equations, namely, uniform circular motions about the origin and motions along a radius from the origin. For uniform circular motion of P along a circle of radius a about the origin we have $r = a$ and

$$x = a \cos \omega t, \qquad y = a \sin \omega t,$$

where ω is a constant. The *period* T of the motion, that is, the time after which P returns to the same position, is $T = 2\pi/\omega$. We find for the velocity components

$$\dot{x} = -a\omega \sin \omega t, \qquad \dot{y} = a\omega \cos \omega t$$

so that the speed of P in its orbit is

$$(52) \qquad v = \sqrt{\dot{x}^2 + \dot{y}^2} = a\omega = \frac{2\pi a}{T}.$$

The acceleration of P has the components

$$\ddot{x} = -a\omega^2 \cos \omega t = -\omega^2 x, \qquad \ddot{y} = -a\omega^2 \sin \omega t = -\omega^2 y.$$

Clearly, the equations of motion (51) are then satisfied if

$$\omega^2 = \frac{\gamma m_0}{a^3}$$

or

$$(53) \qquad a^3 = \frac{\gamma m_0}{\omega^2} = \frac{\gamma m_0}{4\pi^2} T^2.$$

This is just *Kepler's third law* for the special case of circular motion, according to which *the cubes of the distances of the planets from the sun are proportional to the squares of their periods.*

We can give some simple illustrations of Kepler's law for the case where the attracting body is the earth with its mass m_0 and radius R. Observing that here $\gamma m_0 = g R^2$ we have

$$a^3 = \frac{gR^2}{4\pi^2} T^2.$$

For a satellite circling the earth at tree-top level (neglecting, of course, air resistance) we have $a = R \sim 3963$ miles. We find then from our formula for the period of the satellite the value

$$T = 2\pi \sqrt{\frac{R}{g}} \sim 1.4 \text{ hours}$$

and for its velocity in its orbit

$$(54) \qquad v = \frac{2\pi R}{T} = \sqrt{Rg} \sim 27,000 \text{ feet per second.}$$

We can compare the value of T for the satellite circling the earth with the period of 27.32 days of the moon, that is, the time after which the moon returns to the same position among the stars ("sidereal month"). By Kepler's law the ratio of the distance a of the moon to the radius R of the earth should be given by the $\frac{2}{3}$-power of the ratio of the periods. This leads for the distance of the moon from the center of the earth to the value

$$a = \left(\frac{27.32 \times 24}{1.4}\right)^{\frac{2}{3}} R \sim 60R \sim 240,000 \text{ miles,}$$

which agrees well with the actual average value of the distance.

c. Radial Motion—Escape Velocity

The second type of motion we shall consider is that of a particle moving from the center of attraction along a ray, say the x-axis. Here $y = 0$, $x = r$, so that the equations of motion reduce to

$$\ddot{x} = -\frac{\gamma m_0}{x^2}.$$

Following our general procedure for equations of the type $\ddot{s} = f(s)$, we multiply both sides of this equation with $\dot{x}$ and obtain

$$\dot{x}\ddot{x} = -\gamma m_0 \frac{\dot{x}}{x^2}$$

or

$$\frac{d}{dt}\left(\frac{1}{2}\dot{x}^2\right) = \frac{d}{dt}\left(\frac{\gamma m_0}{x}\right).$$

Thus the expression

$$\frac{1}{2}\dot{x}^2 - \frac{\gamma m_0}{x}$$

has a constant value h during the motion. (Later on we shall recognize this fact as an instance of the law of conservation of energy.) If we introduce x instead of t as independent variable, we have then

$$\frac{dt}{dx} = \frac{1}{\dot{x}} = \pm \frac{1}{\sqrt{2h + (2\gamma m_0/x)}},$$

which by integration leads to

$$t = t_0 \pm \int_{x_0}^{x} \frac{d\xi}{\sqrt{2h + (2\gamma m_0/\xi)}}.$$

We shall not bother to carry out the integration which can be performed easily with the help of the methods developed in Chapter 3. For a particle released at the time $t_0 = 0$ at the distance x_0 with initial velocity zero we have $h = -\gamma m_0/x_0$. The time required for such a particle to fall into the attracting particle ($x = 0$) is then

$$t = \int_{0}^{x_0} \frac{d\xi}{\sqrt{2\gamma m_0(1/\xi - 1/x_0)}} = \frac{\pi}{2} \sqrt{\frac{x_0^3}{2\gamma m_0}}.$$

By Kepler's law this is $\sqrt{\frac{1}{32}}$ times the time it would take the particle to circle the center of attraction at the distance x_0 [see Eq. (53), p. 415].

The relation

$$\frac{1}{2}\dot{x}^2 - \frac{\gamma m_0}{x} = h$$

has an interesting consequence when we investigate the circumstances under which a particle can escape to infinity. Since $\frac{1}{2}\dot{x}^2 \geq 0$ we find for $x \to \infty$ that the constant h must be nonnegative, and hence that $\frac{1}{2}\dot{x}^2 - \gamma m_0/x \geq 0$ during the whole motion. In particular, a particle starting at the distance $x = a$ with velocity v can escape to infinity only if $\frac{1}{2}v^2 - \gamma m_0/a \geq 0$. The lowest possible value of the velocity v which will permit a particle to escape to infinity is then $v = \sqrt{2\gamma m_0/a}$. This is the *escape velocity* v_e. For a particle starting at the surface of the earth and escaping to infinity, that is, escaping its gravitational pull, we have $a = R$, $\gamma m_0 = gR^2$, so that

$$v_e = \sqrt{2gR} \sim 37{,}000 \text{ feet per second.}$$

Hence [cf. (54), p. 416] the escape velocity is just $\sqrt{2}$ times the velocity needed to maintain a satellite in a circular orbit near the earth. A meteor falling from infinity onto the earth also would have velocity v_e on impact, if we neglect air resistance and motion of the earth in its orbit.

4.9   Work and Energy

a. Work Done by Forces during a Motion

The *concept of work* throws new light on the considerations of the last section and on many other questions of mechanics and physics.

Let us again think of the particle as moving on a curve under the influence of a force acting along the curve, and let us suppose that its position is specified by the length of arc measured from any fixed initial point. The force acting in the direction of motion itself will then, as a rule, be a function of s. This function will have positive values where the direction of the force is the same as the direction of increasing values of s and negative values where the direction of the force is opposite to that of increasing values of s.

If the magnitude of the force is constant along the path, we mean by the *work done by the force* the product of the force by the distance $(s_1 - s_0)$ traversed, where s_1 denotes the final point and s_0 the initial point of the motion. If the force is not constant, we define the work by means of a limiting process. We subdivide the interval from s_0 to s_1 into n equal or unequal subintervals and notice that if the subintervals are small, the force in each one is nearly constant; if σ_ν is a point chosen arbitrarily in the νth subinterval, then throughout this subinterval the force will be approximately $f(\sigma_\nu)$. If the force throughout the νth subinterval were exactly $f(\sigma_\nu)$, the work done by our force would be exactly

$$\sum_{\nu=1}^{n} f(\sigma_\nu)\,\Delta s_\nu,$$

where Δs_ν as usual denotes the length of the νth subinterval. If we now pass to the limit, letting n increase beyond all bounds while the length of the longest subinterval tends to zero, then by the definition of an integral our sum will tend to

$$W = \int_{s_0}^{s_1} f(s)\,ds,$$

which we naturally call the work done by the force.

If the direction of the force and that of the motion are the same, the work done by the force is positive; we then say that *the force does work*. On the other hand, if the direction of the force and that of the motion are opposed, the work done by the force is negative; we then say that *work is done against the force*.[1]

If we regard the coordinate of position s as a function of the time t, so that the force $f(s) = p$ is also a function of t, then in a plane with rectangular coordinates s and p we can plot the point with coordinates $s = s(t)$, $p = p(t)$ as a function of the time. This point will describe a curve, which may be called the *work diagram* of the motion. If we are dealing with a periodic motion, as in any machine, then after a certain time T (one period) the moving point $(s(t), p(t))$ must return to the same point; that is, the work diagram will be a closed curve. In this case the curve may consist simply of one and the same arc, traversed first forward and then backward; this happens, for instance, in elastic oscillations. However, it is also possible for the curve to be a more general closed curve, enclosing an area; this is the case, for example, with machines in which the pressure on a piston is not the same during the forward stroke as during the backward stroke. The work done in one cycle, that is, in time T, will then be given simply by the negative of the area of the work diagram or, in other words, by the integral

$$\int_{t_0}^{t_0+T} p(t) \frac{ds}{dt}\, dt,$$

where the interval of time from t_0 to $t_0 + T$ represents exactly one period of the motion. If the boundary of the area is positively traversed, the work done is negative, if negatively traversed, the work done is positive. If the curve consists of several loops, some traversed positively and some traversed negatively, the work done is given by the sum of the areas of loops, each with its sign changed.

These considerations are illustrated in practice by the *indicator diagram* of an old-fashioned steam engine. By a suitably designed mechanical device a pencil is made to move over a sheet of paper; the horizontal motion of the pencil relative to the paper is proportional to the distance s of the piston from its extreme position, whereas the vertical motion is proportional to the steam pressure, and hence proportional to the total force p of the steam on the piston. The piston therefore describes the work diagram for the engine on a known scale. The area of this diagram is measured (usually by means of a planimeter), and the work done by the steam on the piston is thus found.

---

[1] Note that here we must carefully characterize the force of which we are speaking. For example, in lifting a weight the work done by the force of gravity is negative: Work is done against gravity. But from the point of view of the person doing the lifting the work done is positive, for the person must exert a force opposed to gravity.

Here we also see that our convention for the sign of an area, as discussed on p. 365 is definitely of practical interest. For it sometimes happens when an engine is running light, that the highly expanded steam at the end of the stroke has a pressure lower than that required to expel it on the return stroke; on the diagram this is shown by a positively traversed loop; the engine itself is drawing energy from the flywheel instead of furnishing energy.

b. Work and Kinetic Energy.   Conservation of Energy

The law of motion

$$m\ddot{s} = f$$

leads to a fundamental relation between the changes in velocity during the motion of a particle along a curve and the work done by the force f in the direction of motion. We apply the same device used already several times in the preceding examples and multiply both sides of the equation of motion by $\dot{s}$:

$$m\ddot{s}\dot{s} = f(s)\dot{s}.$$

Now $m\ddot{s}\dot{s} = (d/dt)\tfrac{1}{2}m\dot{s}^2 = (d/dt)\tfrac{1}{2}mv^2$, where $v(t) = \dot{s}$ is the velocity of the particle. Integrating both sides of the equation with respect to t between the limits t_0 and t_1, we find

$$\frac{1}{2}mv^2(t_1) - \frac{1}{2}mv^2(t_0) = \int_{t_0}^{t_1} f(s)\frac{ds}{dt}\,dt$$

$$= \int_{s_0}^{s_1} f(s)\,ds = W.$$

The quantity $\tfrac{1}{2}mv^2$ is called the *kinetic energy* K of the particle. Hence: *The change in kinetic energy of a particle during the motion equals the work done by the force acting on the particle in the direction of motion.*

The quantity f represented the force acting in the direction of motion or the tangential component of force. For a force $\mathbf{F} = (\rho, \sigma)$ the force in the direction of motion is

$$f = \mathbf{F} \cdot \frac{d\mathbf{R}}{ds} = \rho\,\frac{dx}{ds} + \sigma\,\frac{dy}{ds}.$$

If ρ and σ are known functions of x and y and if the particle is known to move along a curve $x = x(s)$, $y = y(s)$, then f also becomes a known function of s. Hence in order to compute the work

$$(55) \qquad\qquad W = \int_{s_0}^{s_1} f(s)\,ds$$

as the particle moves from one position (x_0, y_0) to another (x_1, y_1), we have to know in general the path along which the particle moves.

In an important class of cases the work W depends only on initial and final position and can be expressed in the form

$$(56) \qquad W = V(x_0, y_0) - V(x_1, y_1)$$

with a suitable function $V(x, y)$ the *potential energy*. The formula expressing that the change in kinetic energy equals the work done by the force then can also be written in the form

$$(57) \qquad \tfrac{1}{2}mv^2(t_1) + V(x_1, y_1) = \tfrac{1}{2}mv^2(t_0) + V(x_0, y_0).$$

Thus the quantity $K + V$, the sum of kinetic and potential mechanical energy, that is, the total energy, does not change during the motion. This is an instance of the general physical law of *conservation of energy*.

A potential energy function V can easily be constructed in some of the motions discussed earlier. Thus for a particle subject to gravity we have $\mathbf{F} = (0, -mg)$ and $f = -mg(dy/ds)$. The work done by the force of gravity as the particle moves from a position (x_0, y_0) to a position (x_1, y_1) is then

$$W = \int_{s_0}^{s_1} -mg \frac{dy}{ds}\, ds = \int_{y_0}^{y_1} -mg\, dy = mgy_0 - mgy_1.$$

We see that W is proportional to the change in altitude between initial and end position. For the potential energy function V we can choose $V = mgy$ (or more generally $V = mgy + c$, where c is any constant). The law of conservation of energy then states that the quantity

$$\tfrac{1}{2}v^2 + gy$$

is constant during the motion. We had noticed this fact already in investigating the motion of a particle sliding down a curve (p. 408).

c. The Mutual Attraction of Two Masses

Another example of a force with which we can associate a potential energy function V is furnished by the gravitational attraction $\mathbf{F}$ exerted by a particle $P_0 = (x_0, y_0)$ of mass m_0 on a particle $P = (x, y)$ of mass m. Here

$$\mathbf{F} = \left[\frac{-\mu(x - x_0)}{r^3}, \frac{-\mu(y - y_0)}{r^3} \right],$$

where $\mu = \gamma m_0 m$ and $r = \sqrt{(x - x_0)^2 + (y - y_0)^2}$. (According to Coulomb's law the same type of formula gives the interaction of two electric charges.)

The force in the direction of motion is then

$$f = -\frac{\mu}{r^3}\left[(x - x_0)\frac{dx}{ds} + (y - y_0)\frac{dy}{ds}\right] = -\frac{\mu}{r^2}\frac{dr}{ds} = \frac{d}{ds}\frac{\mu}{r}$$

since

$$(x - x_0)\frac{dx}{ds} + (y - y_0)\frac{dy}{ds} = \frac{1}{2}\frac{d}{ds}[(x - x_0)^2 + (y - y_0)^2]$$

$$= \frac{1}{2}\frac{dr^2}{ds} = r\frac{dr}{ds}.$$

The work done by the force of attraction when the particle P moves from a position (x_1, y_1) to the position (x_2, y_2) is then

$$W = \int_{s_1}^{s_2}\left(\frac{d}{ds}\frac{\mu}{r}\right)ds = \frac{\mu}{r_2} - \frac{\mu}{r_1} = V(x_1, y_1) - V(x_2, y_2),$$

where $V(x, y) = -\mu/r = -\mu/\sqrt{(x - x_0)^2 + (y - y_0)^2}$ is the potential energy.

If we move the particle from the position (x_1, y_1) to infinity (corresponding to $r_2 = \infty$), the work done by the force of attraction is $-\mu/r_1$. The work done by an opposing force that moves the particle to infinity has the same numerical value but the opposite sign. Hence $\mu/r_1 = -V(x_1, y_1)$ is the work that has to be done *against* the force of attraction in order to move the particle to infinity from the position (x_1, y_1). This important expression is called the *mutual potential* of the two particles. Therefore here the potential is defined as the work required to separate the two attracting masses completely, for example, the work required in order to tear an electron completely away from its atom (ionization potential).

If the attracting mass P_0 is considered as fixed, the law of conservation of energy implies that the attracted particle P moves in such a way that the expression

$$\frac{1}{2}v^2 - \frac{\gamma m_0}{r} = h$$

(the total energy per unit mass m) has a constant value during the motion. We had derived this fact already for the special case of purely radial motion; we see now that it holds for any type of motion under the influence of gravitational attraction. We can conclude again that $h \geq 0$ for a particle escaping to infinity; its orbit is then unbounded (parabola or hyperbola) instead of bounded (ellipse). The escape velocity

$$v_e = \sqrt{\frac{2\gamma m_0}{r}},$$

which corresponds to $h = 0$, is the least velocity which enables the particle to escape to infinity from a given distance r. It does not depend on the *direction* in which the particle is released but only on the distance r from the attracting center.

d. The Stretching of a Spring

As a third example we consider the work done in stretching a spring. Under the assumptions on the elastic properties of the spring made on p. 404, the force acting is $f = -kx$, where k is constant. The work that must be done against this force in order to stretch the spring from the unstretched position $x = 0$ to the final position $x = x_1$ is therefore given by the integral

$$\int_0^{x_1} kx \, dx = \frac{1}{2} kx_1^2.$$

*e. The Charging of a Condenser

The concept of work in other branches of physics can be treated in a similar way. For example, let us consider the charging of a condenser. If we denote the quantity of electricity in the condenser by Q, its capacity by C, and the difference of potential (voltage) across the condenser by V, then we know from physics that $Q = CV$. Moreover, the work done in moving a charge Q through a difference of potential V is equal to QV. Since in the charging of the condenser the difference of potential V is not constant but increases with Q, we perform a passage to the limit exactly analogous to that on p. 418, and as the expression for the work done in charging the condenser we obtain

$$\int_0^{Q_1} V \, dQ = \frac{1}{C} \int_0^{Q_1} Q \, dQ = \frac{1}{2} \frac{Q_1^2}{C} = \frac{1}{2} Q_1 V_1,$$

where Q_1 is the total quantity of electricity passed into the condenser and V_1 is the difference of potential across the condenser at the end of charging process.

Appendix

*A.1   Properties of the Evolute

On p. 359 we defined the evolute E of a curve C as the locus of the centers of curvature of C. If C is represented by: $x = x(s)$, $y = y(s)$, using the arc length s as parameter, then the center of curvature (ξ, η) of the point C with parameter s is given by [cf. (17a), p. 359]

$$(58) \qquad\qquad \xi = x - \rho \dot{y}, \qquad \eta = y + \rho \dot{x},$$

with

$$\kappa = \frac{1}{\rho} = \dot{x}\ddot{y} - \dot{y}\ddot{x}.$$

The quantities κ and $|\rho|$ are, respectively, curvature and radius of curvature of C.

We can deduce some interesting geometrical properties of the evolute from these formulas.

Differentiating the relation $\dot{x}^2 + \dot{y}^2 = 1$ leads to, $\dot{x}\ddot{x} + \dot{y}\ddot{y} = 0$. Since also $\dot{x}\ddot{y} - \dot{y}\ddot{x} = 1/\rho$, we have

$$(59) \qquad\qquad \ddot{x} = -\frac{1}{\rho}\dot{y}, \qquad \ddot{y} = \frac{1}{\rho}\dot{x}.$$

Differentiating the formulas (58) with respect to s

$$\dot{\xi} = \dot{x} - \rho\ddot{y} - \dot{\rho}\dot{y} = -\dot{\rho}\dot{y}, \qquad \dot{\eta} = \dot{y} + \rho\ddot{x} + \dot{\rho}\dot{x} = \dot{\rho}\dot{x},$$

and therefore

$$\dot{\xi}\dot{x} + \dot{\eta}\dot{y} = 0.$$

Since the direction cosines of the normal to the curve are given by $-\dot{y}$ and $\dot{x}$, *the normal to the curve C is tangent to the evolute E at the center of curvature;* or the tangent to the evolute is the normal of the given curve; or *the evolute is the "envelope" of the normals* (cf. Fig. A.1).

If further we denote the length of arc of the evolute, measured from an arbitrary fixed point, by σ, we have, using s as parameter,

$$\dot{\sigma}^2 = \left(\frac{d\sigma}{ds}\right)^2 = \dot{\xi}^2 + \dot{\eta}^2.$$

Since $\dot{x}^2 + \dot{y}^2 = 1$, we obtain from our formulas (59),

$$\dot{\sigma}^2 = \dot{\rho}^2.$$

If we choose the direction in which σ is measured in a suitable way, it follows that

$$\dot{\sigma} = \dot{\rho},$$

provided that $\dot{\rho} \neq 0$. Integration yields

$$\sigma_1 - \sigma_0 = \rho_1 - \rho_0.$$

That is, *the length of arc of the evolute between two points is equal to the difference of the corresponding radii of curvature, provided that $\dot{\rho}$ remains different from zero for the arc under consideration.*

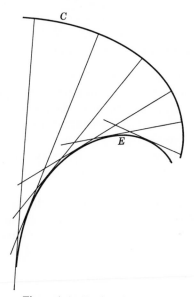

Figure A.1   Evolute (E).

This last condition is not superfluous. For if $\dot{\rho}$ changes sign, then the formula $\dot{\sigma} = \dot{\rho}$ shows that on passing the corresponding point of the evolute the length of arc σ has a maximum or minimum; that is, on passing this point we do not simply continue to reckon σ onward, but we must reverse the sense in which σ is measured. If we wish to avoid this reversal, we must on passing such a point change the sign in the preceding formula, that is, put $\dot{\sigma} = -\dot{\rho}$.

It may also be noted that the centers of curvature which correspond to maxima or minima of the radius of curvature are *cusps of the evolute*. [The proof is omitted here.] (See Figs. A.4, A.6.)

The geometrical relationship just found can be expressed in yet another way: We imagine a flexible inextensible thread laid along an

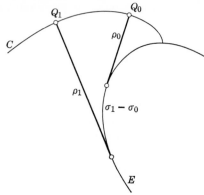

Figure A.2   String construction of the involute C of a curve E: $\rho_1 = \rho_0 + \sigma_1 - \sigma_0$.

arc of the evolute E and stretched so that a part of it extends tangentially away from the curve to it; if in addition the end point Q of this thread lies initially on the original curve C, then as we unwind the thread Q will describe the curve C. This accounts for the name evolute (*evolvere,* to unwind). The curve C is called an *involute* of the evolute E. On the other hand, we may start with an arbitrary curve E and construct its involute C by this unwinding process. Then conversely E is seen to be the evolute of C (Fig. A.2).

For the proof we consider the curve E, which is now the given curve, as given in the form $\xi = \xi(\sigma)$, $\eta = \eta(\sigma)$, where the current rectangular coordinates are denoted by ξ and η and the parameter σ is length of arc on E. The winding is done as indicated in Fig. A.3; when the

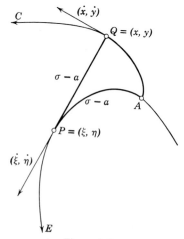

Figure A.3

thread is completely wound on to the evolute E, its end Q coincides with the point A of E corresponding to some arc-length a. If the thread is now unwound until it is tangent to the evolute at the point P, corresponding to the length of arc $\sigma > a$, the length of the segment PQ will be $(\sigma - a)$ and its direction cosines will be $-\dot{\xi}$ and $-\dot{\eta}$, where the dot now denotes differentiation with respect to σ. Thus for the coordinates x, y of the point Q we obtain the expressions

$$(60) \qquad x = \xi - (\sigma - a)\dot{\xi}, \qquad y = \eta - (\sigma - a)\dot{\eta},$$

which give the equations for the involute described by the point Q in terms of the parameter σ. By differentiation with respect to σ we obtain

$$(61) \qquad \begin{aligned} \dot{x} &= \dot{\xi} - \dot{\xi} + (a - \sigma)\ddot{\xi} = (a - \sigma)\ddot{\xi}, \\ \dot{y} &= \dot{\eta} - \dot{\eta} + (a - \sigma)\ddot{\eta} = (a - \sigma)\ddot{\eta}. \end{aligned}$$

Since $\dot{\xi}\ddot{\xi} + \dot{\eta}\ddot{\eta} = 0$, we at once find that

$$\dot{\xi}\dot{x} + \dot{\eta}\dot{y} = 0,$$

which shows that the line PQ is normal to the involute C. We can therefore state that the normals to the curve C are tangent to the curve E. Since the tangent to E has direction cosines $\dot{\xi}$, $\dot{\eta}$ we find for the direction cosines of the tangent of C the expressions

$$(62) \qquad \frac{\dot{x}}{\sqrt{\dot{x}^2 + \dot{y}^2}} = \dot{\eta}, \qquad \frac{\dot{y}}{\sqrt{\dot{x}^2 + \dot{y}^2}} = -\dot{\xi}.$$

Differentiating the relation $\dot{\xi}\dot{x} + \dot{\eta}\dot{y} = 0$ with respect to σ and substituting for $\dot{\xi}$, $\dot{\eta}$, $\ddot{\xi}$, $\ddot{\eta}$, their expressions from the previous equations (61), (62) shows that

$$0 = \ddot{\xi}\dot{x} + \ddot{\eta}\dot{y} + \dot{\xi}\ddot{x} + \dot{\eta}\ddot{y} = \frac{\dot{x}^2 + \dot{y}^2}{a - \sigma} + \frac{-\ddot{x}\dot{y} + \dot{x}\ddot{y}}{\sqrt{\dot{x}^2 + \dot{y}^2}}.$$

Hence the radius of curvature of the curve C corresponding to the point $Q = (x, y)$ turns out to be (see formula (15) on p. 355)

$$\rho = \frac{1}{\kappa} = \frac{(\dot{x}^2 + \dot{y}^2)^{3/2}}{\dot{x}\ddot{y} - \dot{y}\ddot{x}} = \sigma - a.$$

This is also the distance of the point Q from $P = (\xi, \eta)$. Because P also lies on the normal to C at Q, we have in P the center of curvature of C corresponding to the point Q. Thus *every curve E is the evolute of all its involutes.*

Examples. We consider the evolute of the cycloid

$$x = \pi + t + \sin t, \qquad y = -1 - \cos t.$$

By Eq. (17), p. 359, the center of curvature (ξ, η) for a curve referred to an arbitrary parameter t is

$$\xi = x - \dot{y}\, \frac{\dot{x}^2 + \dot{y}^2}{\dot{x}\ddot{y} - \dot{y}\ddot{x}}, \qquad \eta = y + \dot{x}\, \frac{\dot{x}^2 + \dot{y}^2}{\dot{x}\ddot{y} - \dot{y}\ddot{x}}.$$

A short computation yields then for the evolute of the cycloid

$$\xi = \pi + t - \sin t, \qquad \eta = 1 + \cos t.$$

If we put $t = \tau - \pi$, then

$$\xi + \pi = \pi + \tau + \sin \tau, \qquad \eta - 2 = -1 - \cos \tau:$$

these equations show that the evolute is itself a cycloid which is similar to the original curve, and can be obtained from it by translation as indicated in Fig. A.4.

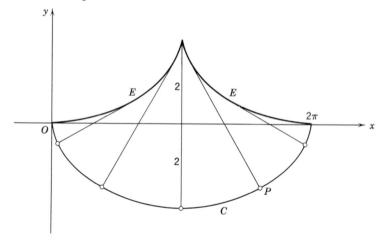

Figure A.4   The cycloidal pendulum.

This gives us a simple method of constructing a *cycloidal pendulum* (see p. 412). If a mass P is attached by a thread of length 4 to one of the cusps of the evolute, then under tension the thread will partly coincide with the evolute and lie along a tangent to the evolute the rest of the way. The mass P is then forced to lie on the involute, that is, on the original cycloid. Under gravity P must describe an isochronous motion over some portion of the cycloid with a period independent of the position at which P begins the motion. (The parameter t to which the cycloid is referred does *not* correspond to the time in the isochronous motion.)

The free straight portion of a pendulum of this type varies in length during the motion (see Fig. A.4).

As a further example we derive the equation for the involute of a circle. We begin with the circle $\xi = \cos\sigma$, $\eta = -\sin\sigma$ and unwind the tangent, as indicated in Fig. A.5. The involute of the circle is then given in the form

$$x = \cos\sigma + \sigma\sin\sigma, \qquad y = -\sin\sigma + \sigma\cos\sigma.$$

(using the equation (60), on p. 427 with $a = 0$).

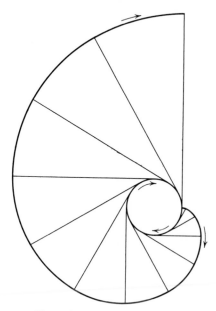

Figure A.5   Involute of the circle.

Finally, we determine the evolute of the ellipse $x = a\cos t$, $y = b\sin t$. We at once have

$$\xi = x - \dot{y}\,\frac{\dot{x}^2 + \dot{y}^2}{\dot{x}\ddot{y} - \dot{y}\ddot{x}} = \frac{a^2 - b^2}{a}\cos^3 t.$$

and

$$\eta = y + \dot{x}\,\frac{\dot{x}^2 + \dot{y}^2}{\dot{x}\ddot{y} - \dot{y}\ddot{x}} = -\frac{a^2 - b^2}{b}\sin^3 t,$$

as parametric representation of the evolute. If from these equations we eliminate t in the usual way, we obtain the equation of the evolute in nonparametric form:

$$(a\xi)^{2/3} + (b\eta)^{2/3} = (a^2 - b^2)^{2/3}.$$

This curve is called an *astroid*. Its graph is given in Fig. A.6. By means of the parametric equations we may readily convince ourselves that the centers of curvature corresponding to the vertices of the ellipse are actually the cusps of the astroid.

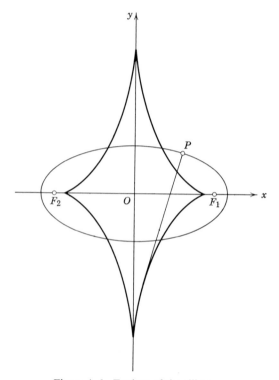

Figure A.6   Evolute of the ellipse.

*A.2   Areas Bounded by Closed Curves.   Indices

In Section 4.2 the oriented area bounded by a closed curve $x = x(t)$, $y = y(t)$, $\alpha \le t \le \beta$, which nowhere intersects itself (a so-called *simple closed curve*), was represented by the integral

$$A = -\int_{\alpha}^{\beta} y(t)\dot{x}(t)\, dt;$$

the value obtained is positive or negative depending on whether the sense in which the boundary is described is counterclockwise or clockwise. This formula remains meaningful as a definition of A if we allow self-intersections of curves. It remains to see how A is related to areas in such cases. Suppose that the curve C, given by the equation $x = x(t)$, $y = y(t)$, intersects itself in a finite number of points, thus dividing the plane into a finite number of portions $R_1, R_2, \ldots$ Suppose further that the derivatives are continuous and that $\dot{x}^2 + \dot{y}^2 \neq 0$, except perhaps for a finite number of jump-discontinuities (which may or may not correspond to corners). Finally, it is assumed that the curve has

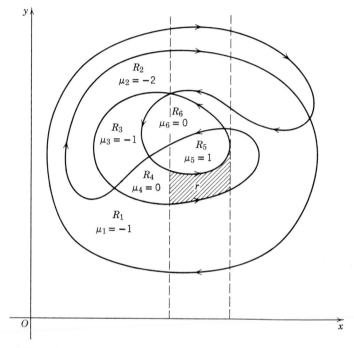

Figure A.7   Indices μ_i of regions R_i formed by oriented closed curve. Figure A.8.

a finite number of *lines of support* $x =$ constant, that is, vertical lines that are either tangent to the curve or pass through a point of self-intersection of the curve.

To each region R_i we then assign an integer, the *index* μ_i, defined in the following way: We choose an arbitrary point Q in R_i, not lying on any line of support, and erect the half-line extending from Q upward in the direction of the positive y-axis. We count the number of times the curve C for increasing t crosses the half-line from right to left, and subtract the number of times the curve C crosses from left to right; the difference is the index μ_i. For example, the interior of the curve illustrated in Fig. 4.17, p. 343, has the index $\mu = +1$; and in Fig. A.7 the regions $R_1, \ldots, R_5$, R_6 have the indexes $\mu_1 = -1$, $\mu_2 = -2$,

$\mu_3 = -1$, $\mu_4 = 0$, $\mu_5 = 1$ and $\mu_6 = 0$. This number μ_i actually depends on the region R_i only and not on the particular point Q chosen in R_i, as we readily see in the following manner. We choose any other point Q' in R_i, not on a line of support, and join Q to Q' by a broken line lying entirely in the region R_i (Fig. A.8). As we proceed along this broken line from Q to Q' the number of right-to-left crossings minus the number of left-to-right crossings is constant; for between lines of support the number of crossings of either type is unchanged, whereas on crossing a line of support the number of crossings of both types

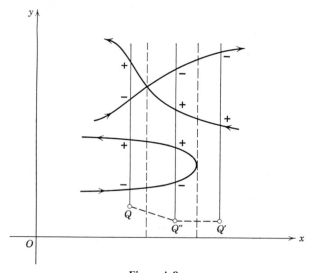

Figure A.8

either stay the same or both numbers increase by one or both decrease by one; in every case, the difference is unaltered. Here a line of support that meets the curve at several different points, say $A, B, \ldots, H$, is considered as several different lines of support, $FA, FB, \ldots, FH$, where F is a point vertically below all the points $A, B, \ldots, H$. Our argument then applies to each of these lines. Hence the number μ_i has the same value whether we use Q or Q' in determining it.

In particular, if our curve does not intersect itself, the interior of the curve consists of a single region R whose index is $+1$ or -1 depending on whether the sense in which the boundary is described is counter-clockwise or clockwise. To see this we draw any vertical line (not a line of support) intersecting the curve; on this line we find the highest point of intersection P with the curve, and in R we choose a point Q

below P and so near it that no point of intersection lies between P and Q. Then above Q there lies one crossing of the curve, which if the curve is traversed in the counterclockwise sense must be a right-to-left crossing, so that $\mu = +1$; otherwise $\mu = -1$. As we have just seen, this same value of μ holds for every other point of R. For such a curve, and, in fact, for all closed curves, one of the regions, the "outside" of the curve, extends unboundedly in all directions; we see immediately that this region has index 0, and ignore it in what follows. Then the relation between the integral A and the areas of the regions R_i is given by the following theorem:

THEOREM. *The value of the integral* $-\displaystyle\int_\alpha^\beta y\dot{x}\,dt$ *is equal to the sum of the absolute areas of the regions* R_i, *each area* R_i *being counted* μ_i *times; in symbols*

$$-\int_\alpha^\beta y\dot{x}\,dt = \sum \mu_i\,|\text{area } R_i|.$$

PROOF. The proof is simple. We assume, as we are entitled to do, that the whole of the curve lies above the x-axis. (Adding a constant to y does not change the value of the integral A for a closed curve.) The lines of support cut R_i into a finite number of portions; let r be one of these portions. Then on taking the integral $-\int y\dot{x}\,dt = -\int y\,dx$ for each single-valued branch of the function $y = y(x)$ and interpreting it as area between the curve and the x-axis, we find that the absolute area of r is counted $+1$ times for each right-to-left branch above r and -1 times for each left-to-right branch above r; in all, μ_i times. The same is true for every other portion of R_i; hence R_i is counted μ_i times. Thus the integral round the complete curve has the value $\Sigma \mu_i\,|\text{area } R_i|$, as stated (cf. Fig. A.7). This formula agrees with what we have found for simple closed curves, as we recognize from the discussion of the values of μ for such curves.

The definition given for the index μ_i has the disadvantage of being stated in terms of a particular coordinate system. As a matter of fact, however, it can be shown that the value of μ_i assigned to a region R_i is independent of the coordinate system and depends solely on the curve. This can be readily seen by identifying μ_i with the total number ν_i of times a point on the curve for t increasing from α to β runs about any fixed point Q_i of R_i in the counterclockwise sense, that is with the number of times C *winds* around Q_i. We shall prove the identity of μ_i and ν_i.

Let C be given parametrically by $x = x(t), y = y(t)$ where $\alpha \leq t \leq \beta$. Let $Q = (\xi, \eta)$ be a point which does not lie on a line of support of C.

We take Q as origin of a system of polar coordinates r, θ in which

$$r = \sqrt{(x - \xi)^2 + (y - \eta)^2}, \qquad \cos\theta = \frac{x - \xi}{r}, \ \sin\theta = \frac{y - \eta}{r}.$$

The polar angle θ is determined only within whole multiples of 2π; however, θ is determined uniquely as a function of t by its value θ_0 for $t = \alpha$ if we require $\theta = \theta(t)$ to vary continuously with t along the curve C. At $t = \beta$ the angle θ will then have a value $\theta(\beta) = \theta_0 + 2\nu\pi$, where ν is an integer. The number

$$\nu = \frac{1}{2\pi}[\theta(\beta) - \theta(\alpha)] = \frac{1}{2\pi}\int_\alpha^\beta \frac{d\theta}{dt}\,dt = \frac{1}{2\pi}\int_C d\theta$$

represents the number of times that the oriented curve C winds around Q.

The curve C crosses the vertical half-line through Q for those values of t for which the expression $(1/2\pi)[\theta(t) - \pi/2]$ has an integral value n. Consider for a fixed n the t-values in the parameter interval for which $(1/2\pi)(\theta - \pi/2) = n$. Let σ_n and τ_n be the number of such t-values for which $d\theta/dt > 0$, respectively $d\theta/dt < 0$. Obviously, the *index* at the point Q is

$$\mu = \sum_n \sigma_n - \sum_n \tau_n = \sum_n (\sigma_n - \tau_n).$$

On the other hand, $\sigma_n - \tau_n$ can only have one of the values 1, 0, -1, for the graph of $\theta(t)$ in the θ, t-plane must cross the line $\theta = \pi/2 + 2n\pi$ alternately from above or below. Actually, we have $\sigma_n - \tau_n = \operatorname{sign}[\theta(\beta) - \theta(\alpha)]$ if $\pi/2 + 2n\pi$ lies between $\theta(\alpha)$ and $\theta(\beta)$ and $\sigma_n - \tau_n = 0$ otherwise.

Consequently, μ equals the number of values of the form $\pi/2 + 2n\pi$ with an integer n that lie between $\theta(\alpha)$ and $\theta(\beta)$ taken with the sign of $\theta(\beta) - \theta(\alpha)$; that is, μ equals the number ν.

Since $\theta = \arctan[(y - \eta)/(x - \xi)]$, we have

$$\frac{d\theta}{dt} = \frac{\dot{y}(x - \xi) - \dot{x}(y - \eta)}{(x - \xi)^2 + (y - \eta)^2}.$$

This yields for the index μ of the oriented closed curve C with respect to the point (ξ, η) the integral representation

$$\mu = \frac{1}{2\pi}\int_\alpha^\beta \frac{\dot{y}(x - \xi) - \dot{x}(y - \eta)}{(x - \xi)^2 + (y - \eta)^2}\,dt.$$

which can be simply written (see p. 367) without referring to the parameter t explicitly:

$$\mu = \frac{1}{2\pi} \int_C \frac{(x - \xi)\, dy - (y - \eta)\, dx}{(x - \xi)^2 + (y - \eta)^2}.$$

The remarkable feature of these results is that the integer μ or ν which describes a *topological* relation between the point Q and the curve C can be determined analytically, from the parameter representation of C, by evaluating an integral.

PROBLEMS

SECTION 4.1c, page 328

1. Sketch the hypocycloid for $a = 4c$ (the astroid) and find its nonparametric equation.

2. Prove that if c/a is rational the general hypocycloid is closed after the moving circle has rotated an integral number of times, whereas if c/a is irrational, the curve has infinitely many points where it meets the circumference of the fixed circle and will not close.

3. Derive the parametric representation

$$x = at - b\sin t, \qquad y = a - b\cos t$$

for ordinary trochoid, that is, for the path of a point P attached to a disc of radius a rolling along a line, P having the distance b from the center of the disc (see Fig. 4.7).

4. Find the parametric equations for the curve $x^3 + y^3 = 3axy$ (the folium of Descartes), choosing as parameter t the tangent of the angle between the x-axis and the ray from the origin to the point (x, y).

SECTION 4.1e, page 343

1. The angle α between two curves at a point of intersection is defined to be the angle between their tangents at the point. Find a formula for $\cos \alpha$ in terms of the parametric representations of the curves.

2. Let $x = f(t)$ and $y = g(t)$. Derive formulas for d^2y/dx^2 and d^3y/dx^3 in terms of derivatives with respect to the parameter t.

3. Find the formula for the angle α between two curves $r = f(\theta)$ and $r = g(\theta)$ in polar coordinates.

4. Find the equations of the curves which everywhere intersect the straight lines through the origin at the same angle α.

5. Prove: if $x = f(t)$ and $y = g(t)$ are continuous on the closed interval $[a, b]$ and differentiable on the open interval (a, b) with $x'^2 + y'^2 > 0$, then there is at least one point on the open arc

$$x = f(t), \qquad y = g(t), \qquad (a < t < b),$$

where the tangent is parallel to the chord joining the end points.

6. Let P be the point of a circle which traces out a cycloid as the circle rolls on a given line. Let Q be the point of contact of the circle with the line. Prove that at any instant, the normal to the cycloid at P passes through Q. What similar property holds for the tangent at P?

7. Prove that the length of the segment of the tangent to the astroid,

$$x = 4c \cos^3 \theta, \qquad y = 4c \sin^3 \theta,$$

cut off by the coordinate axes is constant.

*8. Show that the two families of ellipses and hyperbolas, $(0 < a < b)$

$$\frac{x^2}{a^2 - \lambda^2} + \frac{y^2}{b^2 - \lambda^2} = 1, \quad \text{for } 0 < \lambda < a,$$

$$\frac{x^2}{a^2 - \tau^2} + \frac{y^2}{b^2 - \tau^2} = 1, \quad \text{for } a < \tau < b,$$

are confocal (that is, have the same focii) and intersect at right angles.

9. (*a*) Show for the ellipse that the angle between the two rays from the foci to a point on the curve is bisected by the normal at the point.

(*b*) Show for the hyperbola that the angle is bisected by the tangent.

SECTION 4.1f, page 348

1. Prove that the curve defined by

$$y = \begin{cases} x^2 \sin \dfrac{1}{x}, & 0 < x \le 1 \\ 0, & x = 0 \end{cases}$$

has finite length, but that the continuous curve defined by

$$y = \begin{cases} x \sin \dfrac{1}{x}, & 0 < x \le 1 \\ 0, & x = 0 \end{cases}$$

is not rectifiable.

2. Prove that if the function f is defined and monotone on the closed interval $[a, b]$, then the arc defined by

$$y = f(x), \qquad (a \le x \le b),$$

is rectifiable.

SECTION 4.1g, page 352

1. An *elliptic integral of the second kind* has the form

$$\int_0^\phi \sqrt{1 - k^2 \sin^2 \theta} \, d\theta.$$

(a) Show that the arc length of the ellipse $x = a \cos \theta$, $y = b \sin \theta$ can be expressed in terms of an elliptic integral of the second kind.

(b) Do the same for the trochoid

$$x = at - b \sin t, \qquad y = a - b \cos t.$$

*(c) Show that the arc length of the hyperbola can be expressed in terms of elliptic integrals of the first and second kinds.

SECTION 4.1h, page 354

1. Let P be a point of the rolling circle which generates a cycloid and let Q be the lowest point of the circle at any given instant. Show that Q bisects the segment joining P to the center of the osculating circle of the cycloid at P.

2. Find the center of curvature for $y = x^2$ when $x = 0$. Determine the point of intersection of the normal lines to the curve when $x = 0$ and when $x = \epsilon$. Calculate the distance of the intersection from the center of curvature. Suggest an alternative definition for the center of curvature. Prove that this definition is equivalent to the definition given in the text.

3. Consider the question of whether the osculating circle crosses the curve at the point of contact.

*4. Prove that the circle of curvature at a point P of the curve C is the limit of the circles through three points P, P_1, P_2 as P_1 and P_2 tend to P.

5. Let $r = f(\theta)$ be the equation of a curve in polar coordinates. Prove that the curvature is given by the formula

$$k = \frac{2r'^2 - rr'' + r^2}{(r'^2 + r^2)^{\frac{3}{2}}},$$

where

$$r' = \frac{df}{d\theta}, \qquad r'' = \frac{d^2f}{d\theta^2}.$$

6. The curve for which the length of the tangent intercepted between the point of contact and the y-axis is always equal to 1 is called the tractrix. Find its equation. Show that the radius of curvature at each point of the curve is inversely proportional to the length of the normal intercepted between the point on the curve and the y-axis. Calculate the length of arc of the tractrix and find the parametric equations in terms of the length of arc.

7. Let $x = x(t)$, $y = y(t)$ be a closed curve. A constant length p is measured off along the normal to the curve. The extremity of this segment describes a curve which is called a parallel curve to the original curve. Find the area, the length of arc, and the radius of curvature of the parallel curve.

8. Show that the only curves whose curvature is a fixed constant k are circles of radius $1/k$.

*9. If the curvature of a curve in the xy-plane is a monotonic function of the length of arc, prove that the curve is not closed and that it has no double points.

SECTION 4.1i, page 360

1. Show that the expression for the curvature of a curve $x = x(t)$, $y = y(t)$ is unaltered by rotation of axes and also by change of parameter given by $t = \phi(\tau)$, where $\phi'(\tau) > 0$.

SECTION 4.3d, page 394

1. Prove if the acceleration is always perpendicular to velocity that the speed is constant.

2. The velocity vector, considered as a position vector, traces out a curve known as the hodograph. Show whether or not a particle moving on a closed curve may have a straight line as its hodograph.

3. Assuming the rolling circle moves at constant speed, find the velocity and acceleration of the point P which generates the cycloid.

4. Let A be a fixed point of the plane and suppose that the acceleration vector for a moving point P is always directed toward A and proportional to the $1/|AP|^2$. Prove that the hodograph (cf. Problem 2) is a circle.

5. Let A be a fixed point on a circle. Let P be a point of the circle moving so that the acceleration vector points to A. Prove that the acceleration is proportional to $|AP|^{-5}$.

SECTION 4.5, page 402

1. A particle moves in a straight line subject to a resistance producing the retardation ku^3, where u is the velocity and k a constant. Find expressions for the velocity (u) and the time (t) in terms of s, the distance from the initial position, and v_0, the initial velocity.

2. A particle of unit mass moves along the x-axis and is acted upon by a force $f(x) = -\sin x$.

(*a*) Determine the motion of the point if at time $t = 0$ it is at the point $x = 0$ and has velocity $v_0 = 2$. Show that as $t \to \infty$ the particle approaches a limiting position, and find this limiting position.

(*b*) If the conditions are the same, except that v_0 may have any value, show that if $v_0 > 2$ the point moves to an infinite distance as $t \to \infty$, and that if $v_0 < 2$ the point oscillates about the origin.

3. Choose axes with their origin at the center of the earth, whose radius we shall denote by R. According to Newton's law of gravitation, a particle of unit mass lying on the y-axis is attracted by the earth with a force $-\mu M/y^2$, where μ is the "gravitational constant" and M is the mass of the earth.

(*a*) Calculate the motion of the particle after it is released at the point $y_0 (> R)$; that is, if at time $t = 0$ it is at the point $y = y_0$ and has the velocity $v_0 = 0$.

(*b*) Find the velocity with which the particle in (*a*) strikes the earth.

(*c*) Using the result of (*b*), calculate the velocity of a particle falling to the earth from infinity.[1]

*__4.** A particle perturbed slightly from rest on top of a circle slides downward under the force of gravity. At what point does it fly unconstrained off the circle?

---

[1] This is the same as the least velocity with which a projectile would have to be fired in order that it should leave the earth and never return.

*5. A particle of mass m moves along the ellipse $r = k/(1 - e \cos \theta)$. The force on the particle is cm/r^2 directed toward the origin. Describe the motion of the particle, find its period, and show that the radius vector to the particle sweeps out equal areas in equal times.

SECTION 4A.1, page 424

1. Show that the evolute of an epicycloid (Example, p. 329) is another epicycloid similar to the first, which can be obtained from the first by rotation and contraction.

2. Show that the evolute of a hypocycloid (Example, p. 331) is another hypocycloid, which can be obtained from the first by rotation and expansion.

5

Taylor's Expansion

5.1 Introduction: Power Series

It was a great triumph in the early years of Calculus when Newton and others discovered that many known functions could be expressed as "polynomials of infinite order" or "power series," with coefficients formed by elegant transparent laws. The geometrical series for $1/(1-x)$ or $1/(1+x^2)$

$$(1) \qquad \frac{1}{1-x} = 1 + x + x^2 + \cdots + x^n + \cdots$$

$$(1a) \qquad \frac{1}{1+x^2} = 1 - x^2 + x^4 - x^6 + \cdots + (-1)^n x^{2n} + \cdots$$

valid for the open interval $|x| < 1$, are prototypes (see Chapter 1, p. 67). Similar expansions of the form

$$f(x) = a_0 + a_1 x + \cdots + a_n x^n + \cdots$$
$$= \sum_{v=0}^{\infty} a_v x^v,$$

with numerical coefficients a_v, will be derived in this chapter for many other functions.

The following are striking examples:

$$e^x = 1 + x + \frac{x^2}{2!} + \frac{x^3}{3!} + \cdots + \frac{x^n}{n!} + \cdots;$$

$$\sin x = x - \frac{x^3}{3!} + \frac{x^5}{5!} - + \cdots + \frac{(-1)^n x^{2n+1}}{(2n+1)!} + \cdots;$$

$$\cos x = 1 - \frac{x^2}{2!} + \frac{x^4}{4!} + \cdots + \frac{(-1)^n x^{2n}}{(2n)!} + \cdots$$

These series expansions are valid for all x.

Newton's General Binomial Theorem. The expansion

$$(1 + x)^\alpha = 1 + \frac{\alpha}{1!} x + \frac{\alpha(\alpha - 1)}{2!} x^2 + \cdots$$

$$= \sum_{\nu=0}^{\infty} \binom{\alpha}{\nu} x^\nu$$

is valid for $|x| < 1$ and any exponent α.

To explain the precise meaning of such expansions, we consider the polynomial of order n formed as the sum of the first $n + 1$ terms of the series, *the nth "partial sum,"*

$$S_n = \sum_{\nu=0}^{n} a_\nu x^\nu.$$

The formula

$$f(x) = \sum_{\nu=0}^{\infty} a_\nu x^\nu, \quad \text{for} \quad |x| < a$$

then means: For $n \to \infty$ the sequence S_n tends to the value of the function $f(x)$ at each point x in the interval $|x| < a$. The infinite series is then said to *converge* to $f(x)$ in the interval $|x| < a$. The difference

$$R_n(x) = f(x) - S_n(x),$$

the *"remainder"* of the series, measures the precision with which $f(x)$ is approximated by the polynomial $S_n(x)$ at x. For example,

(1b) $$\frac{1}{1 - x} = 1 + x + x^2 + \cdots + x^n + R_n(x),$$

where the remainder $R_n(x) = x^{n+1}/(1 - x)$ tends to zero for $|x| < 1$ as n increases; thus the infinite geometric series $\sum_{\nu=0}^{\infty} x^\nu = 1/(1 - x)$ results. To find simple manageable estimates for R_n in specific cases is a task of both theoretical and practical importance.

In this chapter we are concerned with such expansions for a wide class of functions, including all the "elementary" transcendental functions. It is a striking fact that in these expansions of transcendental functions the coefficients are elegant expressions in terms of integers. The approach to these expansions will be by *Taylor's theorem*; later in Chapter 7 we shall discuss a different approach by a direct study of power series.

It should be emphasized that often just as for the geometrical series of Eq. (1a), the infinite expansion is not valid outside some interval for x—(in the case of the geometrical series, the interval $x^2 < 1$) even though the function represented by the series is well defined outside this interval.

5.2   Expansion of the Logarithm and the Inverse Tangent

a. The Logarithm

As simple examples we first derive expansions of the logarithmic and the inverse tangent functions by integration, from the geometric series

$$\frac{1}{1 - t} = 1 + t + t^2 + \cdots + t^{n-1} + r_n(t)$$

with $r_n(t) = t^n/(1 - t)$.

We substitute this sum for the integrand in the formula

$$-\log (1 - x) = \int_0^x \frac{dt}{1 - t}$$

and integrate term by term, obtaining for $x < 1$

$$-\log (1 - x) = x + \frac{x^2}{2} + \frac{x^3}{3} + \frac{x^4}{4} + \cdots + \frac{x^n}{n} + R_n(x),$$

with the remainder

$$R_n(x) = \int_0^x r_n \, dt = \int_0^x \frac{t^n}{1 - t} \, dt.$$

Hence for any positive integer n the function $-\log (1 - x)$ is approximated by the polynomial of nth degree,

$$x + \frac{x^2}{2} + \frac{x^3}{3} + \cdots + \frac{x^n}{n},$$

and the remainder R_n indicates the "error" of this approximation.

To appraise the accuracy of this approximation we estimate the remainder R_n. If we at first suppose that $-1 \leq x \leq 0$, then in the entire interval of integration the integrand $t^n/(1 - t)$ in absolute value, nowhere exceeds $|t^n| = (-1)^n t^n$. Thus

$$|R_n| \leq \left| \int_0^x t^n \, dt \right| = \frac{|x|^{n+1}}{n + 1};$$

hence for every value of x in the *closed* interval $-1 \leq x \leq 0$ including $x = -1$ this remainder can be made as small as we wish by choosing n

large enough (cf. p. 61). For $x > 0$ the end point $x = 1$ must be omitted; we have to restrict x to the half-open interval $0 \leq x < 1$; the integrand does not change sign and its absolute value does not exceed $t^n/(1 - x)$; we thus obtain for $0 \leq x < 1$ the estimate

$$|R_n| \leq \frac{1}{1 - x} \int_0^x t^n \, dt = \frac{x^{n+1}}{(1 - x)(n + 1)}.$$

Hence again, if x is fixed, the remainder is arbitrarily small when n is sufficiently large. Of course, the estimate has no meaning for $x = 1$.

Summing up,

$$(2) \qquad \log(1 - x) = -x - \frac{x^2}{2} - \frac{x^3}{3} - \cdots - \frac{x^n}{n} - R_n$$

where the remainder R_n tends to zero as n increases, provided that x lies in the half-open interval $-1 \leq x < 1$.

In fact, this reasoning establishes a "uniform" estimate for the remainder, independent of x and valid for all values of x in the interval $-1 \leq x \leq 1 - h$, where h is any number such that $0 < h \leq 1$; namely, $|R_n| \leq 1/[(n + 1)h]$.

The fact that the remainder R_n tends to zero in the half-open interval $-1 \leq x < 1$ is expressed by saying that in this interval the logarithmic function is given by the infinite series[1]

$$(3) \qquad \log(1 - x) = -x - \frac{x^2}{2} - \frac{x^3}{3} - \frac{x^4}{4} - \cdots.$$

If we insert the particular value $x = -1$ in this series, we obtain the remarkable formula

$$(4) \qquad \log 2 = 1 - \tfrac{1}{2} + \tfrac{1}{3} - \tfrac{1}{4} + - \cdots.$$

This is one of the relations whose discovery made a deep impression on the early pioneers of the calculus.

For the open interval $-1 < x < 1$, we have only to write $-x$ in place of x in (2) in order to obtain

$$(2a) \quad \log(1 + x) = x - \frac{x^2}{2} + \frac{x^3}{3} - \frac{x^4}{4} + - \cdots + (-1)^{n-1} \frac{x^n}{n} - R_n{}',$$

where

$$R_n{}'(x) = \int_0^{-x} \frac{t^n \, dt}{1 - t} = (-1)^{n+1} \int_0^x \frac{t^n \, dt}{1 + t}.$$

---

[1] We leave it as an exercise to the reader to ascertain that for all values of x for which $|x| > 1$ the remainder not only fails to approach zero, but, in fact, that $|R_n|$ increases beyond all bounds as n increases, so that for such values of x the polynomial is not a good approximation of the logarithm and becomes worse with increasing n.

Taking n as even and subtracting (2) from (2a), we have

$$\frac{1}{2} \log \left(\frac{1+x}{1-x}\right) = \operatorname{ar\,tanh} x = x + \frac{x^3}{3} + \frac{x^5}{5} + \cdots + \frac{x^{n-1}}{n-1} + \bar{R}_n,$$

where the remainder $\bar{R}_n$ is given by

$$\bar{R}_n = \frac{1}{2}(R_n - R_n') = \int_0^x \frac{t^n}{1-t^2}\,dt,$$

and where $\operatorname{ar\,tanh} x$ is defined according to p. 233.

Observing that $1/(1-t^2) \le 1/(1-x^2)$, we find by an elementary estimate of the integral that

$$|\bar{R}_n| \le \frac{|x^{n+1}|}{n+1} \cdot \frac{1}{1-x^2} \,;$$

thus the remainder $\bar{R}_n$ tends to zero as n increases, a fact again expressed by writing the expansion as an infinite series:

$$(5) \qquad \frac{1}{2} \log \frac{1+x}{1-x} = \operatorname{ar\,tanh} x = x + \frac{x^3}{3} + \frac{x^5}{5} + \frac{x^7}{7} + \cdots,$$

for all values of x with $|x| < 1$. Incidentally, this result also could be derived directly by integrating the geometric series for $1/(1-x^2)$. It is an advantage of this formula that as x traverses the interval from -1 to 1, the expression $(1+x)/(1-x)$ ranges over all positive numbers. Thus, if the value of x is suitably chosen, the series enables us to calculate the value of the logarithm of any positive number, with an error not exceeding the above estimate for $\bar{R}_n$.

b. The Inverse Tangent

We can treat the inverse tangent in a way similar to that of the logarithm, starting with the formula

$$\frac{1}{1+t^2} = 1 - t^2 + t^4 - + \cdots + (-1)^{n-1} t^{2n-2} + r_n$$

where now $r_n = (-1)^n \frac{t^{2n}}{1+t^2}.$

By integration [see Eq. (14), p. 263], we obtain

$$\operatorname{arc\,tan} x = x - \frac{x^3}{3} + \frac{x^5}{5} - + \cdots + (-1)^{n-1} \frac{x^{2n-1}}{2n-1} + R_n,$$

$$R_n = (-1)^n \int_0^x \frac{t^{2n}}{1+t^2}\,dt;$$

we see at once that in the closed interval $-1 \leq x \leq 1$ the remainder R_n tends to zero as n increases, since

$$|R_n| \leq \int_0^{|x|} t^{2n}\, dt = \frac{|x|^{2n+1}}{2n+1}.$$

From the formula for the remainder we can also easily show that for $|x| > 1$ the absolute value of the remainder increases beyond all bounds as n increases.

We have accordingly deduced the infinite series

(6) $\arc \tan x = x - \dfrac{x^3}{3} + \dfrac{x^5}{5} - + \cdots + (-1)^{n-1} \dfrac{x^{2n-1}}{2n-1} + - \cdots,$

valid for the closed interval $|x| \leq 1$. Since $\arc \tan 1 = \pi/4$, we obtain for $x = 1$, the *Leibnitz-Gregory* series

(7) $\dfrac{\pi}{4} = 1 - \tfrac{1}{3} + \tfrac{1}{5} - + \cdots,$

an expression as remarkable as that found earlier for $\log 2$.

5.3   Taylor's Theorem

Newton's pupil Taylor, observed that the elementary expansion of polynomials lends itself to a wide generalization for nonpolynomial functions, provided that these functions are sufficiently differentiable and that their domain is suitably restricted.

a. Taylor's Representation of Polynomials

This is an entirely elementary algebraic formula concerning a polynomial in x of order n, say

$$f(x) = a_0 + a_1 x + a_2 x^2 + \cdots + a_n x^n.$$

If we replace x by $a + h = b$ and expand each term in powers of h, there results immediately a representation of the form

(8) $f(a + h) = c_0 + c_1 h + c_2 h^2 + \cdots + c_n h^n.$

Taylor's formula is the relation

(8a) $c_v = \dfrac{1}{v!} f^{(v)}(a),$

for the coefficients c_v in terms of f and its derivatives at $x = a$. To prove this fact we consider the quantity $h = b - a$ as the independent variable, and apply the chain rule which shows that differentiation with respect to h is the same as differentiation with respect to $b = a + h$. Thus successively differentiating the formula (8) with respect to h and each time thereafter, substituting $h = 0$ yields successively the results

$$c_0 = f(a), \; c_1 = f'(a), \ldots, \nu! \, c_\nu = f^{(\nu)}(a)$$

and therefore indeed the Taylor formula for polynomials:

$$(9) \quad f(a + h) = f(a) + hf'(a) + \frac{h^2}{2}f''(a) + \ldots + \frac{h^n}{n!}f^{(n)}(a).$$

The $(n + 1)$st derivative vanishes for a polynomial of degree n, and thus our formula (9) naturally terminates.

As stated the formula (9) is nothing but an elementary algebraic rearrangement of a polynomial in powers of $a + h$, into a polynomial in powers of h.

b. Taylor's Formula for Nonpolynomial Functions

Newton and his immediate pupils boldly applied formula (9) to nonpolynomial functions for which the expansion does not automatically stop at the nth term; instead they simply allowed n to increase to infinity, a procedure which for many of the important special functions will be justified later on.

Assuming the function f differentiable at least n times in an interval containing the points a and $a + h$ we certainly can no longer write for $f(a + h)$ an expression as in (9) of a finite number of powers of h, but must account for the discrepancy by an additional "remainder" R_n, writing tentatively

$$(10) \quad f(b) = f(a + h) = f(a) + hf'(a) + \cdots + \frac{h^n}{n!}f^n(a) + R_n;$$

in fact, (10) is nothing but a definition of the corrective *remainder term* R_n and indicates the expectation that R_n might become small and tend to zero for $n \to \infty$. If the remainder indeed tends to zero, then the formula (10) in the limit $n \to \infty$ leads to an expansion

$$(11) \quad f(a + h) = f(a) + hf'(a) + \cdots + \frac{h^n}{n!}f^{(n)}(a) + \cdots$$

of $f(x)$ as an infinite *power series* in h.

The crucial problem, far transcending in difficulty that of the algebraic manipulations in Section 5.3a is then to find estimates for the remainder R_n so that the accuracy of Taylor's representation by the finite *Taylor polynomial* of order n in h

$$(12) \qquad T_n(h) = \sum_0^n \frac{f^{(\nu)}(a)}{\nu!} h^\nu$$

and the passage to the limit for $n \to \infty$, can be rigorously explored. Taylor's polynomial $T_n(h)$ is an approximation to $f(a + h)$ in the sense that at $h = 0$ the functions T_n and f, as well as their derivatives up to order n coincide, so that the difference $R_n = f - T_n$ vanishes at $x = a$ together with its first n derivatives.

5.4   Expression and Estimates for the Remainder

a. Cauchy's and Lagrange's Expressions

A direct representation of the remainder R_n, allowing estimates of its absolute value $|R_n|$, is the core of Taylor's theorem. The results are easily obtained on the basis of the mean value theorem of calculus. They are moreover related to the *linear approximation of functions by differentials* (see p. 179).

Let us first examine again this approximation.

The definition of derivative at the point a states merely that $f(a + h) = f(a) + hf'(a) + h\epsilon$, where $\epsilon \to 0$ for $h \to 0$. We can attain a somewhat sharper approximation by ascertaining that ϵ is in fact of order at least as small as h, provided that not only f' but also f'' exists and is continuous in our interval J. The estimate is obtained if we write again $a + h = b$, introduce a remainder R by

$$(13) \qquad f(b) = f(a) + (b - a)f'(a) + R,$$

and now consider b as fixed and the initial point a as variable; this equation defines R as a function of a in the interval J; then differentiation with respect to the variable a yields zero on the left-hand side since $f(b)$ is constant and the rule for differentiating a product shows that

$$0 = f'(a) - f'(a) + (b - a)f''(a) + R'(a)$$

and hence

$$(14) \qquad -R'(a) = (b - a)f''(a).$$

Now, for $a = b$ we obviously have $R(b) = 0$. By the mean value theorem of calculus $[R(a) - R(b)]/(b - a) = -R'(\xi)$, where ξ is a not otherwise specified value between a and b; because of $R(b) = 0$ we therefore conclude $R(a) = -(b - a)R'(\xi) = -hR'(\xi)$. Now by (14) $R'(\xi) = -(b - \xi)f''(\xi)$ and hence $|R'(\xi)| < h\,|f''(\xi)|$ since $|b - \xi| \leq h$.

Since $|f''(\xi)|$ is bounded in an interval around a, we obtain finally an estimate that shows that the remainder or "error" R_n is small of at least second order in h:

(15) $$|R(a)| < h^2 |f''(\xi)|.$$

We turn from the special case $n = 1$ to that of any order n. The direct characterization of the remainder R_n is achieved by the same device as for $n = 1$. We assume that a and $b = a + h$ are points in an interval J in which $f(x)$ is defined and has continuous derivatives up to the order $n + 1$. We consider a as the independent variable and keep the end point b fixed. In formula (10), p. 446, which defines $R_n(a)$, we write $b - a$ instead of h. Differentiating and taking into account that $f(b)$ is constant, we find from the product rule that almost all terms cancel out, and we are left with the formula.

(16) $$0 = \frac{(b - a)^n}{n!} f^{(n+1)}(a) + R_n{}'(a)$$

for every value a in the interval. Since for $a = b$ the remainder R_n is zero, this direct expression for its derivative as a function of a completely characterizes R_n as the integral $\int_b^a R_n{}'(t)\, dt = -\int_a^b R_n{}'(t)\, dt$ or

(17) $$R_n(a) = \int_a^b \frac{(b - t)^n}{n!} f^{n+1}(t)\, dt.$$

This is an *exact integral representation of the remainder*.

An estimate for R_n similar to the one obtained above for $n = 1$ follows directly by the mean value theorem of calculus applied to (16):

$$\frac{R_n(a) - R_n(b)}{b - a} = \frac{R_n(a)}{b - a} = -R_n{}'(\xi) = \frac{(b - \xi)^n}{n!} f^{n+1}(\xi)$$

or

(18) $$R_n(a) = \frac{(b - a)(b - \xi)^n}{n!} f^{n+1}(\xi),$$

where ξ is a suitable, not specified, intermediate value between a and b. The same estimate can also be obtained by applying to the expression (17) the mean value theorem of integral calculus (Chapter 2, p. 141).

Cauchy's Form of the Remainder. If we define $\xi = a + \theta h$ $= a + \theta(b - a)$ we obtain *Cauchy's formula* for the remainder in Taylor's formula (10)

(19) $$R_n(a) = \frac{h^{n+1}}{n!} (1 - \theta)^n f^{(n+1)}(a + \theta h),$$

where θ is an unspecified quantity between 0 and 1.

We can also apply to the integral (17) for R_n the *generalized mean value theorem* of integral calculus (see p. 142) taking for the "weight function" $p(t)$ the expression $p(t) = (b - t)^n$ which does not change sign throughout the interval of integration.[1] Then

$$(20) \qquad R_n = \frac{1}{n!} f^{(n+1)}(\xi) \int_a^b (b - t)^n \, dt = \frac{(b - a)^{n+1}}{(n + 1)!} f^{(n+1)}(\xi).$$

Lagrange's Form of the Remainder. Setting again $\xi = a + \theta h$ yields *Lagrange's* form for the remainder

$$(21) \qquad R_n(a) = \frac{h^{n+1}}{(n + 1)!} f^{(n+1)}(a + \theta h)$$

with a suitable quantity θ satisfying $0 \leq \theta \leq 1$. Lagrange's form is particularly suggestive, and hence more commonly applied, since it makes the remainder R_n in the formula

$$(22) \quad f(a + h) = f(a) + \frac{h}{1!} f'(a) + \frac{h^2}{2!} f''(a) + \cdots$$
$$+ \frac{h^n}{n!} f^{(n)}(a) + R_n = P_n(h) + R_n$$

look like the term $h^{n+1} f^{(n+1)}(a)/(n + 1)!$ that would arise in the expansion (22) to one order higher, only with the argument a replaced by the intermediate value $a + \theta h$.

For a function f for which $f^{(n+1)}$ is continuous in a closed interval containing the point a, the quantity $|f^{(n+1)}(\xi)|$ has a fixed bound M. Since then

$$|R_n| \leq \frac{h^{n+1}}{(n + 1)!} M$$

the Taylor polynomial $P_n(h)$ gives for fixed n an approximation to the function $f(a + h)$ with an error of order at least $n + 1$ in h.

Our interest will be directed chiefly toward the question whether the remainder R_n tends to zero as n increases; if this is the case, we say that we have expanded the function in an infinite Taylor series

$$(23) \quad f(a + h) = f(a) + \frac{h}{1!} f'(a) + \frac{h^2}{2!} f''(a) + \frac{h^3}{3!} f'''(a) + \cdots;$$

---

[1] The generalized mean value theorem was proved for the case of a positive $p(t)$, but it applies equally well when $p(t)$ is negative throughout the interval of integration.

in particular, if we first put $a = 0$ and then write x in place of h, we obtain the "power series"

$$f(x) = f(0) + \frac{x}{1!}f'(0) + \frac{x^2}{2!}f''(0) + \cdots .$$

We shall discuss examples in Section 5.5.

For applications the finite Taylor expansion (22) for a fixed n with the remainder term is just as important. If we let h tend to zero in this formula in the terminology of Chapter 3, p. 252, the various terms of the series tend to zero with different orders of magnitude in h. The expression $f(a)$ represents the term of zero order in Taylor's series, the expression $hf'(a)$ the term of first order, the expression $h^2 f''(a)/2!$ the term of second order, etc. We see from the form of the remainder that in expanding a function as far as the term of nth order we make an error which tends to zero of order $(n + 1)$ as h tends to zero. The nearer the point $a + h$ lies to the point a, the better is the representation of the function $f(a + h)$ by the approximating polynomial $P_n(h)$; in the cases of greatest interest the approximation in the immediate neighborhood of the point x can be improved by increasing the value of n.

b. An Alternative Derivation of Taylor's Formula

The integral representation (17) for the remainder term R_n in Taylor's theorem was based on formula (16) for $R_n{'}(a)$. Because of the importance of the theorem we give here a different version of the derivation, which leads directly to the expression for R_n by repeated *integration by parts* starting with the formula:

$$(24) \qquad\qquad f(b) - f(a) = \int_a^b f'(t)\, dt.$$

To transform (24) by successive integration by parts, we introduce the functions

$$\phi_1(t),\ \phi_2(t),\ \ldots,\ \phi_\nu(t),\ \ldots,$$

by the relations:

$$(25) \qquad\qquad \phi_0(t) = 1, \qquad \phi_\nu{'}(t) = \phi_{\nu-1}(t)$$

and the conditions

$$(26) \qquad\qquad \phi_\nu(b) = 0, \qquad \text{for} \quad \nu \geq 1$$

where we consider b as a fixed parameter. Clearly, the conditions (25) and (26) determine successively all $\phi_\nu(t)$. As is verified immediately the $\phi_\nu(t)$ are just the polynomials

$$\phi_\nu(t) = \frac{(t-b)^\nu}{\nu!}$$

We note in passing that the functions ϕ_ν originate from each other by successive integration, leaving constants of integration open; therefore the defining conditions (25) could also be satisfied by functions satisfying other side conditions instead of (26) (see p. 189).

Since $\phi_\nu(a) = (-1)^\nu (b-a)^\nu/\nu!$ and $\phi_\nu(b) = 0$, we obtain

$$f(b) - f(a) = \int_a^b \phi_0 f'\, dt = \int_a^b \phi_1' f'\, dt = \phi_1 f' \Big|_a^b - \int_a^b \phi_1 f''\, dt;$$

integrating the last term again by parts, we find

$$f(b) - f(a) = (b-a)f'(a) - \int_a^b \phi_2' f''\, dt$$

$$= (b-a)f'(a) + \frac{(b-a)^2}{2!} f''(a) + \int_a^b \phi_2 f'''\, dt,$$

and repeating the process n times,

$$f(b) - f(a) = (b-a)f'(a) + \frac{(b-a)^2}{2!} f''(a) + \cdots + \frac{(b-a)^n}{n!} f^{(n)}(a)$$

$$+ \int_a^b (-1)^n \phi_n(t) f^{(n+1)}(t)\, dt$$

$$= (b-a)f'(a) + \frac{(b-a)^2}{2!} f''(a) + \cdots + \frac{(b-a)^n}{n!} f^{(n)}(a) + R_n,$$

where, by the definition of ϕ_n,

$$R_n = \int_a^b f^{(n+1)}(t) \frac{(b-t)^n}{n!}\, dt.$$

Thus we have again proved

TAYLOR'S THEOREM. *If a function $f(t)$ has continuous derivatives up to the $(n+1)$th order on a closed interval containing the two points a and b, then:*

$$f(b) = f(a) + (b-a)f'(a) + \cdots + \frac{(b-a)^n}{n!} f^{(n)}(a) + R_n$$

with the remainder R_n, depending on n, a, and b, given by the expression

$$(27) \qquad R_n = \frac{1}{n!} \int_a^b (b - t)^n f^{(n+1)}(t) \, dt.$$

By changes of notation we obtain slightly different expressions of the Taylor formula. Thus, replacing a by x and b by $x + h$, we have

$$(27a) \qquad f(x + h) = f(x) + hf'(x) + \cdots + \frac{h^n}{n!} f^{(n)}(x) + R_n,$$

with

$$R_n = \frac{1}{n!} \int_x^{x+h} (x + h - t)^n f^{(n+1)}(t) \, dt;$$

or with $t = x + \tau$,

$$(27b) \qquad R_n = \frac{1}{n!} \int_0^h (h - \tau)^n f^{(n+1)}(x + \tau) \, d\tau.$$

If we set $x = 0$ and write x in place of h, we obtain[1]

$$(27c) \qquad f(x) = f(0) + \frac{x}{1!} f'(0) + \frac{x^2}{2!} f''(0) + \cdots$$

$$+ \frac{x^n}{n!} f^{(n)}(0) + R_n$$

with the remainder

$$R_n = \frac{1}{n!} \int_0^x (x - t)^n f^{(n+1)}(t) \, dt.$$

Applying the mean value theorem of integral calculus or its generalized form to the integral leads to the Cauchy formula

$$R_n = \frac{(1 - \theta)^n}{n!} x^{n+1} f^{(n+1)}(\theta x)$$

and respectively, the Lagrange formula

$$R_n = \frac{x^{n+1}}{(n + 1)!} f^{(n+1)}(\theta x)$$

for the remainder, as was shown before (p. 448). Here θ is a suitable nonspecified number with $0 \le \theta \le 1$ (not the same in both formulas).

---

[1] This special case of the theorem is sometimes without historical justification, called Maclaurin's theorem. Taylor's general theorem was published in 1715; Maclaurin's special result, in 1742.

As an exercise, the reader should construct functions ϕ_ν satisfying (25) for which the side conditions (26) are replaced by the relations

$$\int_0^1 \phi_\nu(t)\, dt = 0$$

for $\nu \geq 1$ (see Chapter 8, Appendix A).

5.5   Expansions of the Elementary Functions

The preceding general results permit us to expand the simple elementary functions in Taylor series. Expansions of other functions will be discussed in Chapter 7.

a. The Exponential Function

First we expand the exponential function, $f(x) = e^x$. In this case all the derivatives are identical with $f(x)$ and have the value 1 for $x = 0$. Lagrange's form for the remainder (p. 449, Equation (21)), yields at once the formula:

$$e^x = 1 + \frac{x}{1!} + \frac{x^2}{2!} + \frac{x^3}{3!} + \cdots + \frac{x^n}{n!} + \frac{x^{n+1}}{(n+1)!} e^{\theta x}, \qquad 0 < \theta < 1.$$

If we now let n increase beyond all bounds, the remainder R_n tends to zero for any fixed value of x. To prove this we note first that $e^{\theta x} \leq e^{|x|}$ since e^x is a monotone increasing function. Let m be any integer greater than $2|x|$. Then for all $k \geq m$, $|x|/k < \frac{1}{2}$, and

$$\left| \frac{x^{n+1}}{(n+1)!} \right| = \frac{|x^m|}{m!} \cdot \frac{|x|}{m+1} \cdots \frac{|x|}{n+1}$$

$$\leq \frac{|x^m|}{m!} \cdot \frac{1}{2^{n+1-m}} \leq \frac{|2x|^m}{m!} \cdot \frac{1}{2^n}$$

so that

$$|R_n| \leq \frac{|2x|^m}{m!} \cdot e^{|x|} \cdot \frac{1}{2^n}.$$

Since the first two factors on the right are independent of n, whereas $1/2^n \to 0$ for $n \to \infty$ our statement is proved.

The function e^x therefore is represented by the infinite series

$$e^x = 1 + \frac{x}{1!} + \frac{x^2}{2!} + \frac{x^3}{3!} + \cdots$$

$$= \sum_{\nu=0}^{\infty} \frac{x^\nu}{\nu!}.$$

This expansion is *valid for all values of x*. In particular, for $x = 1$ we obtain again the infinite series that served to define the number e in Chapter 1 (cf. p. 77).

Of course, for numerical calculations we must make use of the form of Taylor's theorem with the remainder; for $x = 1$, for example, (compare with similar computation on p. 78) we have

$$e = 1 + 1 + \frac{1}{2!} + \frac{1}{3!} + \cdots + \frac{1}{n!} + \frac{e^\theta}{(n+1)!}.$$

If we wish to calculate e with an error of at most $1/10{,}000$, we need only choose n so large that the remainder is less than $1/10{,}000$, and since this remainder is certainly less[1] than $3/(n+1)!$, it suffices to choose $n = 7$, since $8! > 30{,}000$. We thus obtain the approximate value $e = 2.71825$, with an error less than 0.0001.

b. Expansion of sin x, cos x, sinh x, cosh x

For the functions sin x, cos x, sinh x, cosh x we find the following formulas:

$$
\begin{array}{ccccc}
f(x) = & \sin x & \cos x & \sinh x & \cosh x, \\
f'(x) = & \cos x & -\sin x & \cosh x & \sinh x, \\
f''(x) = & -\sin x & -\cos x & \sinh x & \cosh x, \\
f'''(x) = & -\cos x & \sin x & \cosh x & \sinh x, \\
f^{(4)}(x) = & \sin x & \cos x & \sinh x & \cosh x.
\end{array}
$$

Thus in the approximating polynomials in x for sin x and sinh x, the coefficients of the even powers of x will vanish, whereas in those for cos x and cosh x the coefficients of the odd powers vanish.

---

[1] Here we have made use of the fact that $e < 3$. This follows (cf. p. 78) from our series for e; for it is always true that $1/n! \le 1/2^{n-1}$, and therefore

$$e < 1 + 1 + \tfrac{1}{2} + \tfrac{1}{4} + \cdots = 1 + \frac{1}{1 - \frac{1}{2}} = 3.$$

When we use Lagrange's form of the remainder (21), p. 449, the Taylor series for our functions take the form:

$$\sin x = x - \frac{x^3}{3!} + \frac{x^5}{5!} + \cdots + \frac{(-1)^n x^{2n+1}}{(2n+1)!}$$

$$+ \frac{(-1)^{n+1} x^{2n+3} \cos(\theta x)}{(2n+3)!} \, ;$$

$$\cos x = 1 - \frac{x^2}{2!} + \frac{x^4}{4!} - + \cdots + \frac{(-1)^n x^{2n}}{(2n)!}$$

$$+ \frac{(-1)^{n+1} x^{2n+2} \cos(\theta x)}{(2n+2)!} \, ,$$

$$\sinh x = x + \frac{x^3}{3!} + \frac{x^5}{5!} + \cdots + \frac{x^{2n+1}}{(2n+1)!}$$

$$+ \frac{x^{2n+3} \cosh(\theta x)}{(2n+3)!} \, ,$$

$$\cosh x = 1 + \frac{x^2}{2!} + \frac{x^4}{4!} + \cdots + \frac{x^{2n}}{(2n)!}$$

$$+ \frac{x^{2n+2} \cosh(\theta x)}{(2n+2)!} \, ,$$

where, of course, in each of the four formulas θ denotes a different number in the interval $0 \leq \theta \leq 1$, a number which in addition depends on n and on x. Since in each of these formulas, the remainder tends to zero as n increases, as can be seen by exactly the same argument as in the case of e^x, we can make the approximations as precise as we wish. We thus obtain the four infinite series, valid for all values x:

$$\sin x = x - \frac{x^3}{3!} + \frac{x^5}{5!} - + \cdots = \sum_{\nu=0}^{\infty} \frac{(-1)^\nu x^{2\nu+1}}{(2\nu+1)!} \, ,$$

$$\cos x = 1 - \frac{x^2}{2!} + \frac{x^4}{4!} - + \cdots = \sum_{\nu=0}^{\infty} \frac{(-1)^\nu x^{2\nu}}{(2\nu)!} \, ,$$

$$\sinh x = x + \frac{x^3}{3!} + \frac{x^5}{5!} + \cdots = \sum_{\nu=0}^{\infty} \frac{x^{2\nu+1}}{(2\nu+1)!} \, ,$$

$$\cosh x = 1 + \frac{x^2}{2!} + \frac{x^4}{4!} + \cdots = \sum_{\nu=0}^{\infty} \frac{x^{2\nu}}{(2\nu)!} \, .$$

The last two may also be obtained formally from the series for e^x in accordance with the definitions of the hyperbolic functions (see p. 228).

c. The Binomial Series

We pass over the Taylor series for the functions $\log(1 + x)$ and arc tan x already treated directly in Section 5.2. We shall, however, take up the generalization of the binomial theorem for arbitrary exponents, which was one of the most spectacular of Newton's mathematical discoveries. We wish to expand the function $f(x) = (1 + x)^\alpha$ in a Taylor series where $x > -1$ and α is an arbitrary number, positive or negative, rational or irrational. The function $(1 + x)^\alpha$ is chosen instead of x^α since for the latter at the point $x = 0$ it is not true that all the derivatives are continuous, except in the trivial case of nonnegative integral values of α. We first calculate the derivatives of $f(x)$, obtaining

$$f'(x) = \alpha(1 + x)^{\alpha-1},$$
$$f''(x) = \alpha(\alpha - 1)(1 + x)^{\alpha-2}, \dots,$$
$$\dots\dots\dots\dots\dots\dots\dots\dots\dots\dots\dots\dots\dots\dots\dots$$
$$f^{(\nu)}(x) = \alpha(\alpha - 1)\cdots(\alpha - \nu + 1)(1 + x)^{\alpha-\nu}.$$

In particular, for $x = 0$ we have

$$f'(0) = \alpha, \quad f''(0) = \alpha(\alpha - 1), \dots,$$
$$f^{(\nu)}(0) = \alpha(\alpha - 1)\cdots(\alpha - \nu + 1).$$

Taylor's theorem then states

$$(1 + x)^\alpha = 1 + \alpha x + \frac{\alpha(\alpha - 1)}{2!} x^2 + \cdots$$
$$+ \frac{\alpha(\alpha - 1)(\alpha - 2)\cdots(\alpha - n + 1)}{n!} x^n + R_n.$$

Convergence

We must yet discuss the remainder. This problem is not very difficult, but nonetheless is not quite so simple as the cases previously treated. We shall obtain an estimate for the remainder both directly and also as a special case of a general result of Section A.4. This will permit us to conclude that whenever $|x| < 1$, the remainder R_n for the binomial expansion tends to zero. Thus the expression $(1 + x)^\alpha$ may be expanded in the infinite binomial series

$$(1 + x)^\alpha = 1 + \frac{\alpha}{1!} x + \frac{\alpha(\alpha - 1)}{2!} x^2 + \cdots$$
$$= \sum_{\nu=0}^{\infty} \binom{\alpha}{\nu} x^\nu,$$

where for brevity we have introduced the general binomial coefficients

$$\binom{\alpha}{\nu} = \frac{\alpha(\alpha - 1) \cdots (\alpha - \nu + 1)}{\nu!} \quad \text{(for } \nu > 0\text{)},$$

$$\binom{\alpha}{0} = 1.$$

*To prove directly that the remainder $R_n \to 0$ for $n \to \infty$ in the case where $-1 < x < 1$, we make use of Cauchy's form of the remainder (19), p. 448:

$$R_n = \frac{(1 - \theta)^n}{n!} x^{n+1} f^{(n+1)}(\theta x)$$

$$= \frac{(1 - \theta)^n}{n!} \alpha(\alpha - 1)(\alpha - 2) \cdots (\alpha - n) x^{n+1}(1 + \theta x)^{\alpha - n - 1}$$

$(0 \le \theta \le 1)$. Since $|x| < 1$, we have $0 \le (1 - \theta)/(1 + \theta x) \le 1$ so that

$$|R_n| \le (1 + \theta x)^{\alpha - 1} |\alpha x| \left| \left(1 - \frac{\alpha}{1}\right) x \right| \left| \left(1 - \frac{\alpha}{2}\right) x \right| \cdots \left| \left(1 - \frac{\alpha}{n}\right) x \right|.$$

There exists a number q with $|x| < q < 1$. Then obviously also

$$\left| \left(1 - \frac{\alpha}{m}\right) x \right| < q$$

for all sufficiently large m, say for $m > N$. Thus for $n > N$

$$|R_n| \le (1 + \theta x)^{\alpha - 1} |\alpha| (1 + |\alpha|)^N q^{n - N}.$$

The factor $(1 + x\theta)^{\alpha - 1}$ is bounded (by $2^{\alpha - 1}$ if $\alpha \ge 1$, by $(1 - q)^{\alpha - 1}$ if $\alpha < 1$) so that clearly $R_n \to 0$.

A slightly more general formula gives an expression for $(a + b)^\alpha$. We only have to factor out a^α and apply the binomial expansion with $x = b/a$ to obtain for $a > 0$ and $|b| < a$

$$(a + b)^\alpha = a^\alpha \left(1 + \frac{b}{a}\right)^\alpha = a^\alpha \left(1 + \alpha \frac{b}{a} + \frac{\alpha(\alpha - 1)}{1 \cdot 2} \left(\frac{b}{a}\right)^2 + \cdots\right)$$

$$= a^\alpha + \frac{\alpha}{1} a^{\alpha - 1} b + \frac{\alpha(\alpha - 1)}{1 \cdot 2} a^{\alpha - 2} b^2 + \cdots.$$

5.6 Geometrical Applications

The behavior of a function $f(x)$ in a neighborhood of the point $x = a$, or the behavior of a given curve in a neighborhood of one of its points, can be described in detail by means of Taylor's theorem, since this theorem permits us to resolve the increment of the function on passing to a neighboring point $x = a + h$ into a sum of quantities of the first order, second order, etc., in h.

a. Contact of Curves

Contact of Higher Order

If at a point $x = a$, two curves $y = f(x)$ and $y = g(x)$ intersect and have a common tangent, we say that the curves touch one another or have contact of the first order. In this case the Taylor expansions of the functions $f(a + h)$ and $g(a + h)$ have the same terms of zero order and first order in h. If, in addition, at the point $x = a$ the second derivatives of $f(x)$ and $g(x)$ are also equal to each other, we say that the curves have contact of the second order. Then the terms of second order in the Taylor expansions of f and g will also agree. If we assume that both functions have continuous derivatives of at least the third order, then the difference

$$D(x) = f(x) - g(x)$$

can be expressed in the form

$$D(a + h) = f(a + h) - g(a + h)$$
$$= \frac{h^3}{3!} D'''(a + \theta h) = \frac{h^3}{3!} F(h),$$

where the expression $F(h)$ tends to $f'''(a) - g'''(a)$ as h tends to zero. The difference $D(a + h)$ therefore vanishes to at least the third order with h.

We can proceed in this way and consider the general case where the Taylor series for $f(x)$ and $g(x)$ agree up to terms of the nth order; that is,

$$f(a) = g(a), f'(a) = g'(a), \ldots, f^{(n)}(a) = g^{(n)}(a).$$

We assume that the $(n + 1)$th derivatives are continuous. Under these conditions the curves defined by our two functions are said to have contact of the nth order at the point $x = a$. The difference of the two functions is then of the form

$$D(a + h) = f(a + h) - g(a + h) = \frac{h^{n+1}}{(n + 1)!} D^{(n+1)}(a + \theta h)$$
$$= \frac{h^{n+1}}{(n + 1)!} F(h),$$

where since $0 \le \theta \le 1$ the quantity $F(h) = D^{(n+1)}(a + \theta h)$ tends to $f^{(n+1)}(a) - g^{(n+1)}(a)$ as h tends to zero. We see from this formula that at the point of contact the difference $f(x) - g(x)$ vanishes to at least the $(n + 1)$th order.

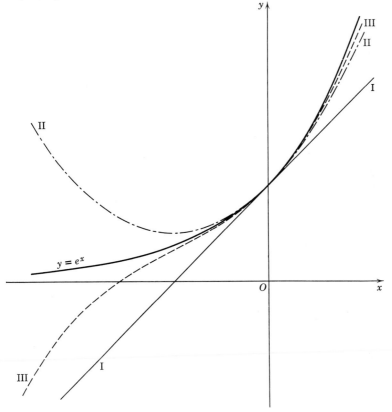

Figure 5.1 Osculating parabolas of e^x.

The Taylor polynomials defined by

$$P_n(x) = f(a) + \frac{x - a}{1!} f'(a) + \cdots + \frac{(x - a)^n}{n!} f^{(n)}(a)$$

are characterized geometrically as the "parabolas" of the nth order having contact of the greatest possible order with the graph of the given function at the given point. Hence these parabolas are sometimes called *osculating parabolas*. (Only for $n = 2$ are these curves "parabolas" in the ordinary sense.)

For the function $y = e^x$, Fig. 5.1 shows the first three osculating parabolas at the point $x = 0$.

Two curves $y = f(x)$ and $y = g(x)$ that have contact of the nth order at a point $x = a$, might possibly have contact of an even higher order, that is, that the equation $f^{(n+1)}(a) = g^{(n+1)}(a)$ might also be true. If this is not the case, that is, if $f^{(n+1)}(a) \neq g^{(n+1)}(a)$, we say that the order of contact is exactly n.[1]

Contact of Even or Odd Order

From our formulas as well as from intuition we can state a remarkable fact often unnoticed by beginners. Let the contact of two curves be exactly of even order; that is, an even number n of derivatives of the two functions have the same value at the point in question, whereas the $(n + 1)$th derivatives differ. Then the preceding formulas show that the difference $f(a + h) - g(a + h)$ has different signs for small positive values of h and for numerically small negative values of h. The two curves then *cross* at the point of contact. This occurs, for instance, in contact of the second order if the third derivatives have different values. In contrast, contact exactly of an odd order, for example, an ordinary contact of the first order, implies that the difference $f(a + h) - g(a + h)$ has the same sign for all numerically small values of h, positive or negative; the two curves therefore do not cross in a neighborhood of the point of contact. The simplest example is the contact of a curve with its tangent. The tangent can cross the curve only at points where the contact is at least of second order; it does actually cross the curve at points where the order of contact is even, for example, an ordinary point of inflection where $f''(x) = 0$ but $f'''(x) \neq 0$. At points where the order of contact is odd the tangent does not cross the curve, as for example, at an ordinary point of the curve where the second derivative is not zero, such as for the curve $y = x^4$ at the origin.

We know from Chapter 4, p. 360, that for the circle of curvature at the point $x = a$ given by the function $y = g(x)$ in a neighborhood of the point $x = a$, we not only have $g(a) = f(a)$ and $g'(a) = f'(a)$, but also $g''(a) = f''(a)$. Hence the circle of curvature is at the same time the osculating circle at the point of the curve under discussion; that is, it is the circle which at that point has contact of the second order with the curve. In the limiting case of a point of inflection, or in general, of a point at which the curvature is zero and the radius of curvature is infinite, the circle of curvature degenerates into the tangent. In ordinary

---

[1] That the order of contact of two curves is a genuine geometrical relation which is unaffected by change of axes is a fact which can be easily confirmed by means of the formulas for change of axes (see Chapter 4, p. 360).

cases, when the contact at the point in question is not of an order higher than the second, the circle of curvature does not merely touch the curve, but also crosses it (cf. Fig. 4.23, p. 359).

In conclusion it should be mentioned that sometimes contact of order exactly m is described by saying: the curves have $m + 1$ infinitely near points in common; of course, the precise meaning of such a statement obviously refers to a limiting process. If the curves have, in fact, $m + 1$ distinct points $P, P_1, \ldots P_m$ in common and if we let all the points P_i tend to P, if necessary modifying one of the curves, then the limiting position might be expected to be that of two curves with a contact of order m. For example, if we draw a circle through three points P, P_1, P_2 on a curve C and then let P_1 and P_2 tend to P, it can be seen that the circle tends to the circle of curvature on C in P. (See Problem 4, p. 437.)

b. On the Theory of Relative Maxima and Minima

As we have already seen in Chapter 3, p. 243, a function $f(x)$, whose first derivative vanishes at $x = a$, has a relative maximum at the point if $f''(a)$ is negative, a minimum if $f''(a)$ is positive. These conditions, therefore, are sufficient conditions for the occurrence of a maximum or minimum. They are by no means necessary; for in the case when $f''(a) = 0$ there are three possibilities open; at the point in question the function may have a maximum or a minimum or neither. Examples of the three possibilities are given by the functions $y = -x^4$, $y = x^4$, and $y = x^3$ at the point $x = 0$. Taylor's theorem at once enables us to make a general statement of sufficient conditions for a maximum or a minimum. We need only to expand the function $f(a + h)$ in powers of h; the essential point is then to find whether the first nonvanishing term contains an even or an odd power of h. In the first case we have a maximum or a minimum depending on whether the coefficient of h is negative or positive; in the second case we have a horizontal inflectional tangent and neither maximum nor minimum. The reader may complete the argument for himself using the formula for the remainder.[1]

---

[1] The necessary and sufficient condition given previously (p. 242), however, is more general and more convenient in applications: provided the first derivative $f'(x)$ vanishes at only a finite number of points, a necessary and sufficient condition for the occurrence of a maximum or minimum at one of these points is that the first derivative $f'(x)$ changes sign as the curve passes through the point.

Appendix I

A.I.1   Example of a Function Which Cannot Be Expanded in a Taylor Series

The possibility of expressing a function by means of a Taylor series with remainder of $(n + 1)$th order depends essentially on the continuity and differentiability of the function at the point in question. For this reason $\log x$ cannot be represented by a Taylor series in powers of x, and the same is true of the function $x^{1/3}$ whose derivative is infinite at $x = 0$.

In order that a function may be capable of being expanded in an infinite Taylor series, all its derivatives must necessarily exist at the point in question; however, this condition is by no means sufficient. A function for which all derivatives exist and are continuous throughout an interval still need not be capable of expansion in a Taylor series; that is, the remainder R_n in Taylor's theorem may fail to tend to zero as n increases, no matter how small the interval is, in which we want to expand the function.

An important simple example of this phenomenon is the function

$$y = f(x) = e^{-1/x^2} \quad \text{for} \quad x \neq 0, \; f(0) = 0,$$

which we have already considered in the Appendix to Chapter 3, p. 255. This function and all its derivatives are continuous in every interval, even at $x = 0$, and as we have seen, at this point all the derivatives vanish, that is, $f^{(n)}(0) = 0$ for every value of n. (Geometrically, this means that the line $y = 0$ has *contact of infinite order* with the curve of the function at the point $x = 0$). Hence in the Taylor expansion

$$f(0) + \frac{x}{1!} f'(0) + \frac{x^2}{2!} f''(0) + \cdots$$

all the coefficients of the approximating polynomials $P_n(x)$ vanish, no matter what value is chosen for n. Thus the remainder remains equal to the function itself, and thus, except for $x = 0$, can not approach zero as n increases, since the function is positive for every other value of x.

Incidentally, this function is useful for the construction of functions exhibiting intuitively unexpected phenomena. For example,

$$g(x) = e^{-1/x^2} \sin (1/x)$$

supplemented by $g(0) = 0$ is again a function with derivatives of all orders, all of which vanish at $x = 0$; the graph of $y = g(x)$ near $x = 0$, intersects the x-axis infinitely many times, and oscillates infinitely often.

A.I.2   Zeros and Infinities of Functions

a. *Zeros of Order n*

The Taylor expansion of a function $f(x)$ allows us to characterize the order to which a function vanishes at a point $x = a$. We say that a function $f(x)$ has an exact n-fold zero at $x = a$ or that it vanishes there exactly of order n, if $f(a) = 0, f'(a) = 0, f''(a) = 0, \ldots, f^{(n-1)}(a) = 0$, and $f^{(n)}(a) \neq 0$. We expressly assume that in the neighborhood the function has continuous derivatives at least to the nth order. By our definition we imply that the Taylor series for the function in the neighborhood of the point can be written in the form

$$(28) \qquad f(a + h) = \frac{h^n}{n!} F(h) = \frac{h^n}{n!} f^{(n)}(a + \theta h), \qquad \theta < 1,$$

in which as h tends to zero the factor $F(h) = n! f(a + h)/h^n$ tends to a limit different from zero, namely, the value $f^{(n)}(a)$. Hence $f(a + h)$ has the same order as h^n for $h \to 0$ or vanishes to order n in the sense defined in Chapter 3, p. 252.

Similarly, expanding the derivatives $f'(x)$, $f''(x), \ldots, f^{(v)}(x)$ by Taylor's theorem with the Lagrange form of the remainder, we obtain a series of expressions

$$f'(a + h) = \frac{h^{n-1}}{(n - 1)!} F_1(h) = \frac{h^{n-1}}{(n - 1)!} f^{(n)}(a + \theta h)$$

$$(29)$$

$$f^v(a + h) = \frac{h^{n-v}}{(n - v)!} F_v(h) = \frac{h^{n-v}}{(n - v)!} f^{(n)}(a + \theta h)$$

in all of which the factors θ *may be different*, whereas the factors $F_1, F_2, \ldots, F_v$ tend *continuously* to $f^{(n)}(a)$ as $h \to 0$. Hence f' vanishes of order $n - 1$, f'' of order $n - 2$, etc.

In these formulas, of course, the assumption is made that $f(x)$ vanishes of order $n \geqslant v$.

b. *Infinity of Order v*

If a function $\phi(x)$ is defined at all points in a neighborhood of the point $x = a$, except perhaps at $x = a$ itself, and if $\phi(x) = f(x)/g(x)$, where at $x = a$ the numerator does not vanish, but the denominator possesses a ν-fold zero, we say that the function $\phi(x)$ becomes infinite of the νth order at the point $x = a$. If at the point $x = a$ the numerator has a μ-fold zero and if $\mu > \nu$, the function has a $(\mu - \nu)$fold zero there; if $\mu < \nu$, the function has a $(\nu - \mu)$fold infinity at the point.

These definitions are in agreement with the conventions already laid down (cf. Section 3.7) regarding the behavior of a function.

A.I.3   Indeterminate Expressions

We now discuss in a more precise manner, the *"indeterminate expressions"* of the form $\phi(x) = f(x)/g(x)$, in which $f(x)$ and $g(x)$ both vanish at the same point $x = a$, such as the function $(\sin x)/x$ at $x = 0$. We shall always assign to such functions the value

$$(30) \qquad \phi(a) = \lim_{h \to 0} \phi(a + h)$$

provided this limit exists.

These limiting values can be characterized by a simple rule, known as *L'Hospital's rule*, for which we assume that all derivatives of f and g that arise are continuous in an interval containing a. We furthermore assume that the denominator $g(x)$ vanishes at $x = a$ to an order ν not higher than that of the numerator $f(x)$, so that the function $\phi(x)$ does not become infinite at $x = a$. Then the rule states

$$(31) \qquad \phi(a) = \frac{f^{(\nu)}(a)}{g^{(\nu)}(a)} \, .$$

By the definition of continuity, the function $\phi(x)$ is then continuous at $x = a$, and being continuous elsewhere, as long as $g(x) \neq 0$, ϕ is continuous in an interval about a.

The proof follows immediately from the results of A.2; applying Eqs. (28) to both f and g, we find the function ϕ is, in a neighborhood of a, given by the relation

$$\phi(a + h) = \frac{f(a + h)}{g(a + h)} = \frac{f^{(\nu)}(a + \theta h)}{g^{(\nu)}(a + \theta_1 h)} \, ,$$

whence the continuity of the numerator and denominator, and the nonvanishing of $g^{(\nu)}(a)$ yield (31). We can express the meaning of the last equations in the following way: if the numerator and denominator of a function $\phi(x) = f(x)/g(x)$ both vanish at $x = a$, we can determine the limiting value as x tends to a by differentiating the numerator and denominator an equal number of times until at least one of the derivatives is not zero at the point. If we encounter a nonvanishing derivative in the denominator before one appears in the numerator, the fraction tends to zero. If a nonvanishing derivative in the numerator is met before one in the denominator, the absolute value of the fraction increases beyond all bounds.

We thus have a method of evaluating the so-called "indeterminate expression" $0/0$, that is, of determining the limiting value of a quotient in which the numerator and denominator tend to zero.

We can arrive at our results in a somewhat different way by basing the proof on the generalized mean value theorem instead of on Taylor's theorem (cf. p. 222). Accordingly, if $g'(x) \neq 0$ in a neighborhood of the point a, we have

$$\frac{f(a+h) - f(a)}{g(a+h) - g(a)} = \frac{f'(a+\theta h)}{g'(a+\theta h)},$$

where θ is the same in both numerator and denominator. Hence, in particular, when $f(a) = 0 = g(a)$,

$$\frac{f(a+h)}{g(a+h)} = \frac{f'(a+\theta h)}{g'(a+\theta h)}.$$

Here θ is a value in the interval $0 < \theta < 1$, and putting $k = \theta h$, we obtain

$$\lim_{h \to 0} \frac{f(a+h)}{g(a+h)} = \lim_{k \to 0} \frac{f'(a+k)}{g'(a+k)},$$

it being assumed that the limit on the right exists.

If $f'(a) = 0 = g'(a)$, we proceed in the same manner until we reach a first index μ for which it is no longer true that simultaneously $f^{(\mu)}(a) = 0 = g^{(\mu)}(a)$. Then

$$\lim_{h \to 0} \frac{f(a+h)}{g(a+h)} = \lim_{l \to 0} \frac{f^{(\mu)}(a+l)}{g^{(\mu)}(a+l)} = \frac{f^{(\mu)}(a)}{g^{(\mu)}(a)},$$

an expression in which we include the case when both sides are infinite.

Examples. The following examples which are significant by themselves, illustrate the application of L'Hospital's rule.

$$\lim_{x \to 0} \frac{\sin x}{x} = \frac{\cos 0}{1} = 1;$$

$$\lim_{x \to 0} \frac{1 - \cos x}{x} = \frac{\sin 0}{1} = 0;$$

$$\lim_{x \to 0} \frac{e^{2x} - 1}{\log(1 + x)} = \lim_{x \to 0} \frac{2e^{2x}}{1/(1 + x)} = 2;$$

$$\lim_{x \to 0} \frac{1 - \cos x}{x^2} = \lim_{x \to 0} \frac{\sin x}{2x} = \lim_{x \to 0} \frac{\cos x}{2} = \frac{1}{2}.$$

Other Indeterminate Forms.   We further note that other so-called indeterminate forms can also be reduced to the case we have considered; for example, the limit of

$$\frac{1}{\sin x} - \frac{1}{x},$$

as x tends to zero, is the limit of the difference of two expressions both of which become infinite, or is an "indeterminate form" $\infty - \infty$.   By the transformation

$$\frac{1}{\sin x} - \frac{1}{x} = \frac{x - \sin x}{x \sin x}$$

we at once arrive at an expression whose limit as x tends to zero is determined by our rule to be

$$\lim_{x \to 0} \frac{1 - \cos x}{x \cos x + \sin x} = \lim_{x \to 0} \frac{\sin x}{2 \cos x - x \sin x} = 0.$$

Derivatives of Indeterminate Forms

The expressions $\phi(x) = f(x)/g(x)$ defined at $x = a$ by our rule are not only continuous but also have continuous derivatives provided that f and g have continuous derivatives of sufficiently high order.

It suffices for us to establish this fact in the case where g vanishes to first order at a, or $g(a) = 0$, $g'(a) \neq 0$.   For $x \neq a$,

$$\phi'(x) = \frac{g(x)f'(x) - f(x)g'(x)}{(g(x))^2} = \frac{z(x)}{N(x)},$$

where again, both numerator and denominator vanish at $x = a$, since $f(a) = g(a) = 0$.   Hence we can determine the limiting value by applying our rule

$$\lim_{x \to a} \phi'(x) = \lim_{x \to 0} \frac{z'(x)}{N'(x)}.$$

Clearly, $d(N(x))/dx = 2g(x)g'(x)$, $d(z(x))/dx = g(x)f''(x) - f(x)g''(x)$, both of which again vanish at $x = a$.   Applying L'Hospital's rule once more, as

$$\lim_{x \to a} \phi'(x) = \lim_{x \to a} \frac{z''(x)}{N''(x)},$$

and noting that $N''(x) = 2g(x)g''(x) + 2(g'(x))^2$, which does not vanish at $x = a$, we find that

$$\lim_{x \to a} \phi'(x) = \frac{g'(a)f''(a) - f'(a)g''(a)}{2(g'(a))^2},$$

and this limit is indeed the derivative of $\phi'(x)$ at $x = a$ (see Chapter 3, p. 261).

Similar rules for indefinite forms hold for $x \to \infty$. Thus let $f(x)$ and $g(x)$ be functions for which $\lim\limits_{x \to \infty} f(x) = \lim\limits_{x \to \infty} g(x) = 0$ while $\lim\limits_{x \to \infty} f'(x)$ and $\lim\limits_{x \to \infty} g'(x)$ exist and are $\neq 0$. Then

$$\lim_{x \to \infty} \frac{f(x)}{g(x)} = \frac{\lim\limits_{x \to \infty} f'(x)}{\lim\limits_{x \to \infty} g'(x)}.$$

The proof follows again from the mean value theorem of differential calculus.

*A.I.4 The Convergence of the Taylor Series of a Function with Nonnegative Derivatives of All Orders

We insert a general theorem concerning the convergence of Taylor's expansion for functions all of whose derivatives are nonnegative.

Consider the class of functions $f(x)$, differentiable to all orders on the closed interval $a \leq x \leq b$, all of whose derivatives are nonnegative on this interval:

$$f^{(\nu)}(x) \geq 0, \qquad \nu = 1, 2, \ldots .$$

We shall show: For every such function the corresponding Taylor expansion of $f(x + h)$ in powers of h converges, and the series represents the value of $f(x + h)$ when x and $\xi = x + h$ lie in the open interval (a, b) and $|h| < b - x$.

For the proof we start with the observation that $f'(x) \geq 0$ by assumption and hence

$$0 \leq f(x) - f(a) = \int_a^x f'(\xi)\, d\xi$$

$$\leq \int_a^b f'(\xi)\, d\xi = f(b) - f(a) = M.$$

Moreover, for x and $\xi = x + h$ in the interval between a and b, we may write

$$f(x + h) - f(x) = hf'(x) + \cdots + \frac{f^{(n)}(x)}{n!} h^n + R_n.$$

Assume first that $h > 0$, or $x < \xi < b$. Then all of the terms on the right-hand side are nonnegative[1] and so each is not greater than the

---

[1] This follows for R_n from the Cauchy or Lagrange formulas and the assumption $f^{(n+1)} \geq 0$.

value of the left-hand side or than M; thus

$$0 \le \frac{f^{(n)}(x)}{n!} \le \frac{M}{h^n} = \frac{M}{(\xi - x)^n}.$$

For $\xi \to b$ it follows that

(32)
$$\frac{f^{(n)}(x)}{n!} \le \frac{M}{(b - x)^n}.$$

Now, using Cauchy's formula ((19), p. 448) for the remainder, we know there exists some θ in the interval $0 \le \theta \le 1$, such that

$$0 \le R_n = \frac{(1 - \theta)^n}{n!} h^{n+1} f^{(n+1)}(x + \theta h)$$

$$\le \frac{h^{n+1}(n + 1)(1 - \theta)^n M}{(b - x - \theta h)^{n+1}}.$$

Since $\xi = x + h < b$, we may choose a positive number p such that

$$0 \le h \le \frac{b - x}{1 + p} \quad \text{or} \quad b - x - \theta h \ge h(1 + p - \theta).$$

We then have

$$0 \le R_n \le \frac{Mh^{n+1}(n + 1)(1 - \theta)^n}{h^{n+1}(1 + p - \theta)^{n+1}}$$

or

$$0 \le R_n \le \frac{M(n + 1)}{(1 + p - \theta)} \left(\frac{1 - \theta}{1 - \theta + p} \right)^n \le \frac{M(n + 1)}{p} \frac{1}{(1 + p)^n}$$

since

$$\frac{1 - \theta}{1 - \theta + p} = \frac{1}{1 + p/(1 - \theta)} \le \frac{1}{1 + p} < 1.$$

We know (Chapter 1, p. 70) that $(n + 1)/(1 + p)^n$ tends to zero as n increases, so that R_n tends to zero as n increases, when $0 \le h < b - x$; thus Taylor's series tends to the function f for $h \ge 0$.

For negative h, the fact that R_n tends to zero with increasing n follows by using the Lagrange form (21), p. 449, for R_n:

$$|R_n| = \frac{1}{(n + 1)!} |h^{n+1}| |f^{(n+1)}(x - \theta |h|)|.$$

Now $f^{(n+2)}$ is nonnegative and hence $f^{(n+1)}$ is monotone nondecreasing. It follows then from the estimate (32) used above that

$$\frac{f^{(n+1)}(x - \theta |h|)}{(n + 1)!} \le \frac{f^{(n+1)}(x)}{(n + 1)!} \le \frac{M}{(b - x)^{n+1}}.$$

Therefore
$$|R_n| \leq \left(\frac{|h|}{b-x}\right)^{n+1} M$$
and so R_n tends to zero as n increases when
$$0 < -h < b - x.$$

Thus for any point x with $a \leq x < b$, the remainder R_n in the Taylor series for $f(x + h)$ in powers of h will tend to zero once $|h| < b - x$ and $h > -(x - a)$.

We note that our result is still true if we assume the inequality $f^{(v)}(x) \geq 0$ only for all sufficiently large v, say for $v > N$ for some integer N, whereas when $v \leq N$ the sign of $f^{(v)}(x)$ may be arbitrary. To prove this we need only replace the function f in our proof by the function
$$g(x) = f(x) + M(x - a + 1)^N,$$
for M some positive constant. Then $g^{(v)}(x) = f^{(v)}(x) \geq 0$ for $v > N$, and $g^{(v)}(x) = f^{(v)}(x) + MN(N-1)\cdots(N-v+1)(x-a+1)^{N-v} \geq f^{(v)}(x) + M$ for $v \leq N$. Thus $g^{(v)}(x) \geq 0$ for all v if M is chosen sufficiently large. This proves that $g(x)$ can be expanded in powers of x, and the same result follows then for the function f, which differs from g only by a polynomial.

The theorem on the binomial series (p. 456) is an immediate consequence of this result: We change the notation slightly and consider first the function $\phi(x) = (1 - x)^\alpha$ in place of $(1 + x)^\alpha$. The derivatives of ϕ are then given by
$$\phi^{(v)}(x) = (-1)^v \binom{\alpha}{v}(1 - x)^{\alpha-v} v!$$

Since the binomial coefficients
$$\binom{\alpha}{v} = \frac{\alpha(\alpha - 1)\cdots(\alpha - v + 1)}{v!}$$
have alternating signs as soon as $\alpha - v$ is negative, we see that either the function $\phi(x)$ or $-\phi(x)$ belongs to the class of functions with nonnegative derivatives from some order on when we limit x to values $x < 1$. Thus for $a = -1$, $b = 1$, $x = 0$, and $|h| < b - x = 1$ our general theorem proves that
$$(1 - h)^\alpha = \sum_{v=0}^{\infty} (-1)^v \binom{\alpha}{v} h^v.$$

If here we write x for $-h$, we obtain the binomial expansion
$$(1 + x)^\alpha = \sum_{v=0}^{\infty} \binom{\alpha}{v} x^v = 1 + \alpha x + \frac{\alpha(\alpha - 1)}{1 \cdot 2} x^2 + \frac{\alpha(\alpha - 1)(\alpha - 2)}{1 \cdot 2 \cdot 3} x^3 + \cdots$$
for any exponent α and any x with $-1 < x < 1$.

Appendix II   Interpolation

*A.II.1   The Problem of Interpolation.   Uniqueness

The Taylor polynomial $P_n(x)$ approximates the function $f(x)$ in such a way that the graphs of $f(x)$ and $P_n(x)$ have contact of order n at a point a, or in such a way that $f(x)$ and $P_n(x)$ coincide at $n + 1$ points "infinitely near" to a. We might "resolve" the point with abscissa a into $n + 1$ distinct points with abscissas $x_0, x_1, \ldots, x_n$ and seek an approximation to $f(x)$ by a polynomial $\phi(x)$ of degree n which coincides with $f(x)$ at these points. This polynomial, as it turns out, is determined uniquely by a system of linear equations. By a passage to the limit $x_i \to x$ for all i we regain the Taylor polynomials. But "interpolation," that is, the approximation by polynomials coinciding with $f(x)$ in distinct points is of great importance in many applications. The following discussion will give a brief account of the theory of interpolation.

We consider the following problem: Determine a polynomial $\phi(x)$ of nth degree, so that it assumes at $n + 1$ given distinct points $x_0, x_1, \ldots, x_n$, the $n + 1$ given values $f_0, f_1, \ldots, f_n$, that is,

$$\phi(x_0) = f_0, \qquad \phi(x_1) = f_1, \ldots, \qquad \phi(x_n) = f_n.$$

If the numbers f_i are the values $f_i = f(x_i)$ assumed by a given (possibly less elementary) function $f(x)$ at the points x_i, then the polynomial $\phi(x)$ will be named the *interpolation polynomial* of nth degree of the function $f(x)$ for the points $x_0, x_1, \ldots, x_n$.

There can at most be one such polynomial of nth degree, for if there were two different such polynomials $\phi(x)$ and $\psi(x)$, then their difference $D(x) = \phi(x) - \psi(x)$ would be a polynomial of mth degree with $0 \leq m \leq n$ having $n + 1$ distinct roots, which is not possible according to elementary algebra.[1]

We can prove the *uniqueness* of the interpolation polynomial by yet another method, based on the

GENERAL THEOREM OF ROLLE.   *If a function $F(x)$ has continuous derivatives of order up to n in an interval, and vanishes at least at $n + 1$*

---

[1] For we would have

$$D(x) = c_0(x - x_1)(x - x_2) \ldots (x - x_m), \qquad c_0 \neq 0,$$

since $x_1, \ldots, x_m$ are zeros of $D(x)$; but then since $D(x_0) = 0$,

$$c_0(x_0 - x_1)(x_0 - x_2) \ldots (x_0 - x_m) = 0$$

contrary to the distinctness of $x_0, x_1, \ldots, x_m$.

distinct points $x_0, x_1, \ldots, x_n$ *of the interval, then there is a point* ξ *in the interior of the interval for which* $F^{(n)}(\xi) = 0$.

PROOF. The general theorem follows easily from the special case $n = 1$ which is the Rolle theorem proved on p. 175. Let the numbers $x_0, x_1, \ldots, x_n$ be arranged in increasing order. Then by the mean value theorem (or by Rolle's theorem) the first derivative $F'(x)$ must vanish at least once within each of the n subintervals (x_i, x_{i+1}). This same consideration applied to $F'(x)$, and the intervals between its zeros tells us that $F''(x)$ vanishes at $n - 1$ points; by applying this argument repeatedly, the assertion is proved.

We now apply this theorem to the difference

$$F(x) = D(x) = \phi(x) - \psi(x)$$
$$= d_0 x^n + d_1 x^{n-1} + \cdots + d_n,$$

which by assumption vanishes at $n + 1$ points. We obtain a point ξ at which the nth derivative vanishes; $D^{(n)}(\xi) = 0$. This is, however, $n! \, d_0$, so that $d_0 = 0$ and the difference is a polynomial of at most degree $n - 1$, vanishing at $n + 1$ points. Again applying the theorem of Rolle, we obtain $d_1 = 0$, etc., or $D(x)$ is identically 0 as we asserted.

These considerations can be extended to the case where the x_i are not all distinct from each other and, perhaps, r of the values x_i agree; that is, $x_0 = x_1 = \cdots = x_{r-1}$. In the interpolation problem we shall then require that $\phi(x)$ and the derivatives $\phi'(x), \ldots, \phi^{(r-1)}(x)$ should assume preassigned values for $x = x_0$, and correspondingly for the other points x_v. The polynomial $D(x)$ then is of the form $c(x - x_0)^r (x - x_r) \cdots$. The general theorem of Rolle and the uniqueness theorem, as well as the proofs, hold unchanged in this case.

A.II.2 Construction of the Solution.
Newton's Interpolation Formula

We shall now construct an interpolation polynomial $\phi(x)$ of nth degree, such that $\phi(x_0) = f_0, \ldots, \phi(x_n) = f_n$. In order to construct it in a stepwise manner, we shall begin with the constant f_0 which is a polynomial $\phi_0(x)$ of 0th order which for all x and, in particular, for $x = x_0$ assumes the value $A_0 = f_0$. To it we add a polynomial of first order, vanishing for $x = x_0$ and therefore of the form $A_1(x - x_0)$; then we determine A_1 such that the sum has for $x = x_1$, the correct value f_1. The resulting polynomial of first degree we name $\phi_1(x)$. Now we add to $\phi_1(x)$ a polynomial of second order which vanishes for $x = x_0$ and $x = x_1$, and is thus of the form $A_2(x - x_0)(x - x_1)$, whose addition thus will not change the behavior at these two points; the factor A_2 is then determined so that the resulting polynomial of second order, $\phi_2(x)$, will also take the assigned value, in this case f_2, at $x = x_2$. This procedure is continued until all points are reached and we obtain the polynomial

$$(33) \quad \phi(x) = \phi_n(x) = A_0 + A_1(x - x_0) + A_2(x - x_0)(x - x_1) + \cdots$$
$$+ A_n(x - x_0) \cdots (x - x_{n-1}).$$

Our method of obtaining the coefficients A_i in the expression for ϕ is made clear by substituting $x = x_0$, $x = x_1$, ..., $x = x_n$ in order, thus obtaining the system of $n + 1$ equations

$$(34) \quad \begin{aligned} f_0 &= A_0 \\ f_1 &= A_0 + A_1(x_1 - x_0) \\ f_2 &= A_0 + A_1(x_2 - x_0) + A_2(x_2 - x_0)(x_2 - x_1) \end{aligned}$$

$$\cdot \quad \cdot \quad \cdot \quad \cdot \quad \cdot \quad \cdot \quad \cdot \quad \cdot \quad \cdot \quad \cdot \quad \cdot \quad \cdot \quad \cdot \quad \cdot$$

$$f_n = A_0 + A_1(x_n - x_0) + \cdots +$$
$$+ A_n(x_n - x_0)(x_n - x_1) \cdots (x_n - x_{n-1}).$$

Clearly, we can determine the coefficients $A_0, A_1, \ldots, A_n$ successively so as to satisfy these equations, and in this way the interpolation polynomial can be constructed.

When the values x_ν are *equidistant*, $x_\nu = x_{\nu-1} + h$, the result can be written explicitly in a more elegant manner. The equations for the A_i now become

$$(35) \quad \begin{aligned} f_0 &= A_0 \\ f_1 &= A_0 + hA_1 \\ f_2 &= A_0 + 2hA_1 + 2!\, h^2 A_2 \\ f_3 &= A_0 + 3hA_1 + 3 \cdot 2h^2 A_2 + 3!\, h^3 A_3 \end{aligned}$$

$$\cdot \quad \cdot \quad \cdot \quad \cdot \quad \cdot \quad \cdot \quad \cdot \quad \cdot \quad \cdot \quad \cdot \quad \cdot \quad \cdot \quad \cdot \quad \cdot$$

$$f_n = A_0 + nhA_1 + \cdots + \frac{n!}{(n - i)!} h^i A_i + \cdots + n!\, h^n A_n.$$

The solutions may easily be expressed as *successive differences* of f: Given any sequence (finite or infinite) of terms $f_0, f_1, f_2, \ldots$, we call the expressions

$$\Delta f_0 = f_1 - f_0, \qquad \Delta f_1 = f_2 - f_1, \qquad \Delta f_2 = f_3 - f_2, \ldots$$

the *first differences* of the f_k. Applying the differencing process again to the sequence of Δf_k, we obtain the expressions

$$\Delta^2 f_0 = \Delta f_1 - \Delta f_0, \qquad \Delta^2 f_1 = \Delta f_2 - \Delta f_1, \qquad \Delta^2 f_2 = \Delta f_3 - \Delta f_2, \ldots,$$

that is,

$$\Delta^2 f_0 = f_2 - 2f_1 + f_0, \qquad \Delta^2 f_1 = f_3 - 2f_2 + f_1, \ldots,$$

which are the *second differences* of the f_k. The nth difference $\Delta^n f_k$ is defined recursively as $\Delta^{n-1} f_{k+1} - \Delta^{n-1} f_k$. When expressed directly in terms of the f_k it is given by the formula

$$(36) \qquad \Delta^n f_k = f_{k+n} - \binom{n}{1} f_{k+n-1} + \binom{n}{2} f_{k+n-2} - \cdots + (-1)^n f_k$$

which follows by a simple inductive argument left to the reader. With this terminology the coefficients A_ν can be written in the form

$$(37) \qquad A_\nu = \frac{1}{\nu!} h^{-\nu} \Delta^\nu f_0$$

as can be verified by induction.[1]

Newton's Interpolation Formula.   Putting $\xi = (x - x_0)/h$ we have $x - x_r = h(\xi - r)$. The expressions $(x - x_0)(x - x_1) \cdots (x - x_\nu)$ assume then the form $\xi(\xi - 1) \cdots (\xi - \nu)h^{\nu+1}$. Thus we obtain for the polynomials $\phi(x)$ from (33), (37), *Newton's interpolation formula*:[2]

$$\phi(x) = \phi(x_0 + \xi h) = f_0 + \binom{\xi}{1} \Delta f_0 + \binom{\xi}{2} \Delta^2 f_0 + \cdots + \binom{\xi}{n} \Delta^n f_0.$$

If $f_0, f_1, f_2, \ldots$ are the values of a function $f(x)$ at the points $x_0, x_1, x_2, \ldots$, where f has continuous derivatives through the nth order, then $\Delta^\nu f_0/h^\nu$ is an

---

[1] We have to verify that the values A_ν given by (37) satisfy the equations (35); that is, for any sequence $f_0, f_1, f_2, \ldots$, the identity

$$f_k = f_0 + \binom{k}{1} \Delta f_0 + \binom{k}{2} \Delta^2 f_0 + \cdots + \binom{k}{k} \Delta^k f_0$$

is satisfied. Assuming that this is true for a certain k, we must show that

$$f_{k+1} = f_1 + \binom{k}{1} \Delta f_1 + \binom{k}{2} \Delta^2 f_1 + \cdots$$

$$= (f_0 + \Delta f_0) + \binom{k}{1}(\Delta f_0 + \Delta^2 f_0) + \binom{k}{2}(\Delta^2 f_0 + \Delta^3 f_0) + \cdots$$

$$= f_0 + \binom{k+1}{1} \Delta f_0 + \binom{k+1}{2} \Delta^2 f_0 + \cdots$$

which is the identity for the case $k + 1$.

[2] As on p. 457 we define here the bionomial coefficients $\binom{\xi}{k}$ for general ξ and position integers k by $\binom{\xi}{k} = \xi(\xi - 1) \cdots (\xi - k + 1)/k!$.

approximation to the derivative $f^{(\nu)}(x_0)$; we shall show on p. 476 that

$$\lim_{h \to 0} \frac{1}{h^\nu} \Delta^\nu f_0 = f^{(\nu)}(x_0).$$

Since also

$$\lim_{h \to 0} h^k \binom{\xi}{k} = \frac{(x - x_0)^k}{k!}$$

we see that in this case $\phi(x)$ tends to the Taylor polynomial $P_n(x)$ when h tends to zero.

We note that the construction of the interpolation polynomial is possible in the same manner, if, perhaps, the first r values $x_0, \ldots, x_{r-1}$ coincide, and corresponding values $f_0, f_0', \ldots, f_0^{(r-1)}$ are preassigned for $\phi(x_0), \phi'(x_0), \ldots, \phi^{(r-1)}(x_0)$, which coincide with the values

$$f(x_0), f'(x_0), \ldots, f^{(r-1)}(x_0),$$

for a given function f. For $\phi(x)$ we write the form

$$\phi(x) = A_0 + A_1(x - x_0) + A_2(x - x_0)^2$$
$$+ \cdots + A_r(x - x_0)^r + A_{r+1}(x - x_0)^r(x - x_r) + \cdots;$$

we then determine the A_ν in order from the equations

$$f_0 = A_0 \qquad f_0' = A_1 \qquad f_0'' = 2A_2$$
$$\cdot \quad \cdot \quad \cdot \quad \cdot \quad \cdot \quad \cdot \quad \cdot \quad \cdot \quad \cdot \quad \cdot \quad \cdot \quad \cdot$$
$$f_0^{(r-1)} = (r - 1)! \, A_{r-1}$$
$$f_r = A_0 + A_1(x_r - x_0) + \cdots + A_r(x_r - x_0)^r$$
$$f_{r+1} = A_0 + A_1(x_{r+1} - x_0) + \cdots$$
$$+ A_r(x_{r+1} - x_0)^r + A_{r+1}(x_{r+1} - x_0)^r(x_{r+1} - x_r)$$
$$\cdot \quad \cdot \quad \cdot \quad \cdot \quad \cdot \quad \cdot \quad \cdot \quad \cdot \quad \cdot \quad \cdot \quad \cdot \quad \cdot \quad \cdot \quad \cdot \quad \cdot \quad \cdot$$

A.II.3   The Estimate of the Remainder

For the foregoing considerations it did not matter how the values $f_0, f_1, \ldots, f_n$ were originally given. For instance, if these values were obtained from physical observations, the problem of constructing the interpolation polynomial could still be completely solved, giving us then in $\phi(x)$ a simple smooth function defined for all x and taking the observed values at the given points, which can be used to "predict" approximate values for $f(x)$ at other x. However, if the function $f(x)$ taking the $n + 1$ given values f_k at the given points x_k is defined also for intermediate values x, we have to face the new problem of estimating the difference $R(x) = f(x) - \phi(x)$, the *error of interpolation*. We know at first only that $R(x_0) = R(x_1) = \cdots R(x_n) = 0$. In order to be able to say more, we must make further assumptions on the behavior of the function $f(x)$, which affect the remainder $R(x)$. We will therefore assume that in the interval under consideration $f(x)$ has continuous derivatives of at least the $(n + 1)$th order.

We note at first that for every choice of the constant c, the function

$$K(x) = R(x) - c(x - x_0)(x - x_1) \cdots (x - x_n)$$

vanishes at the $n + 1$ points $x_0, \ldots, x_n$. Choose now any value y distinct from $x_0, x_1, \ldots, x_n$. We can then determine c so that $K(y) = 0$, that is,

$$c = \frac{R(y)}{(y - x_0)(y - x_1) \cdots (y - x_n)}.$$

Then there are $n + 2$ points at which $K(x)$ vanishes. We apply the generalized Rolle's theorem used earlier to $K(x)$; by this we know there is a value $x = \xi$ between the largest and the smallest of the values $x_0, x_1, \ldots, x_n, y$, such that $K^{(n+1)}(\xi) = 0$. Since $R(x) = f(x) - \phi(x)$, and ϕ, as a polynomial of nth order, has an identically vanishing $(n + 1)$th derivative, we have

$$f^{(n+1)}(\xi) - c(n + 1)! = 0,$$

noting that $(n + 1)!$ is the $(n + 1)$th derivative of $(x - x_0) \cdots (x - x_n)$. Thus we have obtained for c, a second expression $c = f^{(n+1)}(\xi)/(n + 1)!$, containing ξ and depending in some manner on y. We now use the equation $K(y) = 0$, in which y is completely arbitrary and therefore can be replaced by x, and obtain the representation

$$(38) \qquad R(x) = \frac{(x - x_0)(x - x_1) \cdots (x - x_n)}{(n + 1)!} f^{(n+1)}(\xi),$$

where ξ is some value lying between the smallest and the largest of the points $x, x_0, x_1, \ldots, x_n$.

Thus the general problem of interpolation for a given function $f(x)$ is completely solved. We have for $f(x)$ the representation

$$(39) \quad f(x) = A_0 + A_1(x - x_0) + A_2(x - x_0)(x - x_1) + \cdots$$
$$+ A_n(x - x_0)(x - x_1) \cdots (x - x_{n-1}) + R_n,$$

where the coefficients $A_0, A_1, \ldots, A_n$ can be found successively from the values of f at the points $x_0, x_1, \ldots, x_n$ by the recursion formulas (34) on p. 472 and where the remainder R_n is of the form

$$(40) \qquad R_n = \frac{(x - x_0)(x - x_1) \cdots (x - x_n)}{(n + 1)!} f^{(n+1)}(\xi),$$

with a suitable number ξ between the largest and smallest of the values $x, x_0, x_1, \ldots, x_n$.

If we take the corresponding formula (39) for $f(x)$ with n replaced by $n - 1$ and subtract, we obtain

$$A_n(x - x_0)(x - x_1) \cdots (x - x_{n-1}) + R_n - R_{n-1} = 0.$$

For $x = x_n$ we have $R_n = 0$, and hence for the coefficient A_n (using (40) with n replaced by $n - 1$) the representation

$$A_n = \frac{f^{(n)}(\xi)}{n!}$$

where ξ lies between the smallest and largest of the values $x_0, x_1, \ldots, x_n$. Similar representations exist for $A_{n-1}, A_{n-2}, \ldots, A_0$. Thus we recognize that if the points $x_0, x_1, \ldots, x_n$ are tending together to one and the same point, perhaps the origin, then our interpolation formula (39) goes term for term into the Taylor formula (27a), p. 452, with the Lagrange form (21), p. 449, of the remainder. The Taylor formula can thus be considered a limiting case of the Newton interpolation formula.

This formula enables us to give precise meaning to an expression commonly used in geometry. The osculating parabola which meets a given curve at a point, of nth order, is said to have "$(n + 1)$ consecutive points in common" with the given curve at the point. Actually, we obtain this osculating parabola if we find a parabola having $n + 1$ points in common with the curve, and then draw these points together. Analytically, this just corresponds to the transition from the interpolating to the Taylor polynomial. In the same fashion we can characterize the osculation of arbitrary curves. For example, the circle of curvature is that circle which has three consecutive points in common with the given curve.

The interpolation formula can be expected to give the values of a function whose values at some definite points are known, with a high degree of accuracy *between* these points (both $|f^{(v+1)}(\xi)|$ and the $|x - x_i|$ are then bounded). If the value x lies outside the intervals of the points $x_0, x_1, \ldots, x_n$, we speak of *extrapolation*. By means of such an extrapolation we shall obtain good agreement provided the point x is sufficiently near the given points. The Taylor formula corresponds in a sense to *complete extrapolation*; in general, it is suitable for use only in a neighborhood of a point.

A.II.4   The Lagrange Interpolation Formula

In closing, we solve the interpolation problem by a somewhat different formula, due to Lagrange, and differing from Newton's interpolation formula insofar as each individual term contains only one of the given values of the function. Moreover, the formula gives $\phi(x)$ quite explicitly, not requiring any solution of recursive formulas. For brevity we introduce the polynomial of $(n + 1)$th degree

$$\psi(x) = (x - x_0)(x - x_1) \cdots (x - x_n),$$

corresponding to the given points x_ν. Differentiating by the product rule and substituting then successively for x the values $x_0, \ldots, x_n$, we obtain the relations

$$\psi'(x_0) = (x_0 - x_1)(x_0 - x_2) \cdots (x_0 - x_n)$$

. . . .

$$\psi'(x_\nu) = (x_\nu - x_0) \cdots (x_\nu - x_{\nu-1})(x_\nu - x_{\nu+1}) \cdots (x_\nu - x_n),$$

. . . .

$$\psi'(x_n) = (x_n - x_0)(x_n - x_1) \cdots (x_n - x_{n-1}).$$

We note that

$$\frac{\psi(x)}{(x - x_\nu)\psi'(x_\nu)} = \frac{(x - x_0) \cdots (x - x_{\nu-1})(x - x_{\nu+1}) \cdots (x - x_n)}{(x_\nu - x_0) \cdots (x_\nu - x_{\nu-1})(x_\nu - x_{\nu+1}) \cdots (x_\nu - x_n)}$$

is a polynomial of nth degree, having at the point $x = x_\nu$, the value 1, and at the remaining points x_i, the value 0; then it is immediately clear that the expression

(41) $\phi(x) = \psi(x)$

$$\left[\frac{f_0}{(x - x_0)\psi'(x_0)} + \frac{f_1}{(x - x_1)\psi'(x_1)} + \cdots + \frac{f_n}{(x - x_n)\psi'(x_n)} \right]$$

is the desired interpolation polynomial. This is the *interpolation formula of Lagrange*.

PROBLEMS

SECTION 5.4b, page 540

1. Give the complete formal derivation of the remainder formula (27), p. 452, using mathematical induction.

2. (*A Variant of Proof of Taylor's Theorem*)

(*a*) If $g(h)$ has continuous derivatives through the $(n + 1)$th order for $0 \leq h \leq A$, and if $g(0) = g'(0) = \cdots = g^{(n)}(0) = 0$, while $|g^{(n+1)}(h)| \leq M$ on $[0, A]$, for M a constant, show that $|g^{(n)}(h)| \leq Mh$, $|g^{(n-1)}(h)| \leq Mh^2/2!, \ldots, |g^{(n-i)}(h)| \leq Mh^i/i!, \ldots, |g(h)| \leq Mh^n/n!$, for all h in the interval.

(b) Let $f(x)$ be a sufficiently differentiable function on $a \le x \le b$, and $T_n(h)$ be the Taylor polynomial for $f(x)$ at $x = a$. Apply the result of (a) to the function $g(h) = R_n = f(a + h) - T_n(h)$ to obtain Taylor's formula with a rough estimate for the remainder.

3. Let $f(x)$ have a continuous derivative in the interval $a \le x \le b$, and let $f''(x) \ge 0$ for every value of x. Then if ξ is any point in the interval, the curve nowhere falls below its tangent at the point $x = \xi$, $y = f(\xi)$.
(Use the Taylor expansion to three terms.)

4. Deduce the integral formula for the remainder R_n by applying integration by parts to

$$f(x + h) - f(x) = \int_0^h f'(x + \tau)\, d\tau.$$

5. Integrate by parts the formula

$$R_n = \frac{1}{n!} \int_0^h (h - \tau)^n f^{(n+1)}(x + \tau)\, d\tau,$$

and so obtain

$$R_n = f(x + h) - f(x) - hf'(x) - \cdots - \frac{h^n}{n!} f^{(n)}(x).$$

*6. Suppose that in some way a series for the function $f(x)$ has been obtained, namely

$$f(x) = a_0 + a_1 x + a_2 x^2 + \cdots + a_n x^n + R_n(x),$$

where $a_0, a_1, \ldots, a_n$ are constants, $R_n(x)$ is n times continuously differentiable, and $R_n(x)/x^n \to 0$ as $x \to 0$. Show that $a_k = (f^k(0)/k!)$ $(k = 0, \ldots, n)$, that is, that the series is a Taylor series.

SECTION 5.5, page 453

1. Find the first four nonvanishing terms of the Taylor series for the following functions in the neighborhood of $x = 0$:

(a) $x \cot x$

(b) $\dfrac{\sqrt{\sin x}}{\sqrt{x}}$

(c) $\sec x$

(d) $e^{\sin x}$

(e) e^{e^x}

(f) $\log \sin x - \log x$.

2. Find the Taylor series for arc $\sin x$ in the neighborhood of $x = 0$ by using

$$\text{arc } \sin x = \int_0^x \frac{dt}{\sqrt{1 - t^2}}.$$

Compare Section 3.2, Problem 2.

*3. Find the first three nonvanishing terms of the Taylor series for $\sin^2 x$ in the neighborhood of $x = 0$ by multiplying the Taylor series for $\sin x$ by itself. Justify this procedure.

*4. Find the first three nonvanishing terms of the Taylor series for $\tan x$ in the neighborhood of $x = 0$ by using the relation $\tan x = \sin x/\cos x$, and justify the procedure.

*5. Find the first three nonvanishing terms of the Taylor series for $\sqrt{\cos x}$ in the neighborhood of $x = 0$ by applying the binomial theorem to the Taylor series for $\cos x$, and justify the procedure.

*6. Find the Taylor series for $(\arcsin x)^2$. Compare Section 3.2, Problem 2.

7. Find the Taylor series for the following functions in the neighborhood of $x = 0$:

(a) $\sinh^{-1} x$. (b) $\displaystyle\int_0^x e^{-t^2}\, dt$. (c) $\displaystyle\int_0^x \frac{\sin t}{t}\, dt$.

*8. Estimate the error involved in using the first n terms in the series in Problem 7.

9. The elliptic function $s(u)$ has been defined (Section 3.14a) as the inverse of the elliptic integral

$$u(s) = \int_0^s \frac{dx}{\sqrt{(1 - x^2)(1 - k^2 x^2)}}.$$

Find the Taylor expansion of $s(u)$ to the term of degree 5.

10. Evaluate the following limits:

(a) $\displaystyle\lim_{x \to \infty} x\left[\left(1 + \frac{1}{x}\right)^x - e\right]$,

(b) $\displaystyle\lim_{x \to \infty} \left\{\frac{e}{2} x + x^2\left[\left(1 + \frac{1}{x}\right)^x - e\right]\right\}$,

*(c) $\displaystyle\lim_{x \to \infty} x\left[\left(1 + \frac{1}{x}\right)^x - e\log\left(1 + \frac{1}{x}\right)^x\right]$,

(d) $\displaystyle\lim_{x \to 0} \left(\frac{\sin x}{x}\right)^{1/x^2}$,

(e) $\displaystyle\lim_{x \to \infty} \left(\frac{\sin x}{x}\right)^{1/x^2}$.

*11. Find the first three terms of the Taylor series for $[1 + (1/x)]^x$ in powers of $1/x$.

*12. Two oppositely charged particles $+e$, $-e$ situated at a small distance d apart form an electric dipole with moment $M = ed$. Show that the potential energy

(a) At a point situated on the axis of the dipole at a distance r from the center of the dipole is $(M/r^2)(1 + \epsilon)$, where ϵ is approximately equal to $d^2/4r^2$.

(b) At a point situated on the perpendicular bisector of the dipole is 0.

(c) At a point with polar coordinates r, θ relative to the center and axis of the dipole is $[M\cos(\theta/r^2)](1 + \epsilon)$, where ϵ is approximately equal to

$$(d^2/8r^2)(5\cos^2\theta - 3).$$

(The potential energy of a single charge q at a point at a distance r from the charge is q/r; the potential energy of several charges is the sum of the potential energies of the separate charges.)

SECTION 5.6, page 457

1. Prove if $f(a) = 0$ and $f(x)$ has sufficiently many derivatives at $x = a$ that $f(x)^n$ has at least an $(n - 1)$th order contact with the x-axis.

2. The curve $y = f(x)$ passes through the origin O and touches the x-axis at O. Show that the radius of curvature of the curve at O is given by

$$\rho = \lim_{x \to 0} \frac{x^2}{2y}.$$

*3. K is a circle which touches a given curve at a point P and passes through a neighboring point Q of the curve. Show that the limit of the circle K as $Q \to P$ is the circle of curvature of the curve at P.

*4. Show that the order of contact of a curve and its osculating circle is at least three at points where the radius of curvature is a maximum or minimum.

*5. Show that the osculating circle at a point where the radius of curvature is a maximum or minimum does not cross the curve unless the contact is of higher than third order.

*6. Find the maxima and minima of the following functions:

$\qquad$ (*a*) $\cos x \cosh x$ $\qquad$ (*b*) $x + \cos x$

*7. Determine the maxima and minima of the function $y = e^{-1/x^2}$ (see p. 242).

SECTION A.3, page 464

1. Prove if f is continuous on the interval $[0, 1]$ that

$$\lim_{x \to 0} x \int_x^1 \frac{f(z)}{z^2} \, dz = f(0).$$

2. Prove that the function $y = (x^2)^x$, $y(0) = 1$ is continuous at $x = 0$.

6

Numerical Methods

The task of solving an analytical problem always remains uncompleted. The proof of the existence and of some basic properties of the solution is usually considered satisfactory, but relevant questions always remain to be answered. Thus, when the solution is defined by a limit process, for example by an integral, the problem arises of actually finding approximations to this limit and of estimating the accuracy of these approximations. Not only are such questions of basic importance theoretically but they are also inevitable, if we wish to apply analysis to the description and control of natural phenomena which in principle can be described only in an approximate manner.

Accordingly it is a great challenge to carry the solution to the point where numerical answers and estimates of their accuracy come into reach.

Recently, with the advent of high-speed automatic computing machines, theoretical and practical aspects of "numerical analysis" have received a great stimulus; they are presented in a variety of textbooks.[1] For centuries, however, many of the foremost mathematicians, such as Newton, Euler, and, in particular, Gauss, have greatly contributed to numerical methods.

In this volume we cannot present numerical analysis in a comprehensive way, but at least we shall discuss some of the simple classical results.

---

[1] See for example, Hildebrand, *Introduction to Numerical Analysis*, McGraw-Hill Book Co., 1956; Householder, *Principles of Numerical Analysis*, McGraw-Hill Book Co., 1953; and Whittaker and Robinson, *The Calculus of Observations*, Blackie and Sons, Ltd., 1929.

6.1   Computation of Integrals

Although the existence of the integral of a (continuous) function is assured by the theory of Chapter 2, the evaluation of such an integral or "quadrature"[1] cannot be effected by elementary functions except in relatively rare cases. We must therefore devise methods for numerical integration and for estimating the accuracy of the numerical approximation.

To compute approximately the integral

$$(1) \qquad\qquad J = \int_a^b f(x)\,dx$$

with $a < b$, we subdivide the interval $a \leq x \leq b$ into n equal parts, each of length $h = (b - a)/n$ by means of the $n + 1$ points

$$(2) \qquad x_v = a + vh, \qquad nh = b - a, \qquad v = 0, 1, \ldots, n.$$

Then

$$J = \sum_{v=1}^n J_v,$$

where

$$(3) \qquad\qquad J_v = \int_{x_{v-1}}^{x_v} f(x)\,dx;$$

the problem of computing the integral J is reduced to that of obtaining good approximations for the areas J_v of strips of width h into which we have dissected the entire area, represented by J.

a. Approximation by Rectangles

The most direct approximation, paraphrasing the original definition of the integral, yields the relation

$$J = \sum_{v=1}^n J_v$$
$$\approx h(f_1 + f_2 + \cdots + f_n),$$

where for abbreviation we set

$$f_v = f(x_v).$$

---

[1] The word "quadrature" indicates the process of "squaring", that is, of measuring an area inside a curve by finding a square having the same area (as in the problem of "squaring the circle").

Here (and throughout this chapter) the symbol $\approx$ means "approximately equal."

To estimate the accuracy or "error" of this approximation, we assume that $f(x)$ is continuous with a uniformly bounded derivative on the interval $a \leq x \leq b$: $|f'(x)| \leq M_1$. Then it can be proved easily (see Problem 4 p. 507, 6.1) that

$$(4) \qquad |J_\nu - hf_\nu| \leq \frac{M_1 h^2}{2},$$

or therefore

$$\left| J - h\sum_{\nu=1}^n f_\nu \right| \leq n\frac{M_1 h^2}{2}$$

$$(5) \qquad = \tfrac{1}{2}M_1(b-a)h.$$

Thus the accuracy of the approximation of the integral by the finite sum is of the order h of the "mesh width" in the terminology of Chapter 3, p. 252.

b. Refined Approximations—Simpson's Rule

A better approximation is obtained with hardly more effort if we approximate the areas J_ν not by rectangular strips but by the slender trapezoids, as in Fig. 6.1a. The approximation formula (trapezoid formula) is then

$$J \approx \tfrac{1}{2}h(f_0 + f_1) + \tfrac{1}{2}h(f_1 + f_2) + \cdots + \tfrac{1}{2}h(f_{n-1} + f_n)$$

$$(6) \qquad = h(f_1 + f_2 + \cdots + f_{n-1}) + \frac{h}{2}(f_0 + f_n),$$

since every function value except the first and the last appears twice.

An approximation which is generally slightly more precise than that of the trapezoid formula is that in which the νth strip is approximated by a trapezoid bounded above by the tangent to the curve at the midpoint $x_{\nu-1} + h/2$ of the interval $x_{\nu-1} \leq x \leq x_\nu$. The area of this trapezoid is simply

$$hf_{\nu-1/2} = hf\left(x_{\nu-1} + \frac{h}{2}\right),$$

and we obtain by addition the *tangent formula*,

$$(7) \qquad J \approx h(f_{1/2} + f_{3/2} + \cdots + f_{(2n-1)/2}).$$

As we shall see on p. 486, the accuracy of this approximation is of order h^2 when the second derivative of f is continuous in the interval $a \leq x \leq b$ and $|f''(x)| < M_2$, with some constant bound M_2.

Finally, we mention the famous approximation of *Simpson*, which with little additional effort yields a much more accurate approximation

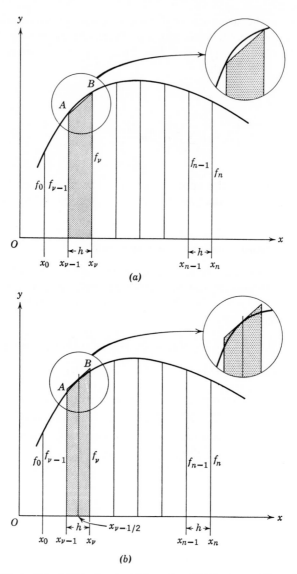

Figure 6.1 (*a*) The trapezoid formula. (*b*) The tangent formula.

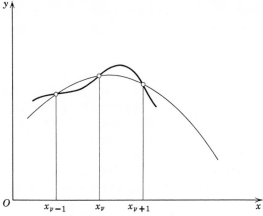

Figure 6.2 Simpson's rule.

if the fourth derivative of f exists and is uniformly bounded in the interval:

$$|f^{(4)}(x)| \leq M_4,$$

with M_4 a constant. Simpson's formula for $n = 2m$ is

$$(8) \quad J \approx \frac{4h}{3}(f_1 + f_3 + f_5 + \cdots + f_{2m-1})$$

$$+ \frac{2h}{3}(f_2 + f_4 + f_6 + \cdots + f_{2m-2}) + \frac{h}{3}(f_0 + f_{2m}).$$

The formula is easily obtained if we approximate the region composed of the νth and $(\nu + 1)$th strips by a strip of width $2h$ bounded above by the parabola which agrees with f at the three abscissae $x_{\nu-1}$, $x_\nu = x_{\nu-1} + h$, and $x_{\nu+1} = x_{\nu-1} + 2h$ (see Fig. 6.2). Newton's interpolation formula (p. 473) yields the equation of this parabola:

$$y = f_{\nu-1} + (x - x_{\nu-1}) \cdot \frac{f_\nu - f_{\nu-1}}{h}$$

$$+ \frac{(x - x_{\nu-1})(x - x_{\nu-1} - h)}{2} \frac{f_{\nu+1} - 2f_\nu + f_{\nu-1}}{h^2};$$

hence we have the approximation

$$J_\nu + J_{\nu+1} \approx \int_{x_{\nu-1}}^{x_{\nu+1}} y\, dx = \int_{x_{\nu-1}}^{x_{\nu-1}+2h} y\, dx$$

$$= 2hf_{\nu-1} + 2h(f_\nu - f_{\nu-1}) + \frac{\frac{8}{3}h - 2h}{2}(f_{\nu+1} - 2f_\nu + f_{\nu-1})$$

$$= \frac{h}{3}(f_{\nu-1} + 4f_\nu + f_{\nu+1}).$$

The formula is now obtained for even $n = 2m$ by adding all these approximate values for $v = 1, 3, 4, \ldots, 2m - 1$ or all the areas of the pairs of strips.

*Accuracy

It is not difficult to estimate the accuracy of our approximations. Each quadrature proceeds by approximating the function $f(x)$ in an interval by an easily integrated function $\phi(x)$ (a polynomial). An estimate for the error in the integration formula can thus be obtained by estimating $|f(x) - \phi(x)|$.

In the tangent formula (p. 483) we replaced $f(x)$ in the interval $[x_{v-1}, x_v]$ by its tangent at the midpoint $x_v - (h/2)$, that is, by

$$\phi(x) = f\left(x_v - \frac{h}{2}\right) + \left(x - x_v + \frac{h}{2}\right) f'\left(x_v - \frac{h}{2}\right).$$

By Taylor's theorem with Lagrange's form of the remainder

$$f(x) = \phi(x) + \frac{1}{2}\left(x - x_v + \frac{h}{2}\right)^2 f''(\xi),$$

where ξ lies between x and $x_v - h/2$. Hence the error corresponding to one strip is estimated by

$$|J_v - h f_{v-\frac{1}{2}}| = \left| \int_{x_{v-1}}^{x_v} [f(x) - \phi(x)]\, dx \right|$$

$$\leq \int_{x_{v-1}}^{x_v} |f(x) - \phi(x)|\, dx \leq M_2 \int_{x_v - h}^{x_v} \frac{1}{2}\left(x - x_v + \frac{h}{2}\right)^2 dx$$

$$= \frac{h^3}{24} M_2.$$

For the total error in the tangent formula contributed by the various intervals[1] we find then the upper bound

$$n \frac{h^3}{24} M_2 = \frac{h^2}{24} M_2(b - a).$$

---

[1] This is the total error inherent in using the approximating formula, the so-called truncation error; in practice, additional error arises because of round off in the computation. The total effect of round-off errors increases most likely with the number of steps taken, that is, with decreasing h, whereas the truncation error decreases.

We use this derivation as a model for estimating the error in the other quadrature formulas. In the trapezoidal rule (6) we approximate $f(x)$ in the interval $[x_{\nu-1}, x_\nu]$ by the linear interpolation polynomial

$$\phi(x) = f_{\nu-1} + (x - x_{\nu-1})\frac{f_\nu - f_{\nu-1}}{h}.$$

From the error estimate for the remainder in the interpolation formula [see p. 475, Eq. (40)] for $n = 1$, we find

$$f(x) - \phi(x) = \tfrac{1}{2}(x - x_{\nu-1})(x - x_\nu)f''(\xi),$$

where ξ lies between $x_{\nu-1}$ and x_ν. Hence the absolute value of the error in the computation of J_ν is at most

$$M_2 \int_{x_{\nu-1}}^{x_\nu} |\tfrac{1}{2}(x - x_{\nu-1})(x - x_\nu)| \, dx = \frac{h^3}{12} M_2,$$

and the total error is then at most n times this quantity:

$$\frac{h^2}{12} M_2(b - a).$$

The same technique can be applied to Simpson's rule (8), taking for $\phi(x)$ the quadratic polynomial agreeing with f in the points $x_{\nu-1}$, x_ν, $x_{\nu+1}$ leading to an error in $J_\nu + J_{\nu+1}$ of the order h^4. Actually, however, the error estimate can be improved by one order of magnitude by using a cubic polynomial $\phi(x)$ that gives a better approximation to f in the interval $[x_{\nu-1}, x_{\nu+1}]$ than the quadratic one, and still has the same integral, thus leading to the same approximation formula (9) for the integral J. We simply use the interpolation polynomial which agrees with $f(x)$ at the points $x_{\nu-1}$, x_ν, $x_{\nu+1}$ and for which $\phi'(x_\nu) = f'(x_\nu)$; it has the form

$$\phi(x) = A_0 + A_1(x - x_{\nu-1}) + A_2(x - x_{\nu-1})(x - x_\nu)$$
$$+ A_3(x - x_{\nu-1})(x - x_\nu)(x - x_{\nu+1}).$$

Here the first three terms represent the quadratic interpolation polynomial agreeing with f at the three points $x_{\nu-1}$, x_ν, $x_{\nu+1}$. The constant A_3 has to be determined from the condition $\phi'(x_\nu) = f'(x_\nu)$.

The last term

$$A_3(x - x_\nu + h)(x - x_\nu)(x - x_\nu - h) = A_3[(x - x_\nu)^2 - h^2] \cdot [x - x_\nu]$$

obviously is an odd function of $x - x_\nu$ and therefore does not contribute to the integral between the limits $x_\nu - h$ and $x_\nu + h$. For the error in the approximation to f we then have the estimate [cf. (40), p. 475, with $n = 3$ and with two of the interpolation points coincident at x_ν].

$$f - \phi = \frac{1}{4!}(x - x_{\nu-1})(x - x_\nu)^2(x - x_{\nu+1})f^{(4)}(\xi).$$

This yields for the error in the computation of $J_\nu + J_{\nu+1}$ the estimate

$$\frac{h^5}{90}M_4,$$

and hence for the total error the estimate

$$\frac{n}{2}\frac{h^5}{90}M_4 = \frac{h^4}{180}(b - a)M_4.$$

Naturally, we may attain higher accuracy by approximation of the function $f(x)$ in a strip by a polynomial of a still higher order.

Examples. We apply these methods to the calculation of

$$\log_e 2 = \int_1^2 \frac{dx}{x}.$$

Dividing the interval $1 \leq x \leq 2$ into ten parts of length $h = \frac{1}{10}$, and using the trapezoidal rule (6), we obtain

$$
\begin{aligned}
x_1 &= 1.1 & f_1 &= 0.90909 \\
x_2 &= 1.2 & f_2 &= 0.83333 \\
x_3 &= 1.3 & f_3 &= 0.76923 \\
x_4 &= 1.4 & f_4 &= 0.71429 \\
x_5 &= 1.5 & f_5 &= 0.66667 \\
x_6 &= 1.6 & f_6 &= 0.62500 \\
x_7 &= 1.7 & f_7 &= 0.58824 \\
x_8 &= 1.8 & f_8 &= 0.55556 \\
x_9 &= 1.9 & f_9 &= 0.52632
\end{aligned}
$$

Sum 6.18773

$$
\begin{aligned}
x_0 &= 1.0 & \tfrac{1}{2}f_0 &= 0.5 \\
x_{10} &= 2.0 & \tfrac{1}{2}f_{10} &= 0.25
\end{aligned}
$$

$$6.93773 \cdot \tfrac{1}{10}$$

$$\log_e 2 \approx 0.69377.$$

Since the graph of the integrand function has its convex side turned towards the x-axis, this value is too large.

Using the tangent rule (7) we have

$$
\begin{array}{ll}
x_0 + \tfrac{1}{2}h = 1.05 & f_{1/2} = 0.95238 \\
x_1 + \tfrac{1}{2}h = 1.15 & f_{3/2} = 0.86957 \\
x_2 + \tfrac{1}{2}h = 1.25 & f_{5/2} = 0.80000 \\
x_3 + \tfrac{1}{2}h = 1.35 & f_{7/2} = 0.74074 \\
x_4 + \tfrac{1}{2}h = 1.45 & f_{9/2} = 0.68966 \\
x_5 + \tfrac{1}{2}h = 1.55 & f_{11/2} = 0.64516 \\
x_6 + \tfrac{1}{2}h = 1.65 & f_{13/2} = 0.60606 \\
x_7 + \tfrac{1}{2}h = 1.75 & f_{15/2} = 0.57143 \\
x_8 + \tfrac{1}{2}h = 1.85 & f_{17/2} = 0.54054 \\
x_9 + \tfrac{1}{2}h = 1.95 & f_{19/2} = 0.51282 \\
\end{array}
$$

$$\overline{ 6.92836 \cdot \tfrac{1}{10}}$$

$$\log_e 2 \approx 0.69284,$$

which, owing to the convexity of the curve, is too small.

For the same subdivision we obtain a much more precise result using Simpson's rule (8). We have

$$
\begin{array}{ll}
x_1 = 1.1 & f_1 = 0.90909 \\
x_3 = 1.3 & f_3 = 0.76923 \\
x_5 = 1.5 & f_5 = 0.66667 \\
x_7 = 1.7 & f_7 = 0.58824 \\
x_9 = 1.9 & f_9 = 0.52632 \\
\end{array}
$$

$$\overline{ \text{Sum } 3.45955 \cdot 4}$$

$$\overline{ 13.83820}$$

$$
\begin{array}{ll}
x_2 = 1.2 & f_2 = 0.83333 \\
x_4 = 1.4 & f_4 = 0.71429 \\
x_6 = 1.6 & f_6 = 0.62500 \\
x_8 = 1.8 & f_8 = 0.55556 \\
\end{array}
$$

$$\overline{ \text{Sum } 2.72818 \cdot 2}$$

$$\overline{ 5.45636}$$
$$13.83820$$

$$
\begin{array}{ll}
x_0 = 1.0 & f_0 = 1.0 \\
x_{10} = 2.0 & f_{10} = 0.5 \\
\end{array}
$$

$$\overline{ 20.79456 \cdot \tfrac{1}{30}}$$

$$\log_e \approx 0.69315.$$

In reality

$$\log_e 2 = 0.693147 \ldots.$$

6.2   Other Examples of Numerical Methods

a.   The "Calculus of Errors"

The "calculus of errors" is simply a numerical application of the basic fact of differential calculus: a function $f(x)$ which is differentiable a sufficient number of times can be represented in the neighborhood of a point by a linear function with an error of higher than the first order, by a quadratic function with an error of higher than the second order, and so on.

Consider the linear approximation to a function $y = f(x)$. If $y + \Delta y = f(x + \Delta x) = f(x + h)$, we have by Taylor's theorem

$$\Delta y = hf'(x) + \frac{h^2}{2} f''(\xi),$$

where $\xi = x + \theta h\,(0 < \theta < 1)$ is an intermediate value which need not be more precisely known. If $h = \Delta x$ is small, we obtain the practical approximation

$$\Delta y \approx hf'(x).$$

Thus we replace the difference quotient by the derivative to which it is approximately equal, and the increment of y by the approximately equal linear expression in h.

This simple fact is used for numerical purposes in the following way. Suppose two physical quantities x and y are related by $y = f(x)$. We then ask what effect an inaccuracy in the measurement of x has on the determination of y. If instead of the "true" value x we use the inaccurate value $x + h$, then the corresponding value of y differs from the true value $y = f(x)$ by the amount $\Delta y = f(x + h) - f(x)$. The error is therefore given approximately by the above relation.

We illustrate the usefulness of such linear approximations by examples.

Examples. (a) In a triangle ABC (cf. Fig. 6.3) suppose that the sides b and c are measured accurately, whereas the angle $\alpha = x$ is only measured to within an error $|\Delta x| < \delta$. What is the corresponding error in the value of the third side $y = a = \sqrt{b^2 + c^2 - 2bc \cos \alpha}$?

We have $\Delta a \approx (bc \sin \alpha\, \Delta \alpha)/a$; the percentage error is therefore

$$\frac{100\, \Delta a}{a} \approx \frac{100\, bc}{a^2} \sin \alpha\, \Delta \alpha.$$

In the special case when $b = 400$ meters, $c = 500$ meters, and $\alpha = 60°$, we have $y = a = 458.2576$ meters, so that

$$\Delta a \approx \frac{200000}{458.2576} \times \tfrac{1}{2}\sqrt{3}\, \Delta\alpha.$$

If $\Delta\alpha$ can be measured to within 10 seconds of arc, that is, if

$$\Delta\alpha = 10'' = 4846 \times 10^{-8} \text{ radians},$$

we find that at worst

$$\Delta a \approx 1.83 \text{ cm};$$

thus the error is at most about 0.004 %.

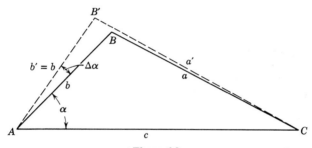

Figure 6.3

(*b*) The following example illustrates the usefulness of the linearization for physical problems.

It is known experimentally that if a metal rod has length l_0 at temperature t_0, then at temperature t its length will be $l = l_0(1 + \alpha(t - t_0))$, where α depends only on t_0 and the material of which the rod is composed. If now a pendulum clock keeps correct time at temperature t_0, how many seconds will it lose per day if the temperature rises to t_1?

For the period $T(l)$ of oscillation we have (see p. 411)

$$T(l) = 2\pi \sqrt{\frac{l}{g}}\,;$$

hence

$$\frac{dT}{dl} = \frac{\pi}{\sqrt{lg}}.$$

If the change of length is Δl, the corresponding change in the period of oscillation is

$$\Delta T \approx \frac{\pi\,\Delta l}{\sqrt{l_0 g}},$$

where $l_1 = l_0(1 + \alpha(t_1 - t_0))$ and $\Delta l = \alpha l_0(t_1 - t_0)$. This is the time lost per oscillation. The time lost per second is $\Delta T/T \approx \Delta l/2l_0$; hence in one day the clock loses $43{,}200\ \Delta l/l_0 = 43{,}200\ \alpha(t_1 - t_0)$ seconds.

In this case and in many other cases where the function under consideration is a product of several factors, we can simplify the calculation by taking the logarithms of both sides before differentiating. In this example we have

$$\log T = \log 2\pi - \tfrac{1}{2} \log g + \tfrac{1}{2} \log l;$$

differentiating, we have

$$\frac{1}{T} \frac{dT}{dl} = \frac{1}{2l}.$$

Replacing dT/dl by $\Delta T/\Delta l$ gives

$$\frac{\Delta T}{T} = \frac{\Delta l}{2l},$$

in agreement with the preceding result.

*b. Calculation of π

A different example, using special artificial devices, is classical, although perhaps made obsolete by modern computers.

Leibnitz's series $\pi/4 = 1 - \tfrac{1}{3} + \tfrac{1}{5} - \tfrac{1}{7} + \cdots$ [Eq. (7), Section 5.2, p. 445], using the series for the inverse tangent, is not suitable for the calculation of π, because of the extreme slowness of its convergence. We may, however, calculate π with comparative ease by the following artifice. If, in the addition theorem for the tangent,

$$\tan (\alpha + \beta) = \frac{\tan \alpha + \tan \beta}{1 - \tan \alpha \tan \beta},$$

we introduce the inverse functions $\alpha = \text{arc tan } u$, $\beta = \text{arc tan } v$, we obtain the formula

$$\text{arc tan } u + \text{arc tan } v = \text{arc tan } \left(\frac{u + v}{1 - uv} \right).$$

Now, choosing u and v so that $(u + v)/(1 - uv) = 1$, we obtain the value $\pi/4$ on the right-hand side, and if u and v are small numbers we can easily calculate the left-hand side by means of known series. If, for example, we put $u = \tfrac{1}{2}$, $v = \tfrac{1}{3}$, as Euler did, we obtain

(9) $$\frac{\pi}{4} = \text{arc tan } \tfrac{1}{2} + \text{arc tan } \tfrac{1}{3}.$$

If we further notice that $\dfrac{\tfrac{1}{3} + \tfrac{1}{7}}{1 - \tfrac{1}{21}} = \dfrac{1}{2}$, we have arc tan $\frac{1}{2}$ = arc tan $\frac{1}{3}$ + arc tan $\frac{1}{7}$, so that by (9),

$$\frac{\pi}{4} = 2 \text{ arc tan } \tfrac{1}{3} + \text{arc tan } \tfrac{1}{7}.$$

Using this formula, Vega calculated the number π to 140 places.

By means of the equation $(\frac{1}{5} + \frac{1}{8})/(1 - \frac{1}{40}) = \frac{1}{3}$, we further obtain

$$\text{arc tan } \tfrac{1}{3} = \text{arc tan } \tfrac{1}{5} + \text{arc tan } \tfrac{1}{8}$$

or

$$\frac{\pi}{4} = 2 \text{ arc tan } \tfrac{1}{5} + \text{arc tan } \tfrac{1}{7} + 2 \text{ arc tan } \tfrac{1}{8}.$$

This expansion is extremely useful for calculating π by means of the series arc tan $x = x - x^3/3 + x^5/5 - \cdots$; for if we substitute for x the value $\frac{1}{5}, \frac{1}{7},$ or $\frac{1}{8}$, we obtain with but few terms a high degree of accuracy, since the terms diminish rapidly.

The reader who is not especially interested in these skilful, yet artificial manipulations, might be satisfied with an understanding of the principle.

*c. Calculation of Logarithms

For the numerical calculation of logarithms we transform the logarithmic series [Eq. (5), p. 444]

$$\tfrac{1}{2} \log \frac{1 + x}{1 - x} = x + \frac{x^3}{3} + \frac{x^5}{5} + \cdots$$

where $0 < x < 1$, by the substitution

$$\frac{1 + x}{1 - x} = \frac{p^2}{p^2 - 1}, \qquad x = \frac{1}{2p^2 - 1}$$

into the series

$$\log p = \tfrac{1}{2} \log (p - 1) + \tfrac{1}{2} \log (p + 1) + \frac{1}{2p^2 - 1}$$
$$+ \frac{1}{3(2p^2 - 1)^3} + \cdots,$$

where $2p^2 - 1 > 1$ or $p^2 > 1$. If p is an integer and $p + 1$ can be resolved into smaller integral factors (for example, if $p + 1$ is even), this last series expresses the logarithm of p by the logarithms of smaller integers plus a series whose terms diminish very rapidly and whose sum can therefore be calculated accurately enough by use of only a few terms. From this series we can therefore calculate successively the logarithms of any prime number, and hence of any number, provided we have already calculated the value of log 2 (for example, by its integral representation, as on p. 489).

The accuracy of this determination of log p can be estimated more easily by means of the geometric series than from the general formula for the remainder. For the remainder R_n of the series, that is, the sum of all the terms following the term $1/n(2p^2 - 1)^n$, we have

$$R_n < \frac{1}{(n+2)(2p^2-1)^{n+2}}\left(1 + \frac{1}{(2p^2-1)^2} + \frac{1}{(2p^2-1)^4} + \cdots\right)$$

$$= \frac{1}{(n+2)(2p^2-1)^n} \cdot \frac{1}{(2p^2-1)^2 - 1},$$

and this formula immediately gives the required estimate of the error.

Let us for example calculate $\log_e 7$ (under the assumption that log 2 and log 3 have already been found numerically), using the first four terms of the series. We have

$$p = 7, \qquad 2p^2 - 1 = 97,$$

$$\log 7 = 2 \log 2 + \tfrac{1}{2} \log 3 + \frac{1}{97} + \frac{1}{3 \cdot 97^3} + \cdots ;$$

$$\tfrac{1}{97} \approx 0.01030928, \qquad \frac{1}{3 \cdot 97^3} \approx 0.00000037,$$

$$2 \log 2 \approx 1.38629436, \qquad \tfrac{1}{2} \log 3 \approx 0.54930614;$$

hence

$$\log_e 7 \approx 1.94591015.$$

Estimation of the error gives

$$R_n < \frac{1}{5 \cdot 97^3} \times \frac{1}{97^2 - 1} < \frac{1}{36 \times 10^9}.$$

However, we note that each of the four numbers which we have added is only given to within an error of 5×10^{-9}, so that the last place in the computed value of log 7 might be wrong by 2. As a matter of fact, however, the last place is also correct.

6.3 Numerical Solution of Equations

We add some remarks about the numerical solution of the equation $f(x) = 0$, where $f(x)$ need not be a polynomial.[1] We start with some tentative first value x_0 of one of the roots and then improve this approximation. How the first approximation for the root is chosen and how good that approximation is may be left open. We may, for example, take a rough guess, or better, obtain a first approximation from the

---

[1] We are, of course, concerned only with the determination of real roots of $f(x) = 0$.

graph of the function $y = f(x)$, whose intersection with the x-axis indicates the required root.

Then we try to improve the approximation by a process or mapping which takes the value x_0 into a "second approximation," and repeat this process. Solving the equation $f(x) = 0$ numerically consists in carrying out such successive approximations repeatedly (or as one says "iterating" the process) with the expectation that the iterated values $x_1, x_2, \ldots, x_n$ converge satisfactorily to the root ξ. We shall consider various such procedures and briefly discuss their accuracy.

a. Newton's Method

Description of Method. Newton's iterative procedure is based on the fundamental principle of the differential calculus—the replacing of a curve by a tangent in the immediate neighborhood of the point of contact. Starting from a first approximate value x_0 for a root ξ of the equation $f(x) = 0$ we consider the point on the graph of the function $y = f(x)$ whose coordinates are $x = x_0$, $y = f(x_0)$. To find a better approximation for the intersection ξ of the curve with the x-axis we determine the point x_1 where the tangent at the point $x = x_0$, $y = f(x_0)$ intersects the x-axis. The abscissa x_1 of this intersection represents a new and, under certain circumstances, a better approximation than x_0 to the required root ξ of the equation.

Figure 6.4 at once gives

$$\frac{f(x_0)}{x_0 - x_1} = f'(x_0);$$

hence the new approximation

$$(10) \qquad x_1 = x_0 - \frac{f(x_0)}{f'(x_0)}.$$

Starting with x_1 as an approximation, we repeat the process to find $x_2 = x_1 - f(x_1)/f'(x_1)$ and so on.

The usefulness of this process depends essentially on the nature of the curve $y = f(x)$. In the situation indicated in Fig. 6.4 the successive approximations x_n converge with increasing accuracy to the required root ξ.

However, Fig. 6.5 shows that with a plausible choice of the original value x_0, our construction need not converge to the required root at all. It is therefore necessary to examine in general the circumstances under which Newton's method furnishes useful approximations to the solution of the equation.

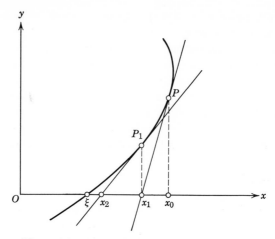

Figure 6.4 Newton's method of approximation.

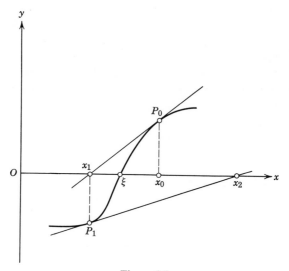

Figure 6.5

Quadratic Convergence of Newton's Method

Assuming that in a sufficiently wide interval about the root ξ the second derivative $f''(x)$ is not "too large" and the first derivative $f'(x)$ not "to small", the main fact concerning Newton's approximation is that the successive "errors"

$$h_1 = \xi - x_1, \qquad h_2 = \xi - x_2, \dots, h_n = \xi - x_n, \dots$$

converge to zero quadratically in the sense that $|h_{n+1}| \leq \mu h_n{}^2$ with a fixed constant μ. This indicates an extremely rapid rate of convergence; if we write the inequality in the form $|h_{n+1}\mu| \leq |h_n\mu|^2$ it implies, for example, that when $|h_n\mu| < 10^{-m}$ we have $|h_{n+1}\mu| < 10^{-2m}$, that is, the number of "significant digits" in μx_n is doubled at each step.

The proof of the quadratic convergence is immediate. From the relations $x_{n+1} = x_n - f(x_n)/f'(x_n)$ and $f(\xi) = 0$ we find that

$$h_{n+1} = \xi - x_{n+1} = \xi - x_n - \frac{f(\xi) - f(x_n)}{f'(x_n)}.$$

By Taylor's formula

$$f(\xi) - f(x_n) = (\xi - x_n)f'(x_n) + \tfrac{1}{2}(\xi - x_n)^2 f''(\eta),$$

where η lies between ξ and x_n. Hence

$$(11) \qquad\qquad h_{n+1} = - \frac{f''(\eta)}{2f'(x_n)} h_n^2.$$

To establish convergence we assume that x_n belongs already to a fixed interval $\xi - \delta < x < \xi + \delta$ in which $|f''|$ has the maximum value M_2, $|f'|$ the positive minimum value m_1, and for which δ is so small that $\tfrac{1}{2}\delta M_2/m_1 < 1$. Putting $\mu = \tfrac{1}{2}M_2/m_1$ we have $\mu\delta < 1$ and

$$|h_{n+1}| \leq \mu |h_n|^2 \leq \mu\delta |h_n| < |h_n|.$$

This inequality shows first of all that x_{n+1} belongs again to the same δ-neighborhood of ξ so that the argument can be repeated. Thus, if only x_0 lies in the δ-neighborhood of ξ, all subsequent x_n will do the same. From $|h_{n+1}| \leq \mu\delta |h_n|$ it follows then that $|h_{n+1}| \leq (\mu\delta)^{n+1} |h_0|$, which implies that $h_n \to 0$ or that $x_n \to \xi$; moreover, the quadratic law of decrease $|h_{n+1}| \leq \mu |h_n|^2$ will hold for the errors. It is clear then that Newton's method will provide us with a sequence x_n which certainly converges toward the solution ξ provided f' and f'' exist, and are continuous near ξ, that $f'(\xi) \neq 0$, and that x_0 is already sufficiently close to ξ. The quadratic character of the approximation is often a decided advantage of Newton's method over others (see p. 503).

*b. The Rule of False Position

Newton's method is the limiting case of an older method, the "rule of false position," in which the secant appears in place of the tangent. Let us assume that we know two points (x_0, y_0) and (x_1, y_1) in the neighborhood of the required intersection with the x-axis. If we replace the curve by the secant joining these two points, the intersection of this

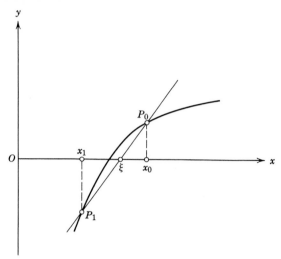

Figure 6.6   The rule of false position.

secant with the x-axis can be an improved approximation to the required root[1] of the equation. For the abscissa ξ of the point of intersection, we have (Fig. 6.6)

(12)
$$\frac{\xi - x_0}{f(x_0)} = \frac{\xi - x_1}{f(x_1)},$$

which leads to

$$\xi = \frac{x_0 f(x_1) - x_1 f(x_0)}{f(x_1) - f(x_0)}$$
$$= \frac{x_0 f(x_1) - x_0 f(x_0) + x_0 f(x_0) - x_1 f(x_0)}{f(x_1) - f(x_0)},$$

or

(13)
$$\xi = x_0 - \frac{f(x_0)}{\dfrac{f(x_1) - f(x_0)}{x_1 - x_0}}.$$

This formula, which determines the further approximation ξ from x_0 and x_1, constitutes the rule of false position. It is useful if one value of the function is positive and the other negative, say as in Fig. 6.6, where $y_0 > 0$ and $y_1 < 0$.

---

[1] This amounts essentially to linear interpolation applied to the inverse function.

The approximation formula of Newton results as a limiting case for $x_1 \to x_0$, for the denominator of the second term on the right-hand side of formula (13) tends to $f'(x_0)$ as x_1 tends to x_0.

Although the rule of false position may be considered more elementary than Newton's method, the latter has the great convenience of requiring only one value of x as initial approximation instead of two values.

c. The Method of Iteration

The Iteration Scheme. We now turn to a far-reaching scheme for solving equations written in the form

$$x = \phi(x),$$

where ϕ is a continuous function with a continuous derivative. The solution of equations of the form $f(x) = 0$ can be reduced to that of $x = \phi(x)$ if we put $\phi(x) = x - c(x)f(x)$ where $c(x)$ is any function different from zero.

In the particularly suggestive *method of iteration*[1] we begin again with a suitably chosen initial approximating value x_0 and then determine a sequence $x_1, x_2, x_3, \ldots$ of values by the conditions

$$x_{n+1} = \phi(x_n), \qquad n = 0, 1, 2, \ldots.$$

If this "iteration" sequence x_n converges to a limit ξ, then $\xi = \phi(\xi)$ is a solution of our equation, since then $\lim_{n \to \infty} x_{n+1} = \xi$ and $\lim_{n \to \infty} \phi(x_n) = \phi(\xi)$ because of the continuity of the function ϕ.

Convergence. The sequence of values x_n in the iteration process converges to a solution under a very general assumption: If the first approximation x_0 lies in an interval[2] J about the solution ξ, in which

$$|\phi'(x)| < q$$

with a constant $q < 1$, then x_n converges to ξ.

For supposing that x_0 lies in J, we have

$$x_1 - \xi = \phi(x_0) - \phi(\xi).$$

---

[1] Sometimes called the method of successive approximation. The method is used in many different mathematical contexts for solving equations of one kind or other.

[2] Although ξ is unknown, we can very often determine such an interval a priori.

By the mean value theorem, the right-hand side of this equation equals $(x_0 - \xi)\phi'(\bar{x})$, where $\bar{x}$ lies in J. Thus by our assumption

$$|x_1 - \xi| \leq q\,|x_0 - \xi|,$$

so that x_1 belongs to J, and then also

$$|x_2 - \xi| \leq q\,|x_1 - \xi\,| \leq q^2\,|x_0 - \xi|.$$

In general, we obtain

$$|x_n - \xi| \leq q^n\,|x_0 - \xi|;$$

since $q^n \to 0$ as $n \to \infty$, our assertion is proved.

We see, moreover, from the preceding, that the iteration sequence x_n does not converge when $\phi'(x) > 1$ in an interval about ξ; if $|\phi'(\xi)| = 1$ we cannot make a general statement.

Attracting and Repelling Fixed Points

It is useful to consider the iteration process in terms of a mapping or transformation. The function $y = \phi(x)$ represents a transformation which maps a point x on the number axis into an image point y of this number axis (see p. 20). The solution ξ is then a point not changed by the transformation ϕ, a so-called *fixed point*, and the problem is thus one of finding a fixed point of the mapping; this problem is solvable by iteration when $|\phi'(\xi)| \leq q < 1$, as we have seen.

The mapping $y = \phi(x)$ of the neighborhood of the root or fixed point ξ has, for $|\phi'(x)| < q < 1$, the property of being *contracting*, that is, diminishing the distance of the original from the fixed point. Such fixed points of contracting mappings are called *attracting fixed points*. Their construction by iteration converges as the terms of a geometric series with the quotient q.

If the root ξ, or the corresponding fixed point of our transformation is in an interval in which $|\phi'(x)| > r$, where r is a constant larger than 1, the *transformation is expanding*, the iteration process diverges, and the *fixed point* is called repelling.

If at the fixed point we have $|\phi'(\xi)| = 1$, no general statements concerning the convergence of the iterations can be made; such fixed points are sometimes called *indifferent*.

The following observation should be stressed: a fixed point ξ of the mapping ϕ is automatically also a fixed point for ψ, the inverse mapping : $\xi = \psi(\xi)$. If $|\phi'(\xi)| > 1$ in a neighborhood of a root ξ and $x = \psi(y)$ is the inverse function of ϕ, then $|\psi'(\xi)| < 1$. Thus ξ is an attracting fixed point for this inverse mapping and it is possible to replace the originally divergent iteration scheme by a convergent one

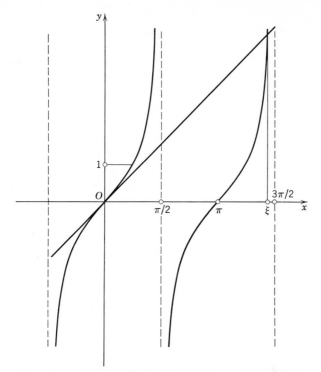

Figure 6.7   Intersection (ξ, ξ) of the curves $y = \tan x$ and $y = x$.

for the inverse mapping. As an example we consider the equation

$$x = \tan x.$$

It is clear from the graphs of the functions $y = x$ and $y = \tan x$ that these intersect somewhere in the interval $\pi < x < \frac{3}{2}\pi$ and that our equation will have a root ξ in that interval (Fig. 6.7). Since

$$\frac{d \tan x}{dx} = \frac{1}{\cos^2 x} > 1,$$

the iteration procedure with any point x_0 in the interval does not converge. However, we obtain a convergent iteration sequence if we write the equation in the inverse form (using the notation arc tan x for the principal branch),

$$x = \text{arc tan } x + \pi.$$

Since here

$$\frac{d}{dx} \text{arc tan } x = \frac{1}{1 + x^2} < 1,$$

the sequence defined by $x_{n+1} = \text{arc tan } x_n + \pi$ and, say, $x_0 = \pi$, converges to ξ.

d. Iterations and Newton's Procedure

As mentioned before the solution of an equation of the form $f(x) = 0$ can be reduced to that of the form $x = \phi(x)$ if we choose for ϕ any expression of the form

$$\phi(x) = x - c(x)f(x)$$

where $c(x)$ is a nonvanishing function. If we want to solve the resulting equation $x = \phi(x)$ by iteration we have to make sure by a suitable choice of $c(x)$ that the fixed point ξ of the mapping ϕ is "attractive", that is, that $|\phi'(\xi)| < 1$. Now for the solution ξ of $f(\xi) = 0$ we have

$$\phi'(\xi) = 1 - c'(\xi)f(\xi) - c(\xi)f'(\xi) = 1 - c(\xi)f'(\xi).$$

The simplest choice is to take for $c(x)$ the expression $1/f'(x)$. Then certainly $|\phi'(\xi)| = 0 < 1$. This choice of $c(x)$ leads to the iteration sequence

$$x_{n+1} = \phi(x_n) = x_n - \frac{f(x_n)}{f'(x_n)},$$

which is just the sequence of approximations (10), p. 495, in Newton's method. For the error $x_n - \xi = h_n$ we have the estimate

$$|h_{n+1}| = |\phi(x_n) - \phi(\xi)| \le qh_n,$$

where q is the maximum of $|\phi'(x)|$ in the interval with end points ξ and x_n. Since here

$$\phi'(x) = \frac{f(x)f''(x)}{f'^2(x)}$$

and $f(x) = f(x) - f(\xi) = f'(\eta)(x - \xi)$, we see that q itself is of the order of h_n, and thus confirm again the quadratic character of the approximation in Newton's method.

Another simple choice for $c(x)$ is to take the constant value $1/f'(x_0)$, leading to the recursion formula

$$x_{n+1} = \phi(x_n) = x_n - \frac{f(x_n)}{f'(x_0)}.$$

Here $\phi'(\xi) = 1 - f'(\xi)/f'(x_0)$. If f' is continuous and different from zero, we will have an attractive fixed point ξ if our initial approximation x_0 is already so close to the solution ξ that

$$|\phi'(\xi)| = \frac{|f'(x_0) - f'(\xi)|}{|f'(x_0)|} < 1.$$

This iteration sequence is somewhat simpler than the one used in Newton's method; however, convergence will be much slower, like that for a geometric progression, as is the case with most iteration schemes.

Examples. As an example we consider the cubic equation

$$f(x) = x^3 - 2x - 5 = 0.$$

Since $f(2) = -1 < 0$, $f(3) = 16 > 0$, a root ξ certainly exists in the interval $2 < x < 3$. Since, moreover, $f'(x) = 3x^2 - 2 > 3(2)^2 - 2 > 0$, the interval contains only one root. By Newton's method we find starting with the approximation $x_0 = 2$ successively

$$x_1 = x_0 - \frac{f(x_0)}{f'(x_0)} = 2 - \frac{-1}{3(2)^2 - 2} = 2.1, \quad f(x_1) = 0.061$$

$$x_2 = x_1 - \frac{f(x_1)}{f'(x_1)} = 2.1 - \frac{0.061}{3(2.1)^2 - 2} = 2.094568.$$

Since $f(2.1) > 0$, $f(2) < 0$, the root ξ lies between 2 and 2.1. In the interval $1.9 < x < 2.2$, and a fortiori then in the interval $\xi - 0.1 < x < \xi + 0.1$, we have the estimates

$$|f''(x)| = |6x| < 6(2.2) = 13.2,$$

$$f'(x) = 3x^2 - 2 > 3(1.9)^2 - 2 = 8.83.$$

It follows [see (11), p. 497] that

$$|\xi - x_{n+1}| \leq \frac{13.2}{2(10.83)} |x_n - \xi|^2 < 0.75 |x_n - \xi|^2$$

provided $|x_n - \xi| < 0.1$. Since $|x_0 - \xi| = |\xi - 2| < 0.1$, we find successively

$$|x_1 - \xi| < (0.75)(0.1)^2 = 0.0061$$

$$|x_2 - \xi| < (0.61)(0.0061)^2 < 0.000042.$$

If this degree of approximation is not sufficient, we obtain a further approximation x_3 with an error $< (0.75)(0.000042)^2 < 0.000\,000\,001\,3$.

All x_n after x_0 must be larger than ξ as is obvious from the fact that f' and f'' are positive, which implies that

$$h_{n+1} = -f''(\eta)h_n^2/2f'(x_n) < 0.$$

Applying instead the rule of false position [(13), p. 498] to the values x_0, x_1 we find for the intersection ξ with the x-axis of the secant joining the points $(x_0, f(x_0))$ and $(x_1, f(x_1))$

$$\xi = x_0 - \frac{f(x_0)(x_1 - x_0)}{f(x_1) - f(x_0)} = 2.09425 \cdots.$$

Since the curve is convex in the interval in question, the secant lies above the curve and the approximation $\bar{\xi}$ must be less than the root ξ.

As a second example, let us solve the equation

$$f(x) = x \log_{10} x - 2 = 0.$$

We have $f(3) = -0.6$ and $f(4) = +0.4$, and therefore use $x_0 = 3.5$ as a first approximation. Using ten-digit logarithmic tables we obtain the successive approximations

$$x_0 = 3.5, \qquad\qquad x_1 = 3.598,$$

$$x_2 = 3.5972849, \qquad x_3 = 3.5972850235.$$

Appendix

*A.1   Stirling's Formula

In many applications, particularly in statistics and the theory of probability, we find it necessary to have a simple approximation to $n!$ as an elementary function of n. Such an expression is given by the following theorem, which bears the name of its discoverer, Stirling (see also Chapter 8, p. 630).

As $n \to \infty$,

$$\text{(14)} \qquad\qquad \frac{n!}{\sqrt{2\pi}\, n^{n+1/2} e^{-n}} \to 1;$$

more exactly,

$$\text{(14a)} \qquad \sqrt{2\pi}\, n^{n+1/2} e^{-n} < n! < \sqrt{2\pi}\, n^{n+1/2} e^{-n}\left(1 + \frac{1}{4n}\right).$$

In other words, the expressions $n!$ and $\sqrt{2\pi}\, n^{n+1/2} e^{-n}$ differ only by a small percentage when the value of n is large—as we say, the two expressions are *asymptotically equal*—and at the same time the factor $1 + 1/4n$ gives us an estimate of the degree of accuracy of the approximation.

We are led to this remarkable formula if we attempt to evaluate the area under the curve $y = \log x$.[1] By integration (p. 276) we find that A_n, the exact area under this curve between the ordinates $x = 1$ and

---

[1] The method used here is a special instance of the Euler MacLaurin formula which will be discussed in Chapter 8, p. 624.

$x = n$, is given by

$$(15) \qquad A_n = \int_1^n \log x \, dx = x \log x - x \Big|_1^n = n \log n - n + 1.$$

If, however, we estimate the area by the trapezoid rule, erecting ordinates at $x = 1$, $x = 2$, $\ldots$, $x = n$ as in Fig. 6.8, we obtain an approximate value T_n for the area [cf. (6), p. 483]

$$(16) \qquad \begin{aligned} T_n &= \log 2 + \log 3 + \cdots + \log (n - 1) + \tfrac{1}{2} \log n \\ &= \log n! - \tfrac{1}{2} \log n. \end{aligned}$$

If we make the reasonable assumption that A_n and T_n are of the same order of magnitude, we find at once that $n!$ and $n^{n+1/2}e^{-n}$ are of the

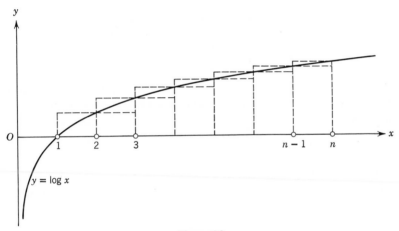

Figure 6.8

same order of magnitude, which is essentially what is stated in Stirling's formula.

To make this argument precise, we first show that the difference $a_n = A_n - T_n$ is bounded, from which it will immediately follow that $T_n = A_n(1 - a_n/A_n)$ is of the same order of magnitude as A_n. The difference $a_{k+1} - a_k$ is the difference between the area under the curve and the area under the secant in the strip $k < x < k + 1$. Since the curve is concave and lies above the secant, $a_{k+1} - a_k$ is positive, and $a_n = (a_n - a_{n-1}) + (a_{n-1} - a_{n-2}) + \ldots + (a_2 - a_1) + a_1$ is monotonic increasing. Moreover, the difference $a_{k+1} - a_k$ is clearly less (cf. Fig. 6.9) than the difference between the area under the tangent

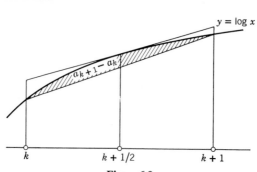

Figure 6.9

at $x = k + \frac{1}{2}$ and the area under the secant; hence we have the inequality

$$a_{k+1} - a_k < \log\left(k + \tfrac{1}{2}\right) - \tfrac{1}{2}\log k - \tfrac{1}{2}\log(k+1)$$

$$= \tfrac{1}{2}\log\left(1 + \frac{1}{2k}\right) - \tfrac{1}{2}\log\left[1 + \frac{1}{2(k + \frac{1}{2})}\right]$$

$$< \tfrac{1}{2}\log\left(1 + \frac{1}{2k}\right) - \tfrac{1}{2}\log\left[1 + \frac{1}{2(k+1)}\right].$$

Adding these inequalities for $k = 1, 2, \ldots, n-1$, we find that all the terms on the right-hand side except two will cancel out, and (since $a_1 = 0$), we have

$$a_n < \tfrac{1}{2}\log\tfrac{3}{2} - \tfrac{1}{2}\log\left(1 + \frac{1}{2n}\right) < \tfrac{1}{2}\log\tfrac{3}{2}.$$

Since a_n is bounded, and in addition monotonic increasing it tends to a limit a as $n \to \infty$. Our inequality for $a_{k+1} - a_k$ now gives us

$$a - a_n = \sum_{k=n}^{\infty}(a_{k+1} - a_k) < \tfrac{1}{2}\log\left(1 + \frac{1}{2n}\right).$$

Since by definition $A_n - T_n = a_n$, we have from (15), (16),

$$\log n! = 1 - a_n + (n + \tfrac{1}{2})\log n - n,$$

or, writing $\alpha_n = e^{1-a_n}$,

$$n! = \alpha_n n^{n+1/2} e^{-n}.$$

The sequence α_n is monotonic *decreasing* and tends to the limit $\alpha = e^{1-a}$; hence

$$1 < \frac{\alpha_n}{\alpha} = e^{a-a_n} < e^{(1/2)\log(1+1/2n)}$$

$$= \sqrt{1 + \frac{1}{2n}} < 1 + \frac{1}{4n}.$$

Hence we have

$$\alpha n^{n+1/2} e^{-n} < n! < \alpha n^{n+1/2} e^{-n} \left(1 + \frac{1}{4n} \right)$$

It only remains for us to find the actual value of the limit a. Here we make use of the formula (80) of Chapter 3, p. 282:

$$\sqrt{\pi} = \lim_{n \to \infty} \frac{(n!)^2 2^{2n}}{(2n)! \sqrt{n}}.$$

Replacing $n!$ by $\alpha_n n^{n+1/2} e^{-n}$ and $(2n)!$ by $\alpha_{2n} 2^{2n+1/2} n^{2n+1/2} e^{-2n}$, we immediately obtain

$$\sqrt{\pi} = \lim_{n \to \infty} \frac{\alpha_n{}^2}{\alpha_{2n} \sqrt{2}}$$

$$= \frac{\alpha^2}{\alpha \sqrt{2}},$$

from which $\alpha = \sqrt{2\pi}$. The proof of Stirling's formula is thus complete.

In addition to its theoretical interest, Stirling's formula is a very useful tool for the numerical calculation of $n!$ when n is large. Instead of multiplying together a large number of integers, we have merely to calculate Stirling's expression by means of logarithms which involves far fewer operations. Thus for $n = 10$ we obtain the value 3598696 for Stirling's expression (using seven-figure tables), whereas the exact value of 10! is 3628800. The percentage error is barely $\frac{5}{6}\%$.

PROBLEMS

SECTION 6.1, page 482

1. Prove if $f''(x) \geq 0$, that the trapezoid rule yields a greater value and the tangent rule a lesser value than the exact integral of f.

2. Estimate the value $h = (b - a)/n$ needed for a calculation by Simpson's rule accurate to p decimal places of

(a) $\log 2 = \int_1^2 \frac{1}{x} \, dx$,

(b) $\pi = 4 \int_0^1 \frac{1}{1 + x^2} \, dx$.

3. Estimate in terms of k and s ($k < 1$ and $s < 1$), the number of points needed to calculate within an error ϵ the elliptic integral

$$u(s) = \int_0^s \frac{dx}{\sqrt{(1 - x^2)(1 - k^2 x^2)}}.$$

4. Let $f(x)$ be a continuous function on the interval $\alpha \leq x \leq \alpha + h$, with a uniformly bounded derivative: $|f'(x)| \leq M_1$ for M_1 a constant. Prove that for any fixed point ξ, $\alpha \leq \xi \leq \alpha + h$, the estimate

$$\left| \int_\alpha^{\alpha+h} f(x)\, dx \; - \; hf(\xi) \right| \leq \frac{M_1 h^2}{2}$$

5. Calculate $\displaystyle\int_\alpha^\infty e^{-x^2}\, dx$ numerically to within $1/100$.

SECTION 6.2, page 490

1. The period of a pendulum is given by

$$T = 2\pi \sqrt{\frac{l}{g}}$$

where l is the length of the pendulum. If the pendulum drives a clock which gains a minute per day determine the necessary correction in l.

2. To measure the height of a hill, a tower 100 meters high on top of the hill is observed from the plain. The angle of elevation of the base of the tower is $42°$ and the tower itself subtends an angle of $6°$. What are the limits of error in the determination of the height if the angle $42°$ is subject to an error of $1°$?

SECTION 6.3, page 494

1. (*a*) To solve the equation $x = f(x)$, show how best to choose the constant a so that the iteration scheme

$$x_{k+1} = x_k + a[x_k - f(x_k)]$$

converges as rapidly as possible in the neighborhood of the solution.

(*b*) Apply this method to solve the equation for $\sqrt{A}$,

$$x = \frac{A}{x}.$$

(*c*) Show if $A \geq 1$ that the number of accurate decimal places is at least doubled at each step of the iteration scheme obtained in (*b*).

2. (*a*) Show how best to choose a polynomial

$$g(x) = a + bx^2$$

so that the iteration scheme for $\sqrt{A}$,

$$x_{k+1} = x_k + g(x_k)\left(x_k - \frac{A}{x_k}\right)$$

converges most rapidly in the neighborhood of the solution.

(*b*) Estimate the rapidity of convergence.

(*c*) Show how to further improve the convergence by suitable choices of polynomials $g(x)$ which are of higher degree.

3. Investigate suitable schemes of the type of Problems 1 and 2 for the calculation of $\sqrt[v]{A}$.

SECTION A.1, page 504

1. Prove that $\lim\limits_{n \to \infty} \dfrac{\sqrt[n]{n!}}{n} = \dfrac{1}{e}$.

*2. By considering $\displaystyle\int_{1/2}^{n+1/2} \log(\alpha + x)\, dx,\ \alpha > 0$, show that

$$\alpha(\alpha + 1) \cdots (\alpha + n) = a_n n!\, n^\alpha,$$

where a_n is bounded below by a positive number. Show that a_n is monotonically decreasing for sufficiently large values of n. [The limit of a_n as $n \to \infty$ is $1/\Gamma(\alpha)$.]

3. Find an approximate expression for $\log \dfrac{n_1!\, n_2! \cdots n_l!}{n!}$, where $n_1 + n_2 + \cdots + n_l = n$.

4. Show that the coefficient of x^n in the binomial expansion of $\dfrac{1}{\sqrt{1-x}}$ is asymptotically given by $\dfrac{1}{\sqrt{\pi n}}$.

7

Infinite Sums and Products

The geometric series, Taylor's series, and a number of examples previously discussed in this book, suggest that we may well study those limiting processes of analysis which involve the summation of *infinite series* from a more general point of view. In principle, any limiting value

$$S = \lim_{n \to \infty} s_n$$

can be written as an infinite series; we need only put $a_n = s_n - s_{n-1}$ for $n > 1$ and $a_1 = s_1$ to obtain

$$s_n = a_1 + a_2 + \cdots + a_n,$$

and the value S thus appears as the limit of s_n, the sum of n terms, as n increases. We express this fact by saying that S is the "sum of the infinite series"

$$a_1 + a_2 + a_3 + \cdots.$$

Such an "infinite sum" is simply a way of representing a limit where each successive approximation is found from the preceding by adding one more term. Thus the expression of a number as a decimal is in principle merely the representation of a number a in the form of an infinite series $a = a_1 + a_2 + a_3 + \cdots$, where, if $0 \leq a \leq 1$, the term a_n is replaced by $\alpha_n \times 10^{-n}$ and α_n is an integer between 0 and 9 inclusive.

Since every limiting value can be written in the form of an infinite series, a special study of series may seem superfluous. However, very often it happens that limiting values occur naturally in the form of such infinite series which exhibit particularly simple laws of formation.

Not *every* series has an easily recognizable law of formation. For example, the number π can certainly be represented as a decimal (which is a series $\Sigma c_\nu \, 10^{-\nu}$), yet we know no simple law enabling us to state the value of an arbitrary digit, say the 7000th, of this decimal. If, however, we consider the Leibnitz-Gregory series for $\pi/4$ instead, we have an expression with a perfectly clear general law of formation [see (7), p. 445].

Analogous to infinite series, in which the approximations to the limit are formed by repeated addition of new terms, are *infinite products*, in which the approximations to the limit arise from repeated multiplication by new factors. We shall not go deeply into the general theory of infinite products, however; the principal subject of this chapter and of Chapter 8 will be infinite series.

7.1 The Concepts of Convergence and Divergence

a. Basic Concepts

Cauchy's Convergence Criterion. We consider an infinite series with the "general term" a_n; the series[1] is then of the form

$$a_1 + a_2 + \cdots = \sum_{\nu=1}^{\infty} a_\nu.$$

The symbol on the right with the summation sign is merely an abbreviated way of writing the expression on the left.

If as n increases, the *nth partial sum*

$$s_n = a_1 + a_2 + \cdots + a_n = \sum_{\nu=1}^{n} a_\nu$$

approaches a limit

$$S = \lim_{n \to \infty} s_n$$

we say that the series is *convergent*; otherwise we say that it is *divergent*. In the first case we call S the *sum of the series*.

We have already encountered many examples of convergent series; for instance, the geometric series $1 + q + q^2 + \cdots$, which converges to the sum $1/(1 - q)$ when $|q| < 1$, the series for log 2, the series for e, and others.

In the language of infinite series, Cauchy's convergence test (cf. Chapter 1, p. 75) is expressed as follows:

---

[1] For formal reasons we include the possibility that certain of the numbers a_n may be zero. If *all* terms from an index N onward (that is, when $n > N$) vanish, we speak of a *terminating series*.

A necessary and sufficient condition for the convergence of a series is that the number

(1)
$$|s_m - s_n| = |a_{n+1} + a_{n+2} + \cdots + a_m|$$

($m > n$), becomes arbitrarily small if m and n are chosen sufficiently large. In other words: *A series converges if, and only if, the following condition is fulfilled: for a given positive number ϵ, it is possible to choose an index $N = N(\epsilon)$, in such a way that the above expression $|s_m - s_n|$ is less than ϵ, provided only that $m > N$ and $n > N$.*

We can illustrate the convergence test by the geometric series for $q = \frac{1}{2}$. If we choose $\epsilon = \frac{1}{10}$, we need only take $N = 4$. For

$$|s_m - s_n| = \frac{1}{2^n} + \cdots + \frac{1}{2^{m-1}}$$

$$= \frac{1}{2^{n-1}}\left(\frac{1}{2} + \frac{1}{2^2} + \cdots + \frac{1}{2^{m-n}}\right) < \frac{1}{2^{n-1}}$$

and
$$\frac{1}{2^{n-1}} < \frac{1}{10} \qquad \text{if } n > 4.$$

If we choose ϵ equal to $\frac{1}{100}$, it is sufficient to take 7 as the corresponding value of N, as may easily be verified.

Obviously, it is a *necessary* condition for the convergence of a series that

$$\lim_{n \to \infty} a_n = 0.$$

Otherwise, the convergence criterion certainly cannot be fulfilled for $m = n + 1$. But this necessary condition is by no means *sufficient* for convergence; on the contrary, it is easy to find infinite series whose general term a_n approaches 0 as n increases, but whose sum does not exist, since the partial sum s_n increases without limit as n increases.

Examples. An example is the series

$$1 + \frac{1}{\sqrt{2}} + \frac{1}{\sqrt{3}} + \cdots + \frac{1}{\sqrt{n}} + \cdots,$$

the general term of which is $1/\sqrt{n}$. We immediately see that

$$s_n > \frac{1}{\sqrt{n}} + \cdots + \frac{1}{\sqrt{n}} = \frac{n}{\sqrt{n}} = \sqrt{n}.$$

The nth partial sum increases beyond all bounds as n increases, and therefore the series diverges.

The same is true for the classic example of the *harmonic series*

$$1 + \frac{1}{2} + \frac{1}{3} + \frac{1}{4} + \cdots .$$

Here

$$a_{n+1} + \cdots + a_{2n} = \frac{1}{n+1} + \cdots + \frac{1}{2n} > \frac{1}{2n} + \cdots + \frac{1}{2n} = \frac{1}{2}.$$

Since n and $m = 2n$ can be chosen to be as large as we please, the series diverges, for Cauchy's test is not fulfilled; in fact, the nth partial sum obviously tends to infinity, since all the terms are positive. On the other hand, the series formed from the same numbers with alternating signs,

$$1 - \frac{1}{2} + \frac{1}{3} - \frac{1}{4} + \frac{1}{5} - + \cdots + \frac{(-1)^{n-1}}{n} + \cdots ,$$

converges [cf. (4) Chapter 5, p. 443], and has the sum log 2.

It is by no means true that in every divergent series s_n tends to $+\infty$ or $-\infty$. Thus in the series

$$1 - 1 + 1 - 1 + 1 + - \cdots ,$$

we see that the partial sum s_n has the values 1 and 0 alternately, and on account of this oscillation backward and forward, neither approaches a definite limit nor increases numerically beyond all bounds.

The following fact, although it is self-evident, is very important and should be noted. *The convergence or divergence of a series is not changed by inserting a finite number of terms or by removing a finite number of terms.* As far as convergence or divergence is concerned, it does not matter in the least whether we begin the series at the term a_0, or a_1, or a_5, or any other term chosen arbitrarily.

b. Absolute Convergence and Conditional Convergence

The harmonic series $1 + \frac{1}{2} + \frac{1}{3} + \frac{1}{4} \cdots$ diverges, but if we change the sign of every other term the resulting series for log 2 converges. On the other hand, the geometric series $1 - q + q^2 - q^3 + - \cdots$ converges and has the sum $1/(1 + q)$, provided that $0 \leq q < 1$, and on making all the signs plus we obtain the series

$$1 + q + q^2 + q^3 + \cdots ,$$

which is also convergent, having the sum $1/(1 - q)$.

Here there appears a distinction which we must examine. With a series whose terms are all positive there are only two possible cases; either it converges or the partial sum increases beyond all bounds as n increases. For the partial sums, being a monotonic increasing sequence, must converge if they remain bounded. Convergence occurs if the individual terms approach zero rapidly enough as n increases; on the other hand, divergence occurs if the terms do not approach zero at all or if they approach zero too slowly. However, in series some terms of which are positive and some negative, it may be that the changes of sign bring about convergence, when too great an increase in the partial sums, due to the positive terms, is compensated by the negative terms, so that as the final result a definite limit is approached.

To understand the possibilities better we consider a series $\sum\limits_{v=1}^{\infty} a_v$ having positive and negative terms and form for comparison the series which has the same terms all with positive signs, that is,

$$|a_1| + |a_2| + \cdots = \sum_{v=1}^{\infty} |a_v|.$$

If this series converges, then for sufficiently large values of n and $m > n$, the expression

$$|a_{n+1}| + |a_{n+2}| + \cdots + |a_m|$$

will certainly be as small as we please; because of the relation

$$|a_{n+1} + \cdots + a_m| \leq |a_{n+1}| + \cdots + |a_m|$$

the expression on the left is also arbitrarily small, and so by the Cauchy test the original series $\sum\limits_{v=1}^{\infty} a_v$ converges. In this case the original series is said to be *absolutely convergent*. Its convergence is due to the absolute smallness of its terms and does not depend on the changes in sign.

If, on the other hand, the series with the terms $|a_n|$ diverges and the original series still converges, we say that the original series is *conditionally convergent*. Conditional convergence results from the terms of opposite signs compensating one another.

Leibnitz's Test. For conditional convergence *Leibnitz's convergence test* is frequently useful:

If the terms of a series are of alternating sign and in addition their absolute values $|a_n|$ tend monotonically to 0 (so that $|a_{n+1}| \leqslant |a_n|$), the series $\sum\limits_{v=1}^{\infty} a_v$ converges. [Example: Leibnitz's series, (7), p. 445.]

For the proof we assume that $a_1 > 0$, which does not limit the generality of the argument, and write our series in the form

$$b_1 - b_2 + b_3 - + \cdots,$$

where all the terms b_n are now positive, b_n tends to zero, and the condition $b_{n+1} \leq b_n$ is satisfied. If we bracket the terms together in the two different ways

$$b_1 - (b_2 - b_3) - (b_4 - b_5) - \cdots$$

and

$$(b_1 - b_2) + (b_3 - b_4) + (b_5 - b_6) + \cdots$$

we see at once that the partial sums $s_n = \sum_1^n a_\nu$ satisfy the following two relations

$$s_1 \geq s_3 \geq s_5 \geq \cdots \geq s_{2n+1} \geq \cdots,$$

$$s_2 \leq s_4 \leq s_6 \leq \cdots \leq s_{2n} \leq \cdots.$$

On the other hand, $s_{2n} \leq s_{2n+1} \leq s_1$ and $s_{2n+1} \geq s_{2n} \geq s_2$. The odd partial sums $s_1, s_3, \ldots$ therefore form a monotonic decreasing sequence,

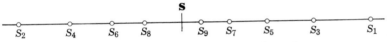

Figure 7.1   Convergence of an alternating series.

which in no case falls below the value s_2; hence this sequence possesses a limit L (p. 73). The even partial sums $s_2, s_4, \ldots$ likewise form a monotonic increasing sequence whose terms in no case exceed the fixed number s_1, and therefore this sequence must have a limiting value L'. Since the numbers s_{2n} and s_{2n+1} differ from one another only by the number b_{2n+1} which approaches 0 as n increases, the limiting values L and L' are equal to one another. That is, the even and the odd partial sums approach the same limit, which we now denote by S (cf. Fig. 7.1). This, however, implies that our series is convergent, as was asserted; its sum is S.

Abel's Test

A test for conditional convergence that includes the Leibnitz test as a special case is *Abel's convergence test*. Let $a_1 + a_2 + \cdots$ be an infinite series whose partial sums $s_n = a_1 + \cdots + a_n$ are bounded independently of n. Let $p_1, p_2, \ldots$ be a sequence of positive numbers decreasing monotonically to the value zero. Then the infinite series

$$(2) \qquad p_1 a_1 + p_2 a_2 + \cdots$$

converges. (For the special series $a_1 + a_2 + \cdots = +1 - 1 + 1 - 1 + - \cdots$ we find that $p_1 - p_2 + p_3 - \cdots$ converges, which is Leibnitz' test.) The proof follows if we apply Cauchy's test using "summation by parts" to estimate

$$|p_{n+1}a_{n+1} + p_{n+2}a_{n+2} + \cdots + p_m a_m|$$

$$= |p_{n+1}(s_{n+1} - s_n) + p_{n+2}(s_{n+2} - s_{n+1}) + \cdots + p_m(s_m - s_{m-1})|$$

$$= |-p_{n+1}s_n + p_m s_m + (p_{n+1} - p_{n+2})s_{n+1} + (p_{n+2} - p_{n+3})s_{n+2} + \cdots$$

$$+ (p_{m-1} - p_m)s_{m-1}|$$

$$\leq p_{n+1}M + p_m M + (p_{n+1} - p_{n+2} + p_{n+2} - p_{n+3} + - \cdots + p_{m-1} - p_m) M$$

$$= 2p_{n+1}M,$$

where M is a bound for the $|s_i|$; since $p_{n+1} \to 0$ the convergence of the series (2) follows by Cauchy's test.

*In conclusion, we make another general remark about the fundamental difference between absolute convergence and conditional convergence. We consider a convergent series $\sum\limits_{\nu=1}^{\infty} a_\nu$. We denote the positive terms of the series by $p_1, p_2, p_3, \ldots$, and the negative terms by $-q_1, -q_2, -q_3, \ldots$. If we form the nth partial sum $s_n = \sum\limits_{\nu=1}^{n} a_\nu$ of the given series, a certain number, say n', of positive terms and a certain number, say n'', of negative terms must appear, where $n' + n'' = n$. Furthermore, if the number of positive terms as well as the number of negative terms in the series is infinite, then the two numbers n' and n'' will increase beyond all bounds as n does. We see immediately that the partial sum s_n is simply equal to the partial sum $\sum\limits_{\nu=1}^{n'} p_\nu$ of the positive terms of the series plus the partial sum $- \sum\limits_{\nu=1}^{n''} q_\nu$ of the negative terms. If the given series converges absolutely, then the series of positive terms $\sum\limits_{\nu=1}^{\infty} p_\nu$ and the series of absolute values of the negative terms $\sum\limits_{\nu=1}^{\infty} q_\nu$ certainly both converge. For as m increases, the partial sums $\sum\limits_{\nu=1}^{m} p_\nu$ and $\sum\limits_{\nu=1}^{m} q_\nu$ are monotonic nondecreasing sequences with the upper bound $\sum\limits_{\nu=1}^{\infty} |a_\nu|$.

The sum of an absolutely convergent series is then simply equal to the sum of the series consisting of the positive terms only, plus the sum of the series consisting of the negative terms only, or, in other words, is equal to the difference of the two series with positive terms.

For $\sum\limits_{\nu=1}^{n} a_\nu = \sum\limits_{\nu=1}^{n'} p_\nu - \sum\limits_{\nu=1}^{n''} q_\nu$; as n increases n' and n'' also increase beyond all bounds, and the limit of the left-hand side must therefore be equal to the difference of the two sums on the right. If the series contains only a finite number of terms of one particular sign, the facts are correspondingly simplified. If, on the other hand, the series does not converge absolutely, but does converge conditionally, then the series $\sum\limits_{\nu=1}^{\infty} p_\nu$ and $\sum\limits_{\nu=1}^{\infty} q_\nu$ must both be divergent. For if both were convergent the series would converge absolutely, contrary to our hypothesis. If only one diverged, say $\sum\limits_{\nu=1}^{\infty} p_\nu$, and the other converged, then separation into positive and negative parts, $s_n = \sum\limits_{\nu=1}^{n'} p_\nu - \sum\limits_{\nu=1}^{n''} q_\nu$, shows that the series could not converge; for as n increases n' and $\sum\limits_{\nu=1}^{n'} p_\nu$ would increase beyond all bounds, whereas the term $\sum\limits_{\nu=1}^{n''} q_\nu$ would approach a definite limit, so that the partial sum s_n would increase beyond all bounds.

We see, therefore, that *a conditionally convergent series cannot be thought of as the difference of two convergent series, the one consisting of its positive terms and the other consisting of the absolute values of its negative terms.*

Closely connected with this fact is another difference between absolutely and conditionally convergent series which we shall now briefly mention.

*c. Rearrangement of Terms

It is a property of finite sums that we can change the order of the terms or, as we say, rearrange the terms at will without changing the value of the sum. The question arises: what is the exact meaning of a change of the order of terms in an infinite series, and does such a rearrangement leave the value of the sum unchanged? Although in finite sums there is no difficulty, for example, in adding the terms in reverse order, in infinite series such a possibility does not exist; there is no last term with which to begin. Now a change of order in an infinite series can only mean this: we say that a series $a_1 + a_2 + a_3 + \cdots$ is transformed by rearrangement into a series $b_1 + b_2 + b_3 + \cdots$, provided that every term a_n of the first series occurs exactly once in the second and conversely. For example, the amount by which a_n is displaced may increase beyond all bounds as n does; the only point is that a_n must appear *somewhere* in the new series. If some of the terms are moved to later positions in the series, other terms must, of course, be moved to earlier positions. For example, the series

$$1 + q + q^2 + q^4 + q^3 + q^8 + q^7 + q^6 + q^5 + q^{16} + \cdots$$

is a rearrangement[1] of the geometric series $1 + q + q^2 + \cdots$.

With regard to change of order there is a fundamental distinction between absolutely convergent series and conditionally convergent series.

In absolutely convergent series rearrangement of the terms does not affect the convergence, and the value of the sum of the series is unchanged, exactly as in finite sums.

In conditionally convergent series, on the other hand, the value of the sum of the series can be changed at will by suitable rearrangement of the series, and the series can even be made to diverge if desired.

The first of these facts, referring to absolutely convergent series, is easily established. Let us assume initially, that our series has positive terms only, and consider the nth partial sum $s_n = \sum_{v=1}^{n} a_v$. All the terms of this partial sum occur in the mth partial sum $t_m = \sum_{v=1}^{m} b_v$ of the rearranged series, provided only that m is chosen large enough. Hence $t_m \geq s_n$. On the other hand, we can determine an index n' so large that the partial sum $s_{n'} = \sum_{v=1}^{n'} a_v$ of the first series contains all the terms $b_1, b_2, \ldots, b_m$. It then follows that $t_m \leq s_{n'} \leq A$, where A is the sum of the first series. Thus for all sufficiently large values of m we have $s_n \leq t_m \leq A$, and since s_n can be made to differ from A by an arbitrarily small amount, it follows that the rearranged series also converges and, in fact, to the same limit A as the original series.

If the absolutely convergent series has both positive and negative terms, we may, in fact, regard it as the difference of two series each of which has positive terms only. Since in the rearrangement of the original series each of these two series merely undergoes rearrangement and therefore converges to the same value as before, the same is true of the original series when rearranged. For by the case just considered the new series is absolutely convergent and is therefore the difference of the two rearranged series of positive terms.

To the beginner the fact just proved may seem a triviality. That it really does require proof, and that in this proof the absolute convergence is essential, can be shown by an example of the opposite behavior of conditionally convergent series. We take the familiar series for log 2, below which we write the result of multiplication by the factor $\frac{1}{2}$,

$$1 - \tfrac{1}{2} + \tfrac{1}{3} - \tfrac{1}{4} + \tfrac{1}{5} - \tfrac{1}{6} + \tfrac{1}{7} - \tfrac{1}{8} + - \cdots = \log 2,$$

$$ \quad \tfrac{1}{2} \phantom{{}+{}} -\tfrac{1}{4} \phantom{{}+{}} +\tfrac{1}{6} \phantom{{}+{}} -\tfrac{1}{8} + - \cdots = \tfrac{1}{2} \log 2,$$

---

[1] For each $n > 0$ the terms q^k with $2^n < k \leq 2^{n+1}$ are written in reverse order.

and add, combining the terms placed in vertical columns.[1] We thus obtain

$$1 + \tfrac{1}{3} - \tfrac{1}{2} + \tfrac{1}{5} + \tfrac{1}{7} - \tfrac{1}{4} + \tfrac{1}{9} + \tfrac{1}{11} - \tfrac{1}{6} + + - \cdots = \tfrac{3}{2}\log 2.$$

This last series can obviously be obtained by rearranging the original series, and yet the value of the sum of the series has been multiplied by the factor $\tfrac{3}{2}$. It is easy to imagine the effect that the discovery of this apparent paradox must have had on the mathematicians of the eighteenth century, who were accustomed to operate with infinite series without regard to their convergence.

*We shall give the proof of the above theorem concerning the change in the sum of a conditionally convergent series $\Sigma\, a_n$ which arises from change of order of the terms, although we shall have no occasion to make use of the result. Let $p_1, p_2, \ldots$ be the positive terms and $-q_1, -q_2, \ldots$ the negative terms of the series. Since the absolute value $|a_n|$ tends to 0 as n increases, the numbers p_n and q_n must also tend to 0 as n increases. As we have already seen, moreover, the sum $\sum_{1}^{\infty} p_\nu$ must diverge, and the same is true of $\sum_{1}^{\infty} q_\nu$.

Now we can easily find a rearrangement of the original series which has an arbitrary number a as sum. Suppose, to be specific, that a is positive. We then add together the first n_1 positive terms, just enough to bring about that the sum $\sum_{1}^{n_1} p_\nu$ is greater than a. Since the sum $\sum_{1}^{n_1} p_\nu$ increases with n_1 beyond all bounds, it is always possible by using enough terms to make the partial sum greater than a. The sum will then differ from the exact value a by p_{n_1}, at most. We now add just enough negative terms $-\sum_{1}^{m_1} q_\nu$ to ensure that the sum $\sum_{1}^{n_1} p_\nu - \sum_{1}^{m_1} q_\nu$ is less than a; this is also possible, as follows from the divergence of the series $\sum_{1}^{\infty} q_\nu$. The difference between this sum and a is now q_{m_1} at most. We now add just enough other positive terms $\sum_{n_1+1}^{n_2} p_\nu$ to make the partial sum again greater than a, as is again possible, since the series of positive terms diverges. The difference between the partial sum and a is now p_{n_2} at most. We again add just enough negative terms $-\sum_{m_1+1}^{m_2} q_\nu$, beginning next after the last one previously used, to make the sum once more less than a, and continue in the same way. The values of the sums thus obtained will oscillate about the number a, and when the process is carried far enough the oscillation will only take place between arbitrarily narrow bounds; for, since the terms p_ν and q_ν themselves tend to 0 when ν is sufficiently large, the length of the interval in which the oscillation takes place will also tend to 0. The theorem is thus proved.

---

[1] For the addition of series see Section 7.1d.

In the same way we can rearrange the series in such a way as to make it diverge: we have only to choose such large numbers of the positive terms as compared with the negative that compensation no longer takes place.

d. Operations with Infinite Series

It is clear that two convergent infinite series $a_1 + a_2 + \cdots = S$ and $b_1 + b_2 + \cdots = T$ can be *added term by term*, that is, that the series formed from the terms $c_n = a_n + b_n$ converges and has the value $S + T$ for its sum.[1] For

$$\sum_{v=1}^{n} c_v = \sum_{v=1}^{n} a_v + \sum_{v=1}^{n} b_v \to S + T.$$

It is also clear that if we multiply each term of a convergent infinite series by the same factor, the series remains convergent, its sum being multiplied by the same factor.

For these operations it is immaterial whether the convergence is absolute or conditional. On the other hand, further study shows that multiplication of two infinite series by the method used in multiplying finite sums does not necessarily lead to a convergent series for the value of the product, unless at least one of the two series is absolutely convergent (cf. Appendix, p. 555).

7.2 Tests for Absolute Convergence and Divergence

In Section 7.1b we have already encountered Leibnitz' useful test for the conditional convergence of series. In the following pages we shall only consider criteria referring to *absolute* convergence.

a. The Comparison Test. Majorants

All such considerations of convergence depend on the comparison of the series in question with a second series; this second series is chosen in such a way that its convergence can readily be tested. The general *comparison test* may be stated as follows:

If the numbers $b_1, b_2, \ldots$ are all positive and the series $\sum_{v=1}^{\infty} b_v$ converges, and if

$$|a_n| \leq b_n$$

for all values of n, then the series $\sum_{n=1}^{\infty} a_n$ is absolutely convergent.

---

[1] This theorem is really nothing more than another statement of the fact (cf. Chapter 1, p. 72) that the limit of the sum of two terms is the sum of their limits.

By Cauchy's test the proof becomes almost trivial. For if $m \geq n$, we have

$$|a_n + \cdots + a_m| \leq |a_n| + \cdots + |a_m| \leq b_n + \cdots + b_m.$$

Since the series $\sum\limits_{n=1}^{\infty} b_n$ converges, the right-hand side is arbitrarily small, provided that n and m are sufficiently large. It follows that for such values of n and m the left-hand side is also arbitrarily small, so that by Cauchy's test the given series converges. The convergence is absolute, since our argument applies equally well to the convergence of the series of absolute values $|a_n|$.

The analogous proof for the following fact can be left to the reader. *If*

$$|a_n| \geq b_n > 0,$$

and the series $\sum\limits_{n=1}^{\infty} b_n$ *diverges, then the series* $\sum\limits_{n=1}^{\infty} a_n$ *is certainly not absolutely convergent.*

Sometimes the above series with the positive terms b_n are called *majorant* and *minorant* series, respectively, for the one with terms a_n.

b. Convergence Tested by Comparison with the Geometric Series

In applications of the test the comparison series most frequently used as a majorant is the geometric series. We at once obtain the following theorem.

THEOREM. *The series* $\sum\limits_{n=1}^{\infty} a_n$ *is absolutely convergent if from a certain term onward a relation of the form*

$$(3) \qquad |a_n| < cq^n$$

holds, where c is a positive number independent of n and q is any fixed positive number less than 1.

Ratio and Root Tests. This test is usually expressed in one of the following weaker forms: the series $\sum\limits_{n=1}^{\infty} a_n$ converges absolutely, if from a certain term onward a relation of the form

$$(4a) \qquad \left|\frac{a_{n+1}}{a_n}\right| < q$$

holds, where q is again a positive number less than 1 and independent of n, or: if from a certain term onward a relation of the form

$$(4b) \qquad \sqrt[n]{|a_n|} < q$$

holds, where q is a positive number less than 1. In particular, the conditions of these tests are satisfied if a relation of the form

$$(5a) \qquad\qquad \lim_{n \to \infty} \left| \frac{a_{n+1}}{a_n} \right| = k < 1$$

or

$$(5b) \qquad\qquad \lim_{n \to \infty} \sqrt[n]{|a_n|} = k < 1$$

is true.

These statements are easily established in the following way.

Let us suppose that the criterion (4a), the *ratio test*, is satisfied from the suffix n_0 onward, that is, when $n > n_0$. For brevity we put $a_{n_0+m+1} = b_m$ and find that

$$|b_1| < q \, |b_0|, \quad |b_2| < q \, |b_1| < q^2 \, |b_0|, \quad |b_3| < q \, |b_2| < q^3 \, |b_0|,$$

and so on; hence

$$|b_m| < q^m \mid b_0|,$$

and then for $n > n_0$, and $c = q^{-n_0-1} \, |b_0|$

$$|a_n| = |b_{n-n_0-1}| < q^{n-n_0-1} \, |b_0|$$

$$= cq^n$$

which establishes our statement. For the criterion (4b), the *root test*, we at once have $|a_n| < q^n$, and our statement follows immediately.

Finally, in order to prove the criteria (5), we consider an arbitrary number q such that $k < q < 1$. Then from a certain n_0 onward, that is, when $n > n_0$, Eqs. 4a, b imply that $\left| \frac{a_{n+1}}{a_n} \right| < q$ and $\sqrt[n]{|a_n|} < q$ respectively, since from a certain term onwards the values of $\left| \frac{a_{n+1}}{a_n} \right|$ or of $\sqrt[n]{|a_n|}$ differ from k by less than $(q - k)$. The statement is then established on the basis of the results already proved.

We stress the point that the four tests 4a, b, 5a, b, derived from the original criterion $|a_n| < cq^n$ are *not* equivalent to one another or to the original, that is, that they cannot be derived from one another in both directions. We shall soon see from examples that if a series satisfies one of the conditions, it need not satisfy all the others.

For completeness it may be pointed out that a series certainly diverges if from a certain term onward

$$|a_n| > c$$

for some positive number c, or if from a certain term onward

$$\sqrt[n]{|a_n|} > 1,$$

or if $\quad \lim\limits_{n \to \infty} \left| \dfrac{a_{n+1}}{a_n} \right| = k, \quad$ or $\quad \lim\limits_{n \to \infty} \sqrt[n]{|a_n|} = k,$

where k is a number greater than 1. For, as we immediately recognize, in such a series the terms cannot tend to zero as n increases; the series must therefore diverge. (In these circumstances the series cannot even be conditionally convergent.)

Our tests furnish *sufficient* conditions for the absolute convergence of a series; that is, when they are satisfied we can conclude that the series converges absolutely. They are definitely not *necessary* conditions, however; that is, absolutely convergent series can be formed which do not satisfy the conditions.

Thus the knowledge that

$$\lim_{n \to \infty} \left| \frac{a_{n+1}}{a_n} \right| = 1 \quad \text{or} \quad \lim_{n \to \infty} \sqrt[n]{|a_n|} = 1$$

does not imply anything about the convergence of the series. Such a series may converge or diverge. For example, the series

$$\sum_{n=1}^{\infty} \frac{1}{n},$$

for which $\lim\limits_{n \to \infty} \sqrt[n]{|a_n|} = 1$ and $\lim\limits_{n \to \infty} \left| \dfrac{a_{n+1}}{a_n} \right| = 1$, is divergent, as we saw on p. 513. On the other hand, as we shall soon see, the series $\sum\limits_{n=1}^{\infty} \dfrac{1}{n^2}$, which satisfies the same relations, is convergent.

As an example of the application of our tests we first consider the series

$$q + 2q^2 + 3q^3 + \cdots + nq^n + \cdots.$$

For this series

$$\lim_{n \to \infty} \sqrt[n]{|a_n|} = |q| \cdot \lim_{n \to \infty} \sqrt[n]{n} = |q|,$$

$$\lim_{n \to \infty} \left| \frac{a_{n+1}}{a_n} \right| = |q| \cdot \lim_{n \to \infty} \frac{n+1}{n} = |q|.$$

That the series converges if $|q| < 1$ follows from the ratio test and from the root test also, even in the weaker form (5).

If, on the other hand, we consider the series

$$1 + 2q + q^2 + 2q^3 + \cdots + q^{2n} + 2q^{2n+1} + \cdots,$$

we can no longer prove convergence by the ratio test when $\frac{1}{2} \leq |q| < 1$; for then $\left| \dfrac{2q^{2n+1}}{q^{2n}} \right| = 2\,|q| \geq 1$. But the root test immediately gives us $\lim\limits_{n\to\infty} \sqrt[n]{|a_n|} = |q|$, and shows that the series converges provided that $|q| < 1$, which, of course, we could also have observed directly.

c. Comparison with an Integral[1]

We now proceed to discuss quite a different method of studying convergence. We shall explain it for the typical, particularly simple and important case of the series

$$\sum_{n=1}^{\infty} \frac{1}{n^\alpha} = 1 + \frac{1}{2^\alpha} + \frac{1}{3^\alpha} + \cdots,$$

where the general term a_n is $1/n^\alpha$, α being a positive number. In order to investigate the convergence or divergence of this series, we consider the graph of the function $y = 1/x^\alpha$ and mark off on the x-axis the integral abscissae $x = 1$, $x = 2, \ldots.$ We first construct the rectangle of height $1/n^\alpha$ over the interval $n - 1 \leq x \leq n$ of the x-axis ($n > 1$), and compare it with the area of the region bounded by the same interval of the x-axis, the ordinates at the ends, and the curve $y = 1/x^\alpha$ (this region is shown shaded in Fig. 7.2). Secondly, we construct the

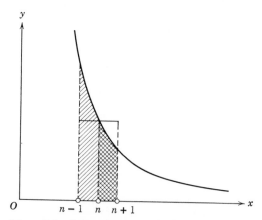

Figure 7.2 Comparison of series with an integral.

---

[1] In this connection see also the Appendix to Chapter 5, p. 505.

rectangle of height $1/n^\alpha$ lying above the interval $n \leq x \leq n + 1$, and similarly compare it with the area of the region lying above the same interval and below the curve (this region is cross-hatched in Fig. 7.2). In the first case the area under the curve is obviously greater than the area of the rectangle; in the second case it is less than the area of the rectangle. In other words,

$$\int_n^{n+1} \frac{dx}{x^\alpha} < \frac{1}{n^\alpha} < \int_{n-1}^n \frac{dx}{x^\alpha}.$$

Writing down these inequalities for $n = 1, 2, 3, \ldots, m$, respectively $n = 2, 3, \ldots, m$, and summing, we obtain the following estimate for the mth partial sum $s_m = \sum_{n=1}^m \frac{1}{n^\alpha}$:

(6)
$$\int_1^{m+1} \frac{dx}{x^\alpha} < s_m < 1 + \int_1^m \frac{dx}{x^\alpha}.$$

Now as m increases the integral $\int_1^m \frac{1}{x^\alpha} dx$ tends to a finite limit or increases without limit depending on whether $\alpha > 1$ or $\alpha \leq 1$. Consequently, the monotonic sequence of numbers s_m is bounded or increases beyond all bounds depending on whether $\alpha > 1$ or $\alpha \leq 1$, and we thus have the following theorem.

THEOREM. *The series of reciprocal powers*

$$\sum_{n=1}^\infty \frac{1}{n^\alpha} = \frac{1}{1^\alpha} + \frac{1}{2^\alpha} + \frac{1}{3^\alpha} + \cdots$$

is convergent if and only if $\alpha > 1$.

For $\alpha = 1$ the divergence of the harmonic series, which we previously proved in a different way, is an immediate consequence; likewise the series

$$\frac{1}{1^2} + \frac{1}{2^2} + \frac{1}{3^2} + \cdots,$$

$$\frac{1}{1^3} + \frac{1}{2^3} + \frac{1}{3^3} + \cdots,$$

converge while the series $\frac{1}{\sqrt{1}} + \frac{1}{\sqrt{2}} + \frac{1}{\sqrt{3}} + \cdots$ diverges.

The convergent series $\sum_{\nu=1}^\infty \frac{1}{\nu^\alpha}$ for $\alpha > 1$ frequently serve as comparison series in investigations of convergence. For example, we see at once that for $\alpha > 1$ the series $\sum\limits_{\nu=1}^{\infty} \dfrac{c_\nu}{\nu^\alpha}$ converges absolutely if the absolute values $|c_\nu|$ of the coefficients remain less than a fixed bound independent of ν.

Euler's Constant.   From the estimate (6) for $\alpha = 1$ it follows at once that the sequence of numbers $C_n = 1 + \dfrac{1}{2} + \dfrac{1}{3} + \cdots + \dfrac{1}{n} - \log n = s_n - \log n > \log (n + 1) - \log n > 0$ is bounded below. Since from the inequality $\dfrac{1}{n + 1} < \displaystyle\int_n^{n+1} \dfrac{dx}{x} = \log (n + 1) - \log n = \dfrac{1}{n + 1} + C_n - C_{n+1}$, we see that the sequence is monotonic decreasing, it must approach a limit

$$\lim_{n \to \infty} C_n = \lim_{n \to \infty} \left(1 + \frac{1}{2} + \frac{1}{3} + \cdots + \frac{1}{n} - \log n\right) = C.$$

The number C whose value is $0.5772 \ldots$, is called *Euler's constant*. In contrast to the other important special numbers of analysis, such as π and e, no other expression with a simple law of formation has been found for Euler's constant. Whether C is rational or irrational is not known to this day.

7.3   Sequences of Functions

As emphasized frequently before, the limit process serves not only to represent known numbers approximately by other, simpler ones, but it also serves to extend the set of known numbers into a wider one. It is of decisive importance in analysis to study limits not only for sequences—or infinite series—of constant numbers, but similarly for sequences of functions, or series whose terms are functions of a variable x, as, for example the Taylor series or power series in general. Not only the approximation of given functions by simpler ones requires such limiting processes but also the definition and analytic description of new functions must frequently be based on the concept of limit of sequences of functions: $f(x) = \lim f_n(x)$ for $n \to \infty$. Equivalently, we may consider $f(x)$ as the sum and the $f_n(x)$ as the partial sums of an infinite series $f(x) = \sum\limits_{r=1}^{\infty} g_r(x)$ of functions $g_n(x)$ where $g_n(x) = f_n(x) - f_{n-1}(x)$ for $n > 1$ and $g_1(x) = f_1(x)$.

We shall now discuss precise definitions and geometrical interpretations.

a. Limiting Processes with Functions and Curves

Definition. The sequence $f_1(x)$, $f_2(x)$, ... converges in the interval $a \leq x \leq b$ to the limit function $f(x)$, if at each point x of the interval the values $f_n(x)$ converge in the usual sense to the value $f(x)$. In this case we write $\lim_{n \to \infty} f_n(x) = f(x)$. According to Cauchy's test (cf. p. 75) we can express the convergence of the sequence without referring to the limit function $f(x)$: The sequence of functions converges to a limit function if and only if at each point x in our interval and for every positive number ϵ, the quantity $|f_n(x) - f_m(x)|$ is less than ϵ, provided that n and m are chosen large enough, that is, larger than a certain number. This number $N = N(\epsilon, x)$ usually depends on ϵ and x and increases beyond all bounds as ϵ tends to zero.

We have frequently met with cases of limits of sequences of functions. We mention only the definition of the power x^α for irrational values of α by the equation

$$x^\alpha = \lim_{n \to \infty} x^{r_n},$$

where $r_1, r_2, \ldots, r_n, \ldots$ is a sequence of rational numbers tending to α; or the equation

$$e^x = \lim_{n \to \infty} \left(1 + \frac{x}{n}\right)^n,$$

where the approximating functions $f_n(x)$ on the right are polymomials of degree n.

The graphical representation of functions by means of curves suggests that we can also speak of limits of sequences of curves, saying, for example, that the graphs of the preceding limit functions x^α and e^x are to be regarded as the limit curves of the graphs of the functions x^{r_n} and $\left(1 + \frac{x}{n}\right)^n$ respectively.

There is, however, a fine distinction between passages to the limit with functions and with curves, not clearly observed until the middle of the nineteenth century. We shall illustrate this point by an example and then discuss it systematically in the next section.

We consider the functions

$$f_n(x) = x^n, \qquad n = 1, 2, \ldots$$

in the interval $0 \leq x \leq 1$. All these functions are continuous, and the limit function $\lim_{n \to \infty} f_n(x) = f(x)$ exists. But this limit function is not continuous. On the contrary, since for all values of n the value of the function $f_n(1) = 1$, the limit

$$f(1) = 1;$$

while, on the other hand, for $0 \le x < 1$, the limit $f(x) = \lim\limits_{n \to \infty} f_n(x) = 0$, as we saw in Chapter 1, p. 65. The function $f(x)$ is therefore a discontinuous function which at $x = 1$ has the value 1 while for all other values of x in the interval has the value 0.

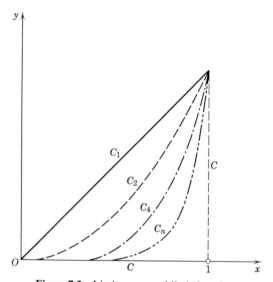

Figure 7.3   Limit curve and limit function.

This discontinuity is geometrically illustrated by the graphs C_n of the functions $y = f_n(x)$. These (cf. Fig. 1.44, p. 66) are continuous curves, all of which pass through the origin and the point $x = 1$, $y = 1$, and which draw in closer and closer to the x-axis as n increases. The *curves* do possess a *limit curve* C which is not discontinuous at all, but consists (cf. Fig. 7.3) of the portion of the x-axis between $x = 0$ and $x = 1$, and the portion of the line $x = 1$ between $y = 0$ and $y = 1$. The *curves* therefore converge to a *continuous* limit curve with a vertical portion, whereas the *functions* converge to a *discontinuous* limit function. We thus recognize that this discontinuity of the limit function expresses itself by the occurrence in the limit curve of a portion perpendicular to the x-axis. This limit curve is *not* the graph of the limit function; for corresponding to the value of x at which the vertical portion occurs the curve gives an infinite number of values of y and the function only one.

Hence the limit of the graphs of the functions $f_n(x)$ is not the same as the graph of the limit $f(x)$ of these function.

Corresponding statements, of course, hold for infinite series as well.

7.4 Uniform and Nonuniform Convergence

a. General Remarks and Definitions

The distinction between the concept of the convergence of functions and that of the convergence of curves is a phenomenon which the

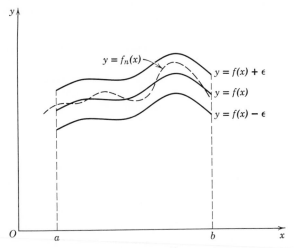

Figure 7.4   To illustrate uniform convergence.

student should clearly grasp. This involves the so-called *nonuniform convergence* of sequences or infinite series of functions which we shall discuss in some detail.

That a function $f(x)$ is the limit of a sequence $f_1(x), f_2(x), \ldots$ in an interval $a \leq x \leq b$ means by definition merely that the usual limit relationship $f(x) = \lim_{n \to \infty} f_n(x)$ holds at each point x of the interval.

Such convergence is a local property of the sequence at the point x. It is, however, natural to require somewhat more than the mere local convergence of our approximations: that if we assign an arbitrary measure of accuracy ϵ, then from a certain index N onward *all* the functions $f_n(x)$ should lie between $f(x) - \epsilon$ and $f(x) + \epsilon$, for *all* values of x, so that their graphs $y = f_n(x)$ lie entirely in the strip shown in Fig. 7.4. If the accuracy of the approximation can be made at least equal to a preassigned positive number ϵ, everywhere in the interval at the same time, that is, by everywhere choosing the same number $N(\epsilon)$ independent of x, we say that the approximation is *uniform*.[1] If $\lim\limits_{n \to \infty} f_n(x) = f(x)$ *uniformly* for $a \leq x \leq b$, there exists for every $\epsilon > 0$ a corresponding number $N = N(\epsilon)$ such that $|f(x) - f_n(x)| < \epsilon$ for all $n > N$ and all x in the interval. Many people were quite surprised when in the middle of the nineteenth century it was noticed by Seidel and others that convergence of functions need not at all be uniform as had been naively assumed.

Examples of Nonuniform Convergence. The concept of uniform convergence is illuminated by examples of nonuniform convergence.

(*a*) The first example occurs for the sequence of functions just considered, $f_n(x) = x^n$; in the interval $0 \leq x \leq 1$ this sequence converges to the limit function $f(x) = 0$ for $0 \leq x < 1$, $f(1) = 1$. Convergence occurs at every point in the interval; that is, if ϵ is any positive number, and if we select any definite fixed value $x = \xi$, the inequality $|\xi^n - f(\xi)| < \epsilon$ certainly holds if n is sufficiently large. Yet this approximation is not uniform. For, if we choose $\epsilon = \frac{1}{2}$, then no matter how large the number n is chosen, we can find a point $x = \eta \neq 1$ at which $|\eta^n - f(\eta)| = \eta^n > \frac{1}{2}$; this is, in fact, true for all points $x = \eta$ where $1 > \eta > \sqrt[n]{\frac{1}{2}}$. It is therefore impossible to choose the number n so large that the difference between $f(x)$ and $f_n(x)$ is less than $\frac{1}{2}$ *throughout the whole interval*.

This behavior becomes intelligible if we refer to the graphs of these functions (Fig. 7.3). We see that no matter how large a value of n we choose, for values of ξ only a little less than 1 the value of the function $f_n(\xi)$ will be very near 1, and therefore cannot be a good approximation to $f(\xi)$, which is 0.

Similar behavior is exhibited by the functions

$$f_n(x) = \frac{1}{1 + x^{2n}}$$

in the neighborhood of the points $x = 1$ and $x = -1$; this can easily be established. Here $f(x) = 1$ for $|x| < 1$, $f(x) = \frac{1}{2}$ for $|x| = 1$ and $f(x) = 0$ for $|x| > 1$.

(*b*) In the above two examples the nonuniformity of the convergence is connected with the fact that the *limit function is discontinuous*. Yet it is also easy to construct a sequence of continuous functions which do converge to a continuous limit function, but not uniformly. We restrict our attention to

---

[1] Compare with the analogous definition at *uniform continuity*, p. 41, where we can choose the same number $\delta(\epsilon)$ independent of x.

the interval $0 \leq x \leq 1$ and make the following definitions for $n \geq 2$:

$$f_n(x) = xn^\alpha \quad \text{for} \quad 0 \leq x \leq \frac{1}{n},$$

$$f_n(x) = \left(\frac{2}{n} - x\right)n^\alpha \quad \text{for} \quad \frac{1}{n} \leq x \leq \frac{2}{n},$$

$$f_n(x) = 0 \quad \text{for} \quad \frac{2}{n} \leq x \leq 1,$$

where to begin with we can choose any value for α, but must then keep this value of α fixed for all terms of the sequence. Graphically, our functions are

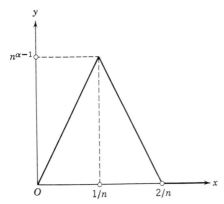

Figure 7.5   To illustrate nonuniform convergence.

represented by a roof-shaped figure made of two line segments lying over the interval $0 \leq x \leq 2/n$ of the x-axis, whereas from $x = 2/n$ onward the graph is the x-axis itself (cf. Fig. 7.5).

If $\alpha < 1$, the altitude of the highest point of the graph, which has in general the value $n^{\alpha-1}$, will tend to zero as n increases; the curves will then tend toward the x-axis, and the functions $f_n(x)$ will converge uniformly to the limit function $f(x) = 0$.

If $\alpha = 1$, the peak of the graph will have the height 1 for every value of n. If $\alpha > 1$, the height of the peak will increase beyond all bounds as n increases.

However, no matter how α is chosen, the sequence $f_1(x), f_2(x), \ldots$ always tends to the limit function $f(x) = 0$. For, if x is positive, we have $2/n < x$, for all sufficiently large values of n so that x is not under the roof-shaped part of the graph and $f_n(x) = 0$; for $x = 0$ all the functional values $f_n(x)$ are equal to 0, so that in either case $\lim\limits_{n \to \infty} f_n(x) = 0$.

The convergence is certainly nonuniform, however, if $\alpha \geq 1$; for it is plainly impossible to choose n so large that the expression $|f(x) - f_n(x)| = f_n(x)$ is less than $\frac{1}{2}$ *everywhere* in the interval.

(c) Exactly similar behavior is exhibited by the sequence of functions

$$f_n(x) = xn^\alpha e^{-nx},$$

where, in contrast with the preceding case, each function of the sequence is represented by a single analytical expression. Here again the equation $\lim_{n\to\infty} f_n(x) = 0$ holds for every positive value of x, since as n increases the

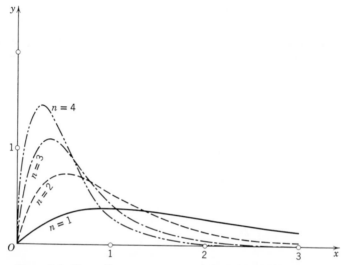

Figure 7.6   Nonuniform convergence of the sequence $f_n(x) = n^2 x e^{-nx}$.

function e^{-nx} tends to zero to a higher order than any power of $1/n$ (cf. Section 3.7b, p. 250). For $x = 0$, we have always $f_n(x) = 0$, and thus

$$f(x) = \lim_{n\to\infty} f_n(x) = 0$$

for every value of x in the interval $0 \le x \le a$, where a is an arbitrary positive number. But here again the convergence to the limit function is not uniform. For at the point $x = 1/n$ [where $f_n(x)$ has its maximum] we have

$$f_n(x) = f_n\left(\frac{1}{n}\right) = \frac{n^{\alpha-1}}{e},$$

and we thus recognize that if $\alpha \ge 1$, the convergence is nonuniform, for every curve $y = f_n(x)$, no matter how large n is chosen, will contain points (namely, the point $x = 1/n$, which varies with n) at which $f_n(x) - f(x) = f_n(x) > 1/2e$ (cf. Fig. 7.6).

(d) The concepts of uniform and nonuniform convergence may, of course, be extended to an infinite series. We say that a series

$$g_1(x) + g_2(x) + \cdots$$

is uniformly convergent, or not, according to the behavior of its partial sums $f_n(x)$. A very simple example of a nonuniformly convergent series is given by

$$f(x) = x^2 + \frac{x^2}{1 + x^2} + \frac{x^2}{(1 + x^2)^2} + \frac{x^2}{(1 + x^2)^3} + \cdots .$$

For $x = 0$ every partial sum $f_n(x) = x^2 + \cdots + x^2/(1 + x^2)^{n-1}$ has the value 0; therefore $f(0) = 0$. For $x \neq 0$ the series is simply a geometric series

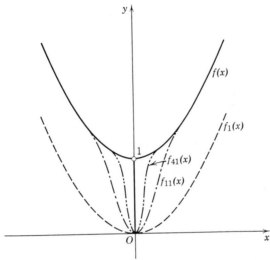

Figure 7.7   Convergence to function with removable jump discontinuity.

with the positive ratio $1/(1 + x^2) < 1$; we can therefore sum it by the elementary rules and thus obtain for every $x \neq 0$ the sum

$$\frac{x^2}{1 - 1/(1 + x^2)} = 1 + x^2.$$

The limit function $f(x)$ is thus given everywhere except at $x = 0$ by the expression $f(x) = 1 + x^2$, whereas $f(0) = 0$; it therefore has a removable discontinuity at the origin.

Here again we have nonuniform convergence in every interval containing the origin. For the difference $f(x) - f_n(x) = r_n(x)$ is always 0 for $x = 0$, whereas it is given by the expression $r_n(x) = 1/(1 + x^2)^{n-1}$ for all other values of x, as the reader may verify for himself. If we require this expression to be less than, say $\frac{1}{2}$, then for each fixed value of x this can be attained by choosing n large enough. But we can find no value of n sufficiently large to ensure that $r_n(x)$ is everywhere less than $\frac{1}{2}$; for if we choose any value of n, no matter how large, we can make $r_n(x)$ greater than $\frac{1}{2}$ by taking x near enough to 0. A uniform approximation to within $\frac{1}{2}$ is therefore impossible. The matter becomes clear if we consider the approximating curves (cf. Fig. 7.7). These curves, except near $x = 0$, lie nearer and nearer to the parabola

$y = 1 + x^2$ as n increases; near $x = 0$, however, the curves send down a narrower and narrower extension to the origin, and as n increases this extension draws in closer and closer to a certain straight line, a portion of the y-axis, so that for the limiting curve we have the parabola plus a linear extension reaching vertically down to the origin.

As a further example of nonuniform convergence we mention the series $\sum_{\nu=0}^{\infty} g_\nu(x)$, where $g_\nu(x) = x^\nu - x^{\nu-1}$ for $\nu \geq 1$, $g_0(x) = 1$, defined in the interval $0 \leq x \leq 1$. The partial sums of this series are the functions x^ν already considered in Example (a), p. 530.

b. A Test of Uniform Convergence

The preceding considerations show us that the uniform convergence of a sequence or series is a special property not possessed by all sequences and series. We now repeat the definition of uniform convergence as it applies to infinite series: the series

$$g_1(x) + g_2(x) + \cdots$$

is uniformly convergent to a function $f(x)$ in an interval if $f(x)$ can be approximated to within a margin of approximation ϵ (where ϵ is an arbitrarily small positive number) by the sum of a fixed and sufficiently large number of terms $g_1(x) + \cdots + g_N(x) = f_N(x)$, independent of x in the interval.

We again have a test (Cauchy's test) for uniform convergence that does not require knowledge of the limit function $f(x)$: the series converges uniformly (or equivalently, the sequence of functions $f_n(x)$ converges uniformly) if and only if the difference $|f_n(x) - f_m(x)|$ can be made less than an arbitrary quantity ϵ everywhere in the interval by choosing n and m larger than a number N independent of x. For, first, if the convergence is uniform, we can make $|f_n(x) - f(x)|$ and $|f_m(x) - f(x)|$ both less than $\epsilon/2$ by choosing n and m greater than a number N independent of x, from which it follows that $|f_n(x) - f_m(x)| < \epsilon$; and secondly, if $|f_n(x) - f_m(x)| < \epsilon$ for all values of x whenever n and m are greater than N, then on choosing any fixed value of $n > N$ and letting m increase beyond all bounds we have the relation

$$|f_n(x) - f(x)| = \lim_{m \to \infty} |f_n(x) - f_m(x)| \leq \epsilon,$$

for every value of x, so that the convergence is uniform.

As we shall see it is just this condition of uniform convergence that makes infinite series and other limiting processes with functions into convenient and useful tools of analysis. Fortunately, in the limiting processes usually encountered in analysis and its applications, non-uniform convergence occurs only at isolated exceptional points and will scarcely trouble us for the present.

Usually, the uniformity of convergence of a series is established by means of the following criterion (comparing the series with a majorant of constant terms):

If the terms of the series $\sum\limits_{v=1}^{\infty} g_v(x)$ satisfy the condition $|g_v(x)| \leq a_v$, where the numbers a_v are positive constants which form a convergent series $\sum\limits_{v=1}^{\infty} a_v$, then the series $\sum\limits_{v=1}^{\infty} g_v(x)$ converges uniformly (and absolutely).

For we then have

$$\left| \sum_{v=n}^{m} g_v(x) \right| \leq \sum_{v=n}^{m} |g_v(x)| \leq \sum_{v=n}^{m} a_v,$$

and since by Cauchy's test the sum $\sum\limits_{v=n}^{m} a_v$ can be made arbitrarily small by choosing n and $m > n$ large enough, this expresses exactly the necessary and sufficient condition for uniform convergence.

A first example is offered by the geometric series $1 + x + x^2 + \cdots$, where x is restricted to the interval $|x| \leq q$, q being any positive number less than 1. The terms of the series are then numerically less than or equal to the terms of the convergent geometric series Σq^v.

A further example is given by the "trigonometric series"

$$\frac{c_1 \sin(x - \delta_1)}{1^2} + \frac{c_2 \sin(x - \delta_2)}{2^2} + \frac{c_3 \sin(x - \delta_3)}{3^2} + \cdots,$$

provided that $|c_n| < c$, where c is a positive constant independent of n. For then we have

$$g_n(x) = \frac{c_n \sin(x - \delta_n)}{n^2}, \quad \text{so that} \quad |g_n(x)| < \frac{c}{n^2}.$$

Hence the uniform and absolute convergence of the trigonometric series follows from the convergence of the series $\sum\limits_{v=1}^{\infty} \frac{c}{n^2}$.

c. Continuity of the Sum of a Uniformly Convergent Series of Continuous Functions

The significance of uniform convergence lies in the fact that a uniformly convergent series in many respects behaves exactly like the sum of a finite number of functions. Thus, for example, the sum of a finite number of continuous functions is itself continuous, and correspondingly we have the following theorem.

THEOREM.   *If a series of continuous terms converges uniformly in an interval, its sum is also a continuous function.*

PROOF. The proof is quite simple.   We subdivide the series

$$f(x) = g_1(x) + g_2(x) + \cdots$$

into the nth partial sum $f_n(x)$ plus the remainder $R_n(x)$. As usual, $f_n(x) = g_1(x) + \cdots + g_n(x)$. If now any positive number ϵ is assigned, we can in virtue of the uniform convergence choose the number n so large that the remainder is less than $\epsilon/4$ throughout the whole interval, and hence

$$|R_n(x + h) - R_n(x)| < \frac{\epsilon}{2}$$

for every pair of numbers x and $x + h$ in the interval.   The partial sum $f_n(x)$ consists of the sum of a finite number of continuous functions and is therefore continuous; for each point x in the interval, therefore, we can choose a positive δ so small that

$$|f_n(x + h) - f_n(x)| < \frac{\epsilon}{2}$$

provided $|h| < \delta$ and the points x and $x + h$ lie in the interval.   It then follows that

$$|f(x + h) - f(x)| = |f_n(x + h) - f_n(x) + R_n(x + h) - R_n(x)|$$
$$\leq |f_n(x + h) - f_n(x)| + |R_n(x + h) - R_n(x)| < \epsilon,$$

which expresses the continuity of our function.

The importance of this theorem becomes clear when we recall that the sums of nonuniformly convergent series of continuous functions are not necessarily continuous from our previous examples. From the preceding theorem we may conclude: if the sum of a convergent series of continuous functions has a point of discontinuity, then in every neighborhood of this point the convergence is nonuniform. Hence every representation of discontinuous functions by series of continuous functions must be based on the use of nonuniformly convergent limiting processes.

d. Integration of Uniformly Convergent Series

A sum of a finite number of continuous functions can be integrated "term by term"; that is, the integral of the sum obtained by integrating each term separately and adding the integrals. In a convergent infinite series of continuous functions the same procedure is permissible, provided that the series converges uniformly in the interval of integration.

A series $\sum_{\nu=1}^{\infty} g_\nu(x) = f(x)$ *which converges uniformly in an interval can be integrated term by term in that interval; or, more precisely, if a and x are two numbers in the interval of uniform convergence, the series* $\sum_{\nu=1}^{\infty} \int_a^x g_\nu(t)\, dt$ *converges, and, in fact, converges uniformly with respect to x, its sum being equal to* $\int_a^x f(t)\, dt$.[1]

To prove this we write as before

$$f(x) = \sum_{\nu=1}^{\infty} g_\nu(x) = f_n(x) + R_n(x).$$

We have assumed that the separate terms of the series are continuous; hence by Section 7.4c the sum is also continuous and therefore integrable. Now if ϵ is any positive number, we can find a number N so large that for every $n > N$ the inequality $|R_n(x)| < \epsilon$ holds for every value of x in the interval. By the mean value theorem of the integral calculus we have

$$\left| \int_a^x [f(t) - f_n(t)]\, dt \right| \leq \epsilon l,$$

where l is the length of the interval of integration. Since the integration of the finite sum $f_n(x)$ can be performed term by term, this gives us

$$\left| \int_a^x f(t)\, dt - \sum_1^n \int_a^x g_\nu(t)\, dt \right| < \epsilon l.$$

But since ϵl can be made as small as we please, this states that

$$\sum_{\nu=1}^{\infty} \int_a^x g_\nu(t)\, dt = \lim_{n \to \infty} \sum_{\nu=1}^n \int_a^x g_\nu(t)\, dt = \int_a^x f(t)\, dt,$$

which was to be proved.

---

[1] Observe that in this theorem we must take *definite* integrals. Thus, for example, the series $\sum_{\nu=1}^{\infty} g_\nu(x)$ with $g_\nu(x) = 0$ converges uniformly; taking the indefinite integral $\int g_\nu(x)\, dx = \text{constant} = c$ of each term, however, leads to the generally divergent series $\sum_{\nu=1}^{\infty} c$.

If, instead of infinite series, we wish to deal with sequences of functions, our result can be expressed in the following way:

If in an interval the sequence of functions $f_1(x)$, $f_2(x)$, ... tends uniformly to the limit function $f(x)$, then

(7) $$\int_a^b f(x)\, dx = \lim_{n \to \infty} \int_a^b f_n(x)\, dx$$

for every pair of numbers a and b lying in the interval; in other words, we can then interchange the order of the operations of integration and passing to the limit.

This fact is not a triviality. From a naive point of view such as prevailed in the eighteenth century it is true that the interchangeability of the two processes is hardly to be doubted; but a glance at the examples in 7.4a shows us that in nonuniform convergence the preceding equation might not hold. We need only consider Example *b*, p. 530, in which the integral of the limit function is 0, whereas the integral of the function $f_n(x)$ over the interval $0 \leq x \leq 1$, that is to say, the area of the triangle in Fig. 7.5, has the value

$$\int_0^1 f_n(x)\, dx = n^{\alpha-2},$$

and when $\alpha \geq 2$ this does not tend to zero. Here we immediately see from the figure that the reason for the difference between $\int_0^1 f(x)\, dx$ and $\lim_{n \to \infty} \int_0^1 f_n(x)\, dx$ lies in the nonuniformity of the convergence.

On the other hand, by considering values of α such that $1 \leq \alpha < 2$, we see that the equation $\lim_{n \to \infty} \int_0^1 f_n(x)\, dx = \int_0^1 f(x)\, dx$ can hold good although the convergence is nonuniform. As a further example, the series $\sum_0^\infty g_n(x)$, where $g_n(x) = x^n - x^{n-1}$ for $n \geq 1$ and $g_0(x) = 1$, can be integrated term by term between the limits 0 and 1, even though it does not converge uniformly. Thus, although uniformity of convergence is a *sufficient* condition for term-by-term integrability, it is by no means a *necessary* condition.

e. Differentiation of Infinite Series

The behavior of uniformly convergent series or sequences with respect to differentiation is quite different from that with respect to integration. For example, the sequence of functions $f_n(x) = \dfrac{\sin n^2 x}{n}$ certainly converges uniformly to the limit function $f(x) = 0$, but the derivative $f_n'(x) = n \cos n^2 x$ certainly does not converge everywhere to the derivative of the limit function $f'(x) = 0$, as we see by considering $x = 0$. In spite of the uniformity of the convergence, therefore, we cannot interchange the processes of differentiation and passage to the limit.

Corresponding statements of course hold for infinite series. For example, the series

$$\sin x + \frac{\sin 2^4 x}{2^2} + \frac{\sin 3^4 x}{3^2} + \cdots$$

is absolutely and uniformly convergent, for its terms are numerically not greater than the terms of the convergent series $\dfrac{1}{1^2} + \dfrac{1}{2^2} + \dfrac{1}{3^2} + \cdots$. If, however, we differentiate the series term by term, we obtain the series

$$\cos x + 2^2 \cos 2^4 x + 3^2 \cos 3^4 x + \cdots,$$

which plainly diverges at $x = 0$.

The only useful criterion which assures us in special cases that term-by-term differentiation is permissible is given by the following theorem.

If, on differentiating a convergent infinite series $\sum\limits_{\nu=0}^{\infty} G_\nu(x) = F(x)$ *term by term, we obtain a uniformly convergent series of continuous terms* $\sum\limits_{\nu=0}^{\infty} g_\nu(x) = f(x)$, *then the sum of this last series is equal to the derivative of the sum of the first series.*

This theorem therefore expressly requires that after differentiating the series term by term we must still investigate whether the result of the differentiation is a uniformly convergent series or not.

The proof of the theorem is almost trivial. For by the theorem in Section 7.4d we can integrate term by term the series obtained by differentiation. Recalling that $g_\nu(t) = G_\nu'(t)$, we obtain

$$\int_a^x f(t)\,dt = \int_a^x \left(\sum_{\nu=0}^{\infty} g_\nu(t) \right) dt = \sum_{\nu=0}^{\infty} \int_a^x g_\nu(t)\,dt = \sum_{\nu=0}^{\infty} (G_\nu(x) - G_\nu(a))$$

$$= F(x) - F(a).$$

This being true for every value of x in the interval of uniform convergence, it follows that

$$f(x) = F'(x),$$

which was to be proved.

7.5   Power Series

Power series occupy a most important position among infinite series. By a power series we mean a series of the type

$$(8) \qquad P(x) = c_0 + c_1 x + c_2 x^2 + \cdots = \sum_{\nu=0}^{\infty} c_\nu x^\nu$$

("power series in x"), or more generally

$$(8a) \quad P(x) = c_0 + c_1 (x - x_0) + c_2 (x - x_0)^2 + \cdots = \sum_{\nu=0}^{\infty} c_\nu (x - x_0)^\nu$$

("power series in $(x - x_0)$"), where x_0 is a fixed number. If in the last series we introduce $\xi = x - x_0$ as a new variable, it becomes a power series $\sum_{\nu=0}^{\infty} c_\nu \xi^\nu$ in the new variable ξ, and we can therefore confine our attention to power series of the more special form $\sum_{\nu=0}^{\infty} c_\nu x^\nu$ without any loss of generality.

In Chapter 5 (p. 446) we considered the approximate representation of functions by polynomials and were thus led to the expansion of functions in Taylor series, which are, in fact, power series. In this section we shall study power series in somewhat greater detail, and shall obtain the expansions of some of the most important functions in series more conveniently than before.

a. Convergence Properties of Power Series—Interval of Convergence

There are power series which converge for *no* value of x except, of course, for $x = 0$, as for example, the series

$$x + 2^2 x^2 + 3^3 x^3 + \cdots + n^n x^n + \cdots.$$

For if $x \neq 0$, we can find an integer N such that $|x| > 1/N$. Then all the terms $n^n x^n$ for which $n > N$ will be greater than 1 in absolute value, and, in fact, as n increases $n^n x^n$ will increase beyond all bounds, so that the series fails to converge.

On the other hand, there are series which converge for *every* value of x; for example, the power series for the exponential function

$$e^x = 1 + x + \frac{x^2}{2!} + \frac{x^3}{3!} + \cdots,$$

whose convergence for every value of x follows at once from the ratio test (criterion 5a, p. 522). The $(n + 1)$th term divided by the nth term gives x/n, and, whatever number x is chosen, this ratio tends to zero as n increases.

The behavior of power series with regard to convergence is expressed in the following fundamental theorem.

If a power series in x converges for a value $x = \xi$, it converges absolutely for every value x such that $|x| < |\xi|$, and the convergence is uniform in every interval $|x| \leq \eta$, where η is any positive number less than $|\xi|$. Here η may lie as near $|\xi|$ as we please.

The proof is simple. If the series $\sum\limits_{v=0}^{\infty} c_v \xi^v$ converges, its terms tend to zero as n increases. From this follows the weaker statement that the terms all lie below a bound M independent of v, that is, $|c_v \xi^v| < M$. If now q is any number such that $0 < q < 1$, and if we restrict x to the interval $|x| \leq q\,|\xi|$, then $|c_v x^v| \leq |c_v \xi^v|\, q^v < Mq^v$. In this interval, therefore, the terms of our series $\sum\limits_{0}^{\infty} c_v x^v$ are smaller in absolute value than the terms of the convergent geometric series $\sum Mq^v$. Hence from the theorem on p. 535 the absolute and uniform convergence of the series in the interval $-q\,|\xi| \leq x \leq q\,|\xi|$ follows.

If a power series does not converge everywhere, that is, if there is a value $x = \xi$ for which it diverges, it must diverge for every value of x such that $|x| > |\xi|$. For if it were convergent for such a value of x, by the theorem above it would have to converge for the numerically smaller value ξ.

From this we recognize that a power series which converges for at least one value of x other than 0 and which diverges for at least one value of x has an *interval of convergence*; that is, a definite positive number ρ exists such that for $|x| > \rho$ the series diverges and for $|x| < \rho$ the series converges. For $|x| = \rho$ no general statement can be made. Here ρ is just the *least upper bound* of the values x for which the series converges (such a least upper bound exists by the theorem on p. 98 since the values x for which the series converges form a bounded set). The limiting cases, those in which the series converges only for $x = 0$ and those in which it converges everywhere, are expressed symbolically by writing $\rho = 0$ and $\rho = \infty$ respectively.[1]

---

[1] It is possible to find this interval of convergence directly from the coefficients c_v of the series. If the limit $\lim\limits_{n \to \infty} \sqrt[n]{|c_n|}$ exists, then

$$\rho = \frac{1}{\lim\limits_{n \to \infty} \sqrt[n]{|c_n|}}.$$

For the general case, see Problem 8, p. 569.

For example, for the geometric series $1 + x + x^2 + \cdots$ we have $\rho = 1$; at the end points of the interval of convergence the series diverges. Similarly, for the series for the inverse tangent (p. 444),

$$\text{arc tan } x = x - \frac{x^3}{3} + \frac{x^5}{5} - + \cdots,$$

we have $\rho = 1$, and at both the end points $x = \pm 1$ of the interval of convergence the series converges, as we recognize at once from Leibnitz's test (p. 514).

From the uniform convergence we derive the important fact that within its interval of convergence (if such an interval exists) the power series represents a continuous function.

b. Integration and Differentiation of Power Series

Because of the uniformity of convergence *it is always permissible to integrate a power series*

$$f(x) = \sum_{\nu=0}^{\infty} c_\nu x^\nu$$

term by term over any closed interval lying entirely within the interval of convergence. We thus obtain the function

$$(9) \qquad\qquad F(x) = c + \sum_{\nu=0}^{\infty} \frac{c_\nu}{\nu+1} x^{\nu+1},$$

for which $F'(x) = f(x) \quad \text{and} \quad F(0) = c.$

We may also differentiate a power series term by term within its interval of convergence, thus obtaining the equation

$$(10) \qquad\qquad f'(x) = \sum_{\nu=1}^{\infty} \nu c_\nu x^{\nu-1}.$$

In order to prove this statement we need only show that the series on the right converges uniformly if x is restricted to an interval lying entirely within the interval of convergence. Suppose then that ξ is a number, lying as close to ρ as we please, for which $\sum_{\nu=1}^{\infty} c_\nu \xi^\nu$ converges; then, as we have seen before, the numbers $|c_\nu \xi^\nu|$ all lie below a bound M independent of ν, so that $|c_\nu \xi^{\nu-1}| < \dfrac{M}{|\xi|} = N.$ Now let q be any number such that $0 < q < 1$; if we restrict x to the interval $|x| \le q\,|\xi|$, the terms of the infinite series (10) are not greater than those of the series $\sum\limits_{\nu=1}^{\infty} |\nu c_\nu\, q^{\nu-1}\, \xi^{\nu-1}|$, and therefore less than those of the series $\sum\limits_{\nu=0}^{\infty} N\nu q^{\nu-1}$. However, in this last series the ratio of the $(n + 1)$th term to the nth term is $q(n + 1)/n$, which tends to q as n increases. Since $0 < q < 1$, it follows [criterion (5a)] that this series converges. Hence the series obtained by differentiation converges uniformly, and by the theorem on p. 539 represents the derivative $f'(x)$ of the function $f(x)$, which proves our statement.

If we apply this result again to the power series

$$f'(x) = \sum_{\nu=1}^{\infty} \nu c_\nu x^{\nu-1},$$

we find on differentiating term by term that

$$f''(x) = \sum_{\nu=2}^{\infty} \nu(\nu - 1)c_\nu x^{\nu-2},$$

and, continuing the process, we arrive at the theorem: *Every function represented by a power series can be differentiated as often as we please within the interval of convergence, and the differentiation can be performed term by term.*[1]

c. Operations with Power Series

The preceding theorems on the behavior of power series are our justification for operating in the same way with power series as with polynomials. It is obvious that two power series can be added or subtracted by adding or subtracting the corresponding coefficients (see p. 520). It is also clear that a power series, like any other convergent series, can be multiplied by a constant factor by multiplying each term by that factor. On the other hand, the multiplication and division of two power series require somewhat more detailed study, for which we

---

[1] As an explicit expression for the kth derivative we obtain

$$f^{(k)}(x) = \sum_{\nu=k}^{\infty} \nu(\nu - 1) \cdots (\nu - k + 1)c\, x^{\nu-k},$$

or in a slightly different form,

$$\frac{f^{(k)}(x)}{k!} = \sum_{\nu=k}^{\infty} \binom{\nu}{k} c_\nu x^{\nu-k} = \sum_{\nu=0}^{\infty} \binom{k + \nu}{k} c_{k+\nu} x^\nu.$$

These two formulas are frequently useful.

refer the reader to the Appendix (p. 555). Here we merely mention without proof that two power series

$$f(x) = \sum_{v=0}^{\infty} a_v x^v$$

and

$$g(x) = \sum_{v=0}^{\infty} b_v x^v$$

can be multiplied together like polynomials. To be specific, we have the following theorems: *Throughout the common part of the intervals of convergence of these two series their product is given by the convergent power series* $\sum_{v=0}^{\infty} c_v x^v$, *where the coefficients* c_v *are given by the formulas*

$$c_0 = a_0 b_0,$$
$$c_1 = a_0 b_1 + a_1 b_0,$$
$$c_2 = a_0 b_2 + a_1 b_1 + a_2 b_0,$$
$$\cdot \quad \cdot \quad \cdot \quad \cdot \quad \cdot \quad \cdot \quad \cdot \quad \cdot \quad \cdot \quad \cdot \quad \cdot$$
$$c_n = a_0 b_n + a_1 b_{n-1} + \cdots + a_n b_0,$$
$$\cdot \quad \cdot \quad \cdot \quad \cdot \quad \cdot \quad \cdot \quad \cdot \quad \cdot \quad \cdot \quad \cdot \quad \cdot$$

d. Uniqueness of Expansion

In the theory of power series the following fact is of importance: if two power series $\sum_{v=0}^{\infty} a_v x^v$ and $\sum_{v=0}^{\infty} b_v x^v$ both converge in an interval which contains the point $x = 0$ in its interior, and if in that interval the two series represent the same function $f(x)$, then they are identical, that is, the equation $a_n = b_n$ is true for every value of n. In other words:

A function $f(x)$ *can be represented by a power series in* x *in only one way, if at all.*

Briefly: the representation of a function by a power series is "unique." For the proof we need only notice that the difference of the two power series, that is, the power series $\phi(x) = \sum_{v=0}^{\infty} c_v x^v$ with coefficients $c_v = a_v - b_v$, represents the function

$$\phi(x) = f(x) - f(x) = 0$$

in the interval; that is, this last power series converges to the limit 0 everywhere in the interval. For $x = 0$, in particular, the sum of the series must be 0; that is, $c_0 = 0$, so that $a_0 = b_0$. We now differentiate the series in the interior of the interval, obtaining $\phi'(x) = \sum_{\nu=1}^{\infty} \nu c_\nu x^{\nu-1}$. However, $\phi'(x)$ is also 0 throughout the interval; hence for $x = 0$, in particular, we have $c_1 = 0$ or $a_1 = b_1$. Continuing this process of differentiating and then putting $x = 0$, we find successively that all the coefficients c_ν are equal to zero, which proves the theorem.

In addition, we can draw the following conclusion from our discussion: if we take the νth derivative of a series $f(x) = \Sigma a_\nu x^\nu$ and then put $x = 0$, we at once obtain

$$a_\nu = \frac{1}{\nu!} f^{(\nu)}(0),$$

that is,

Every power series which converges for points other than $x = 0$ is the Taylor series of the function which it represents.

The uniqueness of the expansion corresponds to the fact that the coefficients can be expressed in terms of the function itself.

*e. Analytic Functions

For functions $f(x)$ which can be expressed by power series, the name "analytic functions" has been used since the importance of such functions was first recognized by Lagrange. Specifically, $f(x)$ is called *analytic* in the neighborhood of $x = a$ if in this neighborhood an expansion of $f(x)$ as a convergent power series in $x - a$ is possible.

While functions which are not at all or not everywhere analytic do play a great role in analysis and applications (See Chapter 8), the analytic functions are particularly important, for they share with polynomials many simple features.

For example, an analytic function which does not vanish identically will have some nonvanishing derivative for $x = a$. Let r be the smallest number for which $f(r)(a) \neq 0$. Then f having a zero of order r at a point $x = a$, can be represented as a product $f(x) = (x - a)^r g(x)$, where $g(x)$ is an analytic function for which $g(a) = \frac{1}{r!} f^{(r)}(a)$ is different from zero. (Compare Chapter 5, p. 463.) Indeed, the possibility of factoring out the power $(x - a)^r$ follows immediately from the convergence of the respective power series.

Also, as is seen from the continuity of the convergent power series for $g(x)$, the factor $g(x)$ cannot vanish in a suitably small neighborhood of $x = a$, or: *the zeros of $f(x)$ are isolated* unless, of course, f vanishes identically.

Since the same is true of the function $f'(x)$ it follows that in a finite interval an analytic function is piecewise monotone, that is, it cannot change its character of monotonicity infinitely often; thus the graph of $y = f(x)$ cannot have infinitely many intersections with a line $y =$ constant (or any straight line) in a finite interval.

One may note that these last statements are not neccessarily true for non-analytic functions, such as for $y = \sin(1/x)\, e^{-1/x^2}$, in the neighborhood of $x = 0$ (see p. 462).

7.6   Expansion of Given Functions in Power Series. Method of Undetermined Coefficients.   Examples

Within its interval of convergence every power series represents a continuous function with continuous derivatives of all orders. We shall now discuss the converse problem of the expansion of a given function in a power series. In theory we can always do this by means of Taylor's theorem; in practice we often meet with difficulties in the actual calculation of the nth derivative and in the estimation of the remainder. But we can often reach our goal more simply by making use of the following device. We first write down tentatively $f(x) = \sum_{\nu=0}^{\infty} c_{\nu} x^{\nu}$, where the coefficients c_{ν} are unknown to begin with. Then by some known property of the function $f(x)$ we determine the coefficients, and then prove the convergence of the series. The series represents a function, and it only remains to prove that this function is identical with $f(x)$. Because of the uniqueness of the expansion in power series we know that no other series than the one just found can be the required expansion. Actually, we have earlier obtaii.ed the series for arc tan x and log $(1 + x)$ by a method related to the idea of this chapter. For we simply integrated term by term the series for the derivatives of these functions, which we knew to be geometric series. We shall now consider some examples of this method.

a. The Exponential Function

As we saw in Chapter 3, Section 4a, p. 223, the function $y = e^x$ is completely characterized by the differential equation $y' = y$ and the initial condition $y = 1$ for $x = 0$. We can use these properties directly to find the power series for the exponential function. Our problem is to find a function $f(x)$ for which $f'(x) = f(x)$ and $f(0) = 1$. If we write tentatively the series with undetermined coefficients

$$f(x) = c_0 + c_1 x + c_2 x^2 + \cdots.$$

and differentiate it, we obtain

$$f'(x) = c_1 + 2c_2 x + 3c_3 x^2 + \cdots.$$

Since by hypothesis these two power series must be identical, we have the equation

$$nc_n = c_{n-1},$$

true for all values of $n \geq 1$. If we observe that because of the relation $f(0) = 1$ the coefficient c_0 must have the value 1, we can calculate all the coefficients successively, and obtain the power series

$$f(x) = 1 + \frac{x}{1!} + \frac{x^3}{2!} + \frac{x^3}{3!} + \cdots.$$

As we easily see by the ratio test, this series converges for all values of x and therefore represents a function for which the relations $f'(x) = f(x)$, $f(0) = 1$ are actually fulfilled. (Here we intentionally avoid making any use of what we have previously learned about the expansion of the exponential function.)

Since only the function e^x possesses these properties we readily deduce that the function $f(x)$ is identical with e^x.

b. The Binomial Series

We can now return to the binomial series (Section 5.5c, p. 456), this time making use of the method of undetermined coefficients. We wish to expand the function $f(x) = (1 + x)^\alpha$ in a power series, and therefore write

$$f(x) = (1 + x)^\alpha = c_0 + c_1 x + c_2 x^2 + \cdots,$$

the coefficients c_ν being undetermined. We now notice that our function obviously satisfies the relation

$$(1 + x)f'(x) = \alpha f(x) = \sum_{\nu=0}^{\infty} \alpha c_\nu x^\nu.$$

On the other hand, if we differentiate the series for $f(x)$ term by term and multiply by $(1 + x)$, we obtain

$$(1 + x)f'(x) = c_1 + (2c_2 + c_1)x + (3c_3 + 2c_2)x^2 + \cdots;$$

and since these two power series for $(1 + x)f'(x)$ must be identical,

$$\alpha c_0 = c_1, \quad \alpha c_1 = 2c_2 + c_1, \quad \alpha c_2 = 3c_3 + 2c_2, \ldots.$$

Now it is certain that $c_0 = 1$, since our series must have the value 1 for $x = 0$, and so we obtain in succession the expressions

$$c_1 = \alpha, \quad c_2 = \frac{(\alpha - 1)\alpha}{2}, \quad c_3 = \frac{(\alpha - 2)(\alpha - 1)\alpha}{3 \cdot 2}, \ldots,$$

for the coefficients, and in general, as is easily established, we have

$$c_\nu = \frac{(\alpha - \nu + 1)(\alpha - \nu + 2) \cdots (\alpha - 1)\alpha}{\nu(\nu - 1) \cdots 2 \cdot 1} = \binom{\alpha}{\nu}.$$

Substituting these values for the coefficients, we have the series $\sum_{v=0}^{\infty} \binom{\alpha}{v} x^v$; we have yet to investigate the convergence of this series and to show that it actually represents $(1 + x)^\alpha$.

By the ratio test we find that when α is not a positive integer, the series converges if $|x| < 1$ and diverges if $|x| > 1$; for then the ratio of the $(n + 1)$th term to the nth term is $\dfrac{\alpha - n + 1}{n} x$, and the absolute value of this expression tends to $|x|$ as n increases beyond all bounds.[1] Hence, if $|x| < 1$ our series represents a function $f(x)$ which satisfies the condition $(1 + x)f'(x) = \alpha f(x)$, as follows from the method of forming the coefficients. Moreover, $f(0) = 1$. Together, these two conditions ensure that the function $f(x)$ is identical with $(1 + x)^\alpha$. For on putting

$$\phi(x) = \frac{f(x)}{(1 + x)^\alpha}$$

we find that

$$\phi'(x) = \frac{(1 + x)^\alpha f'(x) - \alpha(1 + x)^{\alpha-1} f(x)}{(1 + x)^{2\alpha}} = 0;$$

$\phi(x)$ is therefore a constant, and, in fact, is always equal to 1, since $\phi(0) = 1$. We have therefore proved that for $|x| < 1$

$$(1 + x)^\alpha = \sum_{v=0}^{\infty} \binom{\alpha}{v} x^v,$$

which is the binomial series.

Here we note the following special cases of the binomial series; the geometric series

$$\frac{1}{1 + x} = (1 + x)^{-1} = 1 - x + x^2 - x^3 + x^4 - + \cdots$$

$$= \sum_{v=0}^{\infty} (-1)^v x^v;$$

the series

$$\frac{1}{(1 + x)^2} = (1 + x)^{-2} = 1 - 2x + 3x^2 - 4x^3 + - \cdots$$

$$= \sum_{v=0}^{\infty} (-1)^v (v + 1) x^v,$$

---

[1] Here we state, without proof, the exact conditions under which this series converges. If the index α is an integer ≥ 0, the series terminates and is therefore valid for all values of x (becoming the ordinary binomial theorem). For all other values of α the series is absolutely convergent for $|x| < 1$ and divergent for $|x| > 1$. For $x = +1$ the series converges absolutely if $\alpha > 0$, converges conditionally if $-1 < \alpha < 0$, and diverges if $\alpha \leq -1$. Finally, at $x = -1$ the series is absolutely convergent if $\alpha > 0$, divergent if $\alpha < 0$.

which may also be obtained from the geometric series by differentiation; and the series

$$\sqrt{(1+x)} = (1+x)^{1/2} = 1 + \frac{1}{2}x - \frac{1}{2\cdot 4}x^2 + \frac{1\cdot 3}{2\cdot 4\cdot 6}x^3$$
$$- \frac{1\cdot 3\cdot 5}{2\cdot 4\cdot 6\cdot 8}x^4 + - \cdots,$$

$$\frac{1}{\sqrt{(1+x)}} = (1+x)^{-1/2} = 1 - \frac{1}{2}x + \frac{1\cdot 3}{2\cdot 4}x^2 - \frac{1\cdot 3\cdot 5}{2\cdot 4\cdot 6}x^3$$
$$+ \frac{1\cdot 3\cdot 5\cdot 7}{2\cdot 4\cdot 6\cdot 8}x^4 - + \cdots,$$

the first two or three terms of which form useful approximations.

c. The Series for arc sin x

This series can be obtained very easily by expanding the expression $1/\sqrt{(1-t^2)}$ according to the binomial series,

$$(1-t^2)^{-1/2} = 1 + \frac{1}{2}t^2 + \frac{1\cdot 3}{2\cdot 4}t^4 + \cdots.$$

This series converges if $|t| \leq 1$, and so converges uniformly if $|t| \leq q < 1$. On integrating term by term between 0 and x, we obtain

$$\text{arc sin } x = x + \frac{1}{2}\frac{x^3}{3} + \frac{1\cdot 3}{2\cdot 4}\frac{x^5}{5} + \cdots;$$

by the ratio test we find that this converges if $|x| < 1$, and diverges if $|x| > 1$.

The deduction of this series from Taylor's theorem would be decidedly less convenient, owing to the difficulty of estimating the remainder.

d. The Series for ar sinh $x = \log[x + \sqrt{(1+x^2)}]$

We obtain this expansion by a similar method. Using the binomial theorem we write down the series for the derivative of ar sinh x,

$$\frac{1}{\sqrt{1+x^2}} = 1 - \frac{1}{2}x^2 + \frac{1\cdot 3}{2\cdot 4}x^4 - \frac{1\cdot 3\cdot 5}{2\cdot 4\cdot 6}x^6 + - \cdots,$$

and then integrate term by term. We thus obtain the expansion

$$\text{ar sinh } x = x - \frac{1}{2}\frac{x^3}{3} + \frac{1\cdot 3}{2\cdot 4}\frac{x^5}{5} - + \cdots.$$

whose interval of convergence is $-1 < x < 1$.

e. Example of Multiplication of Series

The expansion of the function

$$\frac{\log (1 + x)}{1 + x}$$

is a simple example of the application of the rule for the multiplication of power series. We have only to multiply the logarithmic series

$$\log (1 + x) = x - \frac{x^2}{2} + \frac{x^3}{3} - \frac{x^4}{4} + - \cdots$$

by the geometric series

$$\frac{1}{1 + x} = 1 - x + x^2 - x^3 + x^4 - + \cdots ;$$

as the reader may verify for himself, we obtain the remarkable expansion

$$\frac{\log (1 + x)}{1 + x} = x - (1 + \tfrac{1}{2})x^2 + (1 + \tfrac{1}{2} + \tfrac{1}{3})x^3$$

$$- (1 + \tfrac{1}{2} + \tfrac{1}{3} + \tfrac{1}{4})x^4 + - \cdots$$

for $|x| < 1$.

f. Example of Term-by-Term Integration (Elliptic Integral)

In previous applications pp. 300, 411 we have met with the elliptic integral

$$K = \int_0^{\pi/2} \frac{d\phi}{\sqrt{(1 - k^2 \sin^2 \phi)}}, \qquad \text{for } (k^2 < 1)$$

[the period of oscillation of a pendulum]. In order to evaluate the integral we can first expand the integrand by the binomial theorem, thus obtaining

$$\frac{1}{\sqrt{(1 - k^2 \sin^2 \phi)}} = 1 + \tfrac{1}{2}k^2 \sin^2 \phi + \frac{1 \cdot 3}{2 \cdot 4} k^4 \sin^4 \phi$$

$$+ \frac{1 \cdot 3 \cdot 5}{2 \cdot 4 \cdot 6} k^6 \sin^6 \phi + \cdots .$$

Since $k^2 \sin^2 \phi$ is never greater than k^2 this series converges uniformly for all values of ϕ, and we may integrate term by term:

$$K = \int_0^{\pi/2} \frac{d\phi}{\sqrt{(1 - k^2 \sin^2 \phi)}} = \int_0^{\pi/2} d\phi + \frac{1}{2} k^2 \int_0^{\pi/2} \sin^2 \phi \, d\phi$$

$$+ \frac{1 \cdot 3}{2 \cdot 4} k^4 \int_0^{\pi/2} \sin^4 \phi \, d\phi + \cdots .$$

The integrals occurring here have already been calculated [cf. Eq. (76), p. 279]. If we substitute their values, we have

$$K = \int_0^{\pi/2} \frac{d\phi}{\sqrt{(1 - k^2 \sin^2 \phi)}} = \frac{\pi}{2} \left[1 + \left(\frac{1}{2}\right)^2 k^2 + \left(\frac{1 \cdot 3}{2 \cdot 4}\right)^2 k^4 \right.$$
$$\left. + \left(\frac{1 \cdot 3 \cdot 5}{2 \cdot 4 \cdot 6}\right)^2 k^6 + \cdots \right].$$

7.7 Power Series with Complex Terms

a. Introduction of Complex Terms into Power Series. Complex Representations of the Trigonometric Functions

The similarity between certain power series representing functions which are apparently unrelated led Euler to a purely formal connection between them by giving complex values, in particular, pure imaginary values, to the variable x. We shall first describe Euler's formal, but most striking and fruitful discovery, unhindered by questions of rigor. We shall then indicate a more rigorous justification.

The first relation of this sort is obtained if we replace the quantity x in the series for e^x by a pure imaginary $i\phi$, where ϕ is a real number. If we recall the fundamental equation for the imaginary unit i, that is, $i^2 = -1$, from which $i^3 = -i$, $i^4 = 1$, $i^5 = i, \ldots$ follows, then on separating the real and the imaginary terms of the series, we obtain

$$e^{i\phi} = \left(1 - \frac{\phi^2}{2!} + \frac{\phi^4}{4!} - \frac{\phi^6}{6!} + - \cdots\right)$$
$$+ i\left(\phi - \frac{\phi^3}{3!} + \frac{\phi^5}{5!} - \frac{\phi^7}{7!} + - \cdots\right),$$

or in another form,

(11) $e^{i\phi} = \cos \phi + i \sin \phi.$

This is the well-known and important "Euler formula," a landmark in analysis; as yet it is purely formal.[1] It is consistent with De Moivre's theorem (p. 105), which is expressed by the equation

$$(\cos \phi + i \sin \phi)(\cos \psi + i \sin \psi) = \cos (\phi + \psi) + i \sin (\phi + \psi).$$

By virtue of Euler's formula this equation merely states that the relation

$$e^x \cdot e^y = e^{x+y}$$

continues to hold for pure imaginary values $x = i\phi$, $y = i\psi$.

---

[1] One consequence for $\phi = \pi$ is the formula $e^{\pi i} = -1$, a striking relation between the three most important constants e, π and i.

It should be stated that this Euler formula and the addition theorem $e^{i\varphi}e^{i\psi} = e^{i(\varphi+\psi)}$ may be used rigorously without further justification simply by *defining* $e^{i\phi}$ as the complex number $\cos \phi + i \sin \phi$. This definition is consistent with the ordinary rules for operating with exponentials. In particular, the ordinary rule for multiplying powers of e just furnishes simple concise expressions of the addition theorems of trigonometry as expressed by de Moivre's formula which in turn is of an entirely elementary character. Therefore we are on safe ground when we make use of Euler's relations without the benefit of a more general analysis of functions of a complex variable, as in the next section.

More generally we can define the exponential function for an arbitrary complex exponent $x + iy$ (where x and y are real) by the formula

$$e^{x+iy} = e^x e^{iy} = e^x(\cos y + i \sin y).$$

If we replace the variable x in the power series for $\cos x$ by the pure imaginary ix we at once obtain the series for $\cosh x$; this relation can be expressed by the equation

(12) $$\cosh x = \cos ix.$$

In the same way we obtain

(13) $$\sinh x = \frac{1}{i} \sin ix.$$

Since Euler's formula also gives $e^{-i\phi} = \cos \phi - i \sin \phi$, we arrive at the *exponential expressions for the trigonometric functions*,

(14) $$\sin x = \frac{e^{ix} - e^{-ix}}{2i}, \quad \cos x = \frac{e^{ix} + e^{-ix}}{2}.$$

These are exactly analogous to the exponential expressions for the hyperbolic functions and are, in fact, transformed into them by the relations $\cosh x = \cos ix$, $\sinh x = \frac{1}{i} \sin ix$.

Corresponding formal relations can, of course, be obtained for the functions $\tan x$, $\tanh x$, $\cot x$, $\coth x$, which are connected by the equations $\tanh x = \frac{1}{i} \tan ix$, $\coth x = i \cot ix$.

Finally, similar relations can also be found for the inverse trigonometric and hyperbolic functions. For example, from

$$y = \tan x = \frac{e^{ix} - e^{-ix}}{i(e^{ix} + e^{-ix})} = \frac{e^{2ix} - 1}{i(e^{2ix} + 1)}$$

we immediately find that

$$e^{2ix} = \frac{1 + iy}{1 - iy}.$$

If we take the logarithms of both sides of this equation and then write x instead of y and arc tan x instead of x, we obtain the equation

$$(15) \qquad \text{arc tan } x = \frac{1}{2i} \log \frac{1 + ix}{1 - ix},$$

which expresses a remarkable connection between the inverse tangent and the logarithm. If in the known power series for $\frac{1}{2} \log \frac{1 + x}{1 - x}$ (p. 444) we replace x by ix, we actually obtain the power series for arc tan x,

$$\text{arc tan } x = \frac{1}{i} \left(ix + \frac{(ix)^3}{3} + \frac{(ix)^5}{5} + \cdots \right)$$

$$= x - \frac{x^3}{3} + \frac{x^5}{5} - + \cdots.$$

These relations are as yet of a purely formal character and naturally call for a more exact statement of the meaning they are intended to convey. We have, however, seen above that by using proper definitions these relations acquire a satisfactorily rigorous meaning.

b. A Glance at the General Theory of Functions of a Complex Variable

Although the purely formal point of view indicated in the last Section is in itself free from objection, it is still desirable to recognize in the preceding formulas something more than a mere formal connection. This goal leads to the *general theory of complex functions,* as (for the sake of brevity) we call the general theory of the so-called *analytic functions of a complex variable.* As our starting point we may use a general discussion of the theory of power series with complex variables and complex coefficients. The construction of such a theory of power series offers no difficulty once we define the concept of limit in the domain of complex numbers; in fact, it parallels the theory of real power series almost exactly. However, as we shall not make any use of these matters in what follows we shall content ourselves here by stating certain facts, omitting proofs. It is found that the following generalization of the theorem of Section 7.5a, holds for the complex power series:

If a power series converges for any complex value $x = \xi$ whatever, then it converges absolutely for every value x for which $|x| < |\xi|$; if it diverges for a value $x = \xi$, then it diverges for every value x for which $|x| > |\xi|$. A power series which does not converge everywhere, but does converge for some other point in addition to $x = 0$, possesses a circle of convergence, that is, there exists a number $\rho > 0$ such that the series converges absolutely for $|x| < \rho$ and diverges for $|x| > \rho$.

Having once established the concept of functions of a complex variable represented by power series, and having developed the rules for operating with such functions, we can think of the functions e^x, $\sin x$, $\cos x$, arc $\tan x$, etc., of the *complex* variable x as simply *defined* by the power series which represent them for real values of x.

We shall indicate by two examples how this introduction of complex variables illuminates the behavior of the elementary functions. The geometric series for $1/(1 + x^2)$ ceases to converge when x leaves the interval $-1 \leq x \leq 1$, and so does the series for arc $\tan x$, although there are no peculiarities in the behavior of these functions at the ends of the interval of convergence; in fact, they and all their derivatives are continuous for all real values of x. On the other hand, we can readily understand that the series for $1/(1 - x^2)$ and $\log (1 - x)$ cease to converge as x passes through the value 1, since they become infinite there.

But the divergence of the series for the inverse tangent and the series $\sum_{\nu=0}^{\infty} (-1)^\nu x^{2\nu}$ for $|x| > 1$ immediately becomes clear if we consider complex values of x also. For we find that when $x = i$ the functions become infinite and so cannot be represented by a convergent series. Hence by our theorem about the circle of convergence the series must diverge for all values of x such that $|x| > |i| = 1$; in particular, for real values of x the series diverge outside the interval $-1 \leq x \leq 1$.

Another example is given by the function $f(x) = e^{-1/x^2}$ for $x \neq 0$, $f(0) = 0$ (see p. 462), which, in spite of its completely smooth behavior, cannot be expanded in a Taylor series. As a matter of fact, this function ceases to be continuous if we take pure imaginary values of $x = i\xi$ into account. The function then takes the form e^{1/ξ^2} and increases beyond all bounds as $\xi \to 0$. It is therefore clear that no power series in x can represent this function for all complex values of x in a neighborhood of the origin, no matter how small a neighborhood we choose.

These remarks on the theory of functions and power series of a complex variable must suffice for us here.

Appendix

*A.1   Multiplication and Division of Series

a. Multiplication of Absolutely Convergent Series

Let
$$A = \sum_{\nu=0}^{\infty} a_\nu, \qquad B = \sum_{\nu=0}^{\infty} b_\nu$$

be two absolutely convergent series. Together with these we consider the corresponding convergent series of absolute values

$$\bar{A} = \sum_{\nu=0}^{\infty} |a_\nu| \qquad \text{and} \qquad \bar{B} = \sum_{\nu=0}^{\infty} |b_\nu|.$$

We further put

$$A_n = \sum_{\nu=0}^{n-1} a_\nu, \quad B_n = \sum_{\nu=0}^{n-1} b_\nu, \quad \bar{A}_n = \sum_{\nu=0}^{n-1} |a_\nu|, \quad \bar{B}_n = \sum_{\nu=0}^{n-1} |b_\nu|$$

and
$$c_n = a_0 b_n + a_1 b_{n-1} + \cdots + a_n b_0.$$

We assert that the series $\sum_{\nu=0}^{\infty} c_n$ is absolutely convergent, and that its sum is equal to AB.

To prove this, we write down the series

$$a_0 b_0 + a_1 b_0 + a_1 b_1 + a_0 b_1 + a_2 b_0 + a_2 b_1$$
$$+ a_2 b_2 + a_1 b_2 + a_0 b_2 + \cdots + a_n b_0 + a_n b_1$$
$$+ \cdots + a_n b_n + \cdots + a_1 b_n + a_0 b_n + \cdots,$$

the n^2th partial sum of which is $A_n B_n$, and we assert that it converges absolutely. For the partial sums of the corresponding series with absolute values increase monotonically; the n^2th partial sum is equal to $\bar{A}_n \bar{B}_n$, which is less than $\bar{A}\bar{B}$ (and which tends to $\bar{A}\bar{B}$). The series with absolute values therefore converges, and the series written down above converges absolutely. The sum of the series is obviously AB, since its n^2th partial sum is $A_n B_n$, which tends to AB as $n \to \infty$. We now interchange the order of the terms, which is permissible for absolutely convergent series, and bracket successive terms together. In a convergent series we may bracket successive terms together in as many places as we desire without disturbing the convergence or altering the sum of the series, for if we bracket together, say, all the terms $(a_{n+1} + a_{n+2} + \cdots + a_m)$, then when we form the partial sums we shall omit those partial sums that originally fell between s_n and s_m, which does not affect the convergence or change the value of the limit. Also, if the series was absolutely convergent before the brackets were inserted, it remains absolutely convergent. Since the series

$$\sum_{\nu=0}^{\infty} c_\nu = (a_0 b_0) + (a_0 b_1 + a_1 b_0) + (a_0 b_2 + a_1 b_1 + a_2 b_0) + \cdots$$

is formed in this way from the series written down above, the required proof is complete.

*b. Multiplication and Division of Power Series

The principal use of our theorem is found in the theory of power series. The following assertion is an immediate consequence of it: The product of the two power series

$$\sum_{\nu=0}^{\infty} a_\nu x^\nu \quad \text{and} \quad \sum_{\nu=0}^{\infty} b_\nu x^\nu$$

is represented in the interval of convergence common to the two power series by a third power series $\sum_{\nu=0}^{\infty} c_\nu x^\nu$, whose coefficients are given by

$$c_\nu = a_0 b_\nu + a_1 b_{\nu-1} + \cdots + a_\nu b_0.$$

*As for the division of power series, we can likewise represent the quotient of the two power series above by a power series $\sum_{\nu=0}^{\infty} q_\nu x^\nu$, provided b_0, the constant term in the denominator, does not vanish. (In the latter case such a representation is in general impossible; for it could not converge at $x = 0$ on account of the vanishing of the denominator, whereas on the other hand, every power series must converge at $x = 0$.) The coefficients of the power series

$$\sum_{\nu=0}^{\infty} q_\nu x^\nu$$

can be calculated by remembering that $\sum_{\nu=0}^{\infty} q_\nu x^\nu \cdot \sum_{\nu=0}^{\infty} b_\nu x^\nu = \sum_{\nu=0}^{\infty} a_\nu x^\nu$, so that the following equations must be true:

$$a_0 = q_0 b_0,$$
$$a_1 = q_0 b_1 + q_1 b_0,$$
$$a_2 = q_0 b_2 + q_1 b_1 + q_2 b_0,$$
$$\cdot \; \cdot \; \cdot \; \cdot \; \cdot \; \cdot \; \cdot \; \cdot \; \cdot \; \cdot \; \cdot \; \cdot$$
$$a_\nu = q_0 b_\nu + q_1 b_{\nu-1} + \cdots + q_\nu b_0.$$

From the first of these equations q_0 is readily found, from the second we find the value q_1, from the third (by using the values of q_0 and q_1) we find the value q_2, etc. In order to give strict justification for the expression of the quotient of two power series by the third power series we have to investigate the convergence of the formally-calculated power series $\sum_{\nu=0}^{\infty} q_\nu x^\nu$. However, we

shall make no further use of the result and content ourselves with the statement that the series for the quotient does actually converge in some interval
about the origin. The proof is omitted.

A.2   Infinite Series and Improper Integrals

The infinite series and the concepts developed in connection with
them have simple applications and analogies in the theory of improper
integrals (cf. Chapter 4, p. 301). We confine ourselves to the case of a
convergent integral with an infinite interval of integration, say an
integral of the form $\int_0^\infty f(x)\,dx$. If we divide the interval of integration
by a sequence of numbers $x_0 = 0$, $x_1, \ldots$ tending monotonically to
$+\infty$, we can write the improper integral in the form

$$\int_0^\infty f(x)\,dx = a_1 + a_2 + \cdots,$$

where each term of our infinite series is an integral;

$$a_1 = \int_0^{x_1} f(x)\,dx, \quad a_2 = \int_{x_1}^{x_2} f(x)\,dx, \ldots,$$

and so on. This is true no matter how we choose the points x_ν. We
can therefore relate the idea of a convergent improper integral to that
of an infinite series in many ways.

It is especially convenient to choose the points x_ν in such a way that the
integrand does not change sign within any individual subinterval. The
series $\sum_{\nu=1}^\infty |a_\nu|$ then corresponds to the integral of the absolute value of
our function,

$$\int_0^\infty |f(x)|\,dx.$$

We are thus naturally led to the following concept: *an improper
integral $\int_0^\infty f(x)\,dx$ is said to be absolutely convergent if the integral*
$\int_0^\infty |f(x)|\,dx$ *converges.* Otherwise, if our integral exists at all, we say
that it is *conditionally convergent.*

Some of the integrals considered earlier (pp. 307 to 309), such as

$$\int_0^\infty \frac{1}{1+x^2}\,dx, \quad \int_0^\infty e^{-x^2}\,dx, \quad \Gamma(x) = \int_0^\infty e^{-t}t^{x-1}\,dt,$$

are absolutely convergent.

On the other hand, the important "Dirichlet" integral

$$J = \int_0^\infty \frac{\sin x}{x} \, dx = \lim_{A \to \infty} \int_0^A \frac{\sin x}{x} \, dx,$$

studied on p. 309, is the typical example of a conditionally convergent integral. The simplest proof of convergence is by reduction to an absolutely convergent integral: We write $\sin x = (1 - \cos x)' = 2(\sin^2 x/2)'$ and use integration by parts, transforming J into the absolutely convergent form

$$J = 2 \int_0^\infty \left(\sin^2 \frac{x}{2}\right)\frac{1}{x^2} \, dx.$$

(Note that the new integrand approaches continuously the limit $\frac{1}{2}$ for $x \to 0$ and vanishes of the order x^{-2} for $x \to \infty$.)

*A different proof of the convergence is obtained if we subdivide the interval from 0 to A at the points $x_\nu = \nu\pi (\nu = 0, 1, 2, \ldots, \mu_A)$, where μ_A is the largest possible integer for which $\mu_A \pi \leq A$. We therefore divide the integral into terms of the form

$$a_\nu = \int_{(\nu-1)\pi}^{\nu\pi} \frac{\sin x}{x} \, dx, \quad \text{for} \quad \nu = 1, 2, \ldots,$$

and a remainder R_A of the form

$$\int_{\mu_A \pi}^A \frac{\sin x}{x} \, dx \quad (0 \leq A - \mu_A \pi < \pi).$$

Obviously, the quantities a_ν have alternating signs, since $\sin x$ is alternately positive and negative in consecutive intervals. Moreover, $|a_{\nu+1}| < |a_\nu|$; for on applying the transformation $x = \xi - \pi$, we have

$$|a_\nu| = \int_{(\nu-1)\pi}^{\nu\pi} \frac{|\sin x|}{x} \, dx = \int_{\nu\pi}^{(\nu+1)\pi} \frac{|\sin (\xi - \pi)|}{\xi - \pi} \, d\xi = \int_{\nu\pi}^{(\nu+1)\pi} \frac{|\sin \xi|}{\xi - \pi} \, d\xi$$

$$> \int_{\nu\pi}^{(\nu+1)\pi} \frac{|\sin \xi|}{\xi} \, d\xi = |a_{\nu+1}|.$$

Hence by Leibnitz's test we see that Σa_ν converges. Moreover, the remainder R_A has the absolute value

$$|R_A| = \left| \int_{\mu_A \pi}^A \frac{\sin x}{x} \, dx \right| \leq \int_{\mu_A \pi}^{(\mu_A+1)\pi} \frac{|\sin x|}{x} \, dx$$

$$\leq \frac{1}{\mu_A \pi} \int_{\mu_A \pi}^{(\mu_A+1)\pi} |\sin x| \, dx = \frac{2}{\mu_A \pi},$$

and this tends to 0 as A increases. Thus, if we let A tend to ∞ in the equation

$$\int_\theta^A \frac{\sin x}{x}\, dx = a_1 + a_2 + a_3 + \cdots + a_{\mu_A} + R_A,$$

the right-hand side tends to Σa_ν as a limit, and our integral is convergent. But the convergence is not absolute for

$$|a_\nu| > \int_{(\nu-1)\pi}^{\nu\pi} \frac{|\sin x|}{\nu\pi}\, dx = \frac{2}{\nu\pi}, \qquad \text{so that } \sum |a_\nu| \text{ diverges.}$$

*A.3   Infinite Products

In the introduction to this chapter (p. 511), we stated that infinite series are only *one* way, although a particularly important one, of representing numbers or functions by infinite processes. As an example of another such process, we consider infinite products. No proofs will be given.

On p. 281 we encountered Wallis's product,

$$\frac{\pi}{2} = \frac{2}{1} \cdot \frac{2}{3} \cdot \frac{4}{3} \cdot \frac{4}{5} \cdot \frac{6}{5} \cdot \frac{6}{7} \cdots$$

in which the number $\pi/2$ is expressed as an "infinite product." Generally speaking, by the value of the infinite product

$$\prod_{\nu=1}^{\infty} a_\nu = a_1 \cdot a_2 \cdot a_3 \cdot a_4 \cdots$$

we mean the limit of the sequence of "partial products"

$$a_1, \qquad a_1 \cdot a_2, \qquad a_1 \cdot a_2 \cdot a_3, \qquad a_1 \cdot a_2 \cdot a_3 \cdot a_4, \ldots,$$

provided it exists.

The factors $a_1, a_2, a_3, \ldots$, of course, may also be functions of a variable x. An especially interesting example is the "infinite product" for the function $\sin x$,

$$(16) \qquad \sin \pi x = \pi x \left(1 - \frac{x^2}{1^2}\right)\left(1 - \frac{x^2}{2^2}\right)\left(1 - \frac{x^2}{3^2}\right) \cdots,$$

which we shall obtain in Section 8.5, p. 603.

The infinite product for the *zeta function* plays a very important role in the theory of numbers. In order to retain the notation usual in the theory of numbers we here denote the independent variable by s, and we define the zeta function for $s > 1$, following Riemann, by the expression

$$\zeta(s) = \sum_{n=1}^{\infty} \frac{1}{n^s}.$$

We know (Section 7.2c, p. 525) that the series on the right converges if $s > 1$. If p is any number greater than 1, we obtain the equation

$$\frac{1}{1 - \dfrac{1}{p^s}} = 1 + \frac{1}{p^s} + \frac{1}{p^{2s}} + \frac{1}{p^{3s}} + \cdots$$

by expanding the left-hand side in a geometric series with the quotient p^{-s}. If we imagine this series written down for all the prime numbers $p_1, p_2, p_3, \ldots$ in increasing order of magnitude, and all the equations thus formed multiplied together, we obtain on the left a product of the form

$$\frac{1}{1 - p_1^{-s}} \cdot \frac{1}{1 - p_2^{-s}} \cdots.$$

Without stopping to justify the process, we multiply together the series on the right-hand sides of our equations; we obtain a sum of terms

$$p_1^{-k_1 s} p_2^{-k_2 s} p_3^{-k_3 s} \cdots = (p_1^{k_1} p_2^{k_2} p_3^{k_3} \cdots)^{-s},$$

where $k_1, k_2, k_3, \ldots$ are any nonnegative integers: also we remember that by an elementary theorem each integer $n > 1$ can be expressed in one and only one way as a product of powers of different prime numbers $n = p_1^{k_1} p_2^{k_2} \cdots$. Thus we find that the product on the right is again the function $\zeta(s)$, and so we obtain the remarkable "product form" of Euler

(17) $$\zeta(s) = \frac{1}{1 - p_1^{-s}} \cdot \frac{1}{1 - p_2^{-s}} \cdot \frac{1}{1 - p_3^{-s}} \cdots.$$

This "product form," the derivation of which we have only briefly sketched here, is actually an expression of the zeta function as an infinite product, since the number of prime numbers is infinite.

In the general theory of infinite products one usually excludes the case where the product $a_1 a_2 \cdots a_n$ has the limit zero. Hence it is specially important that none of the factors a_n should vanish. In order that the product may converge, the factors a_n must accordingly tend to 1 as n increases. Since we can if necessary omit a finite number of factors (this has no bearing on the question of convergence), we may assume $a_n > 0$. The following almost trivial theorem applies to this case:

A necessary and sufficient condition for the convergence of the product $\prod_{v=1}^{\infty} a_v$, where $a_v > 0$, is that the series $\sum_{v=1}^{\infty} \log a_v$ should converge. For the partial sums $\sum_{v=1}^{n} \log a_v = \log (a_1 a_2 \cdots a_n)$ of this series will tend to a definite limit if, and only if, the partial products $a_1 a_2 \cdots a_n$ possess a positive limit, as a consequence of the continuity of the logarithm.

In studying convergence the following sufficient condition usually applies, where $a_\nu = 1 + \alpha_\nu$. The product

$$\prod_{\nu=1}^{\infty} (1 + \alpha_\nu)$$

converges, if the series

$$\sum_{\nu=1}^{\infty} |\alpha_\nu|$$

converges and no factor $(1 + \alpha_\nu)$ is zero. In the proof we may assume, after omission of a finite number of factors if necessary, that each $|\alpha_\nu| < \frac{1}{2}$. Then we have $1 - |\alpha_\nu| > \frac{1}{2}$. By the mean value theorem $\log (1 + h) = \log (1 + h) - \log 1 = h/(1 + \theta h)$ with $0 < \theta < 1$.

Therefore

$$|\log (1 + \alpha_\nu)| = \left| \frac{\alpha_\nu}{1 + \theta \alpha_\nu} \right| \le \frac{|\alpha_\nu|}{1 - |\alpha_\nu|} \le 2 |\alpha_\nu|,$$

and so the convergence of the series $\sum_{\nu=1}^{\infty} \log (1 + \alpha_\nu)$ follows from the convergence of $\sum_{\nu=1}^{\infty} |\alpha_\nu|$.

From our criterion it follows that the infinite product (16) above for $\sin \pi x$ converges for all values of x except for $x = 0, \pm 1, \pm 2, \ldots$, where factors of the product are zero. As to the Riemann ζ-function, for $p \ge 2$ and $s > 1$ we readily find that

$$\frac{1}{1 - p^{-s}} = 1 + \frac{1}{p^s - 1}, \qquad 0 < \frac{1}{p^s - 1} < \frac{2}{p^s}.$$

Now if we let p assume all prime values, the series $\Sigma \dfrac{1}{p^s}$ must converge, since its terms form only a part of the convergent series $\sum_{\nu=1}^{\infty} \dfrac{1}{\nu^s}$. The convergence of the product in Eq. (17) for $s > 1$ is thus proved. From the fact that the series for $\zeta(s)$ for $s = 1$ (that is, the harmonic series) diverges, we can draw the remarkable conclusion that the series of reciprocal prime numbers, that is, the series

$$\sum_{k=1}^{\infty} \frac{1}{p_k} = \frac{1}{2} + \frac{1}{3} + \frac{1}{5} + \frac{1}{7} + \frac{1}{11} + \frac{1}{13} + \frac{1}{17} + \frac{1}{19} + \cdots$$

diverges. (Incidentally, this shows that the number of primes is infinite.) Indeed, if the series of reciprocal primes were convergent, then also the series with terms

$$\alpha_k = \frac{1}{1 - p_k^{-1}} - 1 = \frac{p_k^{-1}}{1 - p_k^{-1}}$$

would be convergent, since $p_k \geq 2$ and

$$0 < \alpha_k \leq 2p_k^{-1}.$$

Then, by our test, also the infinite product

$$\prod_{k=1}^{\infty} (1 + \alpha_k) = \prod_{k=1}^{\infty} \frac{1}{1 - p_k^{-1}} = \prod_{k=1}^{\infty} \left(1 + \frac{1}{p_k} + \frac{1}{p_k^2} + \cdots \right)$$

would be convergent; but then clearly the harmonic series would converge as well which is impossible.

*A.4 Series Involving Bernoulli Numbers

So far we have given no expansions in power series for certain elementary functions, for example, $\tan x$. The reason is that the numerical coefficients which occur are not of any simple form. We can express these coefficients, and those in the series for a number of other functions, in terms of the so-called *Bernoulli numbers*. These are curious rational numbers, with a somewhat hidden law of formation, which occur in many parts of analysis. The simplest way to arrive at them is by expanding the function

$$\frac{x}{e^x - 1} = \frac{1}{1 + \dfrac{x}{2!} + \dfrac{x^2}{3!} + \cdots}$$

in a formal power series of the form

$$\frac{x}{e^x - 1} = \sum_{v=0}^{\infty} \frac{B_v^*}{v!} x^v.$$

If we write this equation in the form

$$x = (e^x - 1) \sum_{v=0}^{\infty} \frac{B_v^*}{v!} x^v$$

and substitute on the right the power series for $e^x - 1$, we obtain for the B_n^*, a recurrence relation

$$\binom{n+1}{1} B_n^* + \binom{n+1}{2} B_{n-1}^* + \binom{n+1}{3} B_{n-2}^* + \cdots + \binom{n+1}{n+1} B_0^* = 0$$

for $n > 0$, $B_0{}^* = 1$ from which the $B_i{}^*$ can easily be calculated successively. These rational numbers are called Bernoulli numbers.[1] They are rational since in their formation only rational operations are concerned; as we easily recognize, they vanish for all odd indices other than $\nu = 1$. The first few are

$$B_0{}^* = 1, \quad B_1{}^* = -\tfrac{1}{2}, \quad B_2{}^* = \tfrac{1}{6}, \quad B_4{}^* = -\tfrac{1}{30}, \quad B_6{}^* = \tfrac{1}{42},$$
$$B_8{}^* = -\tfrac{1}{30}, \quad B_{10}{}^* = \tfrac{5}{66}, \dots$$

We must content ourselves with a brief hint as to how these numbers are involved in the power series in question. First, by making use of the transformation

$$1 + \frac{B_2{}^*}{2!} x^2 + \cdots = \frac{x}{e^x - 1} + \frac{x}{2} = \frac{x}{2} \cdot \frac{e^x + 1}{e^x - 1} = \frac{x}{2} \cdot \frac{e^{\frac{1}{2}x} + e^{-\frac{1}{2}x}}{e^{\frac{1}{2}x} - e^{-\frac{1}{2}x}},$$

we obtain

$$\frac{x}{2} \coth \frac{x}{2} = \sum_{\nu=0}^{\infty} \frac{B_{2\nu}{}^*}{(2\nu)!} x^{2\nu}.$$

(This formula proves that $B_{2\nu+1}^* = 0$ for $\nu > 0$, since $(x/2)\coth(x/2)$ is an even function of x.)

If we replace x by $2x$, we have the series

$$x \coth x = \sum_{\nu=0}^{\infty} \frac{2^{2\nu} B_{2\nu}{}^*}{(2\nu)!} x^{2\nu},$$

valid, as can be shown, for $|x| < \pi$, from which, by replacing x by $-ix$, we obtain (cf. p. 552)

$$x \cot x = \sum_{\nu=0}^{\infty} (-1)^{\nu} \frac{2^{2\nu} B_{2\nu}{}^*}{(2\nu)!} x^{2\nu}, \quad |x| < \pi.$$

By means of the equation $2 \cot 2x = \cot x - \tan x$ we now obtain the series

$$\tan x = \sum_{\nu=1}^{\infty} (-1)^{\nu-1} \frac{2^{2\nu}(2^{2\nu} - 1)}{(2\nu)!} B_{2\nu}{}^* x^{2\nu-1},$$

which holds for $|x| < \dfrac{\pi}{2}$.

For further information we refer the reader to Chapter 8 and to more detailed treatises.[2]

---

[1] In a slightly different notation (p. 623), the basic formula will be written

$$\frac{x}{e^x - 1} = 1 - \tfrac{1}{2}x + \sum_{\nu=1}^{\infty} (-1)^{\nu+1} \frac{B_\nu}{(2\nu)!} x^{2\nu}.$$

[2] See, for example, K. Knopp, *Theory and Application of Infinite Series*, p. 183, Blackie & Son, Ltd., 1928 and K. Knopp, *Infinite Sequences and Series*, Dover Publications, 1956.

PROBLEMS

SECTION 7.1, page 511

1. Prove that

$$\sum_{\nu=1}^{\infty} \frac{1}{\nu(\nu + 1)} = \frac{1}{1 \cdot 2} + \frac{1}{2 \cdot 3} + \cdots + = 1$$

[cf. Problems 1.6, 12(a)] and use the result to prove $\sum_{\nu=1}^{\infty} \frac{1}{\nu^2}$ converges.

2. Use the result of Problem 1 to obtain upper and lower bounds for

$$\sum_{\nu=1}^{\infty} \frac{1}{\nu^2}.$$

3. Prove that $\sum_{\nu=0}^{\infty} (-1)^\nu \dfrac{2\nu + 3}{(\nu + 1)(\nu + 2)} = 1$.

4. For what values of α does the series $1 - \dfrac{1}{2^\alpha} + \dfrac{1}{3^\alpha} - \dfrac{1}{4^\alpha} + \cdots$ converge?

5. Prove that if $\sum_{\nu=1}^{\infty} a_\nu$ converges, and $s_n = a_1 + a_2 + \cdots + a_n$, then the sequence

$$\frac{s_1 + s_2 + \cdots + s_N}{N}$$

also converges, and has $\sum_{\nu=1}^{\infty} a_\nu$ as its limit.

6. Is the series $\sum_{n=1}^{\infty} \left(\dfrac{2n}{2n + 1} - \dfrac{2n - 1}{2n} \right)$ convergent?

7. Is the series $\sum_{\nu=1}^{\infty} (-1)^\nu \dfrac{\nu}{\nu + 1}$ convergent?

8. Prove that if $\sum_{\nu=1}^{\infty} a_\nu^2$ converges, so does $\sum_{\nu=1}^{\infty} \dfrac{a_\nu}{\nu}$.

9. (a) If a_n is a monotonic increasing sequence with positive terms, when does the series $\dfrac{1}{a_1} + \dfrac{1}{a_1 a_2} + \dfrac{1}{a_1 a_2 a_3} + \cdots$ converge?

(b) Give an example of a monotone decreasing sequence with $\lim\limits_{n \to \infty} a_n = 1$ for which the series diverges.

(c) Show that if decreasing sequences are allowed, then it is possible to obtain convergent sums even when $\lim\limits_{n \to \infty} a_n = 1$.

10. If the series $\sum_{\nu=1}^{\infty} a_\nu$ with decreasing positive terms converges, then $\lim\limits_{n \to \infty} na_n = 0$.

11. Show that the series $\sum\limits_{\nu=1}^{\infty} \sin \dfrac{\pi}{\nu}$ diverges.

12. Prove that if Σa_ν converges and if $b_1, b_2, b_3, \ldots$ is a bounded monotonic sequence of numbers, then $\Sigma a_\nu b_\nu$ converges. Moreover, prove that if $S = \Sigma a_\nu b_\nu$ and if $\Sigma a_\nu \leq M$, then $|S| \leq Mb_1$.

13. A sequence $\{a_n\}$ is said to be of bounded variation if the series

$$\sum_{i=1}^{\infty} |a_{i+1} - a_i|$$

converges.

(a) Prove that if the sequence $\{a_n\}$ is of bounded variation, then the sequence $\{a_n\}$ converges.

(b) Find a divergent infinite series Σa_i whose elements a_i constitute a sequence which is of bounded variation.

(c) Prove the following generalization of Abel's convergence test (see page 515) due to Dedekind:

The series $\Sigma a_i p_i$ is convergent if Σa_i oscillates between finite bounds and $\{p_i\}$ is a null sequence which is of bounded variation.

(d) Prove the convergence of the following infinite series:

(a) $\sum\limits_{n=2}^{\infty} \dfrac{\sin nx}{\log n} (-1)^n$;

(b) $\sum\limits_{n=2}^{\infty} \dfrac{\cos nx}{\log n} (-1)^n$

for x any fixed real number.

14. Discuss the convergence or divergence of the following series:

(a) $\sum \dfrac{(-1)^\nu}{\nu}$

(d) $\sum \dfrac{\sin \nu\theta}{\nu}$

(b) $\sum \dfrac{(-1)^\nu \cos(\theta/\nu)}{\nu}$

(e) $\sum \dfrac{(-1)^\nu \cos \nu\theta}{\nu}$

(c) $\sum \dfrac{\cos \nu\theta}{\nu}$

(f) $\sum \dfrac{(-1)^\nu \sin \nu\theta}{\nu}$.

15. Find the sums of the following derangements of the series

$$1 - \frac{1}{2} + \frac{1}{3} - \frac{1}{4} + \frac{1}{5} - \frac{1}{6} + \cdots$$

for $\log 2$:

(a) $1 - \frac{1}{2} - \frac{1}{4} + \frac{1}{3} - \frac{1}{6} - \frac{1}{8} + \frac{1}{5} - \frac{1}{10} - \frac{1}{12} + - - \cdots$.

(b) $1 + \frac{1}{3} + \frac{1}{5} - \frac{1}{2} - \frac{1}{4} - \frac{1}{6} + + + \cdots$.

16. Find whether the following series converge or diverge:

(a) $1 + \frac{1}{2} - \frac{1}{3} + \frac{1}{4} + \frac{1}{5} - \frac{1}{6} + \frac{1}{7} + \frac{1}{8} - \frac{1}{9} + + - \cdots$.

(b) $1 + \frac{1}{2} - \frac{2}{3} + \frac{1}{4} + \frac{1}{5} - \frac{2}{6} + \frac{1}{7} + \frac{1}{8} - \frac{2}{9} + + - \cdots$.

SECTION 7.2, page 520

1. Prove that $\sum\limits_{\nu=2}^{\infty} \dfrac{1}{\nu(\log \nu)^\alpha}$ converges when $\alpha > 1$ and diverges when $\alpha \leq 1$.

2. Prove that $\displaystyle\sum_{v=3}^{\infty} \frac{1}{v \log v(\log \log v)^\alpha}$ converges when $\alpha > 1$ and diverges when $\alpha \leq 1$.

3. Prove that if n is an arbitrary integer greater than 1

$$\sum_{v=1}^{\infty} \frac{a_v^{\,n}}{v} = \log n,$$

where $a_v^{\,n}$ is defined as follows:

$$a_v^{\,n} = \begin{cases} 1 \text{ if } n \text{ is not a factor of } v, \\ -(n-1) \text{ if } n \text{ is a factor of } v. \end{cases}$$

4. Show that $\displaystyle\sum_{v=2}^{\infty} \frac{\log(v+1) - \log v}{(\log v)^2}$ converges.

5. Show that $\displaystyle\sum_{v=1}^{\infty} \frac{1 \cdot 2 \cdot 3 \cdots v}{(\alpha+1)(\alpha+2)\cdots(\alpha+v)}$ converges if $\alpha > 1$ and diverges if $\alpha \leq 1$.

*6. By comparison with the series $\displaystyle\sum_{v=1}^{\infty} \frac{1}{v^\alpha}$, prove the following test:

If $\dfrac{\log(1/|a_n|)}{\log n} > 1 + \epsilon$ for some fixed number $\epsilon > 0$ independent of n, and for every sufficiently large n, the series Σa_v converges absolutely; if $\dfrac{\log(1/|a_n|)}{\log n} < 1 - \epsilon$ for every sufficiently large n and some number $\epsilon > 0$ independent of n, the series Σa_v does not converge absolutely.

7. Show that the series $\displaystyle\sum_{v=1}^{\infty} \left(1 - \frac{1}{\sqrt{v}}\right)^v$ converges.

8. For what values of α do the following series converge?

(a) $1 - \dfrac{1}{2^\alpha} + \dfrac{1}{3} - \dfrac{1}{4^\alpha} + \dfrac{1}{5} - \dfrac{1}{6^\alpha} + - \cdots$.

(b) $1 + \dfrac{1}{3^\alpha} - \dfrac{1}{2^\alpha} + \dfrac{1}{5^\alpha} + \dfrac{1}{7^\alpha} - \dfrac{1}{4^\alpha} + - \cdots$.

9. By comparison with the series $\Sigma \dfrac{1}{v(\log v)^\alpha}$, prove the following test:

The series $\Sigma |a_v|$ converges or diverges according as

$$\frac{\log(1/n\,|a_n|)}{\log \log n}$$

is greater than $1 + \epsilon$ or less than $1 - \epsilon$ for every sufficiently large n.

10. Derive the nth root test from the test of Problem 6.

11. Prove the following comparison test: if the series Σb_v of positive terms converges, and

$$\left| \frac{a_{n+1}}{a_n} \right| < \frac{b_{n+1}}{b_n}$$

from a certain term onward, the series Σa_ν is absolutely convergent; if Σb_ν diverges and

$$\left|\frac{a_{n+1}}{a_n}\right| > \frac{b_{n+1}}{b_n}$$

from a certain term onwards, the series Σa_ν is not absolutely convergent.

*12. By comparison with $\sum_{\nu=1}^{\infty} \frac{1}{\nu^\alpha}$, prove "Raabe's" test:

The series $\Sigma |a_\nu|$ converges or diverges according as

$$n\left(\frac{|a_n|}{|a_{n+1}|} - 1\right)$$

is greater than $1 + \epsilon$ or less than $1 - \epsilon$ for every sufficiently large n and for some $\epsilon > 0$ independent of n.

13. By comparison with $\Sigma \frac{1}{\nu(\log \nu)^\alpha}$, prove the following test:

The series $\Sigma |a_\nu|$ converges or diverges according as

$$n \log n \left(\frac{|a_n|}{|a_{n+1}|} - 1 - \frac{1}{n}\right)$$

is greater than $1 + \epsilon$ or less than $1 - \epsilon$ for every sufficiently large n.

14. Prove Gauss's test:

If

$$\frac{|a_n|}{|a_{n+1}|} = 1 + \frac{\mu}{n} + \frac{R_n}{n^{1+\epsilon}},$$

where $|R_n|$ is bounded and $\epsilon > 0$ is independent of n, the $\Sigma |a_\nu|$ converges if $\mu > 1$, diverges if $\mu \leq 1$.

15. Test the following "hypergeometric" series for convergence or divergence:

(a) $\frac{\alpha}{\beta} + \frac{\alpha(\alpha + 1)}{\beta(\beta + 1)} + \frac{\alpha(\alpha + 1)(\alpha + 2)}{\beta(\beta + 1)(\beta + 2)} + \cdots.$

(b) $1 + \frac{\alpha \cdot \beta}{1 \cdot \gamma} + \frac{\alpha(\alpha + 1) \cdot \beta(\beta + 1)}{1 \cdot 2 \cdot \gamma(\gamma + 1)}$

$$+ \frac{\alpha(\alpha + 1)(\alpha + 2) \cdot \beta(\beta + 1)(\beta + 2)}{1 \cdot 2 \cdot 3 \cdot \gamma(\gamma + 1)(\gamma + 2)} + \cdots.$$

SECTION 7.4, page 529

1. The sequence $f_n(x)$, $n = 1, 2, \ldots$, is defined in the interval $0 \leq x \leq 1$ by the equations

$$f_0(x) \equiv 1, \qquad f_n(x) = \sqrt{x f_{n-1}(x)}.$$

(a) Prove that in the interval $0 \leq x \leq 1$ the sequence converges to a continuous limit.

*(b) Prove that the convergence is uniform.

*2. Let $f_0(x)$ be continuous in the interval $0 \leq x \leq a$. The sequence of functions $f_n(x)$ is defined by

$$f_n(x) = \int_0^x f_{n-1}(t)\, dt, \qquad n = 1, 2, \ldots .$$

Prove that in any fixed interval $0 \leq x \leq a$ the sequence converges uniformly to 0.

*3. Let $f_n(x)$, $n = 1, 2, \ldots$, be a sequence of functions with continuous derivatives in the interval $a \leq x \leq b$. Prove that if $f_n(x)$ converges at each point of the interval and the inequality $|f_n'(x)| < M$ (where M is a constant) is satisfied for all values of n and x, then the convergence is uniform.

4. (a) Show that the series $\displaystyle\sum_{\nu=1}^{\infty} \frac{1}{\nu^x}$ converges uniformly for $x \geq 1 + \epsilon$ with $\epsilon > 0$ any fixed number.

(b) Show that the derived series $-\displaystyle\sum \frac{\log \nu}{\nu^x}$ converges uniformly for $x \geq 1 + \epsilon$ with ϵ a fixed positive number.

*5. Show that the series $\displaystyle\sum \frac{\cos \nu x}{\nu^\alpha}$, $\alpha > 0$, converges uniformly for $\epsilon \leq x \leq 2\pi - \epsilon$ with ϵ any small positive value.

6. The series

$$\frac{x-1}{x+1} + \frac{1}{3}\left(\frac{x-1}{x+1}\right)^3 + \frac{1}{5}\left(\frac{x-1}{x+1}\right)^5 + \cdots$$

converges uniformly for $\epsilon \leq x \leq N$ when ϵ, N are fixed positive numbers.

7. Find the regions in which the following series are convergent:

(a) $\displaystyle\sum x^{\nu!}$.

(d) $\displaystyle\sum \frac{a^\nu}{\nu^x}$, $a > 1$.

(b) $\displaystyle\sum \frac{(\nu!)^2 x^\nu}{(2\nu)!}$.

(e) $\displaystyle\sum \frac{\log \nu}{\nu^x}$.

(c) $\displaystyle\sum \frac{a^\nu}{\nu^x}$, $a < 1$.

(f) $\displaystyle\sum \frac{x^\nu}{1 - x^\nu}$.

*8. Prove that if the Dirichlet series $\displaystyle\sum \frac{a_\nu}{\nu^x}$ converges for $x = x_0$, it converges for any $x > x_0$; if it diverges for $x = x_0$, it diverges for any $x < x_0$. Thus there is an "abscissa of convergence" such that for any greater value of x the series converges, and for any smaller value of x the series diverges.

9. If $\displaystyle\sum \frac{a_\nu}{\nu^x}$ converges for $x = x_0$, the derived series $-\displaystyle\sum \frac{a_\nu \log \nu}{\nu^x}$ converges for any $x > x_0$.

SECTION 7.5, page 540

1. If the interval of convergence of the power series $\Sigma a_n x^n$ is $|x| < \rho$, and that of $\Sigma b_n x^n$ is $|x| < \rho'$, where $\rho < \rho'$, what is the interval of convergence of $\Sigma(a_n + b_n)x^n$?

2. If $a_\nu > 0$ and Σa_ν converges, then

$$\lim_{x \to 1-0} \sum a_\nu x^\nu = \sum a_\nu.$$

3. If $a_\nu > 0$ and Σa_ν diverges,

$$\lim_{x \to 1-0} \sum a_\nu x^\nu = \infty.$$

*4. Prove Abel's theorem:
If $\Sigma a_\nu X^\nu$ converges, then $\Sigma a_\nu x^\nu$ converges uniformly for $0 \le x \le X$.

*5. If $\Sigma a_\nu X^\nu$ converges, then $\lim_{x \to X-0} \Sigma a_\nu x^\nu = \Sigma a_\nu X^\nu$.

*6. By multiplication of power series prove that

(*a*) $e^x e^y = e^{x+y}$.   (*b*) $\sin 2x = 2 \sin x \cos x$.

7. Using the binomial series, calculate $\sqrt{2}$ to four decimal places.

8. Let a_n be any sequence of real numbers, and S the set of all limit points of the a_n. We denote the least upper bound p of S by $p = \overline{\lim} \, a_n$. Show that the power series $\sum_{n=0}^{\infty} c_n x^n$ converges for $|x| < \rho$ and diverges for $|x| > \rho$, where

$$\rho = \frac{1}{\overline{\lim} \sqrt[n]{|c_n|}}.$$

APPENDIX, page 555

1. Prove that the power series for $\sqrt{(1-x)}$ still converges when $x = 1$.

2. Prove that for every positive ϵ there is a polynomial in x which represents $\sqrt{(1-x)}$ in the interval $0 \le x \le 1$ with an error less than ϵ.

3. By setting $x = 1 - t^2$ in Problem 2, prove that for every positive ϵ there is a polynomial in t which represents $|t|$ in the interval $-1 \le t \le 1$ with an error less than ϵ.

4. (*a*) Prove that if $f(x)$ is continuous for $a \le x \le b$, then for every $\epsilon > 0$ there exists a polygonal function $\varphi(x)$ (that is, a continuous function whose graph consists of a finite number of rectilinear segments meeting at corners) such that $|f(x) - \varphi(x)| < \epsilon$ for every x in the interval.

(*b*) Prove that every polygonal function $\varphi(x)$ can be represented by a sum $\varphi(x) = a + bx + \Sigma c_i \, |x - x_i|$, where the x_i's are the abscissae of the corners.

5. WEIERSTRASS' APPROXIMATION THEOREM. Prove on the basis of the last statement that if $f(x)$ is continuous in $a \le x \le b$, then for every positive ϵ there exists a polynomial $P(x)$ such that $|f(x) - P(x)| < \epsilon$ for all values of x in the interval $a \le x \le b$.

Hint: Approximate $f(x)$ by linear combinations of the form $(x - x_r) + |x - x_r|$.

6. Prove that the following infinite products converge:

(a) $\displaystyle\prod_{n=1}^{\infty} (1 + (\tfrac{1}{2})^{2n})$;

(b) $\displaystyle\prod_{n=2}^{\infty} \frac{n^3 - 1}{n^3 + 1}$;

(c) $\displaystyle\prod_{n=1}^{\infty} \left(1 - \frac{z^2}{n!}\right)$,

if $|z| < 1$.

7. Prove by the methods of the text that $\displaystyle\prod_{n=1}^{\infty} \left(1 + \frac{1}{n}\right)$ diverges.

8. Prove the identity

$$\prod_{v=1}^{\infty} (1 + x^{2v}) = \frac{1}{1 - x}$$

for $|x| < 1$.

__* 9.__ Consider all the natural numbers which represented in the decimal system have no 9 among their digits. Prove that the sum of the reciprocals of these numbers converge.

10. (a) Prove that for $s > 1$,

$$1 - \frac{1}{2^s} + \frac{1}{3^s} - \frac{1}{4^s} + \cdots = (1 - 2^{1-s})\zeta(s),$$

where $\zeta(s)$ is the Zeta function defined on p. 560.

(b) Use this identity to show that $\displaystyle\lim_{s \to 1+} (s - 1)\zeta(s) = 1$.

11. *Integral test for convergence*

(a) Let $f(x)$ be positive and decreasing for $x \geqslant 1$. Prove that the improper integral $\displaystyle\int_{1}^{\infty} f(x)\, dx$ and the infinite series $\displaystyle\sum_{k=1}^{\infty} f(k)$ either both converge or both diverge.

(b) Prove that in either case the limit

$$\lim_{n \to \infty} \left(\int_{1}^{n} f(x)\, dx - \sum_{k=1}^{n} f(k) \right)$$

exists.

(c) Apply this test to prove that the series

$$\sum_{n=2}^{\infty} \frac{1}{n \log^{\alpha} n}$$

converges for $\alpha > 1$ and diverges for $\alpha \leq 1$.

8

Trigonometric Series

The functions represented by power series, or as Lagrange called them, the *"analytic functions,"* play indeed a central role in analysis. But the class of analytic functions is too restricted in many instances. It was therefore an event of major importance for all of mathematics and for a great variety of applications when Fourier in his "Théorie analytique de la chaleur"[1] observed and illustrated by many examples the fact that convergent *trigonometric series* of the form

$$(1) \qquad f(x) = \frac{a_0}{2} + \sum_{\nu=1}^{\infty} (a_\nu \cos \nu x + b_\nu \sin \nu x)$$

with constant coefficients a_ν, b_ν are capable of representing a wide class of "arbitrary" functions $f(x)$, a class which includes essentially every function of specific interest, whether defined geometrically by mechanical means, or in any other way: even functions possessing jump discontinuities, or obeying different laws of formation in different intervals, can thus be expressed.

Soon after Fourier's dramatic discovery the "Fourier series" were recognized not only as a most powerful tool for physics and mechanics, but just as much as a fruitful source of many beautiful purely mathematical results. Cauchy, and especially Dirichlet, in the years between 1820 and 1830, provided a solid basis for Fourier's somewhat heuristic and incomplete reasoning, making the subject as accessible as it is important.

---

[1] See the translation: *The Analytical Theory of Heat,* by Joseph Fourier, republished, Dover Publications, 1955.

In spite of the "arbitrariness" of the functions expressible by trigono-metrical series they are inherently subjected to the condition of perio-dicity with the period 2π, since each term of the series has this period. But, as we shall see, this restriction is inessential as soon as we consider a function merely in a finite interval from which we can easily extend it as a periodic function.

This chapter provides an elementary introduction to the theory of Fourier series, leaving aside more advanced refinements.

After some preliminary discussion of periodic functions we shall prove the main theorem establishing the validity of the trigonometric expansion for a wide class of functions.

In the subsequent sections we shall discuss somewhat more advanced supplementary topics such as uniform and absolute convergence of the Fourier series and polynomial approximation of arbitrary continuous functions. In the Appendix we shall discuss the theory of Bernoulli's polynomials and their applications.

8.1. Periodic Functions

a. General Remarks. Periodic Extension of a Function

The functions $\sin nx$ and $\cos nx$ are periodic functions of x with the common period 2π; thus any finite or convergent infinite sum of the type (1) is also periodic with period 2π. We now make some general observations concerning periodic functions, amplifying those of Chap-ter 4. p. 336.

Periodicity of a function $f(x)$ with the period T is expressed by the equation

$$(2a) \qquad\qquad f(x + T) = f(x),$$

valid for all values of x.[1] Having the period T implies that $f(x)$ also has the periods $\pm T, \pm 2T, \ldots, \pm mT, \ldots$, and

$$(2b) \qquad\qquad f(x \pm mT) = f(x)$$

for all integers m.

---

[1] In representing periodic functions it is often convenient to think of the independent variable x as a point on the circumference of a circle instead of on a straight line. For a function $f(x)$ with the period 2π, we consider the angle x at the center of a circle of unit radius, included between an arbitrary initial radius and the radius to a variable point on the circumference; then the periodicity of $f(x)$ means that to each point on the circumference there corresponds just one value of the function, although the angle x itself is determined only within multiples of 2π.

In special cases $f(x)$ may also happen to have a shorter period. For example, the function $\sin(4\pi x/T)$ has the period T as well as the smaller period $T/2$.

As we saw already in Chapter 4, p. 337, a function $f(x)$ defined in a closed interval $a \leq x \leq b$, can be extended as a periodic function with period $T = b - a$ for all values of x by defining the function in successive adjacent intervals of length T outside the original interval $a \leq x \leq b$ by the periodicity relation

$$(2c) \qquad f(x + nT) = f(x), \qquad n = \pm 1, \pm 2, \ldots .$$

The extended function is neither defined uniquely nor necessarily continuously at the end points $x = a + nT = b + (n - 1)T$ of our intervals of length T. We must admit functions $f(x)$ with jump discontinuities at points $x = \xi$, which are continuous on either side of ξ but not necessarily defined or continuous at the point ξ itself.

Then the following notations and definition of $f(\xi)$ will be useful throughout this chapter: we denote the right-hand limit and the left-hand limit of $f(x)$ at $x = \xi$ by

$$(3a) \qquad f(\xi + 0) = \lim_{\epsilon \to 0} f(\xi + \epsilon^2),$$

$$(3b) \qquad f(\xi - 0) = \lim_{\epsilon \to 0} f(\xi - \epsilon^2);$$

it is convenient to assign *by definition*, to f as its value at the point of discontinuity ξ itself the *mean value*

$$(4) \qquad f(\xi) = \tfrac{1}{2}[f(\xi + 0) + f(\xi - 0)]$$

disregarding whatever value $f(\xi)$ may have had originally.

With this convention there is no restriction on extending our original function from a closed interval $a \leq x \leq b$ periodically to all values of x even in cases where $f(a) \neq f(b)$. We need pay attention only to the values of $f(x)$ at the jump discontinuities, arising in particular if the originally defined values of $f(a)$ and $f(b)$ do not coincide; to define the periodic extension we have to use the mean value $\tfrac{1}{2}[f(a) + f(b)]$ in place of the values $f(a)$ and $f(b)$.

b. Integrals Over a Period

The graph of a periodic function $f(x)$ clearly has the same shape in any two consecutive intervals corresponding to a period. This implies the important fact that for a periodic function $f(x)$ of period T and for arbitrary a

$$(5) \qquad \int_{-a}^{T-a} f(x)\, dx = \int_0^T f(x)\, dx,$$

or in words: the integral of a periodic function over a period interval of length T always has the same value no matter where the interval lies.

To prove this fact we need only notice that by virtue of the equation $f(\xi - T) = f(\xi)$ the substitution $x = \xi - T$ yields, for any α, β,

$$\int_\alpha^\beta f(x)\,dx = \int_{\alpha+T}^{\beta+T} f(\xi)\,d\xi = \int_{\alpha+T}^{\beta+T} f(x)\,dx.$$

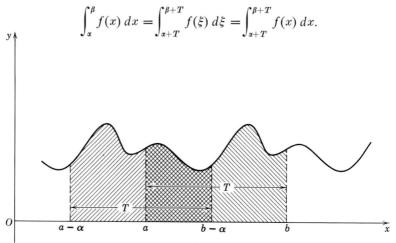

Figure 8.1   To illustrate the integral over a whole period.

In particular, for $\alpha = -a$ and $\beta = 0$

$$\int_{-a}^0 f(x)\,dx = \int_{T-a}^T f(x)\,dx$$

and hence

$$\int_{-a}^{T-a} f(x)\,dx = \int_{-a}^0 f(x)\,dx + \int_0^{T-a} f(x)\,dx$$

$$= \int_{T-a}^T f(x)\,dx + \int_0^{T-a} f(x)\,dx$$

$$= \int_0^T f(x)\,dx,$$

as stated. Recalling the geometrical meaning of the integral, the statement is made obvious by Fig. 8.1.

c. Harmonic Vibrations

The simplest periodic functions from which we shall construct the most general ones are the functions $a \sin \omega x$ and $a \cos \omega x$, or more generally $a \sin \omega(x - \xi)$ and $a \cos \omega(x - \xi)$, where $a \, (\geq 0)$, $\omega \, (> 0)$, and ξ are constants. These functions represent "*sinusoidal vibrations*" or *simple harmonic vibrations* (or *oscillations*).[1] The *period of vibration* is $T = 2\pi/\omega$. The number ω is called the *circular* or *angular frequency* of the vibrations[2]; since $1/T = \omega/2\pi$ is the number of vibrations in unit time, or the frequency, ω is the number of vibrations in the time 2π. The number a is called the *amplitude* of the vibration; it represents the maximum value of the function $a \sin \omega(x - \xi)$ or

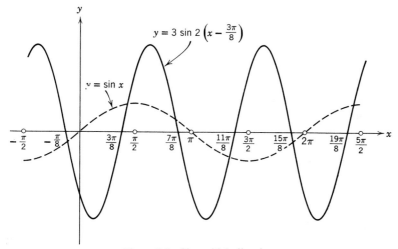

Figure 8.2 Sinusoidal vibrations.

$a \cos \omega(x - \xi)$, since both sine and cosine have the maximum value 1. The number $\omega(x - \xi)$ is called the *phase* and the number $\omega\xi$ is called the phase displacement or *phase shift*.

We obtain the functions $a \sin \omega(x - \xi)$ graphically by stretching the sine curve in the ratios $1 : \omega$ along the x-axis and $a : 1$ along the y-axis, and then translating the curve a distance ξ in the positive direction along the x-axis (cf. Fig. 8.2).

By the addition formulas for the trigonometric functions we can also express harmonic vibrations by $\alpha \cos \omega x + \beta \sin \omega x$ and respectively $\beta \cos \omega x - \alpha \sin \omega x$ where $\alpha = -a \sin \omega\xi$ and $\beta = a \cos \omega\xi$. Conversely, every function of the form $\alpha \cos \omega x + \beta \sin \omega x$ represents a

---

[1] Either of these formulas taken alone (for all values of a and ξ) represents the set of all sinusoidal vibrations; the two formulas are equivalent, since $a \sin \omega(x - \xi) =$

$$a \cos \omega \left[x - \left(\xi + \frac{\pi}{2\omega} \right) \right].$$

[2] Notice that we distinguish between the *frequency* and the *circular frequency*.

sinusoidal vibration $a \sin \omega(x - \xi)$ with the amplitude $a = \sqrt{\alpha^2 + \beta^2}$ and the *phase displacement* $\omega\xi$ given by the equations $\alpha = -a \sin \omega\xi$, $\beta = a \cos \omega\xi$. Using the expression $\alpha \cos \omega x + \beta \sin \omega x$ we immediately can write the sum of two or more such functions with the same circular frequency ω as another vibration with circular frequency ω.

As seen earlier, periodic functions arise when we wish to represent closed curves parametrically. Naturally, they can be used to represent phenomena induced by circular motion, say a process repeated periodically in tune with a flywheel; moreover they are associated with all phenomena of vibration.

8.2   Superposition of Harmonic Vibrations

a. Harmonics.   Trigonometric Polynomials

Although many vibrations are purely sinusoidal (cf. p. 405), most periodic motions have a more complicated character, being obtained by "superposition" of several sinusoidal vibrations. Mathematically, the motion of a point on a line with the coordinate x as a function of the time may be given by a function that is the sum of a number of pure periodic functions of the above type. The harmonic components of the function are then superimposed (that is, their ordinates are added). In this superposition we assume that the circular frequencies (and, of course, the periods) of the superposed vibrations are all different, for the superposition of two sinusoidal vibrations with the same circular frequency yields another sinusoidal vibration with the same circular frequency as shown above.

For the superposition of two sinusoidal vibrations with the different circular frequencies ω_1 and ω_2, there are two fundamentally distinct possibilities, depending on whether ω_1/ω_2 is rational or not, or, as we said, whether the frequencies are *commensurable* or *incommensurable*.

As an example of the first case we assume that the second circular frequency is twice that of the first: $\omega_2 = 2\omega_1$. The period of the second vibration is then half that of the first, $2\pi/2\omega_1 = T_2 = T_1/2$, and so it has not only the period T_2 but also the doubled period T_1, since the function repeats itself after this double period; the function formed by superposition must likewise have the period T_1. The second vibration, with twice the circular frequency and half the period of the first, is called a *first harmonic* of the first vibration (the fundamental).

Corresponding statements are true if we introduce another vibration with circular frequency $\omega_3 = 3\omega_1$. Here again the function $\sin 3\omega_1 x$ necessarily repeats itself with the period $2\pi/\omega_1 = T_1$. Such a vibration is called a *second harmonic* of the given vibration. Similarly, we can consider third, fourth, . . . , $(n - 1)$th harmonics with the circular frequencies $\omega_4 = 4\omega_1$, $\omega_5 = 5\omega_1$, . . . , $\omega_n = n\omega_1$, and, moreover, with any phase displacements we wish. Every such harmonic necessarily repeats itself after the period $T_1 = 2\pi/\omega_1$, and consequently every function obtained by superposing a number of vibrations, each of which is a harmonic of a given fundamental circular frequency ω_1, is itself a periodic function with the period $2\pi/\omega_1 = T_1$. By superposing vibrations with circular frequencies ranging from that of the fundamental to that of the $(n - 1)$th harmonic we obtain a periodic function in the form of a *trigonometric polynomial*

$$(6) \qquad S_n(x) = \frac{a_0}{2} + \sum_{\nu=1}^{n}(a_\nu \cos \nu\omega x + b_\nu \sin \nu\omega x).$$

(The constant $a_0/2$ which does not affect the periodicity is affixed for later convenience.) Since this function contains $2n + 1$ arbitrary constants a_ν, b_ν, we are able to generate curves which may not at all resemble the original sine curves. Figures 8.3 to 8.5 are graphical illustrations.

The term "harmonic" alludes to acoustics,[1] where a fundamental vibration with circular frequency ω corresponds to a tone of a certain pitch, and the first, second, third, etc., harmonics correspond to the sequence of harmonics of the fundamental, that is, to the octave plus fifth, double octave, etc.

In general, for the superposition of vibrations in which the circular frequencies have rational ratios, these circular frequencies can all be represented as integral multiples of a common fundamental frequency.

The superposition of two vibrations having incommensurable circular frequencies ω_1 and ω_2, however, represents a different phenomenon. Here the superposition of sinusoidal vibrations is no longer periodic. Without going into a detailed discussion, we remark that such functions have an "approximately periodic" character or, as we say, are *almost periodic*.

*b. Beats

A final remark on the superposition of sinusoidal vibrations concerns the phenomenon of so-called *beats*. If we superpose two vibrations, each of unit amplitude but having different circular frequencies

---

[1] In acoustics the term overtone is also used.

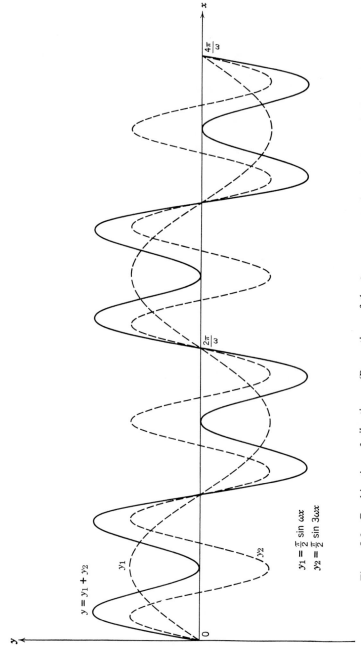

$y = y_1 + y_2$

y_1

y_2

$y_1 = \dfrac{\pi}{2} \sin \omega x$

$y_2 = \dfrac{\pi}{2} \sin 3\omega x$

Figure 8.3 Combination of vibrations. (Proportions of the figure correspond to the assumption $\omega = 1$.)

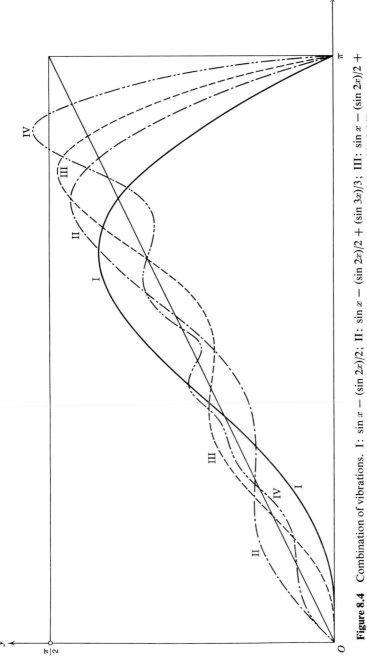

Figure 8.4   Combination of vibrations. I: $\sin x - (\sin 2x)/2$; II: $\sin x - (\sin 2x)/2 + (\sin 3x)/3$; III: $\sin x - (\sin 2x)/2 + (\sin 3x)/3 - (\sin 4x)/4$; IV: $\sin x - (\sin 2x)/2 + (\sin 3x)/3 - (\sin 4x)/4 + (\sin 5x)/5 - (\sin 6x)/6 + (\sin 7x)/7$.

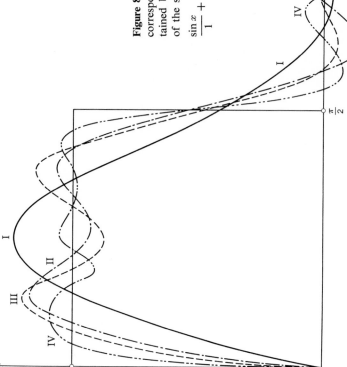

Figure 8.5 Combination of vibrations. The curves correspond to the trigonometrical polynomials obtained by taking 3, 5, 6, and 8 terms respectively of the series

$$\frac{\sin x}{1} + 2\frac{\sin 2x}{2} + \frac{\sin 3x}{3} + \frac{\sin 5x}{5} + 2\frac{\sin 6x}{6}$$

$$+ \frac{\sin 7x}{7} + \frac{\sin 9x}{9} + 2\frac{\sin 10x}{10} + \cdots.$$

ω_1 and ω_2, and if for the sake of simplicity we take the same value of ξ (see p. 575) for both (the generalization to arbitrary phase is left to the reader), then we are concerned with the function

$$y = \sin \omega_1 x + \sin \omega_2 x \qquad (\omega_1 > \omega_2 > 0).$$

By a well-known trigonometrical formula we have

$$y = 2 \cos \left[\tfrac{1}{2}(\omega_1 - \omega_2)x\right] \sin \left[\tfrac{1}{2}(\omega_1 + \omega_2)x\right].$$

This equation represents a phenomenon which we describe as follows: we have a vibration with the circular frequency $\tfrac{1}{2}(\omega_1 + \omega_2)$ and the

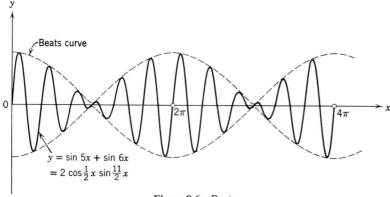

Figure 8.6   Beats.

period $4\pi/(\omega_1 + \omega_2)$. This vibration does not have a constant amplitude but a varying "amplitude" given by the expression

$$2 \cos \left[\tfrac{1}{2}(\omega_1 - \omega_2)x\right]$$

which varies with a longer period $4\pi/(\omega_1 - \omega_2)$. This description is particularly useful when the two circular frequencies ω_1 and ω_2 are relatively large, whereas their difference $(\omega_1 - \omega_2)$ is comparatively small. Then the amplitude $2 \cos \left[\tfrac{1}{2}(\omega_1 - \omega_2)x\right]$ of the vibration with period $4\pi/(\omega_1 + \omega_2)$ varies only slowly compared with the period of vibration, and this change of amplitude repeats itself periodically with the long period $4\pi/(\omega_1 - \omega_2)$. These rhythmic changes of amplitude are called *beats*. Everyone is acquainted with this phenomenon in acoustics and electronics. In radio transmission the circular frequencies ω_1 and ω_2 are, as a rule, far above those which the ear can detect, whereas the difference $\omega_1 - \omega_2$ falls in the range of audible notes. The beats then cause an audible tone, whereas the original vibrations remain imperceptible to the ear.

An example of beats is illustrated graphically in Fig. 8.6.

8.3   Complex Notation

a. General Remarks

Operation with trigonometric functions is often simplified by using complex numbers according to Euler's relation

$$\cos\theta + i\sin\theta = e^{i\theta}$$

or

(7a) $$\cos\theta = \tfrac{1}{2}(e^{i\theta} + e^{-i\theta})$$

(7b) $$\sin\theta = \frac{1}{2i}(e^{i\theta} - e^{-i\theta})$$

(Compare Chapter 7, p. 551.) Accordingly, we can express sinusoidal vibrations in terms of the complex quantities $e^{i\omega x}$, $e^{-i\omega x}$, or $ae^{i\omega(x-\xi)}$, $ae^{-i\omega(x-\xi)}$ respectively, where a, ω, and $\omega\xi$ are amplitude, circular frequency, and phase displacement. Ultimately of course, real vibrations are obtained from the complex expression, simply by separating real and imaginary parts.

One of the conveniences of the complex notation is the fact that the derivatives with respect to the time x are obtained by differentiating the complex exponential function as if i were a real constant; the formula

$$\frac{d}{dx} a[\cos\omega(x-\xi) + i\sin\omega(x-\xi)]$$

$$= a\omega[-\sin\omega(x-\xi) + i\cos\omega(x-\xi)]$$

$$= ia\omega[\cos\omega(x-\xi) + i\sin\omega(x-\xi)]$$

that follows from the formulas for the derivatives of the sine and cosine functions can be written in the concise form

(8) $$\frac{d}{dx} ae^{i\omega(x-\xi)} = ia\omega e^{i\omega(x-\xi)}.$$

The integral of a complex-valued function $\gamma(x)$, say $\gamma(x) = p(x) + iq(x)$, is naturally defined by

$$\int \gamma(x)\,dx = \int p(x)\,dx + i\int q(x)\,dx.$$

Accordingly, for $n \neq 0$

$$\int e^{inx}\,dx = \int \cos nx\,dx + i\int \sin nx\,dx$$

$$= \frac{1}{n}\sin nx - \frac{i}{n}\cos nx = \frac{1}{in}e^{inx}.$$

In particular, for any integer n we have

$$\int_{-\pi}^{\pi} e^{inx}\, dx = \begin{cases} 0 & \text{for } n \neq 0 \\ 2\pi & \text{for } n = 0. \end{cases}$$

More generally, if we remember that $e^{inx} e^{-imx} = e^{i(n-m)x}$, we have for any integers m, n

$$(9) \qquad \int_{-\pi}^{\pi} e^{inx} e^{-imx}\, dx = \begin{cases} 0, & \text{for } n \neq m \\ 2\pi & \text{for } n = m. \end{cases}$$

These relations are merely concise expressions of the orthogonality relations between trigonometric functions (see p. 274).

* b. Application to Alternating Currents

We insert an illustration of these ideas by an important example, denoting the independent variable, the time, by t instead of x.

We consider an electric circuit with resistance R and inductance L, on which an external electromotive force (voltage) E is impressed. In direct current, the voltage E is constant, and the current I is given by Ohm's law, $E = RI$. For an alternating current however, E, and consequently I, is a function of the time t, and *Ohm's law* takes the generalized form (cf. p. 635)

$$(10) \qquad E - L\frac{dI}{dt} = RI.$$

We consider the external electromotive forces E which are sinusoidal with circular frequency ω, given by $\epsilon \cos \omega t$ or $\epsilon \sin \omega t$ and combine both possibilities formally in the complex form

$$E = \epsilon e^{i\omega t} = \epsilon \cos \omega t + i\epsilon \sin \omega t,$$

where ϵ represents the amplitude. Often it is useful to admit *complex* values also for the *amplitude*

$$\epsilon = |\epsilon|\, e^{-i\eta};$$

then

$$E = |\epsilon|\, e^{i(\omega t - \eta)} = |\epsilon|\, \{\cos (\omega t - \eta) + i \sin (\omega t - \eta)\}.$$

We may operate with this "complex voltage" E and the corresponding complex current I as if i were a real parameter. Then the significance of the complex relation between the complex quantities E and I is that the current corresponding to an electromotive force $\epsilon \cos \omega t$ is the real part of I, whereas the current corresponding to an electromotive force $\epsilon \sin \omega t$ is the imaginary part of I. The complex current is given by an expression of the form

$$I = \alpha e^{i\omega t} = \alpha(\cos \omega t + i \sin \omega t)$$

which is also sinusoidal with circular frequency ω. The derivative of I is then given formally by

$$\frac{dI}{dt} = i\,\alpha\omega e^{i\omega t}$$

$$= \alpha\omega(-\sin\omega t + i\cos\omega t) = i\omega I.$$

Substituting these quantities in the *generalized* form of *Ohm's law* (Eq. 10) and dividing by the factor $e^{i\omega t}$, we obtain the equation

$$\epsilon - \alpha Li\omega = R\alpha,$$

or

$$\alpha = \frac{\epsilon}{R + i\omega L},$$

as well as

$$E = (R + i\omega L)I = WI.$$

We may regard this last equation as *Ohm's law* for alternating currents in complex form if we call the quantity

$$W = R + i\omega L$$

the *complex resistance* of the circuit. Ohm's law is then the same as for direct current: the current is equal to the voltage divided by the resistance.

Writing the complex resistance W with $w = |W|$ in the form

$$W = we^{i\delta} = w\cos\delta + iw\sin\delta,$$

where

$$|W| = w = \sqrt{(R^2 + L^2\omega^2)}, \qquad \tan\delta = \frac{\omega L}{R},$$

we obtain

$$I = \frac{\epsilon}{w}e^{i(\omega t - \delta)} = \frac{E}{W}.$$

According to this formula the current has the same period (and circular frequency) as the voltage; the amplitude α of the current is related to the amplitude ϵ of the electromotive force for real ϵ by

$$\alpha = \frac{\epsilon}{w},$$

and, in addition, there is a difference of phase between the current and the voltage. The current reaches its maximum, not at the same time as the voltage but at a time δ/ω later, and the same is, of course, true for the minimum. In electrical engineering the quantity $w = \sqrt{R^2 + L^2\omega^2}$ is frequently called the *impedance* or *alternating current resistance* of the circuit for the circular frequency ω; the phase displacement, usually stated in degrees, is sometimes called the *lag*.

If the amplitude ϵ is complex in the form

$$\epsilon = |\epsilon|\,e^{-i\eta},$$

then nothing essential is changed in the form of Ohm's law, except η is an additional phase shift and we have

$$E = |\epsilon|\, e^{i(\omega t - \eta)},$$

$$I = \frac{E}{W} = \frac{|\epsilon|}{|W|}\, e^{i\omega t} e^{-i(\delta + \eta)}\ .$$

c. Complex Notation for Trigonometrical Polynomials

A compound vibration of the type

$$(11) \qquad S_n(x) = \tfrac{1}{2}a_0 + \sum_{\nu=1}^{n}(a_\nu \cos \nu x + b_\nu \sin \nu x)$$

(for brevity we have taken $\omega = 1$) can be reduced to complex form by substituting

$$\cos \nu x = \tfrac{1}{2}(e^{i\nu x} + e^{-i\nu x}) \qquad \text{and} \quad \sin \nu x = -\tfrac{1}{2}i(e^{i\nu x} - e^{-i\nu x}).$$

This expression then assumes the simpler form

$$(12) \qquad S_n(x) = \sum_{\nu=-n}^{n} \alpha_\nu e^{i\nu x},$$

where the complex numbers α_ν are related to the real numbers a_0, a_ν, and b_ν by the equations

$$(13a) \qquad \begin{cases} \alpha_\nu = \tfrac{1}{2}(a_\nu - ib_\nu), \\ \alpha_{-\nu} = \tfrac{1}{2}(a_\nu + ib_\nu), \qquad \text{for} \quad \nu = 1, 2, \ldots, n, \qquad \text{and} \\ \alpha_0 = \tfrac{1}{2}a_0. \end{cases}$$

Solving these relations for the a_ν and b_ν, we find that

$$(13b) \qquad \begin{aligned} a_\nu &= \alpha_\nu + \alpha_{-\nu}, \\ b_\nu &= i(\alpha_\nu - \alpha_{-\nu}). \end{aligned}$$

(The case $\nu = 0$ is included.)

Conversely, we may regard any arbitrary expression of the form

$$\sum_{\nu=-n}^{n} \alpha_\nu e^{i\nu x}$$

as a function representing the superposition of vibrations written in complex form. The result of this superposition is real if and only if $\alpha_\nu + \alpha_{-\nu}$ is real and $\alpha_\nu - \alpha_{-\nu}$ is pure imaginary; that is, if α_ν and $\alpha_{-\nu}$ are conjugate complex numbers.

d. A Trigonometric Formula

As an application of the complex notation we prove the following identity:

$$\sigma_n(\alpha) = \tfrac{1}{2} + \cos \alpha + \cos 2\alpha + \cdots + \cos n\alpha$$

(14)
$$= \frac{\sin (n + \tfrac{1}{2})\alpha}{2 \sin \tfrac{1}{2}\alpha},$$

which is needed later on in this chapter. The formula makes sense only when $\sin \tfrac{1}{2}\alpha \neq 0$, that is, when α does not have one of the values $0, \pm 2\pi, \pm 4\pi, \ldots$. However, once the formula has been established for $\sin \tfrac{1}{2}\alpha \neq 0$, we conclude that the expression $[\sin (n + \tfrac{1}{2})\alpha]/(2 \sin \tfrac{1}{2}\alpha)$ is a continuous function of α for all α, if we define its value at the exceptional points as that of $\sigma_n(\alpha)$ that is, $n + \tfrac{1}{2}$.

For the proof we replace the cosine function by its exponential expression [see formula (13a) with $a_\nu = 1, b_\nu = 0$]:

$$\sigma_n(\alpha) = \tfrac{1}{2} \sum_{\nu=-n}^{n} e^{i\nu\alpha}.$$

On the right we have a geometric progression with the common ratio $q = e^{i\alpha} = \cos \alpha + i \sin \alpha$. Hence q can have the value 1 only if $\cos \alpha = 1, \sin \alpha = 0$, that is, if α has one of the exceptional values $0, \pm 2\pi, \pm 4\pi, \ldots$. For all other values α the ordinary formula for the sum yields

$$\sigma_n(\alpha) = \tfrac{1}{2} e^{-in\alpha} \frac{1 - q^{2n+1}}{1 - q}$$

$$= \frac{1}{2} \cdot \frac{e^{-in\alpha} - e^{(n+1)i\alpha}}{1 - e^{i\alpha}}.$$

On multiplying the numerator and denominator by $e^{-i\alpha/2}$ we obtain, as stated,

$$\sigma_n(\alpha) = \frac{\sin (n + \tfrac{1}{2})\alpha}{2 \sin \tfrac{1}{2}\alpha}.$$

Integrating $\sigma_n(t)$ on $0 \leq t \leq \pi$, we find the useful result that independently of n

(15)
$$\int_0^\pi \frac{\sin (n + \tfrac{1}{2})t}{2 \sin \tfrac{1}{2}t} \, dt = \int_0^\pi \left(\tfrac{1}{2} + \sum_{\nu=1}^{n} \cos \nu t \right) dt$$

$$= \tfrac{1}{2}\pi$$

since the integral of each term of the series vanishes.

8.4   Fourier Series

a. Fourier Coefficients

Trigonometrical polynomials

(16) $$f(x) = S_n(x) = \tfrac{1}{2}a_0 + \sum_{v=1}^{n}(a_v \cos vx + b_v \sin vx)$$

of order n depend on the $2n + 1$ coefficients a_v and b_v. It is remarkable that these "*Fourier coefficients*" can be expressed simply by the following formulas in terms of the values $f(x)$ of the sum:

(17) $$a_\mu = \frac{1}{\pi}\int_{-\pi}^{\pi} f(x)\cos \mu x\, dx, \quad b_\mu = \frac{1}{\pi}\int_{-\pi}^{\pi} f(x)\sin \mu x\, dx.$$

The proof follows if we multiply (16) by $\cos \mu x$ or $\sin \mu x$ and then integrate. The orthogonality relations (see p. 274) yield the expressions immediately, since only the terms with $v = \mu$ make a nonvanishing contribution.

In the complex terminology

(16a) $$f(x) = S_n(x) = \sum_{v=-n}^{n} \alpha_v e^{ivx},$$

$$a_v = \alpha_v + \alpha_{-v}; \quad b_v = i(\alpha_v - \alpha_{-v}),$$

the corresponding expressions for the complex Fourier coefficients are

(17a) $$\alpha_v = \frac{1}{2\pi}\int_{-\pi}^{\pi} f(x)e^{-ivx}\, dx$$

as is seen also on the basis of the complex orthogonality relations (9), p. 583.

Incidentally, the factor $\tfrac{1}{2}$ in the notation for the constant term $\tfrac{1}{2}a_0$ of (16) serves merely to make the formula (17) valid for $v = 0$.

Now we are led to the main theorem on Fourier series by the natural question of whether, by letting the degree n of the Fourier polynomial (16) tend to infinity, it becomes possible to represent functions $f(x)$ which are periodic with the period 2π but otherwise essentially arbitrary.

Our main result in the next articles will indeed be: *Any periodic function $f(x)$ which is sectionally continuous and has sectionally continuous derivatives of first and second order can be represented by an infinite "Fourier series"*

$$f(x) = \frac{a_0}{2} + \sum_{v=1}^{\infty}(a_v \cos vx + b_v \sin vx)$$

or in complex notation

$$f(x) = \sum_{v=-\infty}^{\infty} \alpha_v e^{ivx}$$

with coefficients given by (17) and (17a).

b. Basic Lemma

We first recall the definition of a *piecewise* or *sectionally continuous* function in an interval, as a function which is continuous except for a finite number of jump discontinuities in the interval.

We further recall that the value of a periodic function $f(x)$ is defined at a point of discontinuity as the mean of the limiting values from the two sides as agreed earlier [Eq. (4), p. 573].

A function $f(x)$ is *sectionally continuous* and has sectionally continuous first and second derivatives, if we can divide the whole interval into a finite number of subintervals, such that f, f', f'' are continuous in each open subinterval and approach definite limits at the end points.

The key to the proof of the main theorem will be a simple fact.

LEMMA. *If a function $k(x)$, and its first derivative $k'(x)$ are sectionally continuous in the interval $a \leq x \leq b$, then the integral*

$$K_\lambda = \int_a^b k(x) \sin \lambda x \, dx$$

tends to zero as $\lambda \to \infty$.

PROOF.   To prove this lemma we use integration by parts. Supposing that k and k' are continuous on $a \leq x \leq b$, we have

$$(18) \quad K_\lambda = \int_a^b k(x) \sin \lambda x \, dx$$

$$= \frac{1}{\lambda} \left[k(a) \cos \lambda a - k(b) \cos \lambda b + \int_a^b k'(x) \cos \lambda x \, dx \right];$$

as λ increases, the right-hand side obviously tends to zero. If $k(x)$ or $k'(x)$ have jump discontinuities ξ in the interval, then we subdivide it into parts by these points ξ, apply our argument to the parts, and add the results.

Omitting the proof, we state that the lemma actually remains true without any assumption about existence of the derivative $k'(x)$, merely using the sectional continuity of k. The proof under these milder conditions relies on the fact that for $\lambda \neq 0$ the function $\sin \lambda x$ is alternately positive and negative in successive intervals of length π/λ. For large values of λ the contributions to the integral from adjacent intervals almost cancel one another because of the continuity of $k(x)$.

c. Proof of $\displaystyle\int_0^\infty \frac{\sin z}{z}\, dz = \frac{\pi}{2}$

As an application of the lemma we evaluate the integral

(19)
$$I = \int_0^\infty \frac{\sin z}{z}\, dz.$$

This improper integral is defined by the relation

$$I = \lim_{M \to \infty} I_M$$

where

$$I_M = \int_0^M \frac{\sin z}{z}\, dz.$$

The convergence of the improper integral I, that is, the existence of the limit of I_M for $M \to \infty$ had been proved already on p. 310. The convergence proof was based on integration by parts and may be restated here. If, say, $0 < M < N$, we have

(20)
$$|I_N - I_M| = \left| \int_M^N \frac{\sin z}{z}\, dz \right|$$
$$= \left| -\frac{\cos z}{z} \Big|_M^N + \int_M^N \frac{\cos z}{z^2}\, dz \right|$$
$$\leq \frac{1}{M} + \frac{1}{N} + \int_M^N \frac{dz}{z^2} = \frac{2}{M}.$$

Since then I_N and I_M differ arbitrarily little if both M and N are sufficiently large, the existence of $I = \lim_{M \to \infty} I_M$ is assured by Cauchy's convergence test. Moreover, letting N tend to infinity in (20) we find an estimate for the rate at which the I_M approach their limit I:

(20a)
$$|I - I_M| \leq \frac{2}{M}.$$

We can rewrite our expression for I in such a way that I appears as a limit of integrals over a fixed finite interval. Let p be an arbitrary positive number; for $M = \lambda p$ the substitution $z = \lambda x$, $dz = \lambda\, dx$ shows that

$$I_{\lambda p} = \int_0^{\lambda p} \frac{\sin z}{z}\, dz = \int_0^p \frac{\sin \lambda x}{x}\, dx.$$

Since $\lambda p \to \infty$ for $\lambda \to \infty$ and fixed positive p, we clearly have

$$I = \lim_{\lambda \to \infty} \int_0^p \frac{\sin \lambda x}{x}\, dx,$$

and more precisely from (20a)

$$\left| I - \int_0^p \frac{\sin \lambda x}{x}\, dx \right| < \frac{2}{\lambda p}.$$

Thus for any positive p the expressions

$$\int_0^p \frac{\sin \lambda x}{x}\, dx$$

approach for $\lambda \to \infty$ one and the same value I; moreover, the convergence is uniform in p as long as we restrict p to values above some fixed positive number P. Indeed, the difference between the integral and the limit I is then less than ϵ for $\lambda > 2/P\epsilon$.

We now apply our lemma of p. 588 to the function

$$k(x) = \frac{1}{x} - \frac{1}{2 \sin (x/2)}.$$

If we define $k(0) = 0$, the function $k(x)$ is continuous and has a continuous first derivative for $0 \le x < 2\pi$ (see p. 466). Hence our lemma shows that

$$\int_0^p \sin \lambda x \left(\frac{1}{x} - \frac{1}{2 \sin (x/2)} \right) dx$$

tends to zero for $\lambda \to \infty$ as long as $0 \le p < 2\pi$. Moreover, by (18) the convergence is uniform for $0 \le p \le \pi$ since $|k(x)|$ and $|k'(x)|$ are bounded in the interval $0 \le x \le \pi$. It follows from our previous result that for any p in the interval $0 < p < 2\pi$

$$\lim_{\lambda \to \infty} \int_0^p \frac{\sin \lambda x}{2 \sin (x/2)}\, dx = I,$$

and also that the convergence is uniform in p for $P \le p \le \pi$, where P is a fixed positive number.

Now for $p = \pi$ and $\lambda = n + \frac{1}{2}$ (where n is an integer) we have evaluated this integral [see formula (15), p. 586] and found that it has the value $\pi/2$ independently of n. Letting λ tend to infinity through values of the form $\lambda = n + \frac{1}{2}$, we find then for I the value $\pi/2$:

$$(21) \qquad \int_0^\infty \frac{\sin z}{z}\, dz = \frac{\pi}{2}.$$

We have proved moreover that

$$(21a) \qquad \lim_{\lambda \to \infty} \int_0^p \frac{\sin \lambda x}{x} \, dx = \frac{\pi}{2},$$

where for a fixed positive P the convergence is uniform for $P \leq p$, and that

$$(21b) \qquad \lim_{\lambda \to \infty} \int_0^p \frac{\sin \lambda x}{2 \sin (x/2)} \, dx = \frac{\pi}{2},$$

where the convergence is uniform for $P \leq p \leq \pi$.

d. Fourier Expansion for the Function $\phi(x) = x$

Our last result leads directly to the Fourier expansion of two related sectionally linear periodic functions $\phi(x)$, $\chi(x)$ defined in the

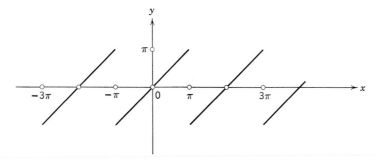

Figure 8.7   The function $\phi(x)$.

interval $-\pi < x < \pi$ by

$$\phi(x) = x$$

and

$$(22) \qquad \chi(x) = \begin{cases} \pi - x & \text{for } x > 0 \\ 0 & \text{for } x = 0 \\ -\pi - x & \text{for } x < 0. \end{cases}$$

(See Figs. 8.7 and 8.8.)

The first function ϕ, periodically extended outside the interval $-\pi < x < +\pi$, has jump discontinuities at the end points, whereas $\chi(x)$ suffers a jump of 2π at $x = 0$. Obviously, the two functions periodically extended are related to each other by

$$\chi(x) = \phi(\pi - x).$$

The Fourier expansion for $\chi(x)$ follows immediately for $0 < x \leq \pi$ from formulas (14), p. 586, and (21*b*), p. 591, for $\lambda = n + \frac{1}{2}$ and $p = x$, by passage to the limit, $n \to \infty$. We find the Fourier series

(23*a*) $$\chi(x) = 2\left(\sin x + \frac{\sin 2x}{2} + \frac{\sin 3x}{3} + \cdots\right).$$

The same holds then also for $-\pi \leq x < 0$, since both sides are odd functions of x. The series is uniformly convergent for $\epsilon < |x| \leq \pi$, with any arbitrarily small, positive value of ϵ. At $x = 0$ all terms of the

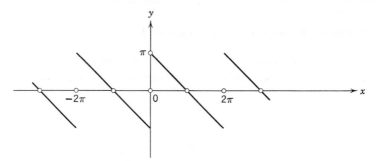

Figure 8.8   The function $\chi(x)$.

series are zero, hence also the sum, in agreement with the definition of $\chi(0)$. Since both sides have period 2π the identity (23*a*) holds then for all x.

That the coefficients of the expansion are indeed the Fourier coefficients defined by formula (17), p. 587, is confirmed easily.

The Fourier expansion for $\phi(x)$ is now obtained directly from $\phi(x) = \chi(\pi - x)$:

(23*b*) $$\phi(x) = 2\sum_{\nu=1}^{\infty}(-1)^{\nu+1}\frac{\sin \nu x}{\nu}$$

$$= 2(\sin x - \tfrac{1}{2}\sin 2x + \tfrac{1}{3}\sin 3x - + \cdots).$$

Here the convergence is uniform as soon as the point x is bounded away from the discontinuity points $x = \pm\pi$ by the condition $|x| < \pi - \epsilon$. For $x = \pi/2$ we obtain again Leibnitz' series

$$\frac{\pi}{2} = 2(1 - \tfrac{1}{3} + \tfrac{1}{5} - \cdots + -).$$

It should be mentioned that the two series for χ and ϕ do not converge absolutely; indeed the absolute values for $x = \pi/2$ form the divergent series

$$2 \sum_{1}^{\infty} \frac{1}{2\nu - 1}.$$

Formula (23b) is remarkable as an example of an infinite series of continuous functions which converges for all x but has as sum a discontinuous function, namely, the piecewise linear function $\phi(x)$. Each partial sum of the series is continuous, since the sum of any finite number of continuous functions must again be continuous. Because a uniformly convergent infinite series of continuous functions has a continuous sum, the Fourier series cannot converge uniformly in a neighborhood of a point x at which ϕ is discontinuous, that is, for $x = \pm\pi, \pm3\pi, \ldots$. Figure 8.4, p. 579 illustrates how the successive partial sums which are trigonometric polynomials and continuous functions approximate the sectionally linear function $\frac{1}{2}\phi(x)$ uniformly in an interval of continuity, but that near the end point the functions change more and more rapidly.

e. The Main Theorem on Fourier Expansion

The Fourier Coefficients. After the preceding preparations, the possibility of expanding a large class of functions can be easily ascertained. The form of such an expansion for a function $f(x)$ with the period 2π is

(24a) $$f(x) = \tfrac{1}{2}a_0 + \sum_{\nu=1}^{\infty}(a_\nu \cos \nu x + b_\nu \sin \nu x),$$

or in complex notation

(24b) $$f(x) = \sum_{\nu=-\infty}^{\infty} \alpha_\nu e^{i\nu x}.$$

We first assume that we have uniformly convergent expansions (24a) or (24b) for the function $f(x)$. We can then determine the coefficients a_ν, b_ν, respectively, α_ν in these expansions by multiplying by $\cos \mu x$, $\sin \mu x$, respectively by $e^{-i\mu x}$, and integrating from $-\pi$ to π, using the orthogonality relations (see pp. 274 and 583)

$$\int_{-\pi}^{\pi} \sin \nu x \sin \mu x \, dx = \int_{-\pi}^{\pi} \cos \nu x \cos \mu x \, dx = \begin{cases} 0 & \text{if } \mu \neq \nu \\ \pi & \text{if } \mu = \nu \neq 0, \end{cases}$$

(25) $$\int_{-\pi}^{\pi} \sin \nu x \cos \mu x \, dx = 0,$$

$$\int_{-\pi}^{\pi} e^{i\nu x} e^{-i\mu x} \, dx = \begin{cases} 0 & \text{if } \mu \neq \nu \\ 2\pi & \text{if } \mu = \nu. \end{cases}$$

Writing t for the variable of integration, we at once obtain the formulas

$$(26a) \quad a_\mu = \frac{1}{\pi} \int_{-\pi}^{\pi} f(t) \cos \mu t \, dt, \qquad b_\mu = \frac{1}{\pi} \int_{-\pi}^{\pi} f(t) \sin \mu t \, dt$$

for $\mu = 0, 1, 2, \ldots$, and

$$(26b) \qquad\qquad \alpha_\mu = \frac{1}{2\pi} \int_{-\pi}^{\pi} f(t) e^{-i\mu t} \, dt$$

for $\mu = 0, \pm 1, \pm 2, \ldots$.

Thus, if $f(x)$ can be expanded at all into a uniformly convergent series $(24a)$ or $(24b)$, then the coefficients can only have the values determined by the formulas $(26a)$ and $(26b)$. But even without a justification for this very tentative procedure, these formulas $(26a)$ or $(26b)$ define sequences of numbers a_ν, b_ν, and α_ν called the *Fourier coefficients* for every function $f(x)$ which is continuous or piecewise continuous in the interval $-\pi \leq x \leq \pi$.

For a given function $f(x)$, we form with the coefficients thus defined by $(26a, b)$ the *Fourier partial sums*

$$S_n(x) = \tfrac{1}{2} a_0 + \sum_{\nu=1}^{n} (a_\nu \cos \nu x + b_\nu \sin \nu x)$$

or

$$S_n(x) = \sum_{\nu=-n}^{\nu=n} \alpha_\nu e^{i\nu}.$$

Our task is to prove that these Fourier sums actually converge for $n \to \infty$ and that the limit is the function $f(x)$.

We now state the

MAIN THEOREM.   *The Fourier series*

$$(27a) \qquad\qquad \tfrac{1}{2} a_0 + \sum_{\nu=1}^{\infty} (a_\nu \cos \nu x + b_\nu \sin \nu x)$$

or

$$(27b) \qquad\qquad \sum_{\nu=-\infty}^{\infty} \alpha_\nu e^{i\nu x}$$

formed with the Fourier coefficients $(26a)$ or $(26b)$ converges to the value $f(x)$ for any sectionally continuous function $f(x)$ of period 2π, which has sectionally continuous derivatives of first and second order.[1] Here the

---

[1] We mention again that this theorem can be proved for much more general classes of functions (see, for example, Section 8.6). The result formulated here, however, amply suffices for most applications.

value of $f(x)$ *at a point of discontinuity must be defined by*

$$(27c) \qquad f(x) = \tfrac{1}{2}[f(x+0) + f(x-0)].$$

PROOF.[1]   For the proof we substitute in the nth "Fourier polynomial"

$$S_n(x) = \tfrac{1}{2}a_0 + \sum_{\nu=1}^{n}(a_\nu \cos \nu x + b_\nu \sin \nu x)$$

the integral expressions (26a) for the coefficients and then interchange the order of integration and summation; we obtain

$$S_n(x) = \frac{1}{\pi} \int_{-\pi}^{\pi} f(t) \left[\tfrac{1}{2} + \sum_{\nu=1}^{n}(\cos \nu t \cos \nu x + \sin \nu t \sin \nu x) \right] dt,$$

or, using the addition theorem for the cosine,

$$S_n(x) = \frac{1}{\pi} \int_{-\pi}^{\pi} f(t) \left[\tfrac{1}{2} + \sum_{\nu=1}^{n} \cos \nu(t - x) \right] dt.$$

By the summation formula (14) of p. 586 therefore,

$$(28) \qquad S_n(x) = \frac{1}{2\pi} \int_{-\pi}^{\pi} f(t) \frac{\sin [(n + \tfrac{1}{2})(t - x)]}{\sin \tfrac{1}{2}(t - x)} dt.$$

Finally, setting $\tau = t - x$ and recalling that periodicity allows us to shift the interval of integration by the quantity x (see p. 574), we obtain

$$(28a) \qquad S_n(x) = \frac{1}{2\pi} \int_{-\pi}^{\pi} f(x + \tau) \frac{\sin (n + \tfrac{1}{2})\tau}{\sin \tfrac{1}{2}\tau} d\tau,$$

where x is, of course, fixed.

We now prove that $S_n(x)$ tends for $n \to \infty$ to $f(x)$; or

$$(29) \qquad \lim_{n \to \infty} S_n(x) = \lim_{n \to \infty} \frac{1}{2\pi} \int_{-\pi}^{\pi} f(x + t) \frac{\sin (n + \tfrac{1}{2})t}{\sin \tfrac{1}{2}t} dt = f(x).$$

Because $f(x) = \tfrac{1}{2}[f(x+0) + f(x-0)]$ for all x, we have [see formula (15), p. 586]

$$S_n(x) - f(x) = \frac{1}{\pi} \int_{0}^{\pi} \frac{[f(x+t) - f(x+0)]}{2 \sin \tfrac{1}{2}t} \sin (n + \tfrac{1}{2})t \, dt$$

$$+ \frac{1}{\pi} \int_{-\pi}^{0} \frac{[f(x+t) - f(x-0)]}{2 \sin \tfrac{1}{2}t} \sin (n + \tfrac{1}{2})t \, dt.$$

---

[1] We give here only the proof for expansion of f in a series (27a). Series (27b) follows then by the substitutions given in Eq. (13b), p. 585.

If we can show now that the functions $[f(x + t) - f(x + 0)]/(2 \sin \frac{1}{2}t)$ and $[f(x + t) - f(x - 0)]/(2 \sin \frac{1}{2}t)$ of the variable t are sectionally continuous together with their first derivatives in the intervals $0 \leq t \leq \pi$ and $-\pi \leq t \leq 0$ respectively, then by our basic lemma (p. 588) both integrals on the right-hand side tend to zero for $n \to \infty$ and formula (29) follows.

Thus the main theorem is proved if we can show that for a fixed x the function of t defined by

$$\phi(t) = \frac{f(x + t) - f(x + 0)}{2 \sin \frac{1}{2}t} \qquad \text{for} \quad 0 < t < \pi$$

$$\phi(t) = \frac{f(x + t) - f(x - 0)}{2 \sin \frac{1}{2}t} \qquad \text{for} \quad -\pi < t < 0$$

is sectionally continuous and has a sectionally continuous first derivative, provided f, f', f'' are sectionally continuous.

To ascertain that these conditions are satisfied for the quotient $\phi(t)$ we first observe that the denominator vanishes only for $t = 0$, and that therefore ϕ and its first derivative are sectionally continuous except possibly near $t = 0$. Only at the singular point $t = 0$ could a loss of differentiability occur. All we have to do, therefore, is to show that $\phi(t)$ and its derivative $\phi'(t)$ approach limits if t tends to zero from positive or negative values respectively. We shall indeed show that these limits exist, and that they have the values

$$\phi(+0) = f'(x + 0), \qquad \phi(-0) = f'(x - 0)$$

respectively

$$\phi'(+0) = \tfrac{1}{2}f''(x + 0), \qquad \phi'(-0) = \tfrac{1}{2}f''(x - 0).$$

For the proof we introduce the function $g(t)$ by $\phi(t) = g(t)h(t)$, where the factor $h(t)$ is defined by

$$h(t) = \frac{t}{2 \sin (t/2)} \qquad \text{for} \quad t \neq 0, \qquad h(0) = 1.$$

We have (see Chapter 5, p. 465) in $h(t)$ a continuous function with a continuous derivative in the whole interval $-\pi \leq t \leq \pi$ with $h(0) = 1$, $h'(0) = 0$; therefore in the limit for $t \to 0$ the values of $g(t)$ and $\phi(t)$ as well as those of $g'(t)$ and $\phi'(t) = gh' + g'h$ coincide.

Now in the interval $0 < t < \pi$ (see Chapter 5, p. 464 for general remarks about indeterminate expressions) by the mean value theorem of calculus

$$g(t) = \frac{f(x+t) - f(x+0)}{t} = f'(x+\xi)$$

with an intermediate value ξ between 0 and t; hence for t, and thus also ξ, tending to zero

$$g(+0) = f'(x+0).$$

For the derivative we obtain

$$g'(t) = \frac{tf'(x+t) + f(x+0) - f(x+t)}{t^2},$$

again an expression where numerator and denominator both tend to zero for $t \to 0$ and have the derivatives $tf''(x+t)$ and $2t$, respectively. To determine the limit for $t \to 0$ we make use of the generalized mean value theorem (cf. p. 222) and find

$$g'(t) = \frac{\eta f''(x+\eta)}{2\eta} = \tfrac{1}{2}f''(x+\eta)$$

with η intermediate between 0 and t. For $t \to 0$ we have $\eta \to 0$, and hence, as said before, $g'(+0) = \phi'(+0) = \tfrac{1}{2}f''(x+0)$.

The same reasoning applies to negative values of t. Consequently, our application of the lemma is justified and the main theorem established.

Again it may be stated that the result obtained is amply sufficient for all needs arising in calculus and its applications. Yet the theoretical interests of mathematicians, starting with the original work of Dirichlet, were frequently aimed at greater generality, that is, at trying to expand functions of a wider class.[1] These efforts have stimulated a more refined analysis of the concepts of function and integral and have led to the development of advanced Fourier analysis as an attractive specialized field, which, however, must remain outside the scope of this book.

---

[1] It might be noted that there are examples of continuous functions which are not expandable in a Fourier series. In addition, there exist examples of functions $f(x)$ represented by a convergent trigonometrical series which, however, are not Fourier series having the expressions (26) as coefficients. Such examples show that for refined investigations a distinction between trigonometric series in general and Fourier series in particular is in order. For us moreover, they illustrate the fact that more restrictive conditions than that of continuity are indeed appropriate, even though the restrictions assumed in our main theorem and in an extension given in Section 8.6 are much more severe than really needed. (See for the general theory, *Trigonometrical Series*, by A. Zygmund, Chelsea Publishing Co., 1952.)

8.5   Examples of Fourier Series

a. Preliminary Remarks

We assume throughout that the period of our functions $f(x)$ is 2π. If $f(x)$ is an even function (cf. p. 29), then clearly $f(x) \sin \nu x$ is odd and $f(x) \cos \nu x$ is even, so that

$$b_\nu = \frac{1}{\pi} \int_{-\pi}^{\pi} f(x) \sin \nu x \, dx = 0,$$

and we obtain a "cosine series." If, on the other hand, the function $f(x)$ is odd, then

$$a_\nu = \frac{1}{\pi} \int_{-\pi}^{\pi} f(x) \cos \nu x \, dx = 0,$$

and we obtain a "sine series."[1]

b. Expansion of the Function $\phi(x) = x^2$

For the even function x^2, we have upon integrating twice by parts

$$a_\nu = \frac{2}{\pi} \int_0^{\pi} x^2 \cos \nu x \, dx = (-1)^\nu \frac{4}{\nu^2}, \qquad (\nu > 0),$$

$$a_0 = \frac{2\pi^2}{3},$$

so that we obtain the expansion

$$(30) \qquad x^2 = \frac{\pi^2}{3} - 4\left(\frac{\cos x}{1^2} - \frac{\cos 2x}{2^2} + \frac{\cos 3x}{3^2} + \cdots \right).$$

Differentiating this series term by term and dividing by 2, we formally recover the series (23b), p. 592, obtained previously for $\phi(x) = x$.

c. Expansion of $x \cos x$

(See Fig. 8.9.) For this odd function we have

$$a_\nu = 0, \quad b_\nu = \frac{2}{\pi} \int_0^{\pi} x \cos x \sin \nu x \, dx.$$

---

[1] Consequently, if the function $f(x)$ is initially given only in the interval $0 < x < \pi$, then we can extend it in the interval $-\pi < x < 0$ either as an odd function or as an even function, and thus for the smaller interval $0 < x < \pi$ either a sine series or a cosine series is obtainable.

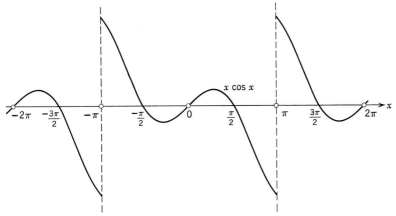

Figure 8.9

Using the formula

$$\int_0^\pi x \sin \mu x \, dx = (-1)^{\mu+1} \frac{\pi}{\mu}, \qquad (\mu = 1, 2, \ldots,),$$

we find

$$b_\nu = \frac{2}{\pi} \int_0^\pi x \cos x \sin \nu x \, dx$$

$$= \frac{1}{\pi} \int_0^\pi x[\sin (\nu + 1)x + \sin (\nu - 1)x] \, dx$$

$$= (-1)^\nu \left(\frac{2\nu}{\nu^2 - 1} \right) \qquad (\nu = 2, 3, \ldots)$$

$$b_1 = -\tfrac{1}{2}.$$

We therefore obtain the series

(31) $$x \cos x = -\tfrac{1}{2} \sin x + 2 \sum_{\nu=2}^{\infty} \frac{(-1)^\nu \nu}{\nu^2 - 1} \sin \nu x.$$

Adding the series (23*b*), p. 592, found for $\phi(x) = x$ yields

(31*a*)

$$x(1 + \cos x) = \tfrac{3}{2} \sin x + 2\left(\frac{\sin 2x}{1 \cdot 2 \cdot 3} - \frac{\sin 3x}{2 \cdot 3 \cdot 4} + \frac{\sin 4x}{3 \cdot 4 \cdot 5} - + \cdots \right).$$

When the function which is equal to $x \cos x$ in the interval $-\pi < x < \pi$ is extended periodically beyond this interval, the same discontinuities (cf. Fig. 8.7) occur as exhibited by the function $\phi(x)$ considered earlier in Section 8.4*d*. On the other hand, the function $x(1 + \cos x)$, periodically extended, remains continuous at the end points of the intervals, and in fact its derivative also remains continuous, since the discontinuities are eliminated by the factor $1 + \cos x$, which together with its derivative vanishes at the end points. This accounts for the fact that the series (31) converges uniformly for all x, as is evident by comparison with the series with constant terms $\dfrac{1}{1^3} + \dfrac{1}{2^3} + \dfrac{1}{3^3} + \cdots$.

d. The Function $f(x) = |x|$

For this even function $b_\nu = 0$, and $a_\nu = \dfrac{2}{\pi} \displaystyle\int_0^\pi x \cos \nu x \, dx$; by integrating by parts we readily obtain

$$\int_0^\pi x \cos \nu x \, dx = \frac{1}{\nu} x \sin \nu x \Big|_0^\pi - \frac{1}{\nu}\int_0^\pi \sin \nu x \, dx$$

$$= \begin{cases} 0, & \text{if } \nu \text{ is even and } \neq 0, \\[2mm] -\dfrac{2}{\nu^2}, & \text{if } \nu \text{ is odd.} \end{cases}$$

Consequently,

$$(32) \qquad |x| = \tfrac{1}{2}\pi - \frac{4}{\pi}\left(\cos x + \frac{\cos 3x}{3^2} + \frac{\cos 5x}{5^2} + \cdots \right).$$

Putting $x = 0$, we obtain the remarkable formula

$$(32a) \qquad \frac{\pi^2}{8} = 1 + \frac{1}{3^2} + \frac{1}{5^2} + \cdots.$$

e. A Piecewise Constant Function

The function defined by the equations

$$f(x) = \operatorname{sgn} x = \begin{cases} -1, & \text{for } -\pi < x < 0, \\[1mm] 0, & \text{for } x = 0, \\[1mm] +1, & \text{for } 0 < x < \pi, \end{cases}$$

as indicated in Fig. 1.22, p. 32, is odd. Hence $a_\nu = 0$ and

$$b_\nu = \frac{2}{\pi}\int_0^\pi \sin \nu x \, dx = \begin{cases} 0 & \text{if } \nu \text{ is even,} \\[2mm] \dfrac{4}{\pi\nu} & \text{if } \nu \text{ is odd,} \end{cases}$$

so that the Fourier series for this function is

(33) $$f(x) = \frac{4}{\pi}\left(\frac{\sin x}{1} + \frac{\sin 3x}{3} + \cdots\right).$$

For $x = \frac{1}{2}\pi$, in particular, this again yields Leibnitz's series.

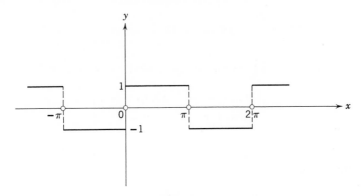

Figure 8.10

The series (33) can be formally derived from that for $|x|$ given in (32), using term-by-term differentiation.

f. The Function $|\sin x|$

The even function $f(x) = |\sin x|$ can be expanded in a cosine series, with the coefficients a_ν given by the following calculations:

$$\frac{1}{2}\pi a_\nu = \int_0^\pi \sin x \cos \nu x \, dx$$

$$= \frac{1}{2}\int_0^\pi [\sin (\nu + 1)x - \sin (\nu - 1)x] \, dx$$

$$= \begin{cases} 0 & \text{if } \nu \text{ is odd,} \\ \dfrac{-2}{\nu^2 - 1} & \text{if } \nu \text{ is even.} \end{cases}$$

We thus obtain, writing 2ν instead of ν,

(34) $$|\sin x| = \frac{2}{\pi} - \frac{4}{\pi}\sum_{\nu=1}^\infty \frac{\cos 2\nu x}{4\nu^2 - 1}.$$

g. Expansion of cos μx. Resolution of the Cotangent into Partial Fractions. The Infinite Product for the Sine

The function $f(x) = \cos \mu x$ for $-\pi < x < \pi$, where μ is not an integer, is even; hence $b_\nu = 0$, whereas

$$\tfrac{1}{2}\pi a_\nu = \int_0^\pi \cos \mu x \cos \nu x \, dx$$

$$= \tfrac{1}{2} \int_0^\pi [\cos (\mu + \nu)x + \cos (\mu - \nu)x] \, dx$$

$$= \tfrac{1}{2} \left[\frac{\sin (\mu + \nu)\pi}{u + \nu} + \frac{\sin (\mu - \nu)\pi}{\mu - \nu} \right]$$

$$= \frac{\mu(-1)^\nu}{\mu^2 - \nu^2} \sin \mu\pi.$$

We thus have

$$(35) \quad \cos \mu x = \frac{2\mu \sin \mu\pi}{\pi} \left(\frac{1}{2\mu^2} - \frac{\cos x}{\mu^2 - 1^2} + \frac{\cos 2x}{\mu^2 - 2^2} + - \cdots \right).$$

This function extended periodically with period 2π from the interval $-\pi < x < \pi$ remains continuous at the points $x = \pm\pi$. Putting $x = \pi$, dividing both sides of the equation by $\sin \mu\pi$, and writing x instead of μ, we obtain the equation

$$(36) \quad \cot \pi x = \frac{2x}{\pi} \left(\frac{1}{2x^2} + \frac{1}{x^2 - 1^2} + \frac{1}{x^2 - 2^2} + \cdots \right).$$

This is the resolution of the cotangent into partial fractions (in analogy to the finite partial fraction resolutions of rational functions discussed in Chapter 3, p. 286), a very important formula of analysis.

We write this series in the form

$$\cot \pi x - \frac{1}{\pi x} = - \frac{2x}{\pi} \left[\frac{1}{1^2 - x^2} + \frac{1}{2^2 - x^2} + \cdots \right].$$

If x lies in an interval $0 \le x \le q < 1$, the nth term on the right is less in absolute value than $2/[\pi(n^2 - q^2)]$. Hence the series converges uniformly in this interval and can be integrated term by term. Multiplying both sides by π and integrating, we obtain

$$\pi \int_0^x \left(\cot \pi t - \frac{1}{\pi t} \right) dt = \log \frac{\sin \pi x}{\pi x} - \lim_{a \to 0} \log \frac{\sin \pi a}{\pi a} = \log \frac{\sin \pi x}{\pi x}$$

on the left and

$$\log\left(1 - \frac{x^2}{1^2}\right) + \log\left(1 - \frac{x^2}{2^2}\right) + \cdots = \lim_{n \to \infty} \sum_{v=1}^{n} \log\left(1 - \frac{x^2}{v^2}\right)$$

on the right. Thus

$$\log\frac{\sin \pi x}{\pi x} = \lim_{n \to \infty} \sum_{v=1}^{n} \log\left(1 - \frac{x^2}{v^2}\right)$$

$$= \lim_{n \to \infty} \log \prod_{v=1}^{n}\left(1 - \frac{x^2}{v^2}\right) = \log \lim_{n \to \infty} \prod_{v=1}^{n}\left(1 - \frac{x^2}{v^2}\right).$$

If we pass from the logarithm to the exponential function we have

$$(36a) \qquad \sin \pi x = \pi x\left(1 - \frac{x^2}{1^2}\right)\left(1 - \frac{x^2}{2^2}\right)\left(1 - \frac{x^2}{3^2}\right) \cdots .$$

We have thus obtained the famous expression for the sine as an infinite product.[1]

From this result, by putting $x = \frac{1}{2}$, we obtain Wallis's product

$$\tfrac{1}{2}\pi = \prod_{v=1}^{\infty} \frac{2v}{2v - 1} \cdot \frac{2v}{2v + 1} = \frac{2}{1} \cdot \frac{2}{3} \cdot \frac{4}{3} \cdot \frac{4}{5} \cdots$$

as derived before on p. 281.

h. Further Examples

By brief calculations similar to the preceding, we obtain further examples of expansions.

The function $f(x)$ defined by the equation $f(x) = \sin \mu x$ for $-\pi < x < \pi$ can be expanded in the series

$$(37) \quad \sin \mu x = -\frac{2 \sin \mu \pi}{\pi}\left(\frac{\sin x}{\mu^2 - 1^2} - \frac{2 \sin 2x}{\mu^2 - 2^2} + \frac{3 \sin 3x}{\mu^2 - 3^2} - + \cdots\right).$$

Putting $x = \frac{1}{2}\pi$ and using the relation $\sin \mu \pi = 2 \sin \frac{1}{2}\mu\pi \cos \frac{1}{2}\mu\pi$ yields the resolution of the secant, that is, of the function $1/\cos \frac{1}{2}\mu\pi$ into partial fractions; this expansion is

$$\pi \sec \pi x = \frac{\pi}{\cos \pi x} = 4 \sum_{v=1}^{\infty} \frac{(-1)^v(2v - 1)}{4x^2 - (2v - 1)^2},$$

where we have written x in place of $\frac{1}{2}\mu$.

---

[1] This formula is particularly interesting because it exhibits directly that the function $\sin \pi x$ vanishes at the points $x = 0, \pm 1, \pm 2, \ldots .$ In this respect it corresponds to the factorization of a polynomial when its zeros are known.

Series analogous to (35) and (37) for the hyperbolic functions $\cosh \mu x$ and $\sinh \mu x \, (-\pi < x < \pi)$ are

$$\cosh \mu x = \frac{2\mu}{\pi} \sinh \mu \pi \left(\frac{1}{2\mu^2} - \frac{\cos x}{\mu^2 + 1^2} + \frac{\cos 2x}{\mu^2 + 2^2} - \frac{\cos 3x}{\mu^2 + 3^2} + \cdots \right),$$

$$\sinh \mu x = \frac{2}{\pi} \sinh \mu \pi \left(\frac{\sin x}{\mu^2 + 1^2} - \frac{2 \sin 2x}{\mu^2 + 2^2} + \frac{3 \sin 3x}{\mu^2 + 3^2} - + \cdots \right).$$

8.6   Further Discussion of Convergence

a. Results

A closer examination of the Fourier coefficients a_ν, b_ν leads easily to the following corollaries to the main theorem of Section 8.4e, p. 593.

(*a*) The Fourier series (27), p. 594, converge to $f(x)$ for all periodic functions under the relaxed condition that $f(x)$ and merely its *first* derivative $f'(x)$ are sectionally continuous or, as we say, that the function is sectionally smooth.

(*b*) If the periodic sectionally smooth function $f(x)$ is continuous, the convergence is absolute and uniform.

(*c*) If the sectionally smooth function $f(x)$ suffers jump discontinuities, the convergence is uniform in each closed interval which does not contain a point of discontinuity.

The proof of (*b*) depends on a simple inequality of Bessel, whereas for the proof of (*a*) and (*c*) the results of Section 8.4d, p. 591, will be used.

b. Bessel's Inequality

This inequality yields bounds for the Fourier coefficients of any piecewise continuous not necessarily differentiable function. It states that

$$(38) \qquad \tfrac{1}{2}a_0^2 + \sum_{\nu=1}^{n}(a_\nu^2 + b_\nu^2) \leq M^2$$

where the bound $M^2 = \dfrac{1}{\pi} \displaystyle\int_{-\pi}^{\pi} f(x)^2 \, dx$ is a number fixed by the function $f(x)$ and depends neither on the individual Fourier coefficients a_ν, b_ν nor the number n. With the complex Fourier coefficients α_ν [see (13a)], p. 585, Bessel's inequality can be immediately written in the form

$$(38a) \qquad \sum_{\nu=-n}^{n} |\alpha_\nu|^2 \leq \frac{1}{2\pi} \int_{-\pi}^{\pi} f(x)^2 \, dx = \tfrac{1}{2}M^2.$$

The inequality is a direct consequence of the obvious fact that

$$\frac{1}{\pi}\int_{-\pi}^{\pi}\left[f(x) - \tfrac{1}{2}a_0 - \sum_{v=1}^{n}(a_v\cos vx + b_v\sin vx)\right]^2 dx \geq 0.$$

We evaluate the integral by expanding the square under the integral sign and observing the orthogonality relations (25), p. 593, as well as the definitions (17), p. 587, of the Fourier coefficients: by integrating the individual terms we immediately obtain Bessel's inequality in the form (38) stated above.

Since the left-hand side of Bessel's inequality increases monotonically with n and the upper bound M^2 is fixed, we can pass to the limit $n \to \infty$ and infer that the inequality

$$(39)\qquad 2\sum_{v=-\infty}^{\infty}|\alpha_v|^2 = \tfrac{1}{2}a_0^2 + \sum_{v=1}^{\infty}(a_v^2 + b_v^2) \leq M^2$$

is valid. The inequality (39) holds for the Fourier coefficients of a piecewise continuous function $f(x)$ even if f should not be represented by the series $(27a)$ or $(27b)$.

Incidentally, we shall show in Section 8.7d that Bessel's inequality (39) remains valid if we replace the inequality sign by that of equality.

*c. Proof of Corollaries (a), (b), and (c)

Assuming $f(x)$ itself to be continuous we apply Bessel's inequality to its piecewise continuous derivative $g(x) = f'(x)$ which has the Fourier coefficients $c_v = +vb_v$, $d_v = -va_v$, as we find immediately using integration by parts (since the integrated terms cancel):

$$c_v = \frac{1}{\pi}\int_{-\pi}^{\pi}f'(x)\cos vx\,dx = +\int_{-\pi}^{\pi}vf(x)\sin vx\,dx = +vb_v,$$

and similarly for d_v. [Here we have made use of the continuity and periodicity of $f(x)$.] We have therefore

$$\sum_{v=1}^{n}v^2(a_v^2 + b_v^2) = \sum_{v=1}^{n}(c_v^2 + d_v^2)$$

$$\leq \frac{1}{\pi}\int_{-\pi}^{\pi}g(x)^2\,dx = \frac{1}{\pi}\int_{-\pi}^{\pi}f'(x)^2\,dx = M^2$$

This result allows us to construct for the Fourier series of $f(x)$ a majorant with constant positive terms, which according to p. 535 assures absolute and uniform convergence as stated in (b). Indeed, we have first for the νth harmonic oscillation by the Cauchy-Schwarz inequality (cf. p. 15)

$$|a_\nu \cos \nu x + b_\nu \sin \nu x|^2 \leq (a_\nu^2 + b_\nu^2)(\cos^2 \nu x + \sin^2 \nu x) = a_\nu^2 + b_\nu^2;$$

then by using the inequality

$$pq \leq \tfrac{1}{2}(p^2 + q^2)$$

for $p = 1/\nu$, $q = \nu\sqrt{a_\nu^2 + b_\nu^2}$, we have for all ν,

$$|a_\nu \cos \nu x + b_\nu \sin \nu x| \leq \frac{1}{\nu}\nu\sqrt{a_\nu^2 + b_\nu^2}$$

$$\leq \frac{1}{2}\left[\frac{1}{\nu^2} + \nu^2(a_\nu^2 + b_\nu^2)\right].$$

Since the sum over ν of the last expression is convergent, we have constructed a majorant. Therefore the Fourier series

$$\tfrac{1}{2}a_0 + \sum_{\nu=1}^{\infty}(a_\nu \cos \nu x + b_\nu \sin \nu x)$$

converges uniformly. It then has a sum $s(x)$ which is a continuous function of x. To show that actually $s(x) = f(x)$ we use an artifice by considering the integrated function

$$F(x) = \int_{-\pi}^{x} (f(t) - \tfrac{1}{2}a_0)\, dt.$$

Clearly, $F(x)$ is continuous for $-\pi \leq x \leq \pi$; moreover, F has the same value at $x = -\pi$ and $x = \pi$, since

$$F(\pi) = \int_{-\pi}^{+\pi} f(t)\, dt - \pi a_0 = 0 = F(-\pi).$$

Hence the periodic extension of F is continuous. Since also the first and second derivatives of F are sectionally continuous, the function F is represented by its Fourier series. By the same argument based on integration by parts as before the Fourier coefficients of F are $-(1/\nu)b_\nu$ and $(1/\nu)a_\nu$ for $\nu \neq 0$, so that

$$F(x) = \tfrac{1}{2}A_0 + \sum_{\nu=1}^{\infty}\frac{1}{\nu}(-b_\nu \cos \nu x + a_\nu \sin \nu x)$$

with some constant coefficient A_0. Now the series obtained by formal term-by-term differentiation is already known to converge uniformly.

Consequently, formal term-by-term differentiation is legitimate (see p. 539), and we obtain the desired relation

$$F'(x) = f(x) - \tfrac{1}{2}a_0 = \sum_{v=1}^{\infty}(a_v \cos vx + b_v \sin vx).$$

To prove the remaining statements for f sectionally continuous and periodic with a sectionally continuous derivative f' we recall that by our previous result they are true for the periodic function $\chi(x)$ of Section 8.4d and hence for the function $\chi(x - \xi)$ which suffers the jump 2π at the point ξ. If now the function $f(x)$ suffers the jumps $\beta_1, \beta_2, \ldots, \beta_m$ at the points $\xi_1, \xi_2, \ldots, \xi_m$, then $f^*(x) = f(x) - \dfrac{1}{2\pi}\sum_{i=1}^{m}\beta_i\chi(x - \xi_i)$ satisfies the conditions of (b) and hence possesses a uniformly convergent Fourier series, thus proving statement (a), (c) for $f(x)$.

d. Order of Magnitude of the Fourier Coefficients. Differentiation of Fourier Series

The preceding discussions of convergence illustrate a general fact: The Fourier coefficients a_v, b_v converge more rapidly to zero as $n \to \infty$, when $f(x)$ is smoother, that is, when more derivatives of the periodic function $f(x)$ are continuous. Correspondingly, the Fourier series converges better as the functions are smoother. We state precisely: If the periodic function $f(x)$ has continuous derivatives up to order k and a piecewise continuous derivative of order $k + 1$, there exists a bound B, depending only on $f(x)$ and k, such that

(40)
$$|a_v|, |b_v| < \frac{B}{v^{k+1}}.$$

The proof is again (see above) almost immediate if we use integration by parts. For brevity we write in complex notation

$$a_v - ib_v = 2\alpha_v$$

and integrate successively by parts until in the integrand the factor $f^{(k+1)}(x)$ appears. Because of the periodicity and continuity of $f(x)$, $f'(x)$, etc., the boundary terms cancel each other and,

$$2\pi\alpha_v = \int_{-\pi}^{\pi} f(x)e^{-ivx}\,dx = -\frac{i}{v}\int_{-\pi}^{\pi} f'(x)e^{-ivx}\,dx$$

$$= \cdots = \left(\frac{-i}{v}\right)^{k+1}\int_{-\pi}^{\pi} f^{(k+1)}(x)e^{-ivx}\,dx.$$

Hence if $\frac{1}{2}B$ is an upper bound for $|f^{(k+1)}(x)|$, then $|\alpha_\nu| \leq \frac{1}{2}B/\nu^{k+1}$, which implies the inequalities (40).

A further remarkable result is that for $k > 2$ the Fourier series can be differentiated term by term $k - 1$ times and then yields the Fourier series for the differentiated function. For the proof we observe that all these differentiated series have the convergent series with $B\sum\limits_{\nu=1}^{\infty} \dfrac{1}{\nu^2}$ as majorant, hence converge absolutely and uniformly themselves (cf. the criteria of Chapter 7, p. 541).

*8.7   Approximation by Trigonometric and Rational Polynomials

a. General Remark on Representations of Functions

In what manner the concept of function should be restricted by demanding the possibility of "explicit expressions" has been a challenging question since the early times of calculus. Functions often are not given analytically, but rather by geometrical or mechanical constructions or by the *geometric* description of their graphs, which could be of a different nature in different intervals.

The discovery of Fourier series in the early nineteenth century was a most illuminating step towards answering the old question; it revealed that indeed "arbitrary" functions, certainly much less restricted than "analytic" ones, can be expressed by convergent Fourier series. Yet even the Fourier series do not cover all continuous functions: as we mentioned without proof, one can define continuous functions for which the Fourier series, formed with the Fourier coefficients, does not converge.

It is all the more remarkable that by giving up the principle of infinite series in which the approximation is achieved by addition of higher order terms only, we can for any continuous function $f(x)$ construct approximating trigonometric or rational polynomials $P_n(x)$ of order n which converge for $n \to \infty$ in a closed interval uniformly to the given function $f(x)$.

b. Weierstrass Approximation Theorem

We prove the following closely related theorems.

(a) If $f(x)$ is a continuous function in a closed interval I, which is contained in the larger interval $-\pi < x < \pi$, then f can in I be uniformly approximated by a trigonometric polynomial of period 2π of sufficiently high order n.

(b) Any function $f(x)$ which is continuous in a closed interval I can be uniformly approximated in I by a polynomial $P(x)$ in x. This statement due to Weierstrass, can be supplemented (see p. 539) by the corollary:

*(c) If $f(x)$ possesses a continuous derivative in I then the approximating polynomials can be so chosen, that the derivative polynomials $P_n{}'(x)$ approximate the derivative $f'(x)$ uniformly.

The proof of (a) is quite direct. We first approximate $f(x)$ by a piecewise linear function whose graph is a polygon $L_n(x)$ inscribed in the graph of $f(x)$ (see Fig. 8.11). Obviously, $L_n(x)$ differs from $f(x)$ absolutely, by less than an arbitrarily small chosen margin $\epsilon/2$, if the

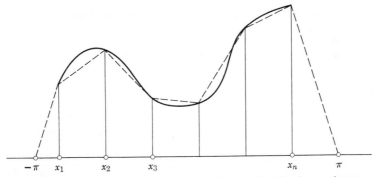

Figure 8.11 Uniform approximation of continuous function by a polygon.

vertices of the polygon are at equally spaced points $x_1, x_2, \ldots, x_n$ and the constant $h = x_{v+1} - x_v$ is chosen sufficiently small, due to the uniform continuity in I of the continuous function f (cf. p. 100).

The next step is to join, as indicated in the figure, the end points $-\pi$ and π of the larger interval by straight lines, and thus extend $L_n(x)$ into a piecewise linear function, again called $L_n(x)$, within the closed interval $-\pi \leq x \leq \pi$: this function, being zero at both end points, can now be extended periodically and, according to Section 8.6a, can be expanded in a uniformly convergent Fourier series whose polynomial section $S_m(x)$ differs from $L_n(x)$ absolutely by less than $\epsilon/2$ if m is sufficiently large. Now $|S_m - f| \leq |S_m - L_n| + |L_n - f| < \epsilon$, and (a) is proved.[1]

To prove (b) we replace in each term of the finite sum $S_m(x)$ according to Section 5.5b, p. 454, the trigonometric functions $\cos vx$ and $\sin vx$ by

---

[1] The same result holds when I is the whole interval $-\pi \leq x \leq +\pi$ if we assume that $f(\pi) = f(-\pi)$. Here we choose an approximating polygon $L_n(x)$ as before, only choosing $L_n(-\pi) = L_n(\pi) = f(-\pi) = f(\pi)$.

Taylor polynomials with a uniformly small remainder; hence, combining these last approximations, we construct a polynomial $P_N(x)$ for which $|P_N(x) - S_m(x)| < \epsilon/2$ where we must choose N large enough to attain the accuracy $\epsilon/2$. Combining, we have certainly in the smaller interval $|P_N(x) - f(x)| < \epsilon$ if m chosen such that $|S_m(x) - f(x)| < \epsilon/2$.

*c. Fejers Trigonometric Approximation of Fourier Polynomials by Arithmetical Means

The theorem (a) of Section 8.7b can be proved very simply by a direct and rather explicit construction of the approximating polynomial, which is provided by the following remarkable theorem of L. Fejer.

THEOREM. *If $S_n(x)$ is the nth Fourier polynomial of a periodic continuous function $f(x)$, then the arithmetical mean*

$$F_n(x) = \frac{S_0(x) + \cdots + S_n(x)}{n + 1}$$

converges uniformly to $f(x)$ for $n \to \infty$.

The theorem guarantees convergence by averaging out whatever disturbing oscillations might occur in the ordinary Fourier approximation.

PROOF. The proof is similar to that of the main theorem of Fourier expansion, but it is simpler because the oscillating kernel $\dfrac{\sin (n + \frac{1}{2})x}{2 \sin \frac{1}{2}x}$ occurring there is replaced here by the positive "Fejer kernel" $s_n(t) = \left(\dfrac{\sin \frac{1}{2}(n + 1)t}{2 \sin \frac{1}{2} t}\right)^2 \cdot \dfrac{2}{n + 1}$. We first note that the function $\sigma_n(\alpha) = \frac{1}{2} + \cos \alpha + \cdots + \cos n\alpha$ of p. 586 can be written in the form

$$\sigma_n(\alpha) = \frac{\sin (n + \frac{1}{2})\alpha}{2 \sin \frac{1}{2}\alpha} = \frac{\sin \frac{1}{2}\alpha \sin (n + \frac{1}{2})\alpha}{2 \sin^2 \frac{1}{2}\alpha}$$

$$= \frac{1}{2} \frac{\cos n\alpha - \cos (n + 1)\alpha}{1 - \cos \alpha},$$

by using the addition formulas for the cosine. We thus obtain the formula

$$\frac{\sigma_0(\alpha) + \sigma_1(\alpha) + \cdots + \sigma_n(\alpha)}{n + 1} = \frac{1}{2(n + 1)} \frac{1 - \cos (n + 1)\alpha}{1 - \cos \alpha}$$

$$= \frac{1}{2(n + 1)} \left(\frac{\sin [(n + 1)\alpha/2]}{\sin (\alpha/2)}\right)^2$$

$$= s_n(\alpha).$$

Since by the definition of the $\sigma_n(\alpha)$, [see (14), p. 586]

$$\frac{1}{\pi} \int_{-\pi}^{\pi} \sigma_k(\alpha) \, d\alpha = 1,$$

it follows that

$$\frac{1}{\pi} \int_{-\pi}^{\pi} s_n(\alpha) \, d\alpha = 1.$$

Now [see (28a), p. 595]

$$S_n(x) = \frac{1}{\pi} \int_{-\pi}^{\pi} f(x + t)\sigma_n(t) \, dt$$

and hence

$$F_n(x) = \frac{1}{\pi(n+1)} \int_{-\pi}^{\pi} f(x+t)[\sigma_0(t) + \cdots + \sigma_n(t)] \, dt$$

$$= \frac{1}{\pi} \int_{-\pi}^{\pi} f(x+t)s_n(t) \, dt.$$

For any positive δ

$$f(x) - F_n(x) = \frac{1}{\pi} \int_{-\pi}^{\pi} [f(x) - f(x+t)]s_n(t) \, dt$$

$$= \frac{1}{\pi} \int_{-\delta}^{\delta} [f(x) - f(x+t)]s_n(t) \, dt$$

$$+ \frac{1}{\pi} \int_{\delta}^{\pi} [f(x) - f(x+t)]s_n(t) \, dt$$

$$+ \frac{1}{\pi} \int_{-\pi}^{-\delta} [f(x) - f(x+t)]s_n(t) \, dt.$$

Now for $f(x)$ continuous the continuity is uniform and we can choose a δ such that $|f(x) - f(x+t)| < \frac{1}{3}\epsilon$ for all x in $[-\pi, \pi]$ and for $|t| < \delta$. Moreover f is bounded, say $|f| < M$. Since from its definition

$$|s_n(t)| \le \frac{1}{2(n+1) \sin^2 (\delta/2)}, \qquad \text{for} \quad \delta \le |t| \le \pi$$

we find using $s_n \ge 0$ that

$$|f(x) - F_n(x)| \le \frac{\epsilon}{3\pi} \int_{-\delta}^{\delta} |s_n(t)| \, dt + \frac{2\pi}{\pi} \int_{\delta}^{\pi} |s_n(t)| \, dt + \frac{2\pi}{\pi} \int_{-\pi}^{-\delta} |s_n(t)| \, dt$$

$$\le \frac{\epsilon}{3\pi} \int_{-\pi}^{\pi} s_n(t) \, dt + \frac{2\pi}{\pi} \cdot \frac{2\pi}{2(n+1) \sin^2 (\delta/2)}$$

$$= \frac{\epsilon}{3} + \frac{2\pi}{(n+1) \sin^2 (\delta/2)}.$$

Clearly,

$$|f(x) - F_n(x)| \leq \epsilon$$

for n sufficiently large, and the theorem is proved.

*d. Approximation in the Mean and Parseval's Relation

The proximity of two functions $g(x)$ and $h(x)$ in a closed interval I in which they are continuous can be measured with a view to uniform convergence by the maximum value of $|g(x) - h(x)|$. Calling the maximum absolute value of a continuous function $\phi(x)$ in I its *maximum norm*, we can express the uniform convergence of a sequence of functions f_n to a function f as $n \to \infty$ by saying that the maximum norm of the difference $f - f_n$ or also of $f_n - f_m$ tends to zero.

For Fourier approximations (as well as for other important mathematical theories outside the scope of this volume) it is natural to consider another measure or "norm" for the deviation between two functions, or what is sufficient, for the "distance" of a function $\phi(x)$ from the function identically zero. This is the "quadratic mean" or the "*mean square norm*" $\mu = \|\phi\|$ defined by an average value

$$\mu^2 = \frac{1}{l} \int_I \phi(x)^2 \, dx = \|\phi\|^2,$$

where l is the length of the interval I. It is a cruder measure than the maximum norm insofar as its smallness does not mean necessarily that the function is small everywhere.

As an example, the norm of x^n over the interval I: $0 \leq x \leq 1$ has the value $(2n + 1)^{-\frac{1}{2}}$ which can be made arbitrarily small by choosing n sufficiently large whereas the function x^n is equal to 1 for $x = 1$.

If the quadratic norm $\|f_n - f\|$ tends to zero as $n \to \infty$, then we say that f_n tends to f in the quadratic mean.

The quadratic norm can be usefully denoted as "distance" because of the so-called triangle inequality, corresponding to that valid for numbers (see p. 14). This inequality $\|f + g\| \leq \|f\| + \|g\|$ for two functions f and g follows immediately: applying the inequality $pq \leq \frac{1}{2}(p^2 + q^2)$ with $p = \dfrac{f(x)}{\|f\|}, q = \dfrac{g(x)}{\|g\|}$, and integrating over I, we find

$$\frac{1}{l} \int_I f(x)g(x) \, dx \leq \|f\| \cdot \|g\|.$$

Now

$$\| f + g \|^2 = \frac{1}{l} \int_I [f(x) + g(x)]^2 \, dx$$

$$= \| f \|^2 + \| g \|^2 + \frac{1}{l} \int_I 2f(x)g(x) \, dx$$

$$\leq (\| f \| + \| g \|)^2$$

or

$$\| f + g \| \leq \| f \| + \| g \|.$$

With these concepts we may illuminate Bessel's inequality of Section 8.6b. We show first: The closest approximation in the mean to a given piecewise continuous function $f(x)$ by a trigonometric polynomial

$$T_n = \frac{c_0}{2} + \sum_{v=1}^{n} c_v \cos vx + d_v \sin vx$$

$$= \sum_{v=-n}^{n} \beta_v e^{ivx},$$

of order n [with $\beta_0 = c_0/2$, $\beta_v + \beta_{-v} = c_v$, $i(\beta_v - \beta_{-v}) = d_v$,] with freedom of the choice of the coefficients c_v, d_v is given by the Fourier polynomial

$$S_n = \frac{a_0}{2} + \sum_{v=1}^{n} a_v \cos vx + b_v \sin vx$$

$$= \sum_{v=-n}^{n} \alpha_v e^{ivx},$$

where a_v, b_v, and α_v are the real and complex Fourier coefficients, determined from f by the formulas (26), p. 594, respectively.

The proof, written in complex notation for brevity, is easily obtained using the orthogonality relations (25) in the interval $I = [-\pi, \pi]$ for the functions e^{ivx}:

$$\frac{1}{l} \int_I \left(f(x) - \sum_{v=-n}^{n} \beta_v e^{ivx} \right)^2 dx$$

$$= \frac{1}{l} \int_I \left[f(x)^2 - 2 \sum_{v=-n}^{n} \beta_v f(x) e^{ivx} + \left(\sum_{v=-n}^{n} \beta_v e^{ivx} \right)^2 \right] dx$$

$$= \| f \|^2 - 2 \sum_{v=-n}^{n} \beta_v \alpha_{-v} + \sum_{v=-n}^{n} \beta_v \beta_{-v}$$

$$= \| f \|^2 - \sum_{v=-n}^{n} \alpha_v \alpha_{-v} + \sum_{v=-n}^{n} (\alpha_v - \beta_v)(\alpha_{-v} - \beta_{-v})$$

$$= \| f \|^2 - \sum_{v=-n}^{n} \alpha_v \bar{\alpha}_v + \sum_{v=-n}^{n} (\alpha_v - \beta_v)(\bar{\alpha}_v - \bar{\beta}_v)$$

$$= \| f \|^2 - \sum_{v=-n}^{n} |\alpha_v|^2 + \sum_{v=-n}^{n} |\alpha_v - \beta_v|^2;$$

clearly, this last expression is minimized when the β_v are chosen as the Fourier coefficients α_v; that is, for $\alpha_v = \beta_v$, or equivalently, $c_v = a_v, d_v = b_v$.

We can now prove *Parseval's theorem*, using the approximation results obtained above.

Bessel's inequality

$$\tfrac{1}{2}a_0^2 + \sum_{v=1}^{n}(a_v^2 + b_v^2) \le \frac{1}{\pi}\int_{-\pi}^{\pi} f(x)^2 \, dx$$

for $n \to \infty$ becomes Parseval's equality

$$\tfrac{1}{2}a_0^2 + \sum_{v=1}^{\infty}(a_v^2 + b_v^2) = \frac{1}{\pi}\int_{-\pi}^{\pi} f(x)^2 \, dx$$

for any function $f(x)$ of period 2π which is continuous for all x.

PROOF.   By the Weierstrass approximation theorem for trigonometrical polynomials we may choose a sequence of polynomials T_n such that $f(x) - T_n(x) \to 0$ uniformly in x. Then also

$$\frac{1}{2\pi}\int_{-\pi}^{\pi}[f(x) - T_n(x)]^2 \, dx \to 0 \qquad \text{as } n \to \infty.$$

However, according to our last result, the Fourier polynomial

$$S_n(x) = \frac{a_0}{2} + \sum_{v=1}^{n}(a_v \cos vx + b_v \sin vx)$$

yields the closest approximation to $f(x)$ in the mean among all nth order Fourier polynomials, so that

$$\frac{1}{2\pi}\int_{-\pi}^{\pi}[f(x) - S_n(x)]^2 \, dx \le \frac{1}{2\pi}\int_{-\pi}^{\pi}[f(x) - T_n(x)]^2 \, dx.$$

It follows that

$$\lim_{n \to \infty}\frac{1}{2\pi}\int_{-\pi}^{\pi}[f(x) - S_n(x)]^2 \, dx = 0.$$

On squaring the integrand as on p. 613, we obtain the Parseval relation.

Finally, we remark that Parseval's relation remains valid even if $f(x)$ has a number of jump discontinuities. The simple proof is omitted here.

Appendix I

*A.I.1   Stretching of the Period Interval.   Fourier's Integral Theorem

The base interval $-\pi \le x \le \pi$ for our periodic functions could be replaced by any interval $-B \le x \le B$. By the transformation

$y = \pi x / B$ this interval of length $2B$ is transformed into the interval $-\pi \leq y \leq \pi$, and a function $f(x)$ with the period $2B$ is transformed into a function $g(y) = f(By/\pi) = f(x)$ with period 2π. The main theorem, written in the complex form [see formula (27b), p. 594], implies

$$g(y) = \sum_{v=-\infty}^{\infty} \frac{1}{2\pi} \int_{-\pi}^{\pi} g(t)e^{-iv(t-y)}\, dt$$

and therefore by this transformation

(41) $$f(x) = \frac{1}{2\pi} \sum_{v=-\infty}^{\infty} \frac{\pi}{B} \int_{-B}^{B} f(s)e^{-iv\pi(s-x)/B}\, ds,$$

where the variable of integration is replaced by $s = Bt/\pi$.

The relation (41) is valid for every function piecewise smooth in $-B \leq x \leq B$.

We set $\pi/B = h$, $v\pi/B = vh = u_v$ and write (41) in the form

$$f(x) = \frac{1}{2\pi} \sum_{v=-\infty}^{\infty} h \int_{-B}^{B} f(s)e^{-iu_v(s-x)}\, ds$$

$$= \frac{1}{2\pi} \sum_{v=-\infty}^{\infty} h e^{iu_v x}\, H_v$$

with $H_v = \int_{-B}^{B} e^{-iu_v s} f(s)ds$. Now the formal passage to the limit for $B \to \infty$ or $\Delta u = h \to 0$ is obvious and yields

(42) $$f(x) = \frac{1}{2\pi} \int_{-\infty}^{\infty} e^{iux}\, du \int_{-\infty}^{\infty} e^{-ius} f(s)\, ds.$$

This is Fourier's integral formula which will be proved rigorously in Volume II for a large class of functions f. This formula can be written in a clearer symmetric form as a pair of reciprocal integral relations between a function $f(x)$ and its "Fourier transform" $F(u)$:

(43) $$F(u) = \frac{1}{\sqrt{2\pi}} \int_{-\infty}^{\infty} f(s)e^{-ius}\, ds$$

(43a) $$f(x) = \frac{1}{\sqrt{2\pi}} \int_{-\infty}^{\infty} F(u)e^{iux}\, dx.$$

Fourier's integral formula (42) can be written in a form which does not involve the use of imaginary exponents. We only have to make use of the expressions

$$e^{iux}e^{-ius} = e^{iu(x-s)} = \cos u(s-x) - i \sin u(s-x).$$

Since $\sin u(s - x)$ is an odd function of u and $\cos u(s - x)$ an even function, integration with respect to u from $-\infty$ to $+\infty$ of the sine term makes no contribution, whereas integrating the cosine term yields twice the value obtained from integrating from 0 to ∞. Hence

$$(43b) \qquad f(x) = \frac{1}{\pi} \int_0^\infty du \int_{-\infty}^{+\infty} f(s) \cos u(s - x)\, ds.$$

*A.I.2   Gibb's Phenomenon at Points of Discontinuity

The nature of the convergence of the Fourier series in the vicinity of a jump discontinuity exhibits a remarkable feature which Gibbs discovered by examining the graphs of the Fourier polynomials

$$S_n(x) = \frac{a_0}{2} + \sum_{v=1}^{n} (a_v \cos vx + b_v \sin vx).$$

As already emphasized in Chapter 7, p. 530 the nonuniform convergence of a convergent sequence in the vicinity of a discontinuity of the limit function can be visualized by the way the continuous graphs of the approximating functions fail to approach the discontinuous graph of the limit function.

In Fourier expansions these graphs do not simply approach the graph of $f(x)$ supplemented by vertical connecting segments $x = \xi$ joining the two end points at the jump position ξ. Instead, the graphs of S_n show waves which near ξ exceed the ordinates $f(\xi + 0)$ and $f(\xi - 0)$ to either side by about 9% of the total height of the jump. Thus the approximating graphs do approximate the graph of $f(x)$ augmented by a vertical line segment at $x = \xi$, not only connecting the two points on the graph of $f(x)$ but overshooting this connecting line segment at both ends; see Figs. 8.4 and 8.5, pp. 579 and 580.

The mathematical analysis of this situation is simple and need be discussed only for the discontinuity of the function $\chi(x)$ of Section 8.4d to which all jump discontinuities were reduced on p. 607.

The function $\frac{1}{2}\chi(x)$ for positive x is [see formula (23a), p. 592], given by

$$\tfrac{1}{2}\chi(x) = \tfrac{1}{2}(\pi - x) = \sum_{v=1}^{\infty} \frac{\sin vx}{v}, \qquad 0 < x < \pi.$$

By integration of formula (14), p. 586, we find that

$$S_n(x) = \sum_{v=1}^{n} \frac{\sin vx}{v} = -\tfrac{1}{2}x + \int_0^x \frac{\sin (n + \tfrac{1}{2})t}{2 \sin \tfrac{1}{2}t}\, dt$$

Hence the remainder $r_n(x) = \frac{1}{2}\chi(x) - S_n(x)$ takes the form

$$r_n(x) = \frac{1}{2}\pi - \int_0^x \frac{\sin (n + \frac{1}{2})t}{t} \, dt + \rho_n(x),$$

where

$$\rho_n(x) = \int_0^x \frac{2 \sin \frac{1}{2}t - t}{2t \sin \frac{1}{2}t} \sin (n + \frac{1}{2})t \, dt.$$

Since the expression $(2 \sin \frac{1}{2}t - t)/2t \sin \frac{1}{2}t$ is sectionally continuous and has a sectionally continuous first derivative the lemma on p. 588 implies that $\rho_n(x)$ for $n \to \infty$ tends to zero uniformly for $0 < x < \pi$. Moreover,

$$\sigma_n(x) = \frac{1}{2}\pi - \int_0^x \frac{\sin (n + \frac{1}{2})t}{t} \, dt = \frac{1}{2}\pi - \int_0^{(n+\frac{1}{2})x} \frac{\sin t}{t} \, dt$$

tends to zero for each individual positive x as $n \to \infty$ (see p. 589). The convergence, however, is not uniform. Clearly, the derivative of $\sigma_n(x)$ vanishes at the points $x_k = 2k\pi/(2n + 1)$ for $k = 1, 2, 3, \ldots$. It is easily seen that more precisely $\sigma_n(x)$ has minima at the points $x_1, x_3, x_5, \ldots$ and maxima at $x_2, x_4, \ldots$. Moreover, the values of σ_n at the minimum points form an increasing sequence. Thus $\sigma_n(x)$ has as its "absolute" minimum for positive x the value

$$\sigma_n(x_1) = \frac{1}{2}\pi - \int_0^\pi \frac{\sin t}{t} \, dt$$

$$= \frac{1}{2}\pi - \int_0^\pi \left(1 - \frac{1}{3!}t^2 + \frac{1}{5!}t^4 - \cdots\right) dt$$

$$= \pi\left(\frac{1}{2} - 1 + \frac{\pi^2}{2 \cdot 3 \cdot 3} - \frac{\pi^4}{2 \cdot 3 \cdot 4 \cdot 5 \cdot 5}\right.$$

$$\left. + \frac{\pi^6}{2 \cdot 3 \cdot 4 \cdot 5 \cdot 6 \cdot 7 \cdot 7} - \cdots\right)$$

$$\approx -0.090 \cdots \pi.$$

For large n the remainder r_n is approximately equal to σ_n. Hence for large n the approximating polynomial S_n exceeds the function χ by about $(9/100)\pi$, that is, by about 9% of the difference of the limiting values of the function at the origin from the right and left. Thus the oscillating branches of the graph of $S_n(x)$ indeed overshoot the height of the graph of $\chi(x)$ and exhibit the limit phenomenon described above.

It is easily seen that the Fejer mean values of the sums $S_n(x)$ are free from Gibb's phenomenon.

*A.I.3   Integration of Fourier Series

In general, as we have seen (p. 536), an infinite series can be integrated term by term if it is uniformly convergent. However, for Fourier series, we have the remarkable result that termwise integration is always possible. We state: *If $f(x)$ is a sectionally continuous function in $-\pi \le x \le \pi$ having the formal Fourier expansion*

$$\tfrac{1}{2}a_0 + \sum_{v=1}^{\infty}(a_v \cos vx + b_v \sin vx),$$

then for any two points x_1, x_2,

$$\int_{x_1}^{x_2} f(x)\,dx = \int_{x_1}^{x_2} \tfrac{1}{2}a_0\,dx + \sum_{v=1}^{\infty}\int_{x_1}^{x_2}(a_v \cos vx + b_v \sin vx)\,dx,$$

or the Fourier series can be integrated termwise. Moreover, the series on the right converges uniformly in x_2 for fixed x_1.

The remarkable part of this theorem is that not only do we not need to assume the uniform convergence of the series but also we do not even need to make use of its convergence.

To prove the theorem, define as on p. 606

$$F(x) = \int_{-\pi}^{x} [f(t) - \tfrac{1}{2}a_0]\,dt.$$

$F(x)$ is continuous and has a sectionally continuous derivative; moreover, it satisfies the condition $F(\pi) = F(-\pi) = 0$, so that it stays continuous when it is periodically extended. Thus the Fourier series

$$\tfrac{1}{2}A_0 + \sum_{v=1}^{\infty}(A_v \cos vx + B_v \sin vx)$$

of $F(x)$ converges uniformly to $F(x)$. Using integration by parts, we obtain for $v \ne 0$, the values

$$A_v = \frac{1}{\pi}\int_{-\pi}^{\pi} F(t)\cos vt\,dt = -\frac{1}{\pi}\int_{-\pi}^{\pi} f(t)\frac{\sin vt}{v}\,dt = -\frac{b_v}{v},$$

$$B_v = \frac{1}{\pi}\int_{-\pi}^{\pi} F(t)\sin vt\,dt = \frac{1}{\pi}\int_{-\pi}^{\pi} f(t)\frac{\cos vt}{v}\,dt = \frac{a_v}{v},$$

for the Fourier coefficients. Therefore the series

$$F(x_2) - F(x_1) = \sum_{v=1}^{\infty}[A_v(\cos vx_2 - \cos vx_1) + B_v(\sin vx_2 - \sin vx_1)]$$

$$= \sum_{v=1}^{\infty}\left[-\frac{b_v}{v}(\cos vx_2 - \cos vx_1) + \frac{a_v}{v}(\sin vx_2 - \sin vx_1)\right]$$

converges uniformly in x. Replacing $F(x)$ by $\int_{\pi}^{x} [f(x) - \frac{1}{2}a_0] \, dx$, we obtain the relation

$$\int_{x_1}^{x_2} [f(x) - \frac{1}{2}a_0] \, dx = \sum_{\nu=1}^{\infty} \int_{x_1}^{x_2} (a_\nu \cos \nu x + b_\nu \sin \nu x) \, dx$$

as was asserted.

Appendix II

*A.II.1 Bernoulli Polynomials and Their Applications

a. Definition and Fourier Expansion

In the derivation of the Taylor series (p. 450) the polynomials $P_n(x) = (x - \xi)^n/n!$, $n \geq 1$ in x with parameter ξ played a role. The sequence of these polynomials is characterized by the conditions that every polynomial P_{n+1} is a primitive function of P_n, that is, $P'_{n+1}(x) = P_n(x)$, and moreover, $P_n(\xi) = 0$ and $P_0(x) \equiv 1$.

We now construct another remarkable sequence of polynomials, by successive integration, the *Bernoulli polynomials*, which we shall then extend as periodic functions and expand in Fourier series.

The Bernoulli polynomials $\phi_n(x)$, for $0 \leq x \leq 1$, are recursively defined by the following relations:

(44a) $\phi_n'(x) = \phi_{n-1}(x)$, $\phi_0(x) = 1$

(44b) $\int_0^1 \phi_n(x) \, dx = 0$,   for $n > 0$.

For known $\phi_0, \phi_1, \ldots, \phi_n$ condition (44a) determines ϕ_n within an arbitrary constant of integration; this constant is then completely fixed by the condition (44b). We see immediately by induction that ϕ_n is a polynomial of the nth order with coefficients that are rational numbers. The first Bernoulli polynomials are easily calculated:

$$\phi_0(x) = 1,$$
$$\phi_1(x) = x - \tfrac{1}{2},$$
$$\phi_2(x) = \tfrac{1}{2}x^2 - \tfrac{1}{2}x + \tfrac{1}{12},$$
$$\phi_3(x) = \tfrac{1}{6}x^3 - \tfrac{1}{4}x^2 + \tfrac{1}{12}x,$$
$$\phi_4(x) = \tfrac{1}{24}x^4 - \tfrac{1}{12}x^3 + \tfrac{1}{24}x^2 - \tfrac{1}{720}.$$

For $n > 1$, we have by (44a, b)

$$\phi_n(1) - \phi_n(0) = \int_0^1 \phi_n'(t) \, dt = 0.$$

Therefore the polynomials ϕ_n may be extended from the basic interval $0 \leq x \leq 1$ to all x as continuous periodic functions $\psi_n(x)$ with the period 1, the so-called Bernoulli functions, whereas the function $\psi_1(x)$ coincides with the discontinuous function $\dfrac{1}{2\pi} \phi(2\pi x - \pi)$ and [see formula (23b), p. 592] can be represented as a Fourier series

$$(45a) \qquad \psi_1(x) = -\frac{1}{\pi} \left(\frac{\sin 2\pi x}{1} + \frac{\sin 4\pi x}{2} + \frac{\sin 6\pi x}{3} + \cdots \right).$$

By means of successive integration, we obtain then

$$(45b) \qquad \psi_n(t) = (-1)^{(n/2)+1} \cdot \frac{2}{(2\pi)^n} \sum_{k=1}^{\infty} \frac{\cos 2\pi k t}{k^n}, \qquad \text{for even } n,$$

$$(45c) \qquad \psi_n(t) = (-1)^{(n+1)/2} \cdot \frac{2}{(2\pi)^n} \sum_{k=1}^{\infty} \frac{\sin 2\pi k t}{k^n}, \qquad \text{for odd } n.$$

In the original interval $0 \leq x \leq 1$ the periodic functions $\psi_n(t)$ are identical with the Bernoulli polynomials $\phi_n(t)$.

For n, even ψ_n is an even function, for n odd ψ_n is odd; equivalently

$$(45d) \qquad\qquad\qquad \psi_n(-x) = (-1)^n \psi_n(x).$$

The constant terms in the successive Bernoulli polynomials form a noteworthy sequence of rational numbers

$$(46a) \qquad\qquad\qquad b_n = \phi_n(0).$$

$$= \begin{cases} \psi_n(0) \text{ for } n \neq 1, \\ -\tfrac{1}{2} \text{ for } n = 1. \end{cases}$$

We obtain immediately from the Fourier expansion

$$(46b) \qquad\qquad\qquad b_n = 0 \qquad \text{for odd } n = 3, 5, \ldots,$$

$$(46c) \quad b_n = (-1)^{(n/2)+1} \cdot \frac{2}{(2\pi)^n} \sum_{k=1}^{\infty} \frac{1}{k^n}, \qquad \text{for even } n = 2, 4, \ldots.$$

Furthermore, evidently for even $n = 2m$, the signs of b_{2m} alternate.

In place of the numbers b_n which decrease rapidly with increasing n, Jacob Bernoulli introduced the following somewhat more suitable numbers:

$$(47) \qquad\qquad\qquad B_m = (-1)^{m-1}(2m)!\, b_{2m},$$

which we call the *Bernoulli numbers*. (That the numbers $B_{2m}{}^* = (-1)^{m-1} B_m$ are identical with the Bernoulli numbers introduced on

p. 562 will become apparent later on.) In particular,

$$B_1 = \frac{1}{6}, \qquad B_2 = \frac{1}{30}, \qquad B_3 = \frac{1}{42}, \qquad B_4 = \frac{1}{30},$$

$$B_5 = \frac{5}{66}, \qquad B_6 = \frac{691}{2730}, \qquad B_7 = \frac{7}{6}, \ldots$$

As a consequence of formula (46c), we have in

$$(48) \qquad \sum_{k=1}^{\infty} \frac{1}{k^{2n}} = (-1)^{n-1}(2\pi)^{2n}\tfrac{1}{2}b_{2n} = \frac{(2\pi)^{2n}}{2(2n)!} B_n$$

an explicit representation of Riemann's ζ-function $\zeta(s)$ for integers $s = 2n$ (see p. 560) by known numbers. For example, we obtain such striking formulas as

$$1 + \frac{1}{2^2} + \frac{1}{3^2} + \frac{1}{4^2} + \cdots = \frac{\pi^2}{6} = \zeta(2)$$

and

$$1 + \frac{1}{2^4} + \frac{1}{3^4} + \frac{1}{4^4} + \cdots = \frac{\pi^4}{90} = \zeta(4).$$

As $n \to \infty$, the numbers b_n and B_n tend to zero and infinity, respectively. For, first of all, we have

$$1 < \sum_{k=1}^{\infty} \frac{1}{k^{2n}} \leq \sum_{k=1}^{\infty} \frac{1}{k^2} = \frac{\pi^2}{6} < 2.$$

Therefore

$$2(2\pi)^{-2n} < |b_{2n}| < 4(2\pi)^{-2n}.$$

Since $2\pi > 1$ and $(2\pi)^{-2n} \to 0$, when $n \to \infty$, we have $b_{2n} \to 0$, whereas $b_{2n+1} = 0$. Furthermore,

$$B_n = (2n)! \, |b_{2n}| > 2(2n)! \, (2\pi)^{-2n};$$

as is seen easily, the right-hand side tends to infinity.

*b. Generating Function; the Taylor Series of the Trigonometric and Hyperbolic Cotangent

The Bernoulli numbers and polynomials lead in an elegant manner to the Taylor expansion of the cotangent and related functions. These expansions follow most easily by means of the so-called *generating function* of the Bernoulli functions, namely, the function

$$(49) \qquad\qquad F(t, z) = \sum_{n=0}^{\infty} \psi_n(t) z^n.$$

This is the power series in z whose coefficients are the Bernoulli functions of the parameter t. On the basis of the Fourier expansion of Eqns. (45) we have the estimate

$$|\psi_n(t)| \leq \left[\frac{2}{(2\pi)^n}\right] \sum_{k=1}^{\infty} \frac{1}{k^n} \leq \left[\frac{2}{(2\pi)^n}\right] \sum_{k=1}^{\infty} \frac{1}{k^2}$$

$$= \frac{\pi^2}{3(2\pi)^n} < \frac{4}{(2\pi)^n}$$

for all t, and $n \geq 2$; hence the absolute value of the nth term of the series for $F(t, z)$ is less than $4(|z|/2\pi)^n$. Thus for all t the radius of convergence of the power series in z is at least 2π, as one sees by comparison with the series

$$4 \sum_{m=1}^{\infty} \left(\frac{|z|}{2\pi}\right)^n$$

Since for a fixed z with $|z| < 2\pi$ the series for $F(z, t)$ has a convergent majorant series, independent of t, it follows from the general theory (see p. 535) that the series converges uniformly for all t. Thus it can be integrated termwise in this domain; it can also be differentiated termwise if the resulting series is also uniformly convergent. We use this fact to determine an explicit formula for $F(t, z)$ (see p. 539). Termwise differentiation with respect to t yields formally for $0 < t < 1$ (for $t = 0$ or $t = 1$, $\psi_1(t)$ has no derivative).

$$\frac{d}{dt} F(t, z) = \sum_{n=1}^{\infty} \psi_{n-1}(t) z^n$$

$$= z \sum_{n=1}^{\infty} \psi_{n-1}(t) z^{n-1}$$

$$= z \sum_{n=0}^{\infty} \psi_n(t) z^n$$

$$= z F(t, z).$$

This series has the same form as the original and is certainly uniformly convergent, so that the termwise differentiation was justified. Hence for every fixed z with $|z| < 2\pi$ and for $0 < t < 1$, the generating function $F(t, z)$ obeys the differential equation $dF/dt = zF(t, z)$. The general solution of this differentiated equation is $F = ce^{zt}$, where c is a factor whose value depends on z as parameter (see p. 223). To determine c, we integrate the series for $F(t, z)$ with respect to t between 0 and 1:

$$\int_0^1 F(t, z)\, dt = c \int_0^1 e^{zt}\, dt$$

$$= c\, \frac{e^z - 1}{z}$$

$$= \int_0^1 \sum_{n=0}^{\infty} z^n \psi_n(t)\, dt$$

$$= 1 + \sum_{n=1}^{\infty} z^n \int_0^1 \psi_n(t)\, dt = 1.$$

Consequently, $c = z/(e^z - 1)$ and so we obtain the final results

$$(50) \qquad\qquad F(t, z) = \frac{z e^{zt}}{e^z - 1}.$$

Letting $t \to 0$ in this expression, we obtain the Taylor series for the function $z/(e^z - 1)$:

$$\lim_{t \to 0} F(t, z) = \frac{z}{e^z - 1} = 1 + \sum_{n=1}^{\infty} b_n z^n.$$

Since $b_1 = -\frac{1}{2}$, adding $\frac{1}{2}z$ to both sides yields

$$(51) \qquad\qquad \frac{z}{e^z - 1} + \frac{z}{2} = 1 + \sum_{n=2}^{\infty} b_n z^n.$$

Incidentally, this formula shows that the numbers $B_n{}^* = n!\, b_n$ are the Bernoulli numbers introduced on p. 562. Since $b_0 = 1$ and $b_n = 0$ for odd n, we have

$$\frac{e^z + 1}{e^z - 1} \cdot \frac{z}{2} = \frac{z}{2} \cdot \frac{e^{z/2} + e^{-z/2}}{e^{z/2} - e^{-z/2}}$$

$$(52) \qquad\qquad = \frac{z}{2} \cdot \frac{2 \cosh \frac{1}{2}z}{2 \sinh \frac{1}{2}z}$$

$$= \tfrac{1}{2}z \coth \tfrac{1}{2}z = \sum_{n=0}^{\infty} b_{2n} z^{2n} = \sum \frac{B_{2n}^*}{(2n)!}\, z^{2n}.$$

Thus we obtain the Taylor series for the hyperbolic cotangent already given on p. 563; the Taylor coefficients are simply related to the Bernoulli numbers; we have proved now that the expansion holds for all $|z| < 2\pi$.

Similarly, we obtain the Taylor series for the ordinary (trigonometric) cotangent. We begin for $|z| < 2\pi$, $0 < t < 1$, with the generating function

(53)
$$G(t, z) = \sum_{n=0}^{\infty} (-1)^n \psi_{2n}(t) z^{2n};$$

differentiating twice, we find that G satisfies the differential equation $d^2G/dt^2 + z^2 G = 0$, whose general solution is $G = a \cos(zt) + b \sin(zt)$, with a and b not depending on t but possibly on z as a parameter. To determine a and b we use two conditions. The first, $\int_0^1 G(t, z)\, dt = 1$, is found through termwise integration. The second,

$$\lim_{t \to 0} \frac{dG(t, z)}{dt} = \tfrac{1}{2} z^2$$

for all z, is found by termwise differentiation, in which we use the fact that for $n > 1$,

$$\psi_{2n}'(0) = \psi_{2n-1}(0) = b_{2n-1} = 0.$$

These conditions imply

$$a = \frac{z}{2} \cot \frac{z}{2}, \qquad b = \frac{z}{2},$$

so that for $|z| < 2\pi$, $0 < t < 1$

$$G(t, z) = \frac{z}{2} \frac{\cos(zt - z/2)}{\sin(z/2)}.$$

We leave the details to the reader.

If we let $t \to 0$ in this formula, we obtain the Taylor expansion of the cotangent (see p. 563) for $|z| < 2\pi$

(54)
$$G(0, z) = \sum_{n=0}^{\infty} (-1)^n b_{2n} z^{2n} = \tfrac{1}{2} z \cot \tfrac{1}{2} z.$$

c. The Euler-Maclaurin Summation Formula

In Section 5.4b we derived Taylor's formula using successive integration by parts. In the following analogous derivation of a famous formula of Euler, the Bernoulli polynomials, or rather their periodic extensions $\psi_n(t)$, take the previous place of the polynomials $(t - b)^n/n!$. (We thus replace a and b from p. 450 by 0 and 1, which is always possible by means of the transformation of the variable t into the variable $s = (t - a)/(b - a)$, and is therefore not an essential change.)

Instead of beginning with the relation

$$(55) \qquad f(1) - f(0) = \int_0^1 f'(t)\, dt$$

which would correspond to our previous derivation of the Taylor formula, we begin with the relation

$$(56) \qquad \int_0^1 f(t)\, dt = \int_0^1 f(t)\psi_0(t)\, dt,$$

which leads to greater symmetry. Since

$$\psi_0(t) = \psi_1'(t), \qquad \psi_1(+0) = -\tfrac{1}{2},$$

and $\psi_1(1-0) = \tfrac{1}{2}$, the formula for integration by parts

$$\int_0^1 u\, dv = uv \Big|_0^1 - \int_0^1 v\, du$$

for $u = f(t)$, $v = \psi_1(t)$, $f(0) = f_0$, $f(1) = f_1$ yields (see also Chapter 3, p. 278).

$$\int_0^1 f(t)\, dt = \tfrac{1}{2}(f_0 + f_1) - \int_0^1 f'(t)\psi_1(t)\, dt,$$

or

$$(57) \qquad \tfrac{1}{2}(f_0 + f_1) = \int_0^1 f(t)\, dt + \int_0^1 f'(t)\psi_1(t)\, dt,$$

an explicit expression for the deviation of the sum on the left from the integral $\int_0^1 f(t)\, dt$.

Since a corresponding formula holds true for every interval between two successive integers due to the periodicity of $\psi_1(t)$, we immediately obtain

$$(58) \quad \tfrac{1}{2}f_0 + f_1 + f_2 + \cdots f_{n-1} + \tfrac{1}{2}f_n$$

$$= \int_0^n f(x)\, dx + \int_0^n f'(x)\psi_1(x)\, dx,$$

or for any interval $a \le x \le b$, with a and b integers,

$$(58a) \quad f_a + f_{a+1} + \cdots + f_{b-1}$$

$$= \int_a^b f(x)\, dx + \int_a^b f'(x)\psi_1(x)\, dx, \ -\tfrac{1}{2}(f_b - f_a).$$

Thus we obtain an exact expression for the difference between the left-hand sum (the area of the inscribed rectangles in the case of an increasing function) and the first term on the right-hand side (the area underneath the curve); formula (58a) is the simplest formulation of the *Euler-Maclaurin summation formula*.

It is natural to improve upon this result by repeating the integration by parts. Integrating the expression $\int_a^b f'(x)\psi_1(x)\,dx$, and setting $u = f'(x)$, $dv = \psi_1(x)\,dx$, we obtain

$$\int_a^b f'(x)\psi_1(x)\,dx = f'(x)\psi_2(x)\Big|_a^b - \int_a^b f''(x)\psi_2(x)\,dx.$$

Since
$$\psi_2(b) = \psi_2(a) = \psi_2(0) = b_2,$$

the first term takes the form

$$b_2[f'(b) - f'(a)];$$

the second term can be again integrated by parts, yielding

$$-b_3[f''(b) - f''(a)] + \int_a^b f'''(x)\psi_3(x)\,dx.$$

Here, since $b_3 = 0$, the first expression vanishes; we again integrate by parts, obtaining

$$b_4[f'''(b) - f'''(a)] - \int_a^b f''''(x)\psi_4(x)\,dx.$$

Repeating this operation, until we reach ψ_{2k}, we obtain the *general form* of the *Euler summation formula*

$$(59) \quad f_a + f_{a+1} + \cdots + f_{b-1} = \int_a^b f(x)\,dx - \tfrac{1}{2}[f(b) - f(a)]$$

$$+ \sum_{n=1}^{k} b_{2n}[f^{(2n-1)}(b) - f^{(2n-1)}(a)] + R_k,$$

where the remainder R_k may be written in one of the two forms

$$(60) \qquad\qquad R_k = -\int_a^b f^{(2k)}(x)\psi_{2k}(x)\,dx,$$

or

$$(60a) \qquad\qquad R_k = \int_a^b f^{(2k+1)}(x)\psi_{2k+1}(x)\,dx.$$

d. Applications. Asymptotic Expressions

Convergent Expansions. Euler's summation formula can be applied in different circumstances. First, if $R_k \to 0$ as $k \to \infty$, then the infinite series

$$\sum_{n=1}^{\infty} b_{2n}[f^{(2n-1)}(b) - f^{(2n-1)}(a)]$$

converges, and the formula gives an important means of expressing the sum of the corresponding series in closed form, or for expressing definite functions as series.

Nonconvergent Expansions.   Secondly, and of more importance, the remainder R_k may not tend to zero as $k \to \infty$; the above series does not necessarily converge. Nevertheless it may happen that at first the absolute values $|R_k|$ decrease with increasing k, and that $|R_k|$ for suitably chosen values of k is very small, whereas $|R_k|$ begins later (for large k) to increase strongly. In this case the summation formula can be an important tool for numerical computations; although it is not possible to obtain arbitrarily high precision, as with convergent series, we can nevertheless compute the value of the left side to within an error which is at most equal to the least value $|R_k|$, which is often a highly satisfactory precision. We shall examine examples of both these phenomena.

Example. Exponential Functions. We consider first the function $f(x) = e^{zx}$ for some fixed z. With $a = 0$ and $b = 1$, we obtain for any number k, the relation

$$f_0 = \int_0^1 f(x)\, dx - \tfrac{1}{2}[f(1) - f(0)]$$
$$+ \sum_{n=1}^{k} b_{2n}[f^{(2n-1)}(b) - f^{(2n-1)}(a)] + R_k.$$

Consequently,

$$1 = \frac{e^z - 1}{z} - \tfrac{1}{2}(e^z - 1) + \sum_{n=1}^{k} b_{2n} z^{2n-1}(e^z - 1) + R_k$$
$$= \frac{e^z - 1}{z} \cdot \left[1 - \frac{z}{2} + \sum_{n=1}^{k} b_{2n} z^{2n}\right] + R_k,$$

where

$$R_k = -\int_0^1 z^{2k} e^{zx} \psi_{2k}(x)\, dx.$$

Since $|\psi_{2k}(x)| \leqslant 4/(2\pi)^{2k}$ (p. 622), it follows that

$$|R_k| \leq |z|^{2k} \cdot e^{|z|} \cdot \frac{4}{(2\pi)^{2k}} = 4e^{|z|}\left(\frac{|z|}{2\pi}\right)^{2k},$$

or $R_k \to 0$, at least for $|z| < 2\pi$. Consequently, for these values of z, we can allow k to grow beyond all bounds in the summation formula, obtaining

(61) $$\frac{z}{e^z - 1} = 1 - \tfrac{1}{2}z + \sum_{n=1}^{\infty} b_{2n} z^{2n}$$

for the function $z/(e^z - 1)$, a formula already found by other methods (p. 623). We note that the interval of convergence is again $|z| < 2\pi$.

e. Sums of Powers; Recursion Formula for Bernoulli Numbers

An even simpler example of a convergent Euler summation formula occurs when the series on the right contains only finitely many terms, especially if $f(x)$ is a polynomial of rth degree with $r \geq 1$, so that $f^{r+1}(x)$ vanishes identically. We choose $f(x) = x^r$, $a = 0$, $b = n$, and $k > \frac{1}{2}r$. For simplification we again introduce the sequence $B_n{}^*$ of Bernoulli numbers, defined previously (p. 623), as $B_n{}^* = n! \, b_n$, for all n. Noting that

$$B_0{}^* = 1, \; B_1{}^* = -\tfrac{1}{2}, \; B_3{}^* = B_5{}^* = B_7{}^* = \cdots = B^*_{2n+1} = 0,$$

we see that Euler's formula (59) takes the form

$$1 + 2^r + 3^r + \cdots + (n-1)^r$$
$$= \int_0^n x^r \, dx + \sum_{v=1}^r \frac{B_v{}^*}{v!} \cdot (f^{(v-1)}(n) - f^{(v-1)}(0))$$
$$= \frac{n^{r+1}}{r+1} + \sum_{v=1}^r \frac{B_v{}^*}{v!} r(r-1) \cdots (r-v+2) n^{r-v+1}$$
$$= \frac{1}{r+1} \left\{ n^{r+1} + \sum_{v=1}^r \binom{r+1}{v} B_v{}^* n^{(r+1)-v} \right\}$$
$$= \frac{1}{r+1} \left\{ \sum_{v=0}^{r+1} \binom{r+1}{v} n^{(r+1)-v} B_v{}^* - B^*_{r+1} \right\}.$$

This formula can be written symbolically as

$$(62) \quad 1 + 2^r + 3^r + \cdots + (n-1)^r = \frac{1}{r+1} \{(n + B^*)^{r+1} - B^{*r+1}\},$$

where the term within the parentheses is to be expanded formally by using the binomial theorem, and each of the "powers" B^{*k} is to be replaced by the corresponding Bernoulli number $B_k{}^*$. For example,

$$1 + 2^2 + 3^2 + \cdots + (n-1)^2 = \tfrac{1}{3}(n^3 + 3n^2 B_1 + 3n B_2)$$
$$= \tfrac{1}{6}(2n^3 - 3n^2 + n)$$
$$1 + 2^3 + 3^3 + \cdots + (n-1)^3 = \tfrac{1}{4}n^2(n-1)^2$$

(cf. p. 58).

By setting $n = 1$, formula (62) assumes the form

$$\frac{1}{r+1} \{(1 + B^*)^{r+1} - B^{*r+1}\} = 0,$$

or

$$(62a) \qquad (1 + B^*)^{r+1} = B^{*r+1} \qquad \text{for all } r = 1.$$

which is just the recursion formula for the $B_k{}^*$ given on p. 562.

f. Euler's Constant and Stirling's Series

An example of an application of the Euler-Maclaurin formula in the second case, that of divergence, is given by the function $f(x) = 1/x$ with $a = 1, b = n$. By (58a)

$$(63) \quad 1 + \tfrac{1}{2} + \tfrac{1}{3} + \cdots + \frac{1}{n-1} = \int_1^n \frac{dx}{x} + \frac{1}{2}\left(1 - \frac{1}{n}\right) - \int_1^n \frac{\psi_1(x)}{x^2}\,dx$$

$$= \log n + \tfrac{1}{2} - \frac{1}{2n} - \int_1^n \frac{\psi_1(x)}{x^2}\,dx$$

or

$$1 + \tfrac{1}{2} + \tfrac{1}{3} + \tfrac{1}{4} + \cdots + \frac{1}{n} - \log n = \tfrac{1}{2} + \frac{1}{2n} - \int_1^n \frac{\psi_1(x)}{x^2}\,dx.$$

For $n \to \infty$ the integral on the right side converges, since $|\psi_1(x)| \leqslant \tfrac{1}{2}$, for all x; thus the absolute value of the integrand is always less than that of the convergent integral $\displaystyle\int_1^\infty dx/x^2$. Hence we obtain in the relation

$$(64) \quad \lim_{n \to \infty}\left[\sum_{k=1}^n \frac{1}{k} - \log n\right] = \tfrac{1}{2} - \int_1^\infty \frac{\psi_1(x)}{x^2}\,dx = C$$

a definite constant C, the Euler constant, already introduced on p. 526.

We have then two results: The harmonic series is of the same order of growth as the logarithm, both diverging to infinity, and there is an explicit expression for the difference between the two

$$\sum_{k=1}^n \frac{1}{k} - \log n - C = R_n = \frac{1}{2n} + \int_n^\infty \frac{\psi_1(x)}{x^2}\,dx.$$

We note that R_n vanishes for $n \to \infty$ at least of first order.

We obtain a more important application when we set $f(x) = \log x$, $a = 1, b = n$ in formula (59), p. 626. Then

$$\log 1 + \log 2 + \cdots + \log(n-1) = n \log n - n + 1 - \tfrac{1}{2}\log n$$

$$- \sum_{m=1}^k b_{2m}(2m-2)!\left(1 - \frac{1}{n^{2m-1}}\right) + \int_1^n \frac{(2k)!}{x^{2k+1}}\,\psi_{2k+1}(x)\,dx.$$

Adding log n to both sides, we obtain

$$(65) \quad \log n! = (n + \tfrac{1}{2}) \log n - n + c_k + \sum_{m=1}^{k} \frac{(2m-2)!}{n^{2m-1}} b_{2m} - r_k(n),$$

where

$$c_k = 1 - \sum_{m=1}^{k} b_{2m}(2m-2)! + \int_1^{\infty} \frac{(2k)!}{x^{2k+1}} \psi_{2k+1}(x) \, dx$$

$$r_k(n) = \int_n^{\infty} \frac{(2k)!}{x^{2k+1}} \psi_{2k+1}(x) \, dx.$$

The improper integrals converge for $k > 0$, since the functions $\psi_{2k+1}(x)$ are periodic, and hence bounded for all x (see p. 307). We can find the value of the constant c_k if we observe that by (65) for $n \to \infty$

$$c_k = \lim_{n \to \infty} \log \left(\frac{n! \, e^n}{n^{n+1/2}} \right).$$

We conclude then from Stirling's formula (14), p. 504 (or directly from Wallis' product for π as on p. 280) that $c_k = \log \sqrt{2\pi}$. If we still express the Bernoulli numbers b_{2m} as $(-1)^{m-1} B_m/(2m)!$ (see formula (47), p. 620), we obtain the so-called *Stirling series*

$$\log \left(\frac{n!}{\sqrt{2\pi} \, n^{n+1/2} e^{-n}} \right) = \sum_{m=1}^{k} \frac{(-1)^{m-1} B_m}{2m(2m-1)n^{2m-1}} - r_k(n).$$

This formula is a refinement of Stirling's formula. For any fixed positive integer k and large n the terms in the sum approach zero respectively of the order of $1/n$, $1/n^3$, $1/n^5$, ... , $1/n^{2k-1}$ The remainder term $r_k(n)$ approaches zero like $1/n^{2k}$, since $\psi_{2k+1}(x)$ is a bounded function. Thus for fixed k and very large n each term in the sum will be very large compared to the following terms, and the remainder will be smaller than all the terms in the sum. We thus obtain an approximation formula of the form

$$(66) \quad \log \left(\frac{n!}{\sqrt{2\pi} n^{n+1/2} e^{-n}} \right) = \frac{B_1}{1 \cdot 2} \frac{1}{n} - \frac{B_2}{3 \cdot 4} \frac{1}{n^3} + \frac{B_3}{5 \cdot 6} \frac{1}{n^5} - + \cdots$$

$$= \frac{1}{12} \frac{1}{n} - \frac{1}{360} \frac{1}{n^3} + \frac{1}{1260} \frac{1}{n^5} - \frac{1}{1680} \frac{1}{n^7} + \frac{1}{1188} \frac{1}{n^9} - + \cdots$$

This expansion must, however, not be considered in the same light as a convergent infinite series. It is only *asymptotically* correct in the sense that if we break off the series after a fixed number of terms, say k terms, then the error r_k is small compared with all the terms kept provided n is sufficiently large. We can never make the error arbitrarily small for a fixed n by taking more and more terms. As a matter of fact the infinite series (66) diverges, as we see immediately from the estimate on p. 621 for the Bernoulli numbers. For a given large n there is an optimum number of terms of the series which one might use. Thus for moderately large n we have the approximation

$$n! \approx \sqrt{2\pi n}\, n^{n+1/2} e^{n+1/12n};$$

for very large n the formula

$$n! \approx \sqrt{2\pi n}\, n^{n+1/2} e^{n+1/12n-1/360n^3}$$

gives a more accurate approximation, etc.

PROBLEMS

SECTION 8.1, page 572

1. The *fundamental period* T of a periodic function f is defined as the greatest lower bound of the positive periods of f. Prove:
(a) If $T \neq 0$, then T is a period.
(b) If $T \neq 0$, then every other period is an integral multiple of T.
(c) If $T = 0$ and if f is continuous at any point, then f is a constant function.

2. Show that if f has incommensurable periods T_1 and T_2, then the fundamental period T is zero. Give an example of a nonconstant function with incommensurable periods.

3. Let f and g have fundamental periods a and b, respectively. If a and b are commensurable, say $a/b = q/p$, where p and q are relatively prime integers, then show by example that $f + g$ can have as its fundamental period any value m/n, where $m = aq = bp$ and n is any natural number.

SECTION 8.5, page 598

1. Obtain the Fourier series for the function $f(x) = \pi x$ on the interval $0 \leq x \leq 1$ as a pure sine series and as a pure cosine series.

2. Show how to represent a function defined on an arbitrary bounded interval as a Fourier series.

3. Obtain the infinite product for the cosine from the relation

$$\cos \pi x = \frac{\sin 2\pi x}{2 \sin \pi x}.$$

4. Using the infinite products for the sine and cosine, evaluate
(a) $\frac{2}{1} \cdot \frac{2}{3} \cdot \frac{6}{5} \cdot \frac{6}{7} \cdot \frac{10}{9} \cdot \frac{10}{11} \cdot \frac{14}{13} \cdots$;
(b) $2 \cdot \frac{2}{3} \cdot \frac{4}{3} \cdot \frac{8}{9} \cdot \frac{10}{9} \cdot \frac{14}{15} \cdot \frac{16}{15}$.

5. Express the hyperbolic cotangent in terms of partial fractions.

6. Determine the special properties of the coefficients of the Fourier expansions of even and odd functions for which $f(x) = f(\pi - x)$.

SECTION 8.6, page 604

1. Investigate the convergence of the Fourier expansion

$$\cos x + \frac{\cos 2x}{2} + \frac{\cos 3x}{3} + \cdots$$

of the function $-\log 2 \left| \sin \dfrac{x}{2} \right|$.

SECTION 8.7, page 608

1. Prove Parseval's equation for a piecewise smooth function f where f may have a number of discontinuities.

APPENDIX II.1, page 619

1. Prove that

$$\phi_n(t) = \frac{1}{n!} \sum_{k=0}^{n} \binom{n}{k} B_k^* t^{n-k}.$$

2. Prove for $n \geqslant 1$ that

$$\phi_n(t) = (-1)^n \phi_n(1 - t).$$

3. Using the expression for the cotangent in partial fractions, expand $\pi x \cot \pi x$ as a power series in x. By comparing this with the series given on p. 625, show that

$$\sum_{\nu=1}^{\infty} \frac{1}{\nu^{2m}} = (-1)^{m-1} \frac{(2\pi)^{2m}}{2 \cdot (2m)!} B_{2m}^*.$$

4. Show that

$$\sum_{\nu=1}^{\infty} \frac{1}{(2\nu - 1)^{2m}} = \frac{(-1)^{m-1}(2^{2m} - 1)\pi^{2m}}{2(2m)!} B_{2m}^*.$$

5. Show that

$$\sum_{\nu=1}^{\infty} \frac{(-1)^\nu}{\nu^{2m}} = \frac{(-1)^m(2^{2m} - 2)\pi^{2m}}{2 \cdot (2m)!} B_{2m}^*.$$

6. Using the infinite products for the sine and cosine, show that

(a) $\log\left(\dfrac{\sin x}{x}\right) = -\displaystyle\sum_{\nu=1}^{\infty} \frac{(-1)^{\nu-1}2^{2\nu-1}B_{2\nu}^*}{(2\nu)! \, \nu} x^{2\nu};$

(b) $\log \cos x = -\displaystyle\sum_{\nu=1}^{\infty} \frac{(-1)^{\nu-1}2^{2\nu-1}(2^{2\nu} - 1)B_{2\nu}^*}{(2\nu)! \, \nu} x^{2\nu}.$

7. Prove that

(a) $\displaystyle\int_0^1 \frac{\log x}{1 - x} \, dx = -\frac{\pi^2}{6};$

(b) $\displaystyle\int_0^1 \frac{\log x}{1 + x} \, dx = -\frac{\pi^2}{12}.$

9

Differential Equations
for the Simplest Types of Vibration

On several previous occasions we have met with *differential equations*, that is, equations from which an unknown function is to be determined and which involve not only this function itself but also its derivatives.

The simplest problem of this type is that of finding the indefinite integral of a given function $f(x)$: to find a function $y = F(x)$ which satisfies the differential equation $y' - f(x) = 0$. Furthermore, in Chapter 3, p. 223, we showed that an equation of the form $y' = \alpha y$ is satisfied by an exponential function $y = ce^{\alpha x}$, and we characterized the trigonometrical functions by differential equations (p. 312). As we saw in Chapter 4 (e.g., p. 405), differential equations arise in connection with the problems of mechanics, and indeed many branches of pure mathematics and most of applied mathematics depend on differential equations. In this chapter, without going into the general theory, we shall consider the differential equations of the simplest types of vibration. These are not only of theoretical value but are also extremely important in applied mathematics.

It will be convenient to bear in mind the following general ideas and definitions. By a *solution* of a differential equation we mean a function which, when substituted in the differential equation, satisfies the equation "identically"; this means for all values of the independent variable that are being considered. Instead of *solution* the term *integral* is often used: first, because the problem is more or less a generalization of the ordinary problem of integration; secondly, because it frequently happens that the solution is actually found by integration.

9.1   Vibration Problems of Mechanics and Physics

a. The Simplest Mechanical Vibrations

The simplest type of mechanical vibration has already been considered in Chapter 4 (p. 404). We there considered a particle of mass m which is free to move on the x-axis and which is brought back to its initial position $x = 0$ by a restoring force. The magnitude of this restoring force we took to be proportional to the displacement x by, in fact, equating it to $-kx$, where k is a positive constant and the negative sign expresses the fact that the force is always directed toward the origin. We shall now assume that there is a frictional force present also and that this frictional force is proportional to the velocity $dx/dt = \dot{x}$ of the particle and opposed to it. This force is then given by an expression of the form $-r\dot{x}$, with a positive frictional constant r. Finally, we shall assume that the particle is also acted on by an external force which is a function $f(t)$ of the time t. Then by Newton's fundamental law the product of the mass m and the acceleration $\ddot{x}$ must be equal to the total force, that is, the elastic force plus the frictional force plus the external force. This is expressed by the equation

$$(1) \qquad\qquad m\ddot{x} + r\dot{x} + kx = f(t).$$

This equation governs the motion of the particle. If we recall the previous examples of differential equations, such as the integration problem for $\dot{x} = dx/dt = f(t)$ solved by $x = \int f(t)\, dt + c$, or the solution of the particular differential equation $m\ddot{x} + kx = 0$ on p. 405, we observe that these problems have an infinite number of distinct solutions. Here too we shall find that there are an infinite number of solutions, a fact expressed in the following way. It is possible to find a *general solution* or *complete integral* $x(t)$ of the differential equation, depending not only on the independent variable t but also on two arbitrary parameters c_1 and c_2, called the *constants of integration*. Assigning special values to these constants we obtain a particular solution, and *every* solution can be found by assigning special values to these constants.

This fact is quite understandable (cf. also p. 404). We cannot expect that the differential equation alone will determine the motion completely. On the contrary, it is plausible that at a given instant, say at the time $t = 0$, we should be able to choose the initial position $x(0) = x_0$ and the initial velocity $\dot{x}(0) = \dot{x}_0$ (in short, the *initial state*) arbitrarily; in other words, at time $t = 0$ we should be able to start the particle from any initial position with any velocity. This being done, we may expect the rest of the motion to be completely determined. The two arbitrary constants c_1 and c_2 in the general solution are just enough to enable us to select the particular solution which fits these initial conditions. In the next section we shall see that this can be done in one way only.

If no external force is present, that is, if $f(t) = 0$, the motion is called a *free motion*. The differential equation is then said to be *homogeneous*. If $f(t)$ is not equal to zero for all values of t, we say that the motion is *forced* and that the differential equation is *nonhomogeneous*. The term $f(t)$ is also occasionally referred to as the *perturbation term*.

b. Electrical Oscillations

A mechanical system of the simple type described can physically be realized only approximately. An example is offered by the pendulum, provided its oscillations are small. The oscillations of a magnetic needle, the oscillations of the centre of a telephone or microphone diaphragm, and other mechanical vibrations can be represented to within a certain degree of accuracy by systems such as described. But there is another type of phenomenon which corresponds with great precision to our differential equation (1). This is the oscillatory electrical circuit.

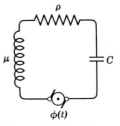

Figure 9.1   Oscillatory electrical circuit.

We consider the circuit sketched in Fig. 9.1, having inductance μ, resistance ρ, and capacity $C = 1/\kappa$. We also suppose that the circuit is acted on by an external electromotive force $\phi(t)$ which is known as a function of the time t, such as the voltage supplied by a dynamo or the voltage due to electric waves. In order to describe the process taking place in the circuit we denote the voltage across the condenser by E and the charge in the condenser by Q. These quantities are then connected by the equation $CE = E/\kappa = Q$. The current I, which like the voltage E is a function of the time, is defined as the rate of change of the charge per unit time, that is, as the rate at which the charge on the condenser diminishes: $I = -\dot{Q} = -dQ/dt = -\dot{E}/\kappa$. Ohm's law states that the product of the current and the resistance is equal to the electromotive force (voltage); that is, it is equal to the condenser voltage E minus the counter electromotive force due to self-induction plus the external electromotive force $\phi(t)$. We thus arrive at the equation $I\rho = E - \mu\dot{I} + \phi(t)$ or $-(\rho/\kappa)\dot{E} = E + (\mu/\kappa)\ddot{E} + \phi(t)$, that is, $\mu\ddot{E} + \rho\dot{E} + \kappa E = -\kappa\phi(t)$, which is satisfied by the voltage in the circuit. We see therefore that we have obtained a differential equation of exactly type (1). Instead of the mass we have the inductance, instead of the frictional force, the resistance, and instead of the elastic constant, the reciprocal of the capacity, whereas the external electromotive force (apart from a constant factor) corresponds to the external force. If the electromotive force is zero, the differential equation is homogeneous.

If we multiply both sides of the differential equation by $-1/\kappa$ and differentiate with respect to the time, we obtain for the current I the corresponding equation

$$\mu \ddot{I} + \rho \dot{I} + \kappa I = \dot{\phi}(t),$$

which differs from the equation for the voltage on the right-hand side only, and for free oscillations ($\phi = 0$) has identically the same form.

9.2 Solution of the Homogeneous Equation. Free Oscillations

a. The Formal Solution

We can easily obtain a solution of the homogeneous equation (1) $m\ddot{x} + r\dot{x} + kx = 0$ in the form of an exponential expression, by determining a constant λ in such a way that the expression $e^{\lambda t} = x$ is a solution. If we substitute this tentative solution and its derivatives $\dot{x} = \lambda e^{\lambda t}$, $\ddot{x} = \lambda^2 e^{\lambda t}$ in the differential equation and remove the common factor $e^{\lambda t}$, we obtain the quadratic equation.

(2) $$m\lambda^2 + r\lambda + k = 0$$

for λ. The roots of this equation are

$$\lambda_1 = -\frac{r}{2m} + \frac{1}{2m}\sqrt{r^2 - 4mk}, \quad \lambda_2 = -\frac{r}{2m} - \frac{1}{2m}\sqrt{r^2 - 4mk}.$$

Each of the two expressions $x = e^{\lambda_1 t}$ and $x = e^{\lambda_2 t}$ is, at least formally, a particular solution of the differential equation, as we see by carrying out the calculations in the reverse direction. Three different cases can now occur:

1. $r^2 - 4mk > 0$. The two roots λ_1 and λ_2 are then real, negative, and unequal, and we have two solutions of the differential equation,

$$u_1 = e^{\lambda_1 t} \quad \text{and} \quad u_2 = e^{\lambda_2 t}.$$

With the help of these two solutions we can at once construct a solution in which two arbitrary constants are present. For after differentiation we see that

(3) $$x = c_1 u_1 + c_2 u_2$$

is also a solution of the differential equation. In Section 9.3 we shall show that this expression is in fact the most general solution of the equation; that is, that we can obtain *every* solution of the equation by substituting suitable numerical values for c_1 and c_2.

2. $r^2 - 4mk = 0$. The quadratic equation has a double root. Thus to begin with we have, apart from a constant factor, only the one solution $x = w_1 = e^{-rt/2m}$. But we easily verify that in this case the function

$$x = w_2 = te^{-rt/2m}$$

is also a solution of the differential equation.[1] For we find that

$$\dot{x} = \left(1 - \frac{r}{2m}t\right)e^{-rt/2m}, \quad \ddot{x} = \left(\frac{r^2}{4m^2}t - \frac{r}{m}\right)e^{-rt/2m},$$

and by substitution we see that the differential equation

$$m\ddot{x} + r\dot{x} + \frac{r^2}{4m}x = m\ddot{x} + r\ddot{x} + kx = 0$$

is satisfied. Then the expression

(4) $$x = c_1 e^{-rt/2m} + c_2 te^{-rt/2m}$$

again gives us a solution of the differential equation with two arbitrary constants of integration c_1 and c_2.

3. $r^2 - 4mk < 0$. We put $r^2 - 4mk = -4m^2v^2$ and obtain two solutions of the differential equation in complex form, given by the expressions $x = u_1 = e^{-rt/2m+ivt}$ and $x = u_2 = e^{-rt/2m-ivt}$. Euler's formula

$$e^{\pm ivt} = \cos vt \pm i \sin vt$$

gives us for the real and imaginary parts of the complex solution u_1, on the one hand the expressions

$$v_1 = e^{-rt/2m}\cos vt, \quad v_2 = e^{-rt/2m}\sin vt,$$

and on the other hand, the representation

$$v_1 = \frac{u_1 + u_2}{2}, \quad v_2 = \frac{u_1 - u_2}{2i}.$$

From the second form of representation we see that v_1 and v_2 are (real) solutions of the differential equation. To verify this directly by differentiation and substitution is a simple exercise.

---

[1] We are led to this solution naturally by the following limiting process: if $\lambda_1 \neq \lambda_2$, then the expression $(e^{\lambda_1 t} - e^{\lambda_2 t})/(\lambda_1 - \lambda_2)$ also represents a solution. If we now let λ_1 tend to λ_2 and write λ instead of λ_1, λ_2, our expression becomes $d(e^{\lambda t})/d\lambda = te^{\lambda t}$.

From our two particular solutions we can again form a general solution

(5) $$x = c_1 v_1 + c_2 v_2 = (c_1 \cos \nu t + c_2 \sin \nu t)e^{-rt/2m}$$

with two arbitrary constants c_1 and c_2. This may also be written in the form

(6) $$x = ae^{-rt/2m} \cos \nu(t - \delta),$$

where we have put $c_1 = a \cos \nu\delta$, $c_2 = a \sin \nu\delta$, and a, δ are two new constants.

We recall that we have already met this solution for the special $r = 0$ (Section 5.4).

b. Interpretation of the Solution

In the two cases $r > 2\sqrt{mk}$ and $r = 2\sqrt{mk}$ the solution is given by the exponential curve or by the graph of the function $te^{-rt/2m}$, which for large values of t resembles the exponential curve, or by the superposition of such curves. In these cases the process is aperiodic; that is, as the time increases the "distance" x approaches the value 0 asymptotically without oscillating about the value $x = 0$. The motion therefore is not oscillatory. The effect of friction or *damping* is so great that it prevents the elastic force from setting up oscillatory motions.

It is quite different for $r < \sqrt{2mk}$, where the damping is so small that complex roots λ_1, λ_2 occur. The expression $x = a \cos \nu(t - \delta)e^{-rt/2m}$ here gives us *damped harmonic oscillations*. These are oscillations which follow the sine law and have the circular frequency $\nu = \sqrt{k/m - r^2/4m^2}$,

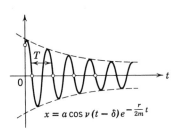

$$x = a \cos \nu (t - \delta) e^{-\frac{r}{2m}t}$$

Figure 9.2 Damped harmonic oscillations.

but whose amplitude, instead of being constant, is given by the expression $ae^{-rt/2m}$. That is, the amplitude diminishes exponentially; the greater the expression $r/2m$ is, the faster is the rate of decrease. In physical literature this damping factor is frequently called the *attenuation constant* of the damped oscillation, the term indicating that the logarithm of the amplitude decreases at the rate $r/2m$. A damped oscillation of this kind is illustrated in Fig. 9.2. As before, we call the quantity $T = 2\pi/\nu$ the period of the oscillation and the quantity $\nu\delta$ the phase displacement. For the special case $r = 0$ we again obtain simple harmonic oscillations with the frequency $\nu_0 = \sqrt{k/m}$, the *natural frequency* of the undamped oscillatory system.

c. Fulfilment of Given Initial Conditions.
Uniqueness of the Solution

We have still to show that the solution with the two constants c_1 and c_2 can be made to fit any preassigned initial state, and also that it represents all the possible solutions of the equation. Suppose that we have to find a solution which at time $t = 0$ satisfies the initial conditions $x(0) = x_0$, $\dot{x}(0) = \dot{x}_0$, where the numbers x_0 and $\dot{x}_0$ can have any values. Then in case 1 of Section 9.2a (p. 636) we must put

$$c_1 + c_2 = x_0,$$
$$c_1\lambda_1 + c_2\lambda_2 = \dot{x}_0.$$

For the constants c_1 and c_2 we accordingly have two linear equations, and these have the unique solutions

$$c_1 = \frac{\dot{x}_0 - \lambda_2 x_0}{\lambda_1 - \lambda_2}, \quad c_2 = \frac{\dot{x}_0 - \lambda_1 x_0}{\lambda_2 - \lambda_1}.$$

In case 2 the same process gives the two linear equations

$$c_1 = x_0,$$
$$\lambda c_1 + c_2 = \dot{x}_0 \quad \left(\lambda = -\frac{r}{2m}\right),$$

from which c_1 and c_2 can again be uniquely determined. Finally, in case 3 the equations determining the constants take the form

$$a \cos \nu\delta = x_0,$$
$$a\left(\nu \sin \nu\delta - \frac{r}{2m} \cos \nu\delta\right) = \dot{x}_0,$$

with the solutions

$$\delta = \frac{1}{\nu} \arccos \frac{x_0}{a}, \quad a = \frac{1}{\nu} \sqrt{\left[\nu^2 x_0^2 + \left(\dot{x}_0 + \frac{r}{2m} x_0\right)^2\right]}.$$

Thus we have shown that the general solutions can be made to fit any arbitrary initial conditions. We have still to show that there is no other solution. For this we need show only that for a given initial state there can never be two different solutions.

If two such solutions $u(t)$ and $v(t)$ existed, for which $u(0) = x_0$, $\dot{u}(0) = \dot{x}_0$ and $v(0) = x_0$, $\dot{v}(0) = \dot{x}_0$, then their difference $w = u - v$ would also be a solution of the differential equation, and we should have $w(0) = 0$, $\dot{w}(0) = 0$. This solution would therefore correspond to an initial state of rest, that is, to a state in which at time $t = 0$ the particle is in its position of rest and has zero velocity. We must show that it can never set itself in motion. To do this we multiply both sides of the differential equation $m\ddot{w} + r\dot{w} + kw = 0$ by $2\dot{w}$ and recall that $2\dot{w}\ddot{w} = (d/dt)\dot{w}^2$ and $2w\dot{w} = (d/dt)w^2$. We thus obtain

$$\frac{d}{dt}(m\dot{w}^2) + \frac{d}{dt}(kw^2) + 2r\dot{w}^2 = 0.$$

If we integrate between the instants $t = 0$ and $t = \tau$ and use the initial conditions $w(0) = 0$, $\dot{w}(0)$, we have

$$m\dot{w}^2(\tau) + kw^2(\tau) + 2r\int_0^\tau \left(\frac{dw}{dt}\right)^2 dt = 0.$$

This equation, however, would yield a contradiction if at any time $\tau > 0$ the function w were different from 0. For then the left-hand side of the equation would be positive, since we have taken m, k, and r to be positive, and the right-hand side is zero. Hence $w = u - v$ is always equal to 0, which proves that the solution is unique.

9.3 The Nonhomogeneous Equation. Forced Oscillations

a. General Remarks. Superposition

Before proceeding to the solution of the problem when an external force $f(t)$ is present, that is, to the solution of the nonhomogeneous equation, we make the following remark.

If w and v are two solutions of the nonhomogeneous equation, the difference $u = w - v$ satisfies the homogeneous equation; this we see at once by substitution. Conversely, if u is a solution of the homogeneous equation and v a solution of the nonhomogeneous equation, then $w = u + v$ is also a solution of the nonhomogeneous equation. Therefore from one solution[1] of the nonhomogeneous equation we obtain *all* its solutions by adding the complete integral of the homogeneous equation. We therefore need find only a *single* solution of the nonhomogeneous equation. Physically this means that if we have a forced oscillation due to an external force, and superpose on it an arbitrary free oscillation, represented by a solution of the homogeneous equation, we obtain a phenomenon which satisfies the same nonhomogeneous equation as the original forced oscillation. If a frictional force is present, the free motion in the case of oscillatory motion must fade out as time goes on because of the damping factor $e^{-rt/2m}$. Hence for a

---

[1] Often called a *particular integral* or particular solution.

given forced vibration with friction it is immaterial what free vibration we superpose; the motion will always tend to the same final state as time goes on.

Second, we notice that the effect of a force $f(t)$ can be split up in the same way as the force itself. By this we mean the following: if $f_1(t), f_2(t)$, and $f(t)$ are three functions such that

$$f_1(t) + f_2(t) = f(t),$$

and if $x_1 = x_1(t)$ is a solution of the differential equation $m\ddot{x} + r\dot{x} + kx = f_1(t)$ and $x_2 = x_2(t)$ is a solution of the equation $m\ddot{x} + r\dot{x} + kx = f_2(t)$, then $x(t) = x_1(t) + x_2(t)$ is a solution of the differential equation

$$(7) \qquad m\ddot{x} + r\dot{x} + kx = f(t).$$

A corresponding statement, of course, holds if $f(t)$ consists of any number of terms. This simple but important fact is called the *principle of superposition*. The proof follows from a glance at the equation itself. By subdividing the function $f(t)$ into two or more terms we can thus split the differential equation into several equations, which in certain circumstances may be easier to manipulate.

The most important case is that of a periodic external force $f(t)$. Such a periodic external force can be resolved into purely periodic components by expansion in a Fourier series, and can therefore[1] be approximated to as closely as we please by a sum of a finite number of purely periodic functions. It is therefore sufficient to find the solution of the differential equation subject to the assumption that the right-hand side has the form

$$a \cos \omega t \qquad \text{or} \qquad b \sin \omega t,$$

where a, b, and ω are arbitrary constants.

Instead of working with these trigonometric functions, we can obtain the solution more simply and neatly if we use complex notation. We put $f(t) = ce^{i\omega t}$, and the principle of superposition shows that we need only consider the differential equation

$$(8) \qquad m\ddot{x} + r\dot{x} + kx = ce^{i\omega t},$$

where by c we mean an arbitrary real or complex constant. Such a differential equation actually represents two real differential equations. For if we split the right-hand side into two terms by taking, for example, $c = 1$ and write $e^{i\omega t} = \cos \omega t + i \sin \omega t$, then x_1 and x_2, the solutions of the two *real* differential equations $m\ddot{x} + r\dot{x} + kx = \cos \omega t$

---

[1] Provided that it is continuous and sectionally smooth (p. 604), which is the most important case in physics.

and $m\ddot{x} + r\dot{x} + kx = \sin \omega t$, combine to form the solution $x = x_1 + ix_2$ of the *complex* differential equation. Conversely, if we first solve the differential equations in complex form, the real part of the solution gives us the function x_1 and the imaginary part the function x_2.

b. Solution of the Nonhomogeneous Equation

We solve Equation (8) by a device suggested naturally by intuition. We assume that c is real and (for the time being) that $r \neq 0$. We now make the guess that a motion will exist which has the same rhythm as the periodic external force, and we accordingly attempt to find a solution of the differential equation in the form

$$\text{(9)} \qquad x = \sigma e^{i\omega t},$$

where we have only to determine the factor σ, which is independent of the time. If we substitute this expression and its derivatives $\dot{x} = i\omega\sigma e^{i\omega t}$, $\ddot{x} = -\omega^2\sigma e^{i\omega t}$ in the differential equation and remove the common factor $e^{i\omega t}$ we obtain the equation

$$-m\omega^2\sigma + ir\omega\sigma + k\sigma = c$$

or

$$\text{(10)} \qquad \sigma = \frac{c}{-m\omega^2 + ir\omega + k}.$$

Conversely, we see that for this value of σ the expression $\sigma e^{i\omega t}$ is actually a solution of the differential equation. To express the meaning of this result clearly, however, we must perform a few transformations.

We begin by writing the complex factor σ in the form

$$\text{(11)} \qquad \sigma = c\,\frac{k - m\omega^2 - ir\omega}{(k - m\omega^2)^2 + r^2\omega^2} = c\alpha e^{-i\omega\delta},$$

where the positive "distortion factor" α and the "phase displacement" $\omega\delta$ are expressed in terms of the given quantities m, r, k, by the equations

$$\alpha^2 = \frac{1}{(k - m\omega^2)^2 + r^2\omega^2}, \quad \sin \omega\delta = r\omega\alpha, \quad \cos \omega\delta = (k - m\omega^2)\alpha.$$

With this notation our solution takes the form

$$x = c\alpha e^{i\omega(t-\delta)},$$

and the meaning of the result is as follows: to the force $c \cos \omega t$ there corresponds the "effect" $c\alpha \cos \omega(t - \delta)$, and to the force $c \sin \omega t$ corresponds the effect $c\alpha \sin \omega(t - \delta)$.

Hence we see that the effect is a function of the same type as the force, that is, an undamped oscillation. This oscillation differs from the oscillation representing the force in that the amplitude is increased in the ratio $\alpha : 1$ and the phase is altered by the angle $\omega\delta$. Of course, it is easy to obtain the same result without using the complex notation, but at the cost of somewhat longer calculations.

According to the remark at the beginning of this section, by finding this one solution we have completely solved the problem; for by superposing any free oscillation we can obtain the most general forced oscillation.

Collecting the results, we state the following:

The complete integral of the differential equation

$$m\ddot{x} + r\dot{x} + kx = ce^{i\omega t}$$

(*where* $x \neq 0$) *is* $x = c\alpha e^{i\omega(t-\delta)} + u$, *where* u *is the complete integral of the homogeneous equation* $m\ddot{x} + r\dot{x} + kx = 0$ and the quantities α and δ are defined by the equations

(12)

$$\alpha^2 = \frac{1}{(k - m\omega^2)^2 + r^2\omega^2}, \quad \sin \omega\delta = r\omega\alpha, \quad \cos \omega\delta = (k - m\omega^2)\alpha.$$

The constants in this general solution leave us the possibility of making the solution suit an arbitrary initial state, that is, for arbitrarily assigned values of x_0 and $\dot{x}_0$ the constants can be chosen in such a way that $x(0) = x_0$ and $\dot{x}(0) = \dot{x}_0$.

c. The Resonance Curve

To acquire a grasp of the solution which we have obtained and of its significance in applications, we shall study the distortion factor α as a function of the "exciting frequency" ω, that is, the function

(13) $$\phi(\omega) = \frac{1}{\sqrt{(k - m\omega^2)^2 + r^2\omega^2}}.$$

Such a detailed investigation is motivated by the fact that for given constants k, m, r, or as we say for a given "oscillatory system," we can think of the system as being acted on by periodic exciting forces of very different circular frequencies, and it is important to consider the solution of the differential equation for these widely different exciting forces. In order to describe the function conveniently we introduce the quantity $\omega_0 = \sqrt{k/m}$. This number ω_0 is the circular frequency which the system would have for free oscillations if the friction r were zero; or, briefly, the *natural frequency of the undamped system* (cf. p. 639). The actual frequency of the free system, owing to the friction r, is not equal to ω_0, but is instead

$$\nu = \sqrt{\frac{k}{m} - \frac{r^2}{4m^2}},$$

where we assume that $4km - r^2 > 0$. (If this is not the case the free system has no frequency; it is aperiodic.)

The function $\phi(\omega)$ tends asymptotically to the value zero as the exciting frequency tends to infinity, and, in fact, it vanishes to the order $1/\omega^2$. Furthermore, $\phi(0) = 1/k$; in other words, an exciting force of frequency zero and magnitude one, that is, a constant force of magnitude one, gives rise to a displacement of the oscillatory system amounting to $1/k$. In the region of positive values of ω the derivative $\phi'(\omega)$ cannot vanish except where the derivative of the expression $(k - m\omega^2)^2 + r^2\omega^2$ vanishes, that is, for a value $\omega = \omega_1 > 0$ for which the equation

$$-4m\omega(k - m\omega^2) + 2r^2\omega = 0$$

holds. In order that such a value may exist we must obviously have $2km - r^2 > 0$; in this case

$$\omega_1 = \sqrt{\frac{k}{m} - \frac{r^2}{2m^2}} = \sqrt{\omega_0^2 - \frac{r^2}{2m^2}}.$$

Since the function $\phi(\omega)$ is positive everywhere, increases monotonically for small values of ω, and vanishes at infinity, this value ω_1 must give a maximum. We call this frequency ω_1 the "resonance frequency" of the system.

By substituting this expression for ω_1 we find that the value of the maximum is

$$\phi(\omega_1) = \frac{1}{r\sqrt{(k/m - r^2/4m^2)}}.$$

As $r \to 0$, this value increases beyond all bounds. For $r = 0$, that is, for an undamped oscillatory system, the function $\phi(\omega)$ has an infinite discontinuity at the value $\omega = \omega_1$. This is a limiting case to which we shall give special consideration later.

The graph of the function $\phi(\omega)$ is called the *resonance curve* of the system. The fact that for $\omega = \omega_1$ (and consequently for small values of r in the neighborhood of the natural frequency) the distortion of amplitude $\alpha = \phi(\omega)$ is particularly large is the mathematical expression of the "phenomenon of resonance," which for fixed values of m and k is more and more evident as r becomes smaller and smaller.

In Fig 9.3 we have sketched a family of resonance curves, all corresponding to the values $m = 1$ and $k = 1$, and consequently to $\omega_0 = 1$, but with
different values of $D = \frac{1}{2}r$. We see that for small values of D well-marked
resonance occurs near $\omega = 1$; in the limiting case $D = 0$ there would be an
infinite discontinuity of $\phi(\omega)$ at $\omega = 1$, instead of a maximum. As D increases

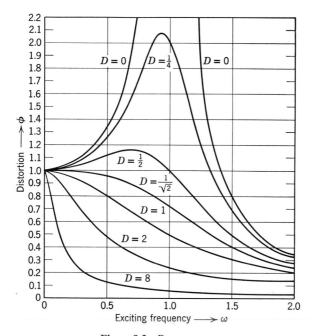

Figure 9.3   Resonance curves.

the maxima move towards the left, and for the value $D = 1/\sqrt{2}$ we have
$\omega_1 = 0$. In this last case the point where the tangent is horizontal has moved
to the origin, and the maximum has disappeared. If $D > 1/\sqrt{2}$ there is no
zero of $\phi'(\omega)$; the resonance curve no longer has a maximum, and resonance
no longer occurs.

In general, the resonance phenomenon ceases as soon as the condition

$$2km - r^2 \leq 0$$

becomes true. In the case of the equality sign, the resonance curve
reaches its greatest height $\phi(0) = 1/k$ at $\omega_1 = 0$; its tangent is horizontal there, and after an initial course which is almost horizontal it
declines towards zero.

d. Further Discussion of the Oscillation

We cannot, however, remain content with the above discussion. To really understand the phenomenon of forced motion an additional point needs to be emphasized. The particular integral $c\alpha e^{i\omega(t-\delta)}$ is to be regarded as a *limiting state* which the complete integral

$$x(t) = c\alpha e^{i\omega(t-\delta)} + c_1 u_1 + c_2 u_2$$

approaches more and more closely *as time goes on*, since the free oscillation $c_1 u_1 + c_2 u_2$ superposed on the particular integral fades away with the passage of time. This fading away will take place slowly if r is small, rapidly if r is large.

Let us suppose, for example, that at the beginning of the motion, that is, at time $t = 0$, the system is at rest, so that $x(0) = 0$ and $\dot{x}(0) = 0$. From this we can determine the constants c_1 and c_2, and we see at once that they are *not* both zero. Even when the exciting frequency is approximately or exactly equal to ω_1, so that resonance occurs, the relatively large amplitude $\alpha = \phi(\omega_1)$ will not at first appear. On the contrary, it will be masked by the function $c_1 u_1 + c_2 u_2$, and will first make its appearance when this function fades away; that is, it will appear more slowly as r grows smaller.

For the undamped system, that is, for $r = 0$, our solution fails when the exciting frequency is equal to the natural circular frequency $\omega_0 = \sqrt{k/m}$, for then $\phi(\omega_0)$ is infinite. We therefore cannot obtain a solution of the equation $m\ddot{x} + kx = e^{i\omega t}$ in the form $\sigma e^{i\omega t}$. We can, however, at once obtain a particular solution in the form $x = \sigma t e^{i\omega t}$. If we substitute this expression in the differential equation, remembering that

$$\dot{x} = \sigma e^{i\omega t}(1 + i\omega t), \quad \ddot{x} = \sigma e^{i\omega t}(2i\omega - t\omega^2),$$

we have

$$\sigma(2im\omega - m\omega^2 t + kt) = 1,$$

and, since $m\omega^2 = k$,

$$\sigma = \frac{1}{2im\omega}.$$

Thus *when resonance occurs in an undamped system we have a solution*

$$x = \frac{t}{2im\omega} e^{i\omega t} = \frac{t}{2i\sqrt{km}} e^{i\omega t}.$$

Using real notation, when $f(t) = \cos \omega t$, we have

$$x = \frac{1}{2} \frac{t}{\sqrt{km}} \sin \omega t$$

and when $f(t) = \sin \omega t$ we have

$$x = -\frac{1}{2}\frac{t}{\sqrt{km}}\cos \omega t.$$

We thus see that we have found a function which may be referred to as an oscillation, but whose amplitude increases proportionally with the time. The superposed free oscillation does not fade away since it is undamped; but it retains its original amplitude and becomes unimportant in comparison with the increasing amplitude of the special forced oscillation. The fact that in this case the solution oscillates backward and forward between positive and negative bounds which continually increase as time goes on represents the real meaning of the infinite discontinuity of the resonance function for an undamped system.

e. Remarks on the Construction of Recording Instruments

In a great variety of applications in physics and engineering the discussion in the previous subsection is of the utmost importance. With many instruments, such as galvanometers, seismographs, oscillatory electrical circuits in radio receivers, and microphone diaphragms, the problem is to record an oscillatory displacement x due to an external periodic force. In such cases the quantity x satisfies our differential equation, at least to a first approximation.

If T is the period of oscillation of the external periodic force, we can expand the force in a Fourier series of the form

$$f(t) = \sum_{l=-\infty}^{\infty} \gamma_l e^{il(2\pi/T)t},$$

or, better still, we can think of it as represented with sufficient accuracy by a trigonometric sum $\sum_{l=-N}^{N} \gamma_l e^{il(2\pi/T)t}$ consisting of a finite number of terms only. By the principle of superposition (p. 641), the solution $x(t)$ of the differential equation, apart from the superposed free oscillation, will be represented by an infinite series[1] of the form

$$x(t) = \sum_{l=-\infty}^{\infty} \sigma_l e^{il(2\pi/T)t},$$

or approximately by a finite expression of the form

$$x(t) = \sum_{l=-N}^{N} \sigma_l e^{il(2\pi/T)t}.$$

---

[1] Questions of convergence will not be discussed here.

By virtue of our previous results

$$\sigma_l = \gamma_l \alpha_l e^{-i\delta_l(2\pi l/T)}$$

and

$$\alpha_l{}^2 = \cfrac{1}{\left(k - ml^2 \cfrac{4\pi^2}{T^2}\right)^2 + r^2l^2 \cfrac{4\pi^2}{T^2}}, \quad \tan \frac{2\pi l}{T}\delta_l = \cfrac{2\pi lr}{T\left(k - m \cfrac{4\pi^2l^2}{T^2}\right)}.$$

We can then describe the action of an arbitrary periodic external force in the following way: if we resolve the exciting force into purely periodic components, the individual terms of the Fourier series, then each component is subject to its own distortion of amplitude and phase displacement, and the separate effects are then superposed additively. If we are interested only in the distortion of amplitude (the phase displacement is only of secondary importance[2] in applications and, moreover, can be discussed in the same way as the distortion of amplitude), a study of the resonance curve gives us complete information about the way in which the motions of the recording apparatus mirror the external exciting force. For very large values of l or $\omega[=(2\pi/T)l]$ the effect of the exciting frequency on the displacement x will be hardly perceptible. On the other hand, all exciting frequencies in the neighborhood of ω_1, the (circular) resonance frequency, will markedly affect the quantity x.

In the construction of physical measuring and recording apparatus the constants m, r, and k are at our disposal, at least within wide limits. These should be chosen so that the shape of the resonance curve is as well adapted as possible to the special requirements of the measurement in question. Here two considerations predominate. First, it is desirable that the apparatus should be as sensitive as possible; that is, for all frequencies ω in question the value of α should be as large as possible. For small values of ω, as we have seen, α is approximately proportional to $1/k$, so that the number $1/k$ is a measure of the sensitiveness of the instrument for small exciting frequencies. The sensitiveness can therefore be increased by increasing $1/k$, that is, by weakening the restoring force.

The other important point is the necessity for *relative freedom from distortion*. Let us assume that the representation $f(t) = \sum\limits_{l=-N}^{N} \gamma_l e^{il(2\pi/T)t}$ is an adequate approximation to the exciting force. We then say that the apparatus records the exciting force $f(t)$ with relative freedom from distortion if for all circular frequencies $\omega \leq N(2\pi/T)$ the distortion factor has approximately the same value. This condition is indispensable if we wish to derive conclusions about the exciting process directly from the behavior of the apparatus; if, for example, a recorder or radio is to reproduce both high and low musical notes with an approximately correct ratio of intensity. The requirement that the reproduction should be relatively "distortionless" can

---

[2] Since, for example, it is imperceptible to the human ear.

never be satisfied exactly, since no portion of the resonance curve is exactly horizontal. We can, however, attempt to choose the constants m, k, r, of the apparatus in such a way that no marked resonance occurs, and also in such a way that the curve has a horizontal tangent at the beginning, so that $\varphi(\omega) = \alpha$ remains approximately constant for small values of ω. As we have learned above, we can do this by putting

$$2km - r^2 = 0.$$

Given a constant m and a constant k, we can satisfy this requirement by adjusting the friction r properly, for example, by inserting a properly chosen resistance in an electrical circuit. The resonance curve then shows us that from the frequency 0 to circular frequencies near the natural circular frequency ω_0 of the undamped system the instrument is nearly distortionless, and that above this frequency the damping is considerable. We therefore obtain relative freedom from distortion in a given interval of frequencies by first choosing m so small and k so large that the natural circular frequency ω_0 of the undamped system is greater than any of the exciting circular frequencies under consideration, and then choosing a damping factor r in accordance with the equation $2km - r^2 = 0$.

List of Biographical Dates

Abel, Niels Henrik (1802–1829)
Archimedes (287?–212 B.C.)
Barrow, Isaac (1630–1677)
Bernoulli, Jakob (1654–1705)
Bernoulli, John (1667–1748)
Bessel, Friedrich Wilhelm (1784–1846)
Bolzano, Bernhard (1781–1848)
Brahe, Tycho (1546–1601)
Briggs, Henry (1556?–1630)
Cantor, Georg (1845–1918)
Cauchy, Augustin (1789–1857)
Coulomb, Charles Augustin de (1736–1806)
Darboux, Gaston (1842–1917)
Dedekind, Richard (1831–1916)
De Moivre, Abraham (1667–1754)
Descartes, (Cartesius) René (1596–1650)
Dirichlet, Gustav Lejeune (1805–1859)
Einstein, Albert (1879–1955)
Euclid (about 300 B.C.)
Euler, Leonhard (1707–1783)
Fejer, Lipot (1880–1959)
Fermat, Pierre de (1601–1665)
Fourier, Joseph (1768–1830)
Fresnel, Augustin (1788–1827)
Gauss, Carl Friedrich (1777–1855)
Gibbs, Josiah Willard (1839–1903)
Gregory, James (1638–1675)
Guldin, Paul (1577–1643)

Hermite, Charles (1822–1901)
Hölder, Otto (1860–1937)
Huygens, Christian (1629–1695)
Jensen, J. L. W. V. (1859–1925)
Kepler, Johannes (1571–1630)
Lagrange, Joseph Louis (1736–1813)
Lambert, Johann Heinrich (1728–1777)
Landau, Edmund (1877–1938)
Leibnitz, Gottfried Wilhelm von (1646–1716)
L'Hôpital, Guillaume, François Antoine de (1661–1704)
Lipschitz, Rudolf Otto (1832–1903)
Lorentz, Hendrik Antoon (1853–1928)
Maclaurin, Colin (1698–1746)
Michelson, Albert Abraham (1852–1931)
Morley, Edmund Williams (1838–1923)
Napier, John (1550–1617)
Newton, Isaac (1642–1727)
Ohm, Georg Simon (1787–1854)
Parseval, Marc Anton (B. ?–1836)
Ptolemy, (Claudius Ptolemaeus) (second Century A.D.)
Raabe, Joseph Ludwig (1801–1859)
Riemann, Bernhard (1826–1866)
Rolle, Michael (1652–1719)
Schwarz, Hermann Amandus (1843–1921)
Seidel, Philipp Ludwig von (1821–1896)
Simpson, Thomas (1710–1761)
Stirling, James (1692–1770)
Taylor, Brook (1685–1731)
Vega, George (1754–1802)
Wallis, John (1616–1703)
Weierstrass, Karl (1815–1897)

Index